D1758965

Diagnostic and Therapeutic Applications of Exosomes in Cancer

Diagnostic and Therapeutic Applications of Exosomes in Cancer

Edited by

Mansoor Amiji
Rajagopal Ramesh

ACADEMIC PRESS
An imprint of Elsevier

Academic Press is an imprint of Elsevier
125 London Wall, London EC2Y 5AS, United Kingdom
525 B Street, Suite 1800, San Diego, CA 92101-4495, United States
50 Hampshire Street, 5th Floor, Cambridge, MA 02139, United States
The Boulevard, Langford Lane, Kidlington, Oxford OX5 1GB, United Kingdom

Notices
Knowledge and best practice in this field are constantly changing. As new research and experience broaden our understanding, changes in research methods, professional practices, or medical treatment may become necessary.

Practitioners and researchers must always rely on their own experience and knowledge in evaluating and using any information, methods, compounds, or experiments described herein. In using such information or methods they should be mindful of their own safety and the safety of others, including parties for whom they have a professional responsibility.

To the fullest extent of the law, neither the Publisher nor the authors, contributors, or editors, assume any liability for any injury and/or damage to persons or property as a matter of products liability, negligence or otherwise, or from any use or operation of any methods, products, instructions, or ideas contained in the material herein.

Library of Congress Cataloging-in-Publication Data
A catalog record for this book is available from the Library of Congress

British Library Cataloguing-in-Publication Data
A catalogue record for this book is available from the British Library

ISBN: 978-0-12-812774-2

For information on all Academic Press publications visit our website at
https://www.elsevier.com/books-and-journals

  Working together
to grow libraries in
developing countries

www.elsevier.com • www.bookaid.org

Publisher: John Fedor
Acquisition Editor: Erin Hill-Parks
Editorial Project Manager: Jennifer Horigan
Production Project Manager: Mohanapriyan Rajendran
Cover Designer: Mark Rogers

Typeset by TNQ Books and Journals

Dedication

I dedicate this book to my family—my lovely wife Tusneem and our three amazing daughters Zahra, Anisa, and Salima. I also dedicate this book to my past and present postdoctoral associates and graduate students, who have made tremendous contribution to the research success of my group for 25 years.

Mansoor M. Amiji

Contents

List of Contributors   xv
Preface   xix
Acknowledgments   xxi

1. Introduction to Exosomes and Cancer   1

   Phillip B. Munson and Arti Shukla

   1. A Brief History   1
   2. Biogenesis   3
   3. Isolation and Characterization   4
   4. Content   5
   5. Biological Function in Cancer   5
       Acknowledgments   7
       References   7

2. Extracellular Vesicle Biogenesis in Cancer   11

   Stephanie N. Hurwitz and David G. Meckes Jr.

   1. Introduction   11
   2. Microvesicle Biogenesis   13
   3. Exosome Biogenesis   14
   4. An Increasing Importance of Examining Diverse Vesicles   17
   5. CD63-Dependent Vesicle Production   19
   6. Vesicle Secretion Is Linked to Intercellular Signaling   20
       References   20

3. Composition, Physicochemical and Biological Properties of
   Exosomes Secreted From Cancer Cells   27

   Scott W. Ferguson, Jake S. Megna and Juliane Nguyen

   1. Introduction   27
   2. Exosome Biogenesis   29

3. Factors Affecting the Release of Exosomes From
    Cancer Cells   34
4. Composition of Exosomes   36
5. Size and Charge of Cancer Exosomes   47
6. Summary and Future Considerations   51
    Acknowledgment   51
    References   51

4. Heterogeneity of Tumor Exosomes – Role in Precision Medicine   59

    Vincent Bernard, Jinjie Ling and Anirban Maitra

1. Introduction   59
2. Double-Stranded Genomic DNA in Circulating Exosomes   60
3. Detection of Mutational Signatures in Genomic DNA-
    Enriched Exosomes   61
4. Exosomes as Agents for Early Detection, Diagnosis, and
    Stratification   61
5. Genomic Molecular Profiling of Exosomal Cargo   63
6. Transcriptomic Characterization of Tumors Through
    Liquid Biopsies   65
7. Conclusions   65
    Acknowledgments   66
    References   66

5. Proteomic Profiling of Tumor Exosomes   69

    Sara Yousuf, Alfred A. Simental and Salma Khan

1. Introduction   69
2. Sources of Exosomes   71
3. Proteomic Content of Exosomes in Cancers   74
4. Conclusion   83
    Acknowledgments   83
    References   83

6. Nucleic Acid Profiling in Tumor Exosomes   93

    Malav S. Trivedi and Maria Abreu

1. Introduction   94
2. Composition of Exosomes   95

3. Nucleic Acid Loading Into Exosomes 95
4. Isolation Techniques for RNA/DNA From Exosomes 97
5. Profiling of Nucleic Acid—RNA 100
6. Single EV Analysis 108
7. Statistical and Bioinformatic Software for evRNA Characterization 109
8. Experimental Artifacts and Contaminants Affecting evRNA Analysis 109
9. Conclusions 110
   References 111

7. Nanotechnology Platforms for Cancer Exosome Analyses 119

   Hyungsoon Im, Katherine S. Yang, Hakho Lee and Cesar M. Castro

1. Introduction 119
2. Nanoplasmonic Sensing 120
3. Electrochemical Sensing 122
4. Immunomagnetic Exosome RNA Analysis 124
5. Conclusions 126
   References 126

8. Exosome RNAs as Biomarkers and Targets for Cancer Therapy 129

   Akhil Srivastava, Narsireddy Amreddy, Rebaz Ahmed, Mohammed A. Razaq, Katherine Moxley, Rheal Towner, Yan D. Zhao, Allison Gillaspy, Ali S. Khan, Anupama Munshi and Rajagopal Ramesh

1. Introduction 130
2. History of RNA in Exosomes 130
3. RNA Species Present in Exosomes 132
4. Sources of Exosomes 144
5. Technology for the Study of Exosomal miRNA 145
6. Challenges and Perspective 151
7. Conclusion 152
   Acknowledgments 152
   References 153
   Further Reading 159

 9. Diagnostic Potential of Tumor Exosomes   161

    Philip Hochendoner, Zheng Zhao and Mei He

    1. Introduction   161
    2. Methods and Tools for Exosome Isolation   163
    3. Tumor Exosomes for Cancer Diagnosis   164
    4. Additional Diagnostic Potential of Exosomes   170
    5. Conclusion/Discussion   170
       References   171

10. Biodistribution of Cancer-Derived Exosomes   175

    Luize G. Lima and Andreas Möller

    1. Introduction   175
    2. Mechanisms Involved in Tumor-Derived Exosomes
       Homing to Specific Organs   177
    3. Impact of Diverse Factors on the Evaluation of Exosome
       Biodistribution In Vivo   179
    4. Conclusions   183
       Acknowledgments   184
       References   184

11. Exosome-Mediated Communication in the Tumor
    Microenvironment   187

    Mei-Ju Su, Neha N. Parayath and Mansoor M. Amiji

    1. Introduction   187
    2. Exosomes and the Tumor Microenvironment   198
    3. Exosome-Mediated Cellular Reprogramming   203
    4. Future Prospects   210
       References   211

12. Exosomes and Tunneling Nanotube Conduits: Synergistic
    Interaction That Facilitates Intercellular Communication
    Between Malignant and Stromal Cells in the Tumor
    Microenvironment   219

    Emil Lou, William Sperduto and Subbaya Subramanian

    1. Introduction   219
    2. Interplay Between Tunneling Nanotubes and Exosomes 220

3. Other Tunneling Nanotube–Like Nanostructures and
Interaction With Exosomes   223

4. Identifying Metabolic Conditions and Physiologic
Factors That Mutually Promote TNT Formation and
Exosome Activity   225

5. Pharmacologic Suppression of Tunneling Nanotubes   226

6. The Search for Specific Molecular and Structural
Biomarkers of TNT Formation in Cancer   227

7. Conclusion and Perspectives   229

References   230

13. Exosomes in Tumor Angiogenesis—Multifunctional
Messengers With Mixed Intentions   235

Liang Zhang

1. Introduction   235

2. Exosomes in Tumor Angiogenesis   237

3. Future Perspectives   241

References   243

14. Role of Exosomes in Development of Premetastatic Niche   247

Sagar Bhayana, Marshleen Yadav and Naduparambil K. Jacob

1. Introduction   247

2. Role of Cancer Cell–Derived Exosomes in Initiation and
Progression of Metastasis   248

3. Role of Exosomal Noncoding RNAs in Initiation and
Priming for Metastasis   250

4. Role of Exosomal Proteins in Priming Metastasis   251

5. Role of Toll-Like Receptors in Exosome-Mediated
Metastasis   252

6. Epithelial–Mesenchymal Transition, a Hallmark of
Premetastatic Niche Formation   253

7. Pharmacologic and Genetic Strategies Targeting
Exosome-Mediated Metastasis   255

References   256

15. Exosomes: Key Supporters of Tumor Metastasis 261

Girijesh K. Patel, Haseeb Zubair, Mohammad A. Khan, Sanjeev K. Srivastava, Aamir Ahmad, Mary C. Patton, Seema Singh, Moh'd Khushman and Ajay P. Singh

1. Introduction 262
2. Exosomes in Metastasis Progression: Events at Primary Tumor Site 262
3. Exosome-Mediated Events at Metastatic Sites 269
4. Strategies Against Exosome-Mediated Metastasis 272
5. Conclusions and Perspective 276
   Acknowledgments 276
   References 277

16. The Emerging Roles and Clinical Potential of Exosomes in Cancer: Drug Resistance 285

Li Min, Cassandra Garbutt, Francis Hornicek and Zhenfeng Duan

1. Introduction 286
2. Mechanisms of Resistance to Current Drug Therapies 287
3. Mechanisms of Exosome in Drug Resistance 291
4. Exosome-Associated Drug Resistance in Cancers 293
5. Conclusion 305
   References 306

17. Exosomes in Cancer Immunotherapy 313

Yuki Takahashi and Yoshinobu Takakura

1. Introduction 313
2. Effect of Tumor Exosomes on T Cells 314
3. Effect of Tumor Exosomes on Natural Killer Cells 316
4. Effect of Tumor Exosomes on Myeloid Lineage Cells 317
5. Other Effects of Exosomes in Cancer Immunotherapies 318
6. Cancer Immunotherapies Using Exosomes 319
7. Conclusion 321
   References 321

18. Extracellular Vesicles as Vehicles of B Cell Antigen
    Presentation: Implications for Cancer Vaccine Therapies   325

    Michael W. Graner

    1. Introduction   325
    2. Extracellular Vesicles: Classifications and Nomenclature 326
    3. B Cells, Extracellular Vesicles, and Associated Impacts   329
    4. Conclusions   336
       References   338

19. Translational Potential of Tumor Exosomes in Diagnosis
    and Therapy   343

    Naureen Javeed and Debabrata Mukhopadhyay

    1. Introduction   343
    2. Exosomes and Cancer   344
    3. Exosomes as Cancer Biomarkers   345
    4. Therapeutic Potential of Exosomes   346
    5. Conclusions   348
       References   349

Index   355

List of Contributors

Maria Abreu Nova Southeastern University, Fort Lauderdale, FL, United States

Aamir Ahmad University of South Alabama-Mitchell Cancer Institute, Mobile, AL, United States

Rebaz Ahmed University of Oklahoma Health Sciences Center, Oklahoma City, OK, United States

Mansoor M. Amiji Northeastern University, Boston, MA, United States

Narsireddy Amreddy University of Oklahoma Health Sciences Center, Oklahoma City, OK, United States

Vincent Bernard Sheikh Ahmed Pancreatic Cancer Research Center, UT MD Anderson Cancer Center, Houston, TX, United States

Sagar Bhayana The Ohio State University Comprehensive Cancer Center, Columbus, OH, United States

Cesar M. Castro Massachusetts General Hospital, Boston, MA, United States

Zhenfeng Duan Massachusetts General Hospital and Harvard Medical School, Boston, MA, United States

Scott W. Ferguson University at Buffalo, The State University of New York, Buffalo, NY, United States

Cassandra Garbutt Massachusetts General Hospital and Harvard Medical School, Boston, MA, United States

Allison Gillaspy University of Oklahoma Health Sciences Center, Oklahoma City, OK, United States

Michael W. Graner University of Colorado Denver, Aurora, CO, United States

Mei He Kansas State University, Manhattan, KS, United States

Philip Hochendoner Kansas State University, Manhattan, KS, United States

Francis Hornicek Massachusetts General Hospital and Harvard Medical School, Boston, MA, United States

Stephanie N. Hurwitz Florida State University College of Medicine, Tallahassee, FL, United States

Hyungsoon Im Massachusetts General Hospital, Boston, MA, United States

Naduparambil K. Jacob The Ohio State University Comprehensive Cancer Center, Columbus, OH, United States

Naureen Javeed Mayo Clinic, Rochester, MN, United States

Ali S. Khan University of Oklahoma Health Sciences Center, Oklahoma City, OK, United States

Mohammad A. Khan University of South Alabama-Mitchell Cancer Institute, Mobile, AL, United States

Salma Khan Loma Linda University, Loma Linda, CA, United States

Moh'd Khushman University of South Alabama-Mitchell Cancer Institute, Mobile, AL, United States

Hakho Lee Massachusetts General Hospital, Boston, MA, United States

Luize G. Lima QIMR Berghofer Medical Research Institute, Herston, QLD, Australia

Jinjie Ling Sheikh Ahmed Pancreatic Cancer Research Center, UT MD Anderson Cancer Center, Houston, TX, United States

Emil Lou University of Minnesota, Minneapolis, MN, United States

Anirban Maitra Sheikh Ahmed Pancreatic Cancer Research Center, UT MD Anderson Cancer Center, Houston, TX, United States

David G. Meckes Jr. Florida State University College of Medicine, Tallahassee, FL, United States

Jake S. Megna University at Buffalo, The State University of New York, Buffalo, NY, United States

Li Min Massachusetts General Hospital and Harvard Medical School, Boston, MA, United States

Andreas Möller QIMR Berghofer Medical Research Institute, Herston, QLD, Australia

Katherine Moxley University of Oklahoma Health Sciences Center, Oklahoma City, OK, United States

Debabrata Mukhopadhyay Mayo Clinic, Jacksonville, FL, United States

Anupama Munshi University of Oklahoma Health Sciences Center, Oklahoma City, OK, United States

Phillip B. Munson   University of Vermont College of Medicine, Burlington, VT, United States

Juliane Nguyen   University at Buffalo, The State University of New York, Buffalo, NY, United States

Neha N. Parayath   Northeastern University, Boston, MA, United States

Girijesh K. Patel   University of South Alabama-Mitchell Cancer Institute, Mobile, AL, United States

Mary C. Patton   University of South Alabama-Mitchell Cancer Institute, Mobile, AL, United States

Rajagopal Ramesh   University of Oklahoma Health Sciences Center, Oklahoma City, OK, United States

Mohammed A. Razaq   University of Oklahoma Health Sciences Center, Oklahoma City, OK, United States

Arti Shukla   University of Vermont College of Medicine, Burlington, VT, United States

Alfred A. Simental   Loma Linda University, Loma Linda, CA, United States

Ajay P. Singh   University of South Alabama-Mitchell Cancer Institute, Mobile, AL, United States

Seema Singh   University of South Alabama-Mitchell Cancer Institute, Mobile, AL, United States

William Sperduto   University of Minnesota, Minneapolis, MN, United States

Akhil Srivastava   University of Oklahoma Health Sciences Center, Oklahoma City, OK, United States

Sanjeev K. Srivastava   University of South Alabama-Mitchell Cancer Institute, Mobile, AL, United States

Subbaya Subramanian   University of Minnesota, Minneapolis, MN, United States

Mei-Ju Su   Northeastern University, Boston, MA, United States

Yuki Takahashi   Kyoto University, Kyoto, Japan

Yoshinobu Takakura   Kyoto University, Kyoto, Japan

Rheal Towner   University of Oklahoma Health Sciences Center, Oklahoma City, OK, United States

Malav S. Trivedi   Nova Southeastern University, Fort Lauderdale, FL, United States

Marshleen Yadav   The Ohio State University Comprehensive Cancer Center, Columbus, OH, United States

Katherine S. Yang   Massachusetts General Hospital, Boston, MA, United States

Sara Yousuf   Loma Linda University, Loma Linda, CA, United States

Liang Zhang   City University of Hong Kong, Kowloon Tong, Hong Kong SAR, China

Yan D. Zhao   University of Oklahoma Health Sciences Center, Oklahoma City, OK, United States

Zheng Zhao   Kansas State University, Manhattan, KS, United States

Haseeb Zubair   University of South Alabama-Mitchell Cancer Institute, Mobile, AL, United States

Preface

The genesis of the book *Diagnostics and Therapeutic Applications of Exosomes in Cancer* originated when the editors met in 2015 at a conference organized by the University of Oklahoma Health Sciences Center in Oklahoma City, OK. Through a series of discussions and engagement with Elsevier Press, the concept of a comprehensive treaty on the role of tumor-derived exosomes became a reality. Exosomes are 30–150 nm cellular structures that are secreted from all cells in the body. In cancer, the potential of exosomes in early-stage diagnostics and affecting communication in the tumor microenvironment as well as at distal metastatic sites is an area of intense investigation. The major objective of this book is to provide scientists and clinicians with the most up-to-date analysis of the field and pose important questions for future scientific research. Each of the chapter was intentionally solicited, and the authors are leaders in their respective areas.

Chapter 1 by Prof. Arti Shukla and her colleagues begins with an overview of tumor-derived exosomes and provides the reader with the background knowledge. The subsequent chapters (Chapters 2, 3, and 4) focus on the biogenesis of tumor-derived exosomes, their properties, and heterogeneity of composition, which play a key role in cell-cell communication and message transmission. Chapters 5 and 6 focus on exosomal proteomics and nucleic acid profiling technologies. Authored by Dr. Hacko Lee and his colleagues, Chapter 7 discusses a novel nanotechnology-based platform for exosome isolation and characterization. Prof. Rajagopal Ramesh and his team address the role of tumor exosomes as biomarkers for prevention and early disease detection in Chapter 8. Further discussion of tumor-derived exosomes as diagnostic agents is provided in Chapter 9 by Prof. Mei He and her colleagues. Prof. Andreas Moller and his team provide an analysis of the biodistribution profile of tumor-derived exosomes in Chapter 10. Cell-cell communication in the tumor microenvironment and the role of exosomes as second messengers affecting other cells such as tumor-associated macrophages and fibroblasts are covered in Chapter 11 by Prof. Mansoor Amiji's group and in Chapter 12 by Prof. Emil Lou's group. In Chapters 13–16, the potential of tumor-derived exosomes to affect angiogenesis, metastasis, and therapeutic resistance is discussed in detail. Exosomes can also affect tumor immunology and have a significant role in cancer immunotherapy. These attributes of exosomes are addressed in Chapters 17 and 18. In Chapter 19, Prof. Dev Mukhopadhyay and his colleagues provide an assessment of the translational potential of exosome science in future clinical practice.

We believe that the book *Diagnostics and Therapeutic Applications of Exosomes in Cancer* is a starting point in furthering the discussions around these unique natural nanosized structures that carry important protein and nucleic acid cargo as vehicles of communication. The field of exosome biology and applications is still at an embryonic stage. As advances in the field continue to occur, the potential of exosomes in cancer is enormous. We hope that this treaty will serve as a catalyst in furthering the science and technology so that one day we may realize the true potential of these structures and improve the lives of cancer patients all over the world.

Mansoor M. Amiji
Boston, MA, USA
Rajagopal Ramesh
Oklahoma City, OK, USA

Acknowledgments

We deeply appreciate the hard work of the chapter authors who provided the contributions for this book. The team at Elsevier Press has been outstanding. Our deepest gratitude to Erin Hills-Park as well as Jennifer Horigan and their team for pulling everything together and making a dream into reality.

Mansoor M. Amiji
Boston, MA, USA
Rajagopal Ramesh
Oklahoma City, OK, USA

1

Introduction to Exosomes and Cancer

Phillip B. Munson, Arti Shukla

UNIVERSITY OF VERMONT COLLEGE OF MEDICINE, BURLINGTON, VT, UNITED STATES

CHAPTER OUTLINE

1. A Brief History 1
2. Biogenesis 3
3. Isolation and Characterization 4
4. Content 5
5. Biological Function in Cancer 5
Acknowledgments 7
References 7

1. A Brief History

One of the most exciting and cutting-edge topics in modern-day science orbits the now dominant theme of extracellular vesicle (EV) research. One area of particular interest is in regard to the subset of EVs referred to as exosomes, which can be seen by the rapid increase of publications over the past 30 years (Fig. 1). Cell-derived vesicles first arrived on the scientific radar in the 1940s when cell-free plasma was discovered to contain a clot-inducing subcellular element (Chargaff & West, 1946). Decades later, the term "platelet dust" was used to describe 20–50 nm vesicles, following the usage of the term microvesicles in 1975 (van der Pol, Boing, Harrison, Sturk, & Nieuwland, 2012; Wolf, 1967). By 1981, the coinage, "exosome," was used by Trams et al. regarding exfoliated vesicles with 5′-nucleotidase activity composed of increased amounts of sphingomyelin and polyunsaturated fatty acids (Trams, Lauter, Salem, & Heine, 1981). Subsequent discoveries over the last decades provided a more comprehensive understanding of the characterization, origin, biogenesis, and functions, which will be briefly summarized in this chapter before linking the ongoing trajectory of exosomes in regard to cancer research.

Currently, the term exosome refers to small (40–140 nm) membrane-bound vesicles of endocytic origin and are not to be confused with the larger (200–1000 nm)

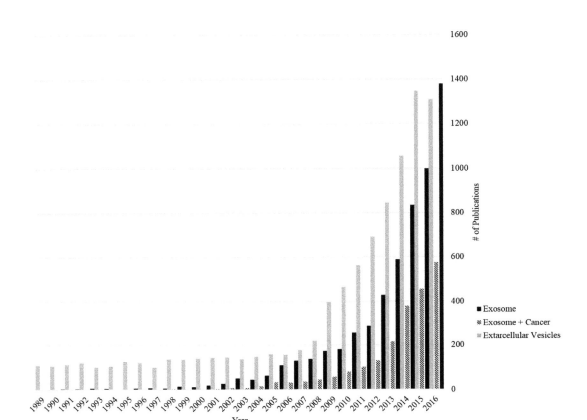

FIGURE 1 Number of publications listed on PubMed using the keywords "exosome," "exosome cancer," or "extracellular vesicles."

microvesicles produced by direct shedding from the plasma membrane. Traditionally, exosomes were deemed as nothing more than a waste mechanism for cells to dispose of unwanted material. However, it has become evidently clear that exosomes serve a much more biologically sophisticated purpose in relaying messages between cells and tissues. Such molecular messages are profoundly adept at altering target cell phenotypes, as scientists have come to discover. Exosomes are known to be present in nearly every body fluid sampled from blood, cerebral spinal fluid, urine, lymph, amniotic fluid, etc., and all mammalian cell types appear to be capable of producing these vesicles. Once released into the extracellular space, exosomes can travel to sites near or distant from their dissemination, thereby providing potential for endocrine, paracrine, and even autocrine signaling. The molecular content of exosomes is of notable intrigue, ranging from biofunctional proteins, RNA species (mRNA and particularly microRNAs), lipids, and some reports indicate the presence of genomic DNA (Keller, Sanderson, Stoeck, & Altevogt, 2006).

2. Biogenesis

The biogenesis of exosomes, as mentioned above, is of endocytic origin and involves a more complicated pathway than that of microvesicles. After the primary invagination of the cell membrane through endocytosis, exosomes begin their creation on the surface of the endosome. A secondary invagination occurs here leading to the deposition of smaller vesicles inside the endosome. These smaller vesicles, at this stage, are referred to as intraluminal vesicles (ILVs), and the endosome containing them is now referred to as either a mature endosome or a multivesicular body (MVB). There are two fates for the ILVs at this stage: one is to be targeted for degradation by the lysosome, which appears to be regulated by lipid composition, cholesterol-poor membranes, and/or the presence of lysobisphosphatidic acid (Wubbolts et al., 2003). The other fate for the MVB, and its content ILVs, is to proceed to fuse with the producer cell's internal surface of the plasma membrane, thereby releasing the contents to the extracellular space. The now released vesicles are exosomes.

The mechanisms by which ILVs/exosomes are produced during the secondary invagination on the endosomal membrane are yet to be fully understood, but certain components have been identified. One mode of exosome biogenesis requires the endosomal sorting complex required for transport (ESCRT) and is known as the ESCRT-dependent pathway, whereas there is also an ESCRT-independent pathway. The ESCRT-dependent pathway utilizes the accessory protein ALIX for the sorting of syndecans through syntenin-mediated interactions and is comprised of four separate complexes: ESCRT-0 for loading ubiquitinated proteins onto the endosomal surface; ESCRT-I and ESCRT-II for budding of the endosomal membrane; and ESCRT-III for separating the membrane. The ESCRT-independent pathway is reported as involving and requiring the lipids sphingosine-1-phosphate and ceramide, the enzyme sphingomyelinase, and tetraspanin-enriched microdomains (Colombo et al., 2013; Piper & Katzmann, 2007). Both of these pathways are targets for inhibition of exosome secretion.

During the process of exosome genesis, lipids are sorted at the site of invagination and molecular cargo is packaged. The exact mechanisms of this process are also unclear, but it is known that there are four underlying requirements for this to occur: cytoskeletal components (actin, microtubules, etc.), molecular motors (kinesin, myosin), molecular switches (predominantly GTPases), and fusion machinery/tethering factors such as SNAREs (Friand, David, & Zimmermann, 2015). Interestingly, exosomes apparently have a diverse range of functions depending on the mode by which they are generated (Blanc & Vidal, 2017; Colombo, Raposo, & Thery, 2014).

Once generated within the MVB, exosome release is mediated by small, vesicular transport regulation GTPases (Rab27A, Rab11, and Rab31), which can work with SNAREs to fuse the MVB membrane to the internal surface of the plasma membrane, and these components are another area of interest for inhibiting exosome secretion (Bobrie, Colombo, Raposo, & Thery, 2011).

3. Isolation and Characterization

To adequately study populations of exosomes, researchers are working toward standardizing isolation techniques. However, more and more techniques are being introduced that make a standard approach increasingly unlikely. Nevertheless, the traditional gold standard for exosome purification is differential ultracentrifugation and, most commonly, exosomes are isolated from conditioned cell culture supernatant supplemented with exosome-free fetal bovine serum, but exosomes are also commonly collected from bodily fluids such as plasma or urine. This technique is widely used because it is less likely to have contaminated protein aggregates; however, it is highly time intensive and requires high volumes of media, as well as yielding low amounts of pelleted exosomes (Lane, Korbie, Anderson, Vaidyanathan, & Trau, 2015). Before ultracentrifugation, samples are cleared of cell and cell debris by shorter, lower speed spins followed by $100,000–150,000 \times g$ spins for about 1 h to clean the sample prior to another high speed spin. Another common technique applauded for clean sample prep is density gradient centrifugation, which utilizes sucrose cushion to separate out vesicles based on size, mass, and density (Chiou & Ansel, 2016; Greening, Xu, Ji, Tauro, & Simpson, 2015).

In addition, exosomes can be captured using various ultrafiltration techniques or size-exclusion chromatography, which have become commercialized to separate preps based on size and molecular weight (Cheruvanky et al., 2007; Rood et al., 2010). Immunoaffinity capture techniques are being employed to separate out exosomes based on their surface protein markers, such as CD63, CD81, and CD9. Another popular method is by exosome precipitation, where samples are cleared of cells/debris and typically a solution of polyethylene glycol is added to insolubilize small exosome-sized vesicles and after an overnight incubation can be spun out at low speed (Helwa et al., 2017; Li, Kaslan, Lee, Yao, & Gao, 2017).

There are multiple means, after isolation, by which to characterize exosome preparations. The most agreed-upon standards for defining an exosome now are membrane-bound vesicles with a diameter of 40–140 nm coming from endocytic origin. Because of this, size characterization and Western blot analysis for exosome markers such as certain tetraspanins (CD9, CD63, and CD81), MVB proteins (ALIX and TSG101), and heat shock proteins (Hsc70, Hsp 90) are most commonly employed (Vlassov, Magdaleno, Setterquist, & Conrad, 2012). Size characterization includes multiple avenues, most useful of which is nanoparticle tracking analysis, which not only provides size characterization but also particle concentration and, in many circumstances, zeta potential. Dynamic light scattering can be used to determine exosome size populations, but to less effectiveness as nanoparticle tracking analysis. Additionally, transmission electron microscopy (TEM) can allow size characterization along with morphological characterization that is useful for defining pure exosome isolation, and recently scanning electron microscopy has been employed (Wu, Deng, & Klinke, 2015). TEM allows visualization of the membrane structure of exosomes, along with a notable cup-shaped morphology that is indicative of exosomes in TEM. Additional techniques are also being developed and used, such as high-resolution flow cytometry (Hoen et al., 2012) and microfluidic-based systems (Yang, Liao, Tian, & Li, 2017).

4. Content

Exosomes have been reported to be released from nearly every cell type and bodily fluid studied and can interact with myriad target cell/tissue types. The molecular cargo of these exosomes dictates the diverse range of effects they may have on target cells, either near or far from their production. Such content ranges from proteins (surface or cytoplasmic), RNA species, functional lipids, and occasionally reported genomic DNA (Coumans et al., 2017; Skotland, Sandvig, & Llorente, 2017). The surface protein content of exosomes resembles the surface composition and topology of its producer cell, and these surface molecules are determinants for potential target cell interaction, and such interactions are pertinent to normal physiology as well as disease processes (Vyas & Dhawan, 2017). It should be noted here that there are multiple ways in which an exosome may interact with a target cell: docking to cell surface protein and conducting a signaling cascade within the cell; the exosome may be endocytosed; or there may be a direct fusion of the exosome to the target cell membrane.

Beyond proteins, RNA species enriched in exosomes are of robust functional importance to the capacity of exosomes to elicit a biological effect. There have been many RNAs identified in exosomes to date (mRNA, piRNA, snoRNA, scaRNA, Y RNA, siRNA, tRNA fragments, vault RNA), but having gained the most attention is undoubtedly miRNAs (Janas, Janas, Sapon, & Janas, 2015).

The focus on miRNAs is because these small (around 22 nucleotides in length) noncoding RNA strands are now known to regulate up to half of the genes within the human body (Chen et al., 2014) and are highly enriched in exosomes. These miRNAs are transferred to target cells via exosomes, maintaining functionality to posttranscriptionally regulate gene expression, and are therefore of great interest when studying exosome components and signaling.

5. Biological Function in Cancer

The many roles of exosomes in normal physiology are vast: immune function and surveillance; neural plasticity and brain function; tissue repair; stem cell maintenance and function; blood coagulation; heart function and cardio protection; and the list could go on (Isola & Chen, 2017; Lee, El Andaloussi, & Wood, 2012). Conversely, exosomes also play an outsize role in many disease states, including pathogenesis of viruses such as HIV-1 and parasites such as malaria (Madison & Okeoma, 2015; Sampaio, Cheng, & Eriksson, 2017); heart diseases (Boulanger, Loyer, Rautou, & Amabile, 2017); kidney diseases, diabetes, and metabolic disorders (Campion, Sanchez-Ferras, & Batchu, 2017;

Martinez & Andriantsitohaina, 2017); disorders of pregnancy (Cuffe, Holland, Salomon, Rice, & Perkins, 2017); central nervous system diseases such as Parkinson's, Alzheimer's, and multiple sclerosis (Selmaj, Mycko, Raine, & Selmaj, 2017; Wu, Zheng, & Zhang, 2017); and, as this book will focus on, exosomes are enormously important in the biology of cancer (Brinton, Sloane, Kester, & Kelly, 2015).

Because these small vesicles are such big players in human physiology and cancer, it is fundamental that researchers use them to uncover the mechanisms of healthy and abnormal physiologies, while also utilizing the molecular cargo as quarries of biomarkers. The dynamic role of exosomes secreted by cancer cells is substantial and is becoming understood to be involved in tumorigenesis, tumor growth/progression, metastasis, angiogenesis, extracellular matrix (ECM) remodeling, immune evasion, chemoresistance, and the establishment of the premetastatic niche (Andaloussi, Mager, Breakefield, & Wood, 2013; Peinado et al., 2012). On the front of biomarkers, it is important to note that neoplastic cells secrete exosomes with content that is markedly dissimilar to that of their noncancerous counterparts, and in many cases it has been reported that cancer cells secrete a larger volume of exosomes altogether when compared with noncancerous cells. The unique signature associated with cancer exosomes is an exciting front in excavating for biomarkers to diagnose cancer, provide better prognostics, and identify therapeutic targets.

Proteomic identification and analysis of miRNAs as exosomal biomarkers of cancer is of great interest and importance to the field. Protein profiling from exosomes has led to very intriguing finds that may lead to clinical use, such as the discovery that Glypican-1 in exosomes can identify pancreatic cancer (Melo et al., 2015). Differential expression of miRNAs in exosomes not only provides insight for biomarkers (Yan et al., 2017), but the array of functionality imparted by transferred exosomal miRNAs to tumor cell function phenotypically change target cells (Takahashi, Prieto-Vila, Hironaka, & Ochiya, 2017). Interestingly, along with sending protumor miRNAs to targets, it is being uncovered that some tumors even shuttle tumor suppressor miRNAs away from themselves via exosomes to prevent their antitumorigenic effects (Kanlikilicer et al., 2016; Rashed et al., 2017). Identifying such biomarkers and their mechanistic effects is of utmost importance in the realm of understanding cancer, and the field of exosomes has significant potential.

Tumor-derived exosomes carry their epigenetic cargo to other tumor cells to aid in their progression and also to nontumor cells for the purpose of phenotypically altering them to aid in tumor growth and spread. The alterations caused by tumor exosomes to nontumor cells can be that they dampen the immune response against the malignancy, reprogram surrounding cells in the tumor microenvironment to aid the tumor, or even convert nonmalignant cells to become cancerous. The established communicatory link between tumor cells and host cells via exosomes turns out to be a dynamic system that promotes tumor survival. Tumor exosomes are capable of directly targeting immune cells to aid in the tumor's evasion from the immune system by carrying immunoinhibitory signals to immune cells (Whiteside, 2017). In addition, tumor exosomes are adept at establishing cells in the tumor microenvironment to make the location more favorable for the tumor. The cells that can be targeted in this setting include fibroblasts, stromal cells, endothelial, and other inflammatory immune cells, and the vasculature surrounding the tumor. Altering and repaving the framework of the ECM by this route is accomplished by exosome signals to these cells and also by exosomal secretion of ECM metalloproteinases (Joyce & Pollard, 2009; Millimaggi et al., 2007; Webber, Steadman, Mason, Tabi, & Clayton, 2010). Furthermore, tumor exosomes are reportedly capable of leading to the well-characterized herald of tumorigenesis known as epithelial to mesenchymal transition, suggesting an extra layer of exosomes' role in coordinating the spread of cancer (Vella, 2014). Exosomes promote angiogenesis and are also capable of conferring chemotherapeutic drug resistance to cancers by allowing the transfer of genetic cargo that more quickly allow the tumor cells to adapt and become resistant, but the exosomes are also used by the cancer cells to spit out the chemotherapeutics that are internalized (Bach, Hong, Park, & Lee, 2017; Lobb et al., 2017).

The therapeutic approaches using exosomes and what is newly being discovered is expanding greatly. Not only are the molecules present in tumor exosomes useful as therapeutic targets but exosomes themselves can be engineered as therapeutic delivery agents, and other treatment approaches include the inhibition of exosome secretion from tumor cells. The therapeutic potentials for exosomes in cancer include the direct targeting of exosomes that tumor cells produce and may be the progenitors of its progression, using them as drug delivery devices and as diagnostic and prognostic indicators of tumorigenesis based on biomarker discovery (Luan et al., 2017; Moore, Kosgodage, Lange, & Inal, 2017; Shahabipour et al., 2017; Sterzenbach et al., 2017; Syn, Wang, Chow, Lim, & Goh, 2017).

As the field of cancer exosomes expands, we are likely to uncover fascinating insights into the biology of cancer and the sophisticated mechanisms by which cancer develops, grows, and spreads. In these efforts, it is becoming clear that exosomes have enormous potential to biomarker discovery and therapeutic options that will shift the paradigm by which we understand, diagnose, and treat cancer.

Acknowledgments

We acknowledge financial support from the Department of Defense (W81XWH-13-PRCRP-IA) and NIH (RO1 ES021110) to AS. PM received Department of Pathology and Laboratory Medicine, UVM, Graduate Student Fellowship.

References

Andaloussi, E. L., Mager, I., Breakefield, X. O., & Wood, M. J. (2013). Extracellular vesicles: Biology and emerging therapeutic opportunities. *Nature Reviews. Drug Discovery*, *12*(5), 347–357. https://doi.org/10.1038/nrd3978.

Bach, D. H., Hong, J. Y., Park, H. J., & Lee, S. K. (2017). The role of exosomes and miRNAs in drug-resistance of cancer cells. *International Journal of Cancer*, *141*(2), 220–230. https://doi.org/10.1002/ijc.30669.

Blanc, L., & Vidal, M. (2017). New insights into the function of Rab GTPases in the context of exosomal secretion. *Small GTPases*, 1–12. https://doi.org/10.1080/21541248.2016.1264352.

Bobrie, A., Colombo, M., Raposo, G., & Thery, C. (2011). Exosome secretion: Molecular mechanisms and roles in immune responses. *Traffic*, *12*(12), 1659–1668. https://doi.org/10.1111/j.1600-0854.2011.01225.x.

Boulanger, C. M., Loyer, X., Rautou, P. E., & Amabile, N. (2017). Extracellular vesicles in coronary artery disease. *Nature Reviews Cardiology, 14*(5), 259–272. https://doi.org/10.1038/nrcardio.2017.7.

Brinton, L. T., Sloane, H. S., Kester, M., & Kelly, K. A. (2015). Formation and role of exosomes in cancer. *Cellular and Molecular Life Sciences: CMLS, 72*(4), 659–671. https://doi.org/10.1007/s00018-014-1764-3.

Campion, C. G., Sanchez-Ferras, O., & Batchu, S. N. (2017). Potential role of serum and urinary biomarkers in diagnosis and prognosis of diabetic nephropathy. *Canadian Journal of Kidney Health and Disease, 4*. https://doi.org/10.1177/2054358117705371. 2054358117705371.

Chargaff, E., & West, R. (1946). The biological significance of the thromboplastic protein of blood. *Journal of Biological Chemistry, 166*(1), 189–197.

Chen, W. X., Liu, X. M., Lv, M. M., Chen, L., Zhao, J. H., Zhong, S. L., ... Tang, J. H. (2014). Exosomes from drug-resistant breast cancer cells transmit chemoresistance by a horizontal transfer of microRNAs. *PLoS One, 9*(4), e95240. https://doi.org/10.1371/journal.pone.0095240.

Cheruvanky, A., Zhou, H., Pisitkun, T., Kopp, J. B., Knepper, M. A., Yuen, P. S. T., & Star, R. A. (2007). Rapid isolation of urinary exosomal biomarkers using a nanomembrane ultrafiltration concentrator. *American Journal of physiology. Renal Physiology, 292*(5), F1657–F1661. https://doi.org/10.1152/ajprenal.00434.2006.

Chiou, N.-T., & Ansel, K. M. (2016). Improved exosome isolation by sucrose gradient fractionation of ultra-centrifuged crude exosome pellets.

Colombo, M., Moita, C., van Niel, G., Kowal, J., Vigneron, J., Benaroch, P., ... Raposo, G. (2013). Analysis of ESCRT functions in exosome biogenesis, composition and secretion highlights the heterogeneity of extracellular vesicles. *Journal of Cell Science, 126*(Pt 24), 5553–5565. https://doi.org/10.1242/jcs.128868.

Colombo, M., Raposo, G., & Thery, C. (2014). Biogenesis, secretion, and intercellular interactions of exosomes and other extracellular vesicles. *Annual Review of Cell and Developmental Biology, 30*, 255–289. https://doi.org/10.1146/annurev-cellbio-101512-122326.

Coumans, F. A. W., Brisson, A. R., Buzas, E. I., Dignat-George, F., Drees, E. E. E., El-Andaloussi, S., ... Nieuwland, R. (2017). Methodological guidelines to study extracellular vesicles. *Circulation Research, 120*(10), 1632–1648. https://doi.org/10.1161/circresaha.117.309417.

Cuffe, J. S. M., Holland, O., Salomon, C., Rice, G. E., & Perkins, A. V. (2017). Review: Placental derived biomarkers of pregnancy disorders. *Placenta, 54*, 104–110. https://doi.org/10.1016/j.placenta.2017.01.119.

Friand, V., David, G., & Zimmermann, P. (2015). Syntenin and syndecan in the biogenesis of exosomes. *Biology of the cell.* https://doi.org/10.1111/boc.201500010.

Greening, D. W., Xu, R., Ji, H., Tauro, B. J., & Simpson, R. J. (2015). A protocol for exosome isolation and characterization: Evaluation of ultracentrifugation, density-gradient separation, and immunoaffinity capture methods. *Methods in Molecular Biology, 1295*, 179–209. https://doi.org/10.1007/978-1-4939-2550-6_15.

Helwa, I., Cai, J., Drewry, M. D., Zimmerman, A., Dinkins, M. B., Khaled, M. L., ... Liu, . (2017). A comparative study of serum exosome isolation using differential ultracentrifugation and three commercial reagents. *PLoS One, 12*(1), e0170628. https://doi.org/10.1371/journal.pone.0170628.

Hoen, E. N. M.N-t., van der Vlist, E. J., Aalberts, M., Mertens, H. C. H., Bosch, B. J., Bartelink, W., ... Wauben, M. H. M. (2012). Quantitative and qualitative flow cytometric analysis of nanosized cell-derived membrane vesicles. *Nanomedicine: Nanotechnology, Biology and Medicine, 8*(5), 712–720. https://doi.org/10.1016/j.nano.2011.09.006.

Isola, A. L., & Chen, S. (2017). Exosomes: The messengers of health and disease. *Current Neuropharmacology, 15*(1), 157–165.

Janas, T., Janas, M. M., Sapon, K., & Janas, T. (2015). Mechanisms of RNA loading into exosomes. *FEBS Letters, 589*(13), 1391–1398. https://doi.org/10.1016/j.febslet.2015.04.036.

Joyce, J. A., & Pollard, J. W. (2009). Microenvironmental regulation of metastasis. *Nature Reviews Cancer, 9*(4), 239–252.

Kanlikilicer, P., Rashed, M. H., Bayraktar, R., Mitra, R., Ivan, C., Aslan, B., … Lopez-Berestein, G. (2016). Ubiquitous release of exosomal tumor suppressor miR-6126 from ovarian cancer cells. *Cancer Research, 76*(24), 7194–7207. https://doi.org/10.1158/0008-5472.can-16-0714.

Keller, S., Sanderson, M. P., Stoeck, A., & Altevogt, P. (2006). Exosomes: From biogenesis and secretion to biological function. *Immunology Letters, 107*(2), 102–108. https://doi.org/10.1016/j.imlet.2006.09.005.

Lane, R. E., Korbie, D., Anderson, W., Vaidyanathan, R., & Trau, M. (2015). Analysis of exosome purification methods using a model liposome system and tunable-resistive pulse sensing. *Scientific Reports, 5*, 7639. https://doi.org/10.1038/srep07639.

Lee, Y., El Andaloussi, S., & Wood, M. J. (2012). Exosomes and microvesicles: Extracellular vesicles for genetic information transfer and gene therapy. *Human Molecular Genetics, 21*(R1), R125–R134. https://doi.org/10.1093/hmg/dds317.

Li, P., Kaslan, M., Lee, S. H., Yao, J., & Gao, Z. (2017). Progress in exosome isolation techniques. *Theranostics, 7*(3), 789–804. https://doi.org/10.7150/thno.18133.

Lobb, R. J., van Amerongen, R., Wiegmans, A., Ham, S., Larsen, J. E., & Moller, A. (2017). Exosomes derived from mesenchymal non-small cell lung cancer cells promote chemoresistance. *International Journal of Cancer, 141*(3), 614–620. https://doi.org/10.1002/ijc.30752.

Luan, X., Sansanaphongpricha, K., Myers, I., Chen, H., Yuan, H., & Sun, D. (2017). Engineering exosomes as refined biological nanoplatforms for drug delivery. *Acta Pharmacologica Sinica, 38*(6), 754–763. https://doi.org/10.1038/aps.2017.12.

Madison, M. N., & Okeoma, C. M. (2015). Exosomes: Implications in HIV-1 pathogenesis. *Viruses, 7*(7), 4093–4118. https://doi.org/10.3390/v7072810.

Martinez, M. C., & Andriantsitohaina, R. (2017). Extracellular vesicles in metabolic syndrome. *Circulation Research, 120*(10), 1674–1686. https://doi.org/10.1161/circresaha.117.309419.

Melo, S. A., Luecke, L. B., Kahlert, C., Fernandez, A. F., Gammon, S. T., Kaye, J., … Kalluri, R. (2015). Glypican-1 identifies cancer exosomes and detects early pancreatic cancer. *Nature*. https://doi.org/10.1038/nature14581.

Millimaggi, D., Mari, M., D'Ascenzo, S., Carosa, E., Jannini, E. A., Zucker, S., … Dolo, V. (2007). Tumor vesicle-associated CD147 modulates the angiogenic capability of endothelial cells. *Neoplasia, 9*(4), 349–357.

Moore, C., Kosgodage, U., Lange, S., & Inal, J. M. (2017). The emerging role of exosome and microvesicle- (EMV-) based cancer therapeutics and immunotherapy. *International Journal of Cancer, 141*(3), 428–436. https://doi.org/10.1002/ijc.30672.

Peinado, H., Aleckovic, M., Lavotshkin, S., Matei, I., Costa-Silva, B., Moreno-Bueno, G., … Lyden, D. (2012). Melanoma exosomes educate bone marrow progenitor cells toward a pro-metastatic phenotype through MET. *Nature Medicine, 18*(6), 883–891. https://doi.org/10.1038/nm.2753.

Piper, R. C., & Katzmann, D. J. (2007). Biogenesis and function of multivesicular bodies. *Annual Review of Cell and Developmental Biology, 23*, 519–547. https://doi.org/10.1146/annurev.cellbio.23.090506.123319.

van der Pol, E., Boing, A. N., Harrison, P., Sturk, A., & Nieuwland, R. (2012). Classification, functions, and clinical relevance of extracellular vesicles. *Pharmacological Reviews, 64*(3), 676–705. https://doi.org/10.1124/pr.112.005983.

Rashed, M. H., Kanlikilicer, P., Rodriguez-Aguayo, C., Pichler, M., Bayraktar, R., Bayraktar, E., … Berestein, G. L. (2017). Exosomal miR-940 maintains SRC-mediated oncogenic activity in cancer cells: A possible role for exosomal disposal of tumor suppressor miRNAs. *Oncotarget, 8*(12), 20145–20164. https://doi.org/10.18632/oncotarget.15525.

Rood, I. M., Deegens, J. K., Merchant, M. L., Tamboer, W. P., Wilkey, D. W., Wetzels, J. F., & Klein, J. B. (2010). Comparison of three methods for isolation of urinary microvesicles to identify biomarkers of nephrotic syndrome. *Kidney International, 78*(8), 810–816. https://doi.org/10.1038/ki.2010.262.

Sampaio, N. G., Cheng, L., & Eriksson, E. M. (2017). The role of extracellular vesicles in malaria biology and pathogenesis. *Malaria Journal, 16*(1), 245. https://doi.org/10.1186/s12936-017-1891-z.

Selmaj, I., Mycko, M. P., Raine, C. S., & Selmaj, K. W. (2017). The role of exosomes in CNS inflammation and their involvement in multiple sclerosis. *Journal of Neuroimmunology, 306,* 1–10. https://doi.org/10.1016/j.jneuroim.2017.02.002.

Shahabipour, F., Barati, N., Johnston, T. P., Derosa, G., Maffioli, P., & Sahebkar, A. (2017). Exosomes: Nanoparticulate tools for RNA interference and drug delivery. *Journal of Cellular Physiology, 232*(7), 1660–1668. https://doi.org/10.1002/jcp.25766.

Skotland, T., Sandvig, K., & Llorente, A. (2017). Lipids in exosomes: Current knowledge and the way forward. *Progress in Lipid Research, 66,* 30–41. https://doi.org/10.1016/j.plipres.2017.03.001.

Sterzenbach, U., Putz, U., Low, L. H., Silke, J., Tan, S. S., & Howitt, J. (2017). Engineered exosomes as vehicles for biologically active proteins. *Molecular Therapy: The Journal of the American Society of Gene Therapy, 25*(6), 1269–1278. https://doi.org/10.1016/j.ymthe.2017.03.030.

Syn, N. L., Wang, L., Chow, E. K., Lim, C. T., & Goh, B. C. (2017). Exosomes in cancer nanomedicine and immunotherapy: Prospects and challenges. *Trends in Biotechnology, 35*(7), 665–676. https://doi.org/10.1016/j.tibtech.2017.03.004.

Takahashi, R. U., Prieto-Vila, M., Hironaka, A., & Ochiya, T. (2017). The role of extracellular vesicle microRNAs in cancer biology. *Clinical Chemistry and Laboratory Medicine, 55*(5), 648–656. https://doi.org/10.1515/cclm-2016-0708.

Trams, E. G., Lauter, C. J., Salem, N., Jr., & Heine, U. (1981). Exfoliation of membrane ecto-enzymes in the form of micro-vesicles. *Biochimica et Biophysica Acta, 645*(1), 63–70.

Vella, L. J. (2014). The emerging role of exosomes in epithelial-mesenchymal-transition in cancer. *Frontiers in Oncology, 4,* 361. https://doi.org/10.3389/fonc.2014.00361.

Vlassov, A. V., Magdaleno, S., Setterquist, R., & Conrad, R. (2012). Exosomes: Current knowledge of their composition, biological functions, and diagnostic and therapeutic potentials. *Biochimica et Biophysica Acta (BBA) - General Subjects, 1820*(7), 940–948. https://doi.org/10.1016/j.bbagen.2012.03.017.

Vyas, N., & Dhawan, J. (2017). Exosomes: Mobile platforms for targeted and synergistic signaling across cell boundaries. *Cellular and Molecular Life Sciences: CMLS, 74*(9), 1567–1576. https://doi.org/10.1007/s00018-016-2413-9.

Webber, J., Steadman, R., Mason, M. D., Tabi, Z., & Clayton, A. (2010). Cancer exosomes trigger fibroblast to myofibroblast differentiation. *Cancer Research, 70*(23), 9621–9630. https://doi.org/10.1158/0008-5472.CAN-10-1722.

Whiteside, T. L. (2017). Exosomes carrying immunoinhibitory proteins and their role in cancer. *Clinical and Experimental Immunology.* https://doi.org/10.1111/cei.12974.

Wolf, P. (1967). The nature and significance of platelet products in human plasma. *British Journal of Haematology, 13*(3), 269–288. https://doi.org/10.1111/j.1365-2141.1967.tb08741.x.

Wubbolts, R., Leckie, R. S., Veenhuizen, P. T., Schwarzmann, G., Mobius, W., Hoernschemeyer, J., … Stoorvogel, W. (2003). Proteomic and biochemical analyses of human B cell-derived exosomes. Potential implications for their function and multivesicular body formation. *Journal of Biological Chemistry, 278*(13), 10963–10972. https://doi.org/10.1074/jbc.M207550200.

Wu, X., Zheng, T., & Zhang, B. (2017). Exosomes in Parkinson's disease. *Neuroscience Bulletin, 33*(3), 331–338. https://doi.org/10.1007/s12264-016-0092-z.

Wu, Y., Deng, W., & Klinke, D. J. (2015). Exosomes: Improved methods to characterize their morphology, RNA content, and surface protein biomarkers. *Analyst, 140*(19), 6631–6642. https://doi.org/10.1039/c5an00688k.

Yan, S., Han, B., Gao, S., Wang, X., Wang, Z., Wang, F., … Sun, B. (2017). Exosome-encapsulated microRNAs as circulating biomarkers for colorectal cancer. *Oncotarget.* https://doi.org/10.18632/oncotarget.18557.

Yang, F., Liao, X., Tian, Y., & Li, G. (2017). Exosome separation using microfluidic systems: Size-based, immunoaffinity-based and dynamic methodologies. *Biotechnology Journal, 12*(4). https://doi.org/10.1002/biot.201600699.

2

Extracellular Vesicle Biogenesis in Cancer

Stephanie N. Hurwitz, David G. Meckes Jr.

FLORIDA STATE UNIVERSITY COLLEGE OF MEDICINE, TALLAHASSEE, FL, UNITED STATES

CHAPTER OUTLINE

1. Introduction 11

2. Microvesicle Biogenesis 13

3. Exosome Biogenesis... 14

4. An Increasing Importance of Examining Diverse Vesicles........................ 17

5. CD63-Dependent Vesicle Production 19

6. Vesicle Secretion Is Linked to Intercellular Signaling 20

References 20

1. Introduction

It is an exciting time for the field of extracellular vesicles (EVs) as researchers have rapidly produced novel insights into pathological processes that offer huge potential for clinical applications. Of great interest is the utility of EVs in the diagnosis and treatment of many types of human cancer. Nearly every cell examined to date produces secreted vesicles harboring proteins, lipids, and nucleic acids for cell-to-cell transmission, and a plethora of growing evidence has suggested that EV secretion is enhanced in tumorigenic cells and after cellular transformation (Al-Nedawi et al., 2008; Ceccarelli et al., 2007; Dvorak et al., 1981; Ji et al., 2008). Furthermore, circulating plasma levels of EVs appear to be increased in patients with active malignant disease (Baran et al., 2010; Kim et al., 2003; Logozzi et al., 2009; Silva et al., 2012; Taylor & Gercel-Taylor, 2008). Numerous studies have found distinct EV cargo released from cancer cells using a variety of genomic, proteomic, and lipidomic techniques. Our recent findings suggest that overall EV protein cargo reliably reflect the protein content of their progenitor tumor cells and thereby provide a wealth of potential membrane-protected biomarkers for clinical utility (Hurwitz, Rider, et al., 2016). A multitude of *in vitro* and preclinical studies have offered further understanding of the contents of cell-derived and circulating EVs with insights into their use in cancer diagnostics. For instance, utility of urinary vesicles may provide novel noninvasive diagnostic and grading tools for prostate cancer (McKiernan et al., 2016).

Although many studies have focused on the appeal of using EVs to reflect ongoing disease processes, much remains to be understood with regard to the role of secreted vesicles in disease pathogenesis. Vesicles have largely been described as enhancers of tumor growth and metastasis through promotion of migratory phenotypes, angiogenesis, and tumor microenvironment formation (Costa-Silva et al., 2015; Hoshino et al., 2015; Melo et al., 2014; Park et al., 2010; Peinado et al., 2012). Manipulation of stromal cell phenotype in a future tumor microenvironment is also a likely role of EVs in the context of cancer (Cho et al., 2011; Cho, Park, Lim, & Lee, 2012; Hoshino et al., 2015; Webber, Steadman, Mason, Tabi, & Clayton, 2010; Webber et al., 2015). Furthermore, the activity of EVs in immunological response has been well described, although not completely understood, and may enhance both tumorigenic and antitumorigenic responses through a number of manners, including tumor antigen presentation or T cell cytotoxicity (André et al., 2002; Chaput et al., 2004; Schartz, Chaput, André, & Zitvogel, 2002; Taylor & Gerçel-Taylor, 2005; Zitvogel et al., 1998).

It is increasingly evident that secreted vesicles likely play profound roles in cancer development and progression and may further aid in clinical diagnostic tools for early detection. However, perhaps of even greater significance is the potential therapeutic value of EVs. Indeed, a recent *Nature* study demonstrated the potential efficacy of engineered vesicles in combating specific cancer mutations, a large step toward personalized cancer therapy (Kamerkar et al., 2017). The authors of the study showed that electroporated exosomes loaded with a short interfering RNA (siRNA) targeting a specific mutated *KRAS* gene in cancer cells suppressed pancreatic tumor growth and metastatic behavior in mice. In the past two decades, appreciable advances have been made in understanding mechanisms of vesicle biogenesis and secretion from a variety of cell types. Clearly, a more complete knowledge of EV formation will be necessary for therapeutic application of EVs. A major question remains: what underlies the mechanism of increased vesicle secretion observed in tumorigenesis? Ongoing research into the pathways involved in cancer EV biogenesis will pave the way for interventions such as blockade of pathogenic EV secretion, inhibition of tumorigenic cargo packaging and cell-to-cell transmission, or promotion of healthy, tumor-suppressing EV populations. In this chapter, we will review important discoveries made, as well as the gaps remaining, in identifying mechanisms of distinct vesicle population formation and the interaction of intracellular organelle systems in these processes. We also highlight the key pathways and proteins known to be involved in tumor EV biogenesis and cargo sorting, and the use of secreted exogenous tumor virus protein in further understanding EV production in the context of cancer.

2. Microvesicle Biogenesis

EVs are diverse membrane-enclosed vesicles released from cells that have historically been classified by their size or subcellular origin (Fig. 1). Originally believed to be simply an artifact of cell culture or a cellular waste mechanism to discard unwanted material, it is now well established that these tiny vesicles participate in broad cell-to-cell communication events in normal and pathological conditions. EVs generally refer to a broad diversity of secreted vesicles, including microvesicles (MVs), exosomes, apoptotic bodies, and more recently viral particles. Outward budding and fission at the plasma membrane generates vesicles ranging from 150 to 500 nm in size that are termed shedding MVs or ectosomes. However, vesicles less than 100 nm in size have been observed to bud from the plasma membrane and likely copurify with smaller

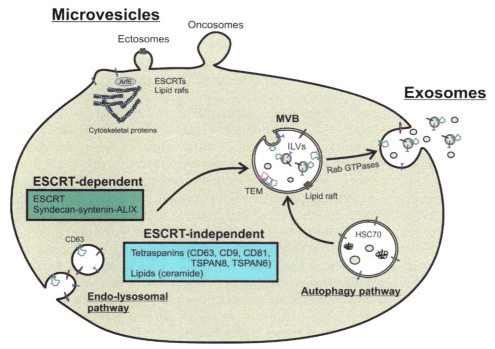

FIGURE 1 Extracellular Vesicle Biogenesis Pathways. Microvesicles (MVs) are a class of extracellular vesicles (EVs) also referred to as ectosomes that are shed from the plasma membrane after budding and fusion events. Additionally, larger MVs, termed oncosomes, have been observed to be released from cancer cells. MVs are believed to require cytoskeletal protein interaction with other *endosomal sorting complexes required for transport* (ESCRT), lipid raft–associated proteins, and GTPases such as Arf6. EVs derived from internal endosomal membranes of multivesicular bodies (MVBs) are called exosomes. Exosomes can be formed through ESCRT-dependent or ESCRT-independent pathways. ESCRT complexes and syndecan–syntenin–Alix interactions can guide endosomal cargo sorting and intraluminal (ILV) budding into MVBs. Alternatively, ESCRT-independent mechanisms requiring tetraspanin proteins or ceramide can be involved in exosome biogenesis. Recent evidence also suggests convergence of endo/lysosomal and autophagy pathways that contribute to cargo sorting into ILVs and exosome formation. *TEM*, tetraspanin-enriched microdomain.

EV populations (Gould, Booth, & Hildreth, 2003). Additionally, large MVs over 500 nm in size, termed oncosomes have been found to be released from certain cancer cells (Di Vizio et al., 2009; Di Vizio et al., 2012; Minciacchi, Freeman, & Di Vizio, 2015; Minciacchi, You et al., 2015). MV biogenesis is poorly understood but likely utilizes similar cellular machinery described for virus budding from the plasma membrane (Gould et al., 2003; Meckes & Raab-Traub, 2011). For instance, the recruitment of *endosomal sorting complexes required for transport* (ESCRT) components to the plasma membrane results in MV release (Nabhan, Hu, Oh, Cohen, & Lu, 2012). Likewise, viruses that bud from the plasma membrane, such as HIV, use similar ESCRT machinery (Meckes & Raab-Traub, 2011).

Another mechanism of MV biogenesis appears to involve alterations in phospholipid microdomains at the plasma membrane, resulting in increased negative membrane curvature and budding that may require the cytoskeletal actin–myosin machinery (Akers, Gonda, Kim, Carter, & Chen, 2013). Membrane alterations can include phosphatidylserine moving from the inner to the outer leaflet of the plasma membrane through the enzymatic activity of flippases, floppases, and scramblases (Akers et al., 2013). Similarly, the lipid metabolism enzyme acid sphingomyelinase controls large vesicle release from the plasma membrane of glial cells (Bianco et al., 2009). Moreover, the membranes of MVs are enriched in lipid rafts, important microdomain organizing centers within the cell that have been demonstrated to associate with numerous cytoskeletal proteins (Del Conde, Shrimpton, Thiagarajan, & López, 2005; Harder, Scheiffele, Verkade, & Simons, 1998; Lingwood & Simons, 2010; Meckes, Menaker, & Raab-Traub, 2013; Simons & Sampaio, 2011). Lipid rafts contain high levels of cholesterol and depletion of cholesterol inhibits MV formation, suggesting the importance of these microdomains in MV biogenesis (Del Conde et al., 2005). In platelets, the calcium-dependent enzyme calpain cleaves cytoskeletal proteins and controls MV formation, pointing to the importance of cytoskeletal remodeling in MV biogenesis (Pasquet, Toti, Nurden, & Dachary-Prigent, 1996). For example, Arf6, an important GTPase that regulates myosin and actin polymerization, has been implicated as a key component of the MV biogenesis machinery in tumor cells (Muralidharan-Chari et al., 2009). In the aforementioned study, inhibition of Arf6 or its downstream targets resulted in fewer MVs released from the plasma membrane. Finally, hypoxia has been shown to induce MV biogenesis and release that is dependent upon the GTPase Rab22a (Wang et al., 2014). Taken together, MV biogenesis requires complex interactions between lipids and cytoskeletal components to drive membrane budding. The diversity of mechanisms described in the literature suggests that multiple cellular pathways may guide the formation of distinct MV populations with unique cargo and functions. Multiple mechanisms of formation could also reflect cell type–specific biogenesis pathways.

3. Exosome Biogenesis

Exosomes are generally accepted as small EVs less than 150 nm in size. Unlike MVs, exosomes originate at internal endosomal membranes of multivesicular bodies (MVBs) (Fig. 1). Budding and fusion events into the lumen of MVBs result in the accumulation of intraluminal vesicles (ILVs). Upon fusion of the MVB with the plasma membrane, ILVs are expelled into the extracellular milieu, now termed exosomes. Despite the broad and rapid ongoing work investigating exosome biogenesis, there is still only a limited understanding of the molecular mechanisms driving the formation and sorting of cargo into exosomes. Adding to the complexity is the existence of exosome subpopulations that copurify together and are likely generated by distinct mechanisms. The most established mechanisms of exosome biogenesis is dependent on the ESCRT machinery, consisting of proteins that assemble into four complexes called ESCRT-0, -I, -II, and -III. Ubiquitinated cargo is recognized by ESCRT-0 protein Hrs on endosomal membranes that recruits ESCRT-I and -II to drive the cargo sorting and intraluminal membrane budding at MVBs (Babst, Katzmann, Snyder, Wendland, & Emr, 2002; Henne, Buchkovich, & Emr, 2011; Raiborg, Malerød, Pedersen, & Stenmark, 2008; Raiborg & Stenmark, 2009). ESCRT-III, comprised of the charged multivesicular body proteins (CHMPs), is subsequently activated at the endosomal membrane by recruitment to ESCRT-I and -II complexes through interactions with Alix and Tsg101. ESCRT-III drives the final steps in ILV formation and promotes deubiquitination of cargo proteins, perhaps once these proteins have been successfully sequestered within an ILV (Henne et al., 2011). Finally, the AAA-ATPase Vps4 is recruited to facilitate membrane fission and dissociation of ESCRT-III, promoting vesicle formation and recycling of the endosomal sorting machinery (Bishop & Woodman, 2001; Liégeois, Benedetto, Garnier, Schwab, & Labouesse, 2006; Wollert et al., 2009).

A large siRNA screen recently performed in HeLa cervical cancer cells has shed new light on the importance of specific ESCRT components and accessory proteins in exosome formation and content (Colombo et al., 2013). In the study by Colombo and colleagues, knock-down of ESCRT-0 components Hrs and Stam1 or the ESCRT-1 component Tsg101 resulted in decreased EV secretion. In particular, EVs produced by ESCRT knock-down cells contained less CD63 and MHC class II. Surprisingly, depletion of Vps4B led to an observable increase in exosome secretion without altering the EV components tested, whereas depletion of Alix resulted in an increase in the number of MHC-II-containing vesicles without affecting total EV levels. Similar experiments performed in MCF7 breast cancer cells found decreased exosome secretion after Tsg101 depletion, but increased exosome production after Vps4 and Alix knock-down (Baietti et al., 2012). These studies highlight the complex cell type–specific exosome biogenesis pathways that exist in cancer cells.

Another ESCRT-dependent mechanism of exosome biogenesis containing syndecan, syntenin, and Alix has been described (Baietti et al., 2012). As indicated in this study, syntenin was found to interact directly with Alix through LYPX(n)L motifs similar to those found in viral late domains required for retrovirus, rhabdovirus, and filovirus budding (Dilley, Gregory, Johnson, & Vogt, 2010; Freed, 2002; Parent et al., 1995; Wills et al., 1994). Syntenin functions as an adapter protein to connect syndecans to Alix on endosomal membranes, whereas Alix acts as the connection to the ESCRT machinery to drive ILV budding. The digestion of heparin sulfate chains on oligomerized syndecans by heparanase induces the release of exosomes containing syntenin-1, syndecan, and CD63, suggesting the importance of syndecans in ILV production (Roucourt, Meeussen, Bao, Zimmermann, & David, 2015). In another study, syntenin-mediated exosome formation was found to be controlled by Arf6 and its effector phospholipase D2 (PLD2) (Ghossoub et al., 2014). These data describe a new role of Arf6 in exosome production that is distinct from its function in MV biogenesis.

Despite the importance of ESCRT components in exosome production, budding of ILVs into MVBs continues to occur in the absence of these complexes, indicating the existence of ESCRT-independent mechanisms of exosome formation (Stuffers, Sem Wegner, Stenmark, & Brech, 2009). The first ESCRT-independent mechanism of ILV formation was described in 2008 when Trajkovic and colleagues found elevated levels of ceramide in EV preparations enriched in exosomes (Trajkovic et al., 2008). The authors further demonstrated a decrease in the release of EVs after inhibition of neutral sphingomyelinase 2 (nSMase2), an enzyme responsible for the conversion of sphingomyelin to ceramide. These results were independent of ESCRT function. Interestingly, the addition of sphingomyelinase alone to large synthetic vesicles containing dioleoylphosphatidylcholine, sphingomyelin, and cholesterol generated *in vitro* was sufficient to drive budding and formation of smaller vesicles. Other lipids implicated in EV biogenesis and cargo sorting include cholesterol, phosphatidic acid (PA), and sphingosine 1-phosphate (SP1) (Kajimoto, Okada, Miya, Zhang, & Nakamura, 2013; Laulagnier et al., 2004; Möbius et al., 2003). It has been suggested that ILV budding likely occurs at raft-based microdomains containing high levels of sphingolipids that are converted to ceramide, as lipid raft microdomains are present in EVs and contain proteins implicated in ILV formation (Dubois, Ronquist, Ek, Ronquist, & Larsson, 2015). Lipid rafts have been found to be important for cargo sorting into exosomes (de Gassart, Geminard, Fevrier, Raposo, & Vidal, 2003; Valapala & Vishwanatha, 2011), and exosomes are enriched in classical lipid raft markers such as flotillins (Keerthikumar et al., 2015). Lipid rafts may also be involved in protein sorting and biogenesis of MVs, as they are enriched on the plasma membrane and implicated in the formation of MV containing tissue factor stomatin, synexin, and sorcin (Del Conde et al., 2005; Salzer, Hinterdorfer, Hunger, Borken, & Prohaska, 2002; Simons & Sampaio, 2011).

Similar to lipid rafts, tetraspanin-enriched microdomains (TEMs) have been postulated to drive EV cargo sorting and biogenesis. Proteomics analyses have revealed EV-associated proteins that interact with CD81 and are devoid in EVs isolated from CD81-deficient animals (Perez-Hernandez et al., 2013). The tetraspanins CD9 and CD82 also direct secretion of cargo including B-catenin into exosomes through a ceramide-dependent pathway (Chairoungdua, Smith, Pochard, Hull, & Caplan, 2010). Likewise, CD9 was found to interact with metalloprotease CD10 and facilitate its exosomal release (Mazurov, Barbashova, & Filatov, 2013). After CD9 knock-out, vesicle production was decreased from mouse-derived dendritic cells (Mazurov et al., 2013). The tetraspanin protein Tspan8 also plays a role in exosome production; when overexpressed, Tspan8 enhances vesicular release of its interacting proteins CD106 and CD49d (Nazarenko et al., 2010). Recent work from our laboratory and the Pegtel laboratory has revealed an important function of CD63 in the trafficking of the Epstein–Barr virus latent membrane protein 1 (LMP1) protein into the exosomal pathway (Hurwitz et al., 2017; Verweij et al., 2011). Interestingly, LMP1 expression enhances exosome production through a CD63-dependent mechanism, providing evidence for a role of CD63 in exosome biogenesis (Hurwitz et al., 2017). Like LMP1, efficient sorting of PMEL to ILVs and exosomal secretion requires CD63 in melanocytes (van Niel et al., 2011). In a recent study, we described CD63-dependent versus CD63-independent packaging of many EV proteins, highlighting a definite role of the tetraspanin in sorting cargo involved in MAPK and NF-kB signaling, protein transport, macroautophagy, and MVB assembly (Hurwitz et al., 2018). Finally, the tetraspanin 6 (Tspan6) has recently been reported to recruit syntenin and mediate secretion of exosomes containing amyloid precursor protein fragments (Guix et al., 2017). Together, these data point to the importance of tetraspanins in EV cargo sorting and exosome biogenesis.

In addition, autophagy pathways have recently been implicated in endosomal cargo trafficking and exosome secretion (Papandreou & Tavernarakis, 2017). The induction of macroautophagy by nutrient deprivation or other cellular stressors such as viral infection results in the engulfment of cytoplasmic material into membrane-enclosed autophagosomes. Autophagosomes can deliver their cargo to lysosomes for degradation or fuse with MVBs for secretion in a process termed secretory autophagy (Ponpuak et al., 2015). An alternative pathway called microautophagy involves the incorporation of cytoplasmic components into endo/lysosomal membranes through the adapter protein Hsc70 (Baixauli, López-Otín, & Mittelbrunn, 2014; Sahu et al., 2011). Interestingly, Hsc70 is abundantly present in EVs, suggesting microautophagy processes may regulate trafficking of protein cargo to exosomes (Hurwitz, Rider, et al., 2016; Hurwitz et al., 2018). Together, macro- and microautophagy likely represent two important mechanisms of protein sorting through the endosomal system, which influence the production and components of exosomes. The intersection of autophagic processes and exosome biogenesis clearly warrants further investigation.

4. An Increasing Importance of Examining Diverse Vesicles

With a growing appreciation of vesicle diversity across cell types, tissue origins, and even subcellular entities, examination of a broad range of EV sources has been challenging but necessary. The collection of human cancer cells compiled by the National Cancer Institute (NCI-60 panel) has played a large role in providing initial characterizations of cancer vesicle contents. Using a proteomics approach, Keerthikumar and coauthors used a subset of the NCI-60 panel to demonstrate the enrichment of oncogenic proteins packaged into exosomes versus MVs, including the differential expression of many Rab GTPases in smaller exosome-sized vesicles (Keerthikumar et al., 2015). Analysis of the protein cargo of NCI-60 ovarian cancer cell lines also highlighted the sorting of MAPK and PI3K signaling proteins into EVs, supporting a role of EVs in cell-to-cell signaling and potential tumor growth and migration (Sinha, Ignatchenko, Ignatchenko, Mejia-Guerrero, & Kislinger, 2014). Recently, multiple laboratories have reported on differential expression of proteins and RNAs in subsets of the NCI-60 cells compared with human primary cells, providing early evidence of novel markers of aggressive or metastatic cell behavior (Hoshino et al., 2015; Ji et al., 2013; Ren et al., 2017).

Of perhaps more broad relevance, proteomic analyses of cancer cells from subsets of the NCI-60 panel have contributed to recognition of regulated cargo sorting into EVs and general mechanisms of vesicle formation and secretion (Kosaka et al., 2013; Phuyal, Hessvik,

Skotland, Sandvig, & Llorente, 2014). For instance, Staubach and colleagues revealed major overlap between intracellular lipid raft and vesicular lipid raft contents, indicating a likely involvement of these detergent-resistant membranes in protein and lipid trafficking (Staubach, Razawi, & Hanisch, 2009). In 2016, we conducted a large-scale quantification of vesicle release from the NCI-60 panel by nanoparticle tracking, standardizing the isolation method and cell confluency to describe the number of vesicles produced by diverse cell types and tissue origins (Hurwitz, Conlon, Rider, Brownstein, & Meckes, 2016). In attempts to generate specific associations between the number of EVs secreted by a given cell type and the progenitor cell genetic profile, we applied an algorithmic bioinformatics approach comparing gene transcript expression levels as they correlated to greater vesicle production. We further grossly stratified vesicles by size: small exosome-sized vesicles (less than or equal to 150 nm) and larger microvesicle-sized EVs (greater than 150 nm). Interestingly, a wide variation was observed in vesicle populations secreted by diverse cell lines. Larger cells produced a greater number of EVs in culture, which we hypothesized could be due to increased surface area and endosomal machinery, although it is possible that conversely the increased EV production resulted in subsequent cell growth. Moreover, a subset of cells were observed to secrete predominately small-sized vesicles, whereas others produced much larger EV populations. These cell lines may prove to be useful in studying biogenesis mechanisms of differential EV subtypes. Furthermore, the observations from this study highlight the necessity of characterizing large numbers of diverse cell types and tissue origins in the process of initial EV population and content description.

We hypothesized that EV biogenesis is controlled by specific genes whose levels of expression may reflect differences in EV secretion. Indeed, our genetic bioinformatics analysis across the NCI-60 panel revealed many genes to be positively associated with increasing vesicle production. In particular, small vesicle release was correlated with higher levels of tetraspanin CD63 and GTPase protein transcripts including Rab17, Rab9A, Rab5B, MRas, and ERas. Cholesterol trafficking protein Npc2, ESCRT protein Hrs, and ATP-binding protein ATP6V0D2 were also highlighted for their potential involvement in small vesicle production. Strikingly, although several gene transcripts were similarly associated with larger vesicle production, additional involvement of actin-, myosin-, and dynein-binding proteins were suggested to facilitate large EV formation. Altogether, our findings supported the general consensus that many overlapping mechanisms of EV subpopulation biogenesis likely exist, but that unique pathways may also control the formation of independent vesicle types. It is possible that shared molecular government of budding or scission and release guides both ILV formation at the endosome and MVB or microvesicle release at the plasma membrane (Booth et al., 2006; Raposo & Stoorvogel, 2013). Indeed these mechanisms are also implicated in enveloped virus release, highlighting the similarities between viral particles and EVs, and the utility of virology research in understanding cellular vesicle production (Gould et al., 2003; Meckes, 2015; Meckes, Gunawardena, et al., 2013; Meckes & Raab-Traub, 2011; Meckes et al., 2010).

We later conducted a large-scale proteomic analysis of vesicle content secreted from cells across the NCI-60 panel (Hurwitz, Rider, et al., 2016), identifying a core of 213 conserved vesicular proteins across all cell types. Of these common proteins, 25 positively predicted total vesicle secretion from each cell type. Among the core predictive proteins were tetraspanin CD81, vesicle trafficking protein Sec22b, vesicle-associated membrane protein 3 (VAMP3), and a number of Ras GTPase family members: HRas, NRas, Rab5c, Rab7a, RALb. Other Rab family members including Rab1a, Rab1b, Rab2a, Rab6a, Rab8a, Rab10, Rab11b, and Rab14 were found in vesicles from all cells, though not necessarily correlated with secretion quantity. Together with previous evidence supporting the functions of many Ras GTPases in vesicle trafficking and secretion (Chavrier & Goud, 1999; Hsu et al., 2010; Hutagalung & Novick, 2011; Li et al., 2014; Ostrowski et al., 2010; Savina, Fader, Damiani, & Colombo, 2005; Savina, Vidal, & Colombo, 2002; Stenmark, 2009; Wang et al., 2014), this study suggested the conserved secretion of this family of proteins across cell types. Similar to Arf6 that has been demonstrated to contribute to mechanisms of both MV and exosome biogenesis, Arf4 was identified in this study as a conserved and predictive protein of vesicle release. Several heat shock proteins were identified in all cell-derived vesicle isolates, including Hspa1b and Hspb1, indicating a potential conserved role of autophagic processes in the formation of EVs. Finally, in line with a study by Hoshino and colleagues (Hoshino et al., 2015), integrin beta-1 was identified as a core vesicle protein as well. Altogether, examination of vesicle quantity and content across a broad range of cells has provided important insight into the conserved molecular mechanisms governing EV biogenesis.

5. CD63-Dependent Vesicle Production

Although the use of large-scale "omics" approaches are being increasingly employed in the EV field to characterize content, focused mechanistic studies into EV biogenesis still remain scarce. In our previous studies, we found the historical exosome tetraspanin marker CD63 to be positively correlated with vesicle secretion in a large population of cells (Hurwitz, Conlon, et al., 2016; Hurwitz, Rider, et al., 2016). In the initial study, we proceeded to efficiently knock out CD63 using a CRISPR/Cas9 system and confirmed a reduction in small vesicle release (Hurwitz, Conlon, et al., 2016). Others' work similarly demonstrated a reduction in EV protein secretion after knock-down of CD63 (Verweij et al., 2011). Interestingly, CD63 was demonstrated to interact with an Epstein–Barr virus oncoprotein called LMP1 that traffics through the endosomal pathway and is secreted efficiently into EVs (Verweij et al., 2011). We hypothesized that recruitment of CD63 to LMP1 on endosomal membranes directs packaging of the viral protein into exosomes. Fascinatingly, we found EV production to be significantly increased after introduction of LMP1 into a variety of epithelial and B cells and that this augmentation of EV secretion was dependent on the presence of CD63 in cells (Hurwitz et al., 2017). This observation was consistent with the multitude of previous studies demonstrating EV enhancement in the context of tumorigenesis and may provide a unique means to understand how vesicle production is increased in cancer. Furthermore, LMP1 was efficiently packaged into CD63-positive vesicles and trafficking of LMP1 into EVs required CD63. Even intracellular localization of LMP1 appeared disrupted in the absence of CD63. Altogether, these data point to an important role of CD63 in vesicle cargo sorting and EV production. Surprisingly, we found that CD63 did not localize to intracellular lipid raft microdomains, and LMP1 trafficking to lipid rafts remained intact after CD63 knock-out. These findings suggest a lipid raft-independent mechanism of LMP1 and CD63 cotrafficking into EVs, and we instead propose the protein–protein interaction may occur in separate TEM compartments in cells. Given the known enrichment of TEMs in secreted EVs, and the overlapping lipid and protein composition of TEMs and EVs, this trafficking mechanism should be further explored (Le Naour, André, Boucheix, & Rubinstein, 2006; Perez-Hernandez et al., 2013).

6. Vesicle Secretion Is Linked to Intercellular Signaling

Importantly, an inverse association between EV secretion and intracellular signaling was described by Verweij and coauthors in the context of the EBV oncoprotein LMP1 (Verweij et al., 2011). More recently, we expanded on this observation, demonstrating specific increases in noncanonical NF-κB, MAPK/ERK, and mTOR signaling in the absence of CD63-dependent LMP1 secretion (Hurwitz et al., 2017; Hurwitz et al., 2018). These findings implicate a role of EVs in downregulating oncogenic signaling, likely through extracellular secretion of signal transduction molecules. However, packaging of oncogenic molecules may explain the increasingly evident impact of EV secretion on the surrounding tumor microenvironment and metastasis to distal sites. It is therefore clear that EVs represent a complex regulatory function of cells to monitor intracellular homeostatic signaling but simultaneously have innumerous impacts on surrounding cells and throughout the body. Through our work, we have demonstrated the utility in examining an exogenous oncogenic viral protein encoded by EBV to understand vesicular trafficking and signal transduction mechanisms within cells. As multiple studies point to a close connection of virus formation with molecular mechanisms governing EV biogenesis, similar models of viral protein endosomal trafficking will likely prove insightful. Altogether, a further understanding of ESCRT-, ceramide-, and tetraspanin-dependent EV biogenesis mechanisms will shed insight into important implications of secreted vesicles in tumor cell growth and metastasis.

References

Akers, J. C., Gonda, D., Kim, R., Carter, B. S., & Chen, C. C. (2013). Biogenesis of extracellular vesicles (EV): Exosomes, microvesicles, retrovirus-like vesicles, and apoptotic bodies. *Journal of Neuro-Oncology*, *113*(1), 1–11. https://doi.org/10.1007/s11060-013-1084-8.

Al-Nedawi, K., Meehan, B., Micallef, J., Lhotak, V., May, L., Guha, A., & Rak, J. (2008). Intercellular transfer of the oncogenic receptor EGFRvIII by microvesicles derived from tumour cells. *Nature Cell Biology*, *10*(5), 619–624. https://doi.org/10.1038/ncb1725.

André, F., Schartz, N. E., Chaput, N., Flament, C., Raposo, G., Amigorena, S., … Zitvogel, L. (2002). Tumor-derived exosomes: A new source of tumor rejection antigens. *Vaccine*, *20*(suppl 4), A28–A31.

Babst, M., Katzmann, D. J., Snyder, W. B., Wendland, B., & Emr, S. D. (2002). Endosome-associated complex, ESCRT-II, recruits transport machinery for protein sorting at the multivesicular body. *Developmental Cell*, *3*(2), 283–289.

Baietti, M. F., Zhang, Z., Mortier, E., Melchior, A., Degeest, G., Geeraerts, A., ... David, G. (2012). Syndecan-syntenin-ALIX regulates the biogenesis of exosomes. *Nature Cell Biology*, *14*(7), 677–685. https://doi.org/10.1038/ncb2502.

Baixauli, F., López-Otín, C., & Mittelbrunn, M. (2014). Exosomes and autophagy: Coordinated mechanisms for the maintenance of cellular fitness. *Frontiers in Immunology*, *5*, 403. https://doi.org/10.3389/fimmu.2014.00403.

Baran, J., Baj-Krzyworzeka, M., Weglarczyk, K., Szatanek, R., Zembala, M., Barbasz, J., ... Szczepanik, A. (2010). Circulating tumour-derived microvesicles in plasma of gastric cancer patients. *Cancer Immunology, Immunotherapy*, *59*(6), 841–850. https://doi.org/10.1007/s00262-009-0808-2.

Bianco, F., Perrotta, C., Novellino, L., Francolini, M., Riganti, L., Menna, E., ... Verderio, C. (2009). Acid sphingomyelinase activity triggers microparticle release from glial cells. *The EMBO Journal*, *28*(8), 1043–1054. https://doi.org/10.1038/emboj.2009.45.

Bishop, N., & Woodman, P. (2001). TSG101/mammalian VPS23 and mammalian VPS28 interact directly and are recruited to VPS4-induced endosomes. *Journal of Biological Chemistry*, *276*(15), 11735–11742. https://doi.org/10.1074/jbc.M009863200.

Booth, A. M., Fang, Y., Fallon, J. K., Yang, J. M., Hildreth, J. E., & Gould, S. J. (2006). Exosomes and HIV Gag bud from endosome-like domains of the T cell plasma membrane. *The Journal of Cell Biology*, *172*(6), 923–935. https://doi.org/10.1083/jcb.200508014.

Ceccarelli, S., Visco, V., Raffa, S., Wakisaka, N., Pagano, J. S., & Torrisi, M. R. (2007). Epstein-Barr virus latent membrane protein 1 promotes concentration in multivesicular bodies of fibroblast growth factor 2 and its release through exosomes. *International Journal of Cancer*, *121*(7), 1494–1506. https://doi.org/10.1002/ijc.22844.

Chairoungdua, A., Smith, D. L., Pochard, P., Hull, M., & Caplan, M. J. (2010). Exosome release of β-catenin: A novel mechanism that antagonizes Wnt signaling. *The Journal of Cell Biology*, *190*(6), 1079–1091. https://doi.org/10.1083/jcb.201002049.

Chaput, N., Taïeb, J., Schartz, N. E., André, F., Angevin, E., & Zitvogel, L. (2004). Exosome-based immunotherapy. *Cancer Immunology, Immunotherapy*, *53*(3), 234–239. https://doi.org/10.1007/s00262-003-0472-x.

Chavrier, P., & Goud, B. (1999). The role of ARF and Rab GTPases in membrane transport. *Current Opinion in Cell Biology*, *11*(4), 466–475. https://doi.org/10.1016/S0955-0674(99)80067-2.

Cho, J. A., Park, H., Lim, E. H., Kim, K. H., Choi, J. S., Lee, J. H., ... Lee, K. W. (2011). Exosomes from ovarian cancer cells induce adipose tissue-derived mesenchymal stem cells to acquire the physical and functional characteristics of tumor-supporting myofibroblasts. *Gynecologic Oncology*, *123*(2), 379–386. https://doi.org/10.1016/j.ygyno.2011.08.005.

Cho, J. A., Park, H., Lim, E. H., & Lee, K. W. (2012). Exosomes from breast cancer cells can convert adipose tissue-derived mesenchymal stem cells into myofibroblast-like cells. *International Journal of Oncology*, *40*(1), 130–138. https://doi.org/10.3892/ijo.2011.1193.

Colombo, M., Moita, C., van Niel, G., Kowal, J., Vigneron, J., Benaroch, P., ... Raposo, G. (2013). Analysis of ESCRT functions in exosome biogenesis, composition and secretion highlights the heterogeneity of extracellular vesicles. *Journal of Cell Science*, *126*(Pt 24), 5553–5565. https://doi.org/10.1242/jcs.128868.

Costa-Silva, B., Aiello, N. M., Ocean, A. J., Singh, S., Zhang, H., Thakur, B. K., ... Lyden, D. (2015). Pancreatic cancer exosomes initiate pre-metastatic niche formation in the liver. *Nature Cell Biology*, *17*(6), 816–826. https://doi.org/10.1038/ncb3169.

de Gassart, A., Geminard, C., Fevrier, B., Raposo, G., & Vidal, M. (2003). Lipid raft-associated protein sorting in exosomes. *Blood*, *102*(13), 4336–4344. https://doi.org/10.1182/blood-2003-03-0871.

Del Conde, I., Shrimpton, C. N., Thiagarajan, P., & López, J. A. (2005). Tissue-factor-bearing microvesicles arise from lipid rafts and fuse with activated platelets to initiate coagulation. *Blood*, *106*(5), 1604–1611. https://doi.org/10.1182/blood-2004-03-1095.

Di Vizio, D., Kim, J., Hager, M. H., Morello, M., Yang, W., Lafargue, C. J., … Freeman, M. R. (2009). Oncosome formation in prostate cancer: Association with a region of frequent chromosomal deletion in metastatic disease. *Cancer Research*, *69*(13), 5601–5609. https://doi.org/10.1158/0008-5472.CAN-08-3860.

Di Vizio, D., Morello, M., Dudley, A. C., Schow, P. W., Adam, R. M., Morley, S., … Freeman, M. R. (2012). Large oncosomes in human prostate cancer tissues and in the circulation of mice with metastatic disease. *The American Journal of Pathology*, *181*(5), 1573–1584. https://doi.org/10.1016/j.ajpath.2012.07.030.

Dilley, K. A., Gregory, D., Johnson, M. C., & Vogt, V. M. (2010). An LYPSL late domain in the gag protein contributes to the efficient release and replication of Rous sarcoma virus. *Journal of Virology*, *84*(13), 6276–6287. https://doi.org/10.1128/JVI.00238-10.

Dubois, L., Ronquist, K. K., Ek, B., Ronquist, G., & Larsson, A. (2015). Proteomic profiling of detergent resistant membranes (lipid rafts) of prostasomes. *Molecular & Cellular Proteomics*, *14*(11), 3015–3022. https://doi.org/10.1074/mcp.M114.047530.

Dvorak, H. F., Quay, S. C., Orenstein, N. S., Dvorak, A. M., Hahn, P., Bitzer, A. M., & Carvalho, A. C. (1981). Tumor shedding and coagulation. *Science*, *212*(4497), 923–924.

Freed, E. O. (2002). Viral late domains. *Journal of Virology*, *76*(10), 4679–4687.

Ghossoub, R., Lembo, F., Rubio, A., Gaillard, C. B., Bouchet, J., Vitale, N., … Zimmermann, P. (2014). Syntenin-ALIX exosome biogenesis and budding into multivesicular bodies are controlled by ARF6 and PLD2. *Nature Communications*, *5*, 3477. https://doi.org/10.1038/ncomms4477.

Gould, S. J., Booth, A. M., & Hildreth, J. E. (2003). The Trojan exosome hypothesis. *Proceedings of the National Academy of Sciences of the United States of America*, *100*(19), 10592–10597. https://doi.org/10.1073/pnas.1831413100.

Guix, F. X., Sannerud, R., Berditchevski, F., Arranz, A. M., Horré, K., Snellinx, A., … De Strooper, B. (2017). Tetraspanin 6: A pivotal protein of the multiple vesicular body determining exosome release and lysosomal degradation of amyloid precursor protein fragments. *Molecular Neurodegeneration*, *12*(1), 25. https://doi.org/10.1186/s13024-017-0165-0.

Harder, T., Scheiffele, P., Verkade, P., & Simons, K. (1998). Lipid domain structure of the plasma membrane revealed by patching of membrane components. *The Journal of Cell Biology*, *141*(4), 929–942.

Henne, W. M., Buchkovich, N. J., & Emr, S. D. (2011). The ESCRT pathway. *Developmental Cell*, *21*(1), 77–91. https://doi.org/10.1016/j.devcel.2011.05.015.

Hoshino, A., Costa-Silva, B., Shen, T. L., Rodrigues, G., Hashimoto, A., Tesic Mark, M., … Lyden, D. (2015). Tumour exosome integrins determine organotropic metastasis. *Nature*, *527*(7578), 329–335. https://doi.org/10.1038/nature15756.

Hsu, C., Morohashi, Y., Yoshimura, S., Manrique-Hoyos, N., Jung, S., Lauterbach, M. A., … Simons, M. (2010). Regulation of exosome secretion by Rab35 and its GTPase-activating proteins TBC1D10A-C. *The Journal of Cell Biology*, *189*(2), 223–232. https://doi.org/10.1083/jcb.200911018.

Hurwitz, S. N., Conlon, M. M., Rider, M. A., Brownstein, N. C., & Meckes, D. G., Jr. (2016). Nanoparticle analysis sheds budding insights into genetic drivers of extracellular vesicle biogenesis. *Journal of Extracellular Vesicles*, *5*, 31295.

Hurwitz, S. N., Nkosi, D., Conlon, M. M., York, S. B., Liu, X., Tremblay, D. C., & Meckes, D. G., Jr. (2017). CD63 regulates Epstein-Barr virus LMP1 exosomal packaging, enhancement of vesicle production, and noncanonical NF-κB signaling. *Journal of Virology*, *91*(5). https://doi.org/10.1128/JVI.02251-16.

Hurwitz, S. N., Rider, M. A., Bundy, J. L., Liu, X., Singh, R. K., & Meckes, D. G., Jr. (2016). Proteomic profiling of NCI-60 extracellular vesicles uncovers common protein cargo and cancer type-specific biomarkers. *Oncotarget*. https://doi.org/10.18632/oncotarget.13569.

Hurwitz, S.N., Cheerathodi, M.R., Nkosi, D., York, S.B., Meckes, D.G., Jr. (2018). Tetraspanin CD63 bridges autophagic and endosomal processes to regulate exosomal secretion and intracellular signaling of Epstein-Barr virus LMP1. *Journal of Virology*, *92*(5), 223–232. https://doi.org/10.1128/JVI.01969-17.

Hutagalung, A. H., & Novick, P. J. (2011). Role of Rab GTPases in membrane traffic and cell physiology. *Physiological Reviews*, *91*(1), 119–149. https://doi.org/10.1152/physrev.00059.2009.

Ji, H., Erfani, N., Tauro, B. J., Kapp, E. A., Zhu, H. J., Moritz, R. L., … Simpson, R. J. (2008). Difference gel electrophoresis analysis of Ras-transformed fibroblast cell-derived exosomes. *Electrophoresis*, *29*(12), 2660–2671. https://doi.org/10.1002/elps.200800015.

Ji, H., Greening, D. W., Barnes, T. W., Lim, J. W., Tauro, B. J., Rai, A., … Simpson, R. J. (2013). Proteome profiling of exosomes derived from human primary and metastatic colorectal cancer cells reveal differential expression of key metastatic factors and signal transduction components. *Proteomics*, *13*(10–11), 1672–1686. https://doi.org/10.1002/pmic.201200562.

Kajimoto, T., Okada, T., Miya, S., Zhang, L., & Nakamura, S. (2013). Ongoing activation of sphingosine 1-phosphate receptors mediates maturation of exosomal multivesicular endosomes. *Nature Communications*, *4*, 2712. https://doi.org/10.1038/ncomms3712.

Kamerkar, S., LeBleu, V. S., Sugimoto, H., Yang, S., Ruivo, C. F., Melo, S. A., … Kalluri, R. (2017). Exosomes facilitate therapeutic targeting of oncogenic KRAS in pancreatic cancer. *Nature*, *546*(7659), 498–503. https://doi.org/10.1038/nature22341.

Keerthikumar, S., Gangoda, L., Liem, M., Fonseka, P., Atukorala, I., Ozcitti, C., … Mathivanan, S. (2015). Proteogenomic analysis reveals exosomes are more oncogenic than ectosomes. *Oncotarget*, *6*(17), 15375–15396.

Kim, H. K., Song, K. S., Park, Y. S., Kang, Y. H., Lee, Y. J., Lee, K. R., … Kim, S. (2003). Elevated levels of circulating platelet microparticles, VEGF, IL-6 and RANTES in patients with gastric cancer: Possible role of a metastasis predictor. *European Journal of Cancer*, *39*(2), 184–191.

Kosaka, N., Iguchi, H., Hagiwara, K., Yoshioka, Y., Takeshita, F., & Ochiya, T. (2013). Neutral sphingomyelinase 2 (nSMase2)-dependent exosomal transfer of angiogenic microRNAs regulate cancer cell metastasis. *Journal of Biological Chemistry*, *288*(15), 10849–10859. https://doi.org/10.1074/jbc.M112.446831.

Laulagnier, K., Grand, D., Dujardin, A., Hamdi, S., Vincent-Schneider, H., Lankar, D., … Record, M. (2004). PLD2 is enriched on exosomes and its activity is correlated to the release of exosomes. *FEBS Letters*, *572*(1–3), 11–14. https://doi.org/10.1016/j.febslet.2004.06.082.

Le Naour, F., André, M., Boucheix, C., & Rubinstein, E. (2006). Membrane microdomains and proteomics: Lessons from tetraspanin microdomains and comparison with lipid rafts. *Proteomics*, *6*(24), 6447–6454. https://doi.org/10.1002/pmic.200600282.

Liégeois, S., Benedetto, A., Garnier, J. M., Schwab, Y., & Labouesse, M. (2006). The V0-ATPase mediates apical secretion of exosomes containing Hedgehog-related proteins in *Caenorhabditis elegans*. *The Journal of Cell Biology*, *173*(6), 949–961. https://doi.org/10.1083/jcb.200511072.

Li, W., Hu, Y., Jiang, T., Han, Y., Han, G., Chen, J., & Li, X. (2014). Rab27A regulates exosome secretion from lung adenocarcinoma cells A549: Involvement of EPI64. *Acta Pathologica, Microbiologica et Immunologica Scandinavica: Acta Pathologica, Microbiologica, et Immunologica Scandinavica*, *122*(11), 1080–1087. https://doi.org/10.1111/apm.12261.

Lingwood, D., & Simons, K. (2010). Lipid rafts as a membrane-organizing principle. *Science*, *327*(5961), 46–50. https://doi.org/10.1126/science.1174621.

Logozzi, M., De Milito, A., Lugini, L., Borghi, M., Calabrò, L., Spada, M., … Fais, S. (2009). High levels of exosomes expressing CD63 and caveolin-1 in plasma of melanoma patients. *PLoS One*, *4*(4), e5219. https://doi.org/10.1371/journal.pone.0005219.

Mazurov, D., Barbashova, L., & Filatov, A. (2013). Tetraspanin protein CD9 interacts with metalloprotease CD10 and enhances its release via exosomes. *The FEBS Journal*, *280*(5), 1200–1213. https://doi.org/10.1111/febs.12110.

McKiernan, J., Donovan, M. J., O'Neill, V., Bentink, S., Noerholm, M., Belzer, S., … Carroll, P. (2016). A novel urine exosome gene expression assay to predict high-grade prostate cancer at initial biopsy. *JAMA Oncol*, *2*(7), 882–889. https://doi.org/10.1001/jamaoncol.2016.0097.

Meckes, D. G., Jr. (2015). Exosomal communication goes viral. *Journal of Virology, 89*(10), 5200–5203. https://doi.org/10.1128/JVI.02470-14.

Meckes, D. G., Jr., Gunawardena, H. P., Dekroon, R. M., Heaton, P. R., Edwards, R. H., Ozgur, S., … Raab-Traub, N. (2013). Modulation of B-cell exosome proteins by gamma herpesvirus infection. *Proceedings of the National Academy of Sciences of the United States of America, 110*(31), E2925–E2933. https://doi.org/10.1073/pnas.1303906110.

Meckes, D. G., Jr., Menaker, N. F., & Raab-Traub, N. (2013). Epstein-Barr virus LMP1 modulates lipid raft microdomains and the vimentin cytoskeleton for signal transduction and transformation. *Journal of Virology, 87*(3), 1301–1311. https://doi.org/10.1128/JVI.02519-12.

Meckes, D. G., Jr., & Raab-Traub, N. (2011). Microvesicles and viral infection. *Journal of Virology, 85*(24), 12844–12854. https://doi.org/10.1128/JVI.05853-11.

Meckes, D. G., Jr., Shair, K. H., Marquitz, A. R., Kung, C. P., Edwards, R. H., & Raab-Traub, N. (2010). Human tumor virus utilizes exosomes for intercellular communication. *Proceedings of the National Academy of Sciences of the United States of America, 107*(47), 20370–20375. https://doi.org/10.1073/pnas.1014194107.

Melo, S. A., Luecke, L. B., Kahlert, C., Fernandez, A. F., Gammon, S. T., Kaye, J., … Kalluri, R. (2015). Glypican-1 identifies cancer exosomes and detects early pancreatic cancer. *Nature, 523*(7559), 177–182. https://doi.org/10.1038/nature14581.

Melo, S. A., Sugimoto, H., O'Connell, J. T., Kato, N., Villanueva, A., Vidal, A., … Kalluri, R. (2014). Cancer exosomes perform cell-independent microRNA biogenesis and promote tumorigenesis. *Cancer Cell, 26*(5), 707–721. https://doi.org/10.1016/j.ccell.2014.09.005.

Minciacchi, V. R., Freeman, M. R., & Di Vizio, D. (2015). Extracellular vesicles in cancer: Exosomes, microvesicles and the emerging role of large oncosomes. *Seminars in Cell & Developmental Biology, 40*, 41–51. https://doi.org/10.1016/j.semcdb.2015.02.010.

Minciacchi, V. R., You, S., Spinelli, C., Morley, S., Zandian, M., Aspuria, P. J., … Di Vizio, D. (2015). Large oncosomes contain distinct protein cargo and represent a separate functional class of tumor-derived extracellular vesicles. *Oncotarget, 6*(13), 11327–11341. https://doi.org/10.18632/oncotarget.3598.

Möbius, W., van Donselaar, E., Ohno-Iwashita, Y., Shimada, Y., Heijnen, H. F., Slot, J. W., & Geuze, H. J. (2003). Recycling compartments and the internal vesicles of multivesicular bodies harbor most of the cholesterol found in the endocytic pathway. *Traffic, 4*(4), 222–231.

Muralidharan-Chari, V., Clancy, J., Plou, C., Romao, M., Chavrier, P., Raposo, G., & D'Souza-Schorey, C. (2009). ARF6-regulated shedding of tumor cell-derived plasma membrane microvesicles. *Current Biology: CB, 19*(22), 1875–1885. https://doi.org/10.1016/j.cub.2009.09.059.

Nabhan, J. F., Hu, R., Oh, R. S., Cohen, S. N., & Lu, Q. (2012). Formation and release of arrestin domain-containing protein 1-mediated microvesicles (ARMMs) at plasma membrane by recruitment of TSG101 protein. *Proceedings of the National Academy of Sciences of the United States of America, 109*(11), 4146–4151. https://doi.org/10.1073/pnas.1200448109.

Nazarenko, I., Rana, S., Baumann, A., McAlear, J., Hellwig, A., Trendelenburg, M., … Zöller, M. (2010). Cell surface tetraspanin Tspan8 contributes to molecular pathways of exosome-induced endothelial cell activation. *Cancer Research, 70*(4), 1668–1678. https://doi.org/10.1158/0008-5472.CAN-09-2470.

Ostrowski, M., Carmo, N. B., Krumeich, S., Fanget, I., Raposo, G., Savina, A., … Thery, C. (2010). Rab27a and Rab27b control different steps of the exosome secretion pathway. *Nature Cell Biology, 12*(1), 19–30. https://doi.org/10.1038/ncb2000. sup 11–13.

Papandreou, M. E., & Tavernarakis, N. (2017). Autophagy and the endo/exosomal pathways in health and disease. *Biotechnology Journal, 12*(1). https://doi.org/10.1002/biot.201600175.

Parent, L. J., Bennett, R. P., Craven, R. C., Nelle, T. D., Krishna, N. K., Bowzard, J. B., … Wills, J. W. (1995). Positionally independent and exchangeable late budding functions of the Rous sarcoma virus and human immunodeficiency virus Gag proteins. *Journal of Virology, 69*(9), 5455–5460.

Park, J. E., Tan, H. S., Datta, A., Lai, R. C., Zhang, H., Meng, W., … Sze, S. K. (2010). Hypoxic tumor cell modulates its microenvironment to enhance angiogenic and metastatic potential by secretion of proteins and exosomes. *Molecular & Cellular Proteomics, 9*(6), 1085–1099. https://doi.org/10.1074/mcp.M900381-MCP200.

Pasquet, J. M., Toti, F., Nurden, A. T., & Dachary-Prigent, J. (1996). Procoagulant activity and active calpain in platelet-derived microparticles. *Thrombosis Research, 82*(6), 509–522.

Peinado, H., Alečković, M., Lavotshkin, S., Matei, I., Costa-Silva, B., Moreno-Bueno, G., … Lyden, D. (2012). Melanoma exosomes educate bone marrow progenitor cells toward a pro-metastatic phenotype through MET. *Nature Medicine, 18*(6), 883–891. https://doi.org/10.1038/nm.2753.

Perez-Hernandez, D., Gutiérrez-Vázquez, C., Jorge, I., López-Martín, S., Ursa, A., Sánchez-Madrid, F., … Yáñez-Mó, M. (2013). The intracellular interactome of tetraspanin-enriched microdomains reveals their function as sorting machineries toward exosomes. *Journal of Biological Chemistry, 288*(17), 11649–11661. https://doi.org/10.1074/jbc.M112.445304.

Phuyal, S., Hessvik, N. P., Skotland, T., Sandvig, K., & Llorente, A. (2014). Regulation of exosome release by glycosphingolipids and flotillins. *The FEBS Journal, 281*(9), 2214–2227. https://doi.org/10.1111/febs.12775.

Ponpuak, M., Mandell, M. A., Kimura, T., Chauhan, S., Cleyrat, C., & Deretic, V. (2015). Secretory autophagy. *Current Opinion in Cell Biology, 35*, 106–116. https://doi.org/10.1016/j.ceb.2015.04.016.

Raiborg, C., Malerød, L., Pedersen, N. M., & Stenmark, H. (2008). Differential functions of Hrs and ESCRT proteins in endocytic membrane trafficking. *Experimental Cell Research, 314*(4), 801–813. https://doi.org/10.1016/j.yexcr.2007.10.014.

Raiborg, C., & Stenmark, H. (2009). The ESCRT machinery in endosomal sorting of ubiquitylated membrane proteins. *Nature, 458*(7237), 445–452. https://doi.org/10.1038/nature07961.

Raposo, G., & Stoorvogel, W. (2013). Extracellular vesicles: Exosomes, microvesicles, and friends. *The Journal of Cell Biology, 200*(4), 373–383. https://doi.org/10.1083/jcb.201211138.

Ren, J., Zhou, Q., Li, H., Li, J., Pang, L., Su, L., … Liu, B. (2017). Characterization of exosomal RNAs derived from human gastric cancer cells by deep sequencing. *Tumor Biologyogy: the Journal of the International Society for Oncodevelopmental Biology and Medicine, 39*(4). https://doi.org/10.1177/1010428317695012.

Roucourt, B., Meeussen, S., Bao, J., Zimmermann, P., & David, G. (2015). Heparanase activates the syndecan-syntenin-ALIX exosome pathway. *Cell Research, 25*(4), 412–428. https://doi.org/10.1038/cr.2015.29.

Sahu, R., Kaushik, S., Clement, C. C., Cannizzo, E. S., Scharf, B., Follenzi, A., … Santambrogio, L. (2011). Microautophagy of cytosolic proteins by late endosomes. *Developmental Cell, 20*(1), 131–139. https://doi.org/10.1016/j.devcel.2010.12.003.

Salzer, U., Hinterdorfer, P., Hunger, U., Borken, C., & Prohaska, R. (2002). Ca(++)-dependent vesicle release from erythrocytes involves stomatin-specific lipid rafts, synexin (annexin VII), and sorcin. *Blood, 99*(7), 2569–2577.

Savina, A., Fader, C. M., Damiani, M. T., & Colombo, M. I. (2005). Rab11 promotes docking and fusion of multivesicular bodies in a calcium-dependent manner. *Traffic, 6*(2), 131–143. https://doi.org/10.1111/j.1600-0854.2004.00257.x.

Savina, A., Vidal, M., & Colombo, M. I. (2002). The exosome pathway in K562 cells is regulated by Rab11. *Journal of Cell Science, 115*(Pt 12), 2505–2515.

Schartz, N. E., Chaput, N., André, F., & Zitvogel, L. (2002). From the antigen-presenting cell to the antigen-presenting vesicle: The exosomes. *Current Opinion in Molecular Therapeutics, 4*(4), 372–381.

Silva, J., Garcia, V., Rodriguez, M., Compte, M., Cisneros, E., Veguillas, P., … Bonilla, F. (2012). Analysis of exosome release and its prognostic value in human colorectal cancer. *Genes Chromosomes & Cancer, 51*(4), 409–418.

Simons, K., & Sampaio, J. L. (2011). Membrane organization and lipid rafts. *Cold Spring Harbor Perspectives in Biology, 3*(10), a004697. https://doi.org/10.1101/cshperspect.a004697.

Sinha, A., Ignatchenko, V., Ignatchenko, A., Mejia-Guerrero, S., & Kislinger, T. (2014). In-depth proteomic analyses of ovarian cancer cell line exosomes reveals differential enrichment of functional categories compared to the NCI 60 proteome. *Biochemical and Biophysical Research Communications, 445*(4), 694–701. https://doi.org/10.1016/j.bbrc.2013.12.070.

Staubach, S., Razawi, H., & Hanisch, F. G. (2009). Proteomics of MUC1-containing lipid rafts from plasma membranes and exosomes of human breast carcinoma cells MCF-7. *Proteomics, 9*(10), 2820–2835. https://doi.org/10.1002/pmic.200800793.

Stenmark, H. (2009). Rab GTPases as coordinators of vesicle traffic. *Nature Reviews. Molecular Cell Biology, 10*(8), 513–525. https://doi.org/10.1038/nrm2728.

Stuffers, S., Sem Wegner, C., Stenmark, H., & Brech, A. (2009). Multivesicular endosome biogenesis in the absence of ESCRTs. *Traffic, 10*(7), 925–937. https://doi.org/10.1111/j.1600-0854.2009.00920.x.

Taylor, D. D., & Gercel-Taylor, C. (2008). MicroRNA signatures of tumor-derived exosomes as diagnostic biomarkers of ovarian cancer. *Gynecologic Oncology, 110*(1), 13–21. https://doi.org/10.1016/j.ygyno.2008.04.033.

Taylor, D. D., & Gerçel-Taylor, C. (2005). Tumour-derived exosomes and their role in cancer-associated T-cell signalling defects. *British Journal of Cancer, 92*(2), 305–311. https://doi.org/10.1038/sj.bjc.6602316.

Trajkovic, K., Hsu, C., Chiantia, S., Rajendran, L., Wenzel, D., Wieland, F., … Simons, M. (2008). Ceramide triggers budding of exosome vesicles into multivesicular endosomes. *Science, 319*(5867), 1244–1247. https://doi.org/10.1126/science.1153124.

Valapala, M., & Vishwanatha, J. K. (2011). Lipid raft endocytosis and exosomal transport facilitate extracellular trafficking of annexin A2. *Journal of Biological Chemistry, 286*(35), 30911–30925. https://doi.org/10.1074/jbc.M111.271155.

van Niel, G., Charrin, S., Simoes, S., Romao, M., Rochin, L., Saftig, P., … Raposo, G. (2011). The tetraspanin CD63 regulates ESCRT-independent and -dependent endosomal sorting during melanogenesis. *Developmental Cell, 21*(4), 708–721. https://doi.org/10.1016/j.devcel.2011.08.019.

Verweij, F. J., van Eijndhoven, M. A., Hopmans, E. S., Vendrig, T., Wurdinger, T., Cahir-McFarland, E., … Pegtel, D. M. (2011). LMP1 association with CD63 in endosomes and secretion via exosomes limits constitutive NF-κB activation. *The EMBO Journal, 30*(11), 2115–2129. https://doi.org/10.1038/emboj.2011.123.

Wang, T., Gilkes, D. M., Takano, N., Xiang, L., Luo, W., Bishop, C. J., … Semenza, G. L. (2014). Hypoxia-inducible factors and RAB22A mediate formation of microvesicles that stimulate breast cancer invasion and metastasis. *Proceedings of the National Academy of Sciences of the United States of America, 111*(31), E3234–E3242. https://doi.org/10.1073/pnas.1410041111.

Webber, J. P., Spary, L. K., Sanders, A. J., Chowdhury, R., Jiang, W. G., Steadman, R., … Clayton, A. (2015). Differentiation of tumour-promoting stromal myofibroblasts by cancer exosomes. *Oncogene, 34*(3), 290–302. https://doi.org/10.1038/onc.2013.560.

Webber, J., Steadman, R., Mason, M. D., Tabi, Z., & Clayton, A. (2010). Cancer exosomes trigger fibroblast to myofibroblast differentiation. *Cancer Research, 70*(23), 9621–9630. https://doi.org/10.1158/0008-5472.CAN-10-1722.

Wills, J. W., Cameron, C. E., Wilson, C. B., Xiang, Y., Bennett, R. P., & Leis, J. (1994). An assembly domain of the Rous sarcoma virus Gag protein required late in budding. *Journal of Virology, 68*(10), 6605–6618.

Wollert, T., Yang, D., Ren, X., Lee, H. H., Im, Y. J., & Hurley, J. H. (2009). The ESCRT machinery at a glance. *Journal of Cell Science, 122*(Pt 13), 2163–2166. https://doi.org/10.1242/jcs.029884.

Zitvogel, L., Regnault, A., Lozier, A., Wolfers, J., Flament, C., Tenza, D., … Amigorena, S. (1998). Eradication of established murine tumors using a novel cell-free vaccine: Dendritic cell-derived exosomes. *Nature Medicine, 4*(5), 594–600.

3

Composition, Physicochemical and Biological Properties of Exosomes Secreted From Cancer Cells

Scott W. Ferguson, Jake S. Megna, Juliane Nguyen

UNIVERSITY AT BUFFALO, THE STATE UNIVERSITY OF NEW YORK, BUFFALO, NY, UNITED STATES

CHAPTER OUTLINE

1. Introduction 27

2. Exosome Biogenesis 29

3. Factors Affecting the Release of Exosomes From Cancer Cells 34

4. Composition of Exosomes 36
   4.1 Proteins in Cancer Exosomes 36
   4.2 Exosomal RNA in Cancer 41
   4.3 Lipids in Exosomes 45

5. Size and Charge of Cancer Exosomes 47

6. Summary and Future Considerations 51

Acknowledgment 51

References 51

1. Introduction

It is now well established that cells use exosomes for cellular communication and as a means to modulate their microenvironment. Thus, characterizing the molecular composition and physicochemical properties of cancer exosomes is not only important for diagnostics or therapeutics but also to fully understand their biological functions. Cancer exosomes promote numerous procancer hallmarks including tumor cell proliferation, extracellular matrix (ECM) degradation, epithelial-to-mesenchymal transition (EMT) and increased cellular motility, increased metastatic potential, promotion of

Diagnostic and Therapeutic Applications of Exosomes in Cancer. https://doi.org/10.1016/B978-0-12-812774-2.00003-1

angiogenesis under hypoxia, immune suppression, and preparation of premetastatic niches in specific and distant organ sites (Azmi, Bao, & Sarkar, 2013; Fong et al., 2015; Harris et al., 2015; Liu et al., 2016; Tadokoro, Umezu, Ohyashiki, Hirano, & Ohyashiki, 2013; Vella, 2014; Ye et al., 2014; Zhang et al., 2015; Zhou et al., 2014; Zhuang et al., 2012a, 2012b). Although there has been some progress in establishing how exosomes exert these pleiotropic effects, many questions still remain. In particular, few studies have attributed specific phenotypic effects to specific molecular cargoes. Studies that go further to explore how these molecular cargoes are selected during cancer exosome formation are even scarcer. Complicating these efforts is the likelihood that the complex communication achieved by exosomes is itself necessarily pleiotropic. Understanding the network of interactions between all of these molecular participants requires detailed knowledge of the exosomal components (proteins, RNAs, lipids) and their respective function.

Information about the molecular composition of cancer exosomes versus those derived from healthy cells is also highly useful for the development of clinical diagnostics to find predictors of disease occurrence or progression through highly specific molecular markers. In one example of a potential exosome-based diagnostic, it was shown that exosome-associated glypican1 (GPC1) was a specific protein marker for pancreatic cancer (Melo et al., 2015). Healthy patients had no GPC1 on their exosomes, whereas patients with pancreatic cancer had circulating GPC1-containing exosomes. Furthermore, the amount of circulating exosomal GPC1 correlated with tumor burden and survival. Findings such as these have immense clinical utility, and there are intensive ongoing efforts to uncover additional "cancer-specific" diagnostic markers either in the form of exo-RNAs or exoproteins.

In addition to their molecular composition, the physicochemical properties of exosomes play an important role in exosome interactions with recipient cells. Exosome size and surface charge affect their adhesion and uptake into neighboring cells and thus how they mediate their phenotypic effects. The surface charge is controlled by the nature of the lipids in the exosome membrane and the incorporated proteins.

With this backdrop in mind, this chapter focuses on the biogenesis of exosomes and the general composition of cancer exosomes, which includes proteins, RNAs, and lipids, and their physicochemical properties such as charge and size. Recent publications indicate preferential enrichment of specific proteins, lipids, and RNAs into cancer exosomes compared with exosomes derived from healthy cells. Although additional studies are required to obtain the full picture of the composition and function of cancer exosomes (current studies are limited to only a few selected cell types), the adaptability of biogenic pathways to specific cargo incorporation (with the goal of promoting phenotypic effects that are beneficial to cancer propagation) are worthy of greater scrutiny. Despite differences in enrichment of specific components, it seems that the biogenesis pathways available to healthy and cancerous cells are largely conserved at some level. Indeed, many of the common exosomal proteins such as tetraspanins are present in cancer exosomes and in exosomes derived from healthy cells.

2. Exosome Biogenesis

Exosomes are generally described as forming within the endosomal pathway in an organelle known as the "multivesicular body" (MVB) (Raposo & Stoorvogel, 2013). The dynamic nature of the endosomal pathway has, however, infused a high degree of complexity into exosome biogenesis research (Grant & Donaldson, 2009). The membrane structures associated with endosomes are involved in internalization of ligands from the extracellular space through numerous pathways mediated by a host of adaptor proteins such as clathrin or caveolin (Mulcahy, Pink, & Carter, 2014).

Within the canonical endocytic pathway, these internalized vesicles are first delivered into early endosomes. Early endosomes proceed toward late endosomes where they accumulate, hence the term "multivesicular" (Luzio et al., 2000; Mellman, 1996). A typical feature of MVBs is cytosolic sampling, where proteins and RNAs are incorporated through inward budding of the endosomal membrane. Some MVBs are destined for lysosomes, which typically concludes the endosomal pathway because the high acidity and presence of hydrolases within this compartment efficiently degrade the contained proteins and RNAs (Luzio et al., 2000).

However, it is increasingly apparent that endosome biogenesis proceeds less along a stepwise pathway and more within a complex network. MVBs can also form from the Golgi complex, and MVBs are not always destined to lysosomes (Campanella et al., 2012; Kwon, Oh, Nacke, Mostov, & Lipschutz, 2016; Ostrowski et al., 2010). Indeed, some MVBs with markers of late endosomes are trafficked to the plasma membrane (PM) where they are released extracellularly to become exosomes. In cancer, MVBs with markers of early endosomes are also released (Gross, Chaudhary, Bartscherer, & Boutros, 2012).

MVBs are morphologically heterogeneous in different cell lines under different conditions. In fact, even within the same cell, different types of MVBs are often present containing intraluminal vesicles (ILVs) of various sizes and molecular composition (White, Bailey, Aghakhani, Moss, & Futter, 2006). It is likely that the heterogeneity of released exosomes and extracellular vesicles at least in part reflects the complex biology of membrane-sorting organelles within cells.

MVBs destined for PM fusion and exosome release have mostly been studied within the context of endosomal sorting complexes required for transport (ESCRT)-dependent biogenesis. However, recent work examining protein and RNA sorting into exosomes has revealed that many cargoes present in exosomes are loaded in an ESCRT-independent fashion. It was demonstrated that knocking out specific components of the ESCRT machinery resulted in differential effects on the total exosomes secreted and the population of released exosomes in terms of subset-specific markers (Colombo et al., 2013). Furthermore, these differences varied between nontumor- and tumor-derived cell lines.

Therefore, the molecular composition of exosomes recovered from various sources is important to characterize and consider to better understand the totality of pathways available to cells when generating distinct sets of released vesicles. Early results suggest that cancer cells have numerous biogenic pathways at their disposal to generate different exosomes and extracellular vesicles (EVs) with diverse biological functions.

The ESCRT machinery is a well-described set of protein complexes conserved across eukaryotes. ESCRT was mostly studied in the context of ubiquitin-mediated protein degradation, an important process that enables cells to recognize and destroy misfolded proteins before they exert any pathological effect. The ESCRT machinery, in general, is important for membrane protein sorting and membrane abscission as vesicles bud from the endosomal membrane and pinch off into the intraluminal space (Schmidt & Teis, 2012).

The involvement of ESCRT in exosome biogenesis was first reported in 2001, when Thery et al. noted that the ESCRT-associated proteins TSG101 and ALG-2-interacting protein X (ALIX) were present in the exosomes derived from various cell lines (Thery et al., 2001). Early studies in reticulocytes also posited a role for ALIX in exosome biogenesis because it was shown to bind to the cytoplasmic domain of the transferrin receptor (TfR). A few years later a competition model was proposed in which binding of ALIX to TfR replaced the heat shock protein 70 (HSC70) and resulted in the receptor being sorted into exosomes (Geminard, De Gassart, Blanc, & Vidal, 2004). Many studies in both cancer and noncancer cell lines have implicated a role for ESCRT-related proteins in exosome formation. In particular the ESCRT-0 protein, HRS, was found to be important for exosome formation in dendritic cells, HEK-293 cells, and tumor cells (Gross et al., 2012; Hoshino et al., 2013; Tamai et al., 2010). It is also possible that a subset of exosomes released from cells is ESCRT-dependent. For instance, dendritic cell exosome release after HRS knockdown only decreased after antigen stimulation and not at equilibrium. Given that MHC-II-bearing dendritic cell exosomes play an important role in antigen presentation, it is possible these vesicles are only released under specific physiological states, i.e., specific vesicle biogenesis and release is a function of the physiological state of the cell. The altered physiology of cancer cells opens up opportunities for looking in more depth at all the possible exosome biogenesis pathways in cancer.

Baietti et al. demonstrated a role for syndecan, syntenin, and ALIX in tumor exosome biogenesis. Syntenin overexpression resulted in increased ALIX-dependent exosome release and knockdown of any of the three proteins impaired exosome release (Baietti et al., 2012). Their studies also showed that the formation of exosomes containing these proteins was dependent on ESCRT-II, ESCRT-III, and the vacuolar sorting protein, VSP4. Syntenin interacted with ALIX directly through LYPX(n)L motifs similar to retroviral budding, highlighting how interdisciplinary studies in the retroviral field may be a useful approach to discovering general rules for protein sorting in membranes and the physical rules governing membrane budding.

In a particularly thought-provoking study, a more general model of exosome biogenesis that accounted for heterogeneity in released exosomes and other membrane vesicles was proposed (Fang et al., 2007). The researchers suggested that some exosome populations are released in an ESCRT-independent manner. In their model, higher-order protein oligomerization, which induces membrane curvature, may be sufficient across divergent protein types to cause budding of endosomal membranes away from the cytoplasm and generate ILVs. They described a process independent of type E VSPs in Jurkat cells whereby any membrane-anchored protein prone to such higher-order oligomerizations seemed sufficient to drive membrane budding and "exosome" formation. At the time, they referred to exosomes as both "immediate" (vesicles budding directly from the PM) and "delayed" (vesicles first formed at the MVB). Because it is still not possible to fully delineate the origin of the vesicles they termed exosomes, it may be that these more general definitions are still the de facto reality in the field of EV-mediated communication.

The prototypical ESCRT biogenesis mechanism has since been reassessed as new evidence has emerged. A closer look at the Colombo report illuminates the complexity of the ESCRT pathway in exosome biogenesis (Colombo et al., 2013). Their work interrogated ESCRT-dependent biogenesis by systematic knockdown of 23 individual ESCRT components in HeLa cells. When HGS (encoding HRS), STAM1 (both components of ESCRT-0), or TSG101 (an ESCRT-I component) were silenced, fewer exosomes were secreted (~50% reduction). These proteins seemed to be especially important for the biogenesis of CD63- and MHC-II-bearing exosomes because the remaining released vesicles were significantly depleted of these markers. When VPS4B was silenced, secretion actually increased (by ~150%) without difference in composition. Silencing of ALIX resulted in an increase of MHC-II in both the cells and exosomes, possibly indicating that ALIX is important for MHC-II degradation and not for the release of exosomes per se (ALIX depletion did not influence secretion in HeLa cells). Conversely, in dendritic cells, silencing ALIX resulted in a ~50% reduction of exosome release. These results not only share at least some similarities with those reported by Baietti et al. (2012) but also contain some disagreements. For example, even though both studies used cancer cells (HeLa and MCF-7), Baietti reported that ALIX knockdown reduced exosome secretion (Table 1). In oligodendrocytes, GPI-anchored proteolipoprotein (PLP) release by exosomes and flotilin-containing exosomes were not affected by depleting any of the tested ESCRT proteins (TSG101, ALIX, or VPS4B). However, in these cells, anthrax toxin–containing exosomes were depleted on knockdown of ALIX or TSG101 (Abrami et al., 2013).

We recently reviewed the importance of exosomal molecular heterogeneity in relation to translating exosomes and EVs into therapeutics (Ferguson & Nguyen, 2016). There are now considerable efforts being made to find accurate methods to describe the subsets of released vesicles, and such efforts need to be married to biogenesis studies in cancer research. The relationship between ESCRT-dependent exosome formation and their cargo is still not fully understood.

Progress is, however, being made in delineating EV subtypes within the heterogeneous population of released vesicles. In 2016, Kowal et al. published a detailed proteomic study of novel markers to characterize heterogeneous populations of EV subtypes, with tetraspanins proving particularly useful for demarcating EV subsets (Kowal et al., 2016). These findings seemed to support the hypothesis that exosomes are generated through tetraspanin networks (Rana & Zoller, 2011). For further reading on tetraspanins and their role in exosome biogenesis, we direct interested readers to a review on exosome biogenesis that describes tetraspanin-enriched microdomains (TeMs), which seem to be implicated in cancer exosome biogenesis (Andreu & Yanez-Mo, 2014).

It was (Perez-Hernandez et al., 2013) shown that certain ESCRT-independent sorting of proteins such as Rac GTPase were dependent on tetraspanins such as CD81. Their proteomic analysis of the TeM interactome accounted for nearly half of the exosomal proteins.

Table 1   Proteins Involved in Exosome Secretion From Cancer Cells and Noncancer Cells. Effects on Secretion and Exosomal Composition Derived From Studies Where These Proteins Were Knocked Down

| | | Cancer Cells | | | | Noncancer Cells | | | |
| | | HeLa | | MCF-7 | | Dendritic Cells | | Oligodendrocytes | |
		Exosome Secretion	Exosome Composition	Exosome Secretion	Exosome Composition	Exosome Secretion	Exosome Composition	Exosome Secretion	Exosome Composition
ESCRT-0	HRS	↓	Decrease in CD63 and MHCII+exosomes			↓	Decrease in ubiquitinated protein cargo		
	STAM1	↓	Decrease in CD63 and MHCII+exosomes						
ESCRT-1	TSG101	↓	Decrease in CD63 and MHCII+exosomes	↓	Strong general decrease in endosomal trafficking			↔,↓	No change in proteolipoprotein- (PLP) or flotilin-containing exosome secretion, decrease in anthrax toxin–containing exosomes
Accessory proteins	VPS4B	↑	No composition changes	↓	General decrease in exosome release			↕	No change in PLP- or flotilin-containing exosome secretion
	ALIX	↕	Increase in MHCII (in exosomes and in cells)	↓	Decrease in exosomal syndecan, CD63 and HSP70			↔,↓	No change in PLP- or flotilin-containing exosome secretion, decrease in anthrax toxin–containing exosomes

TeMs may serve as a platform for exosome protein incorporation through binding motifs such as PDZ (syntenin, for example, is a PDZ-binding protein).

Another exosome biogenesis pathway is the lipid-dependent ceramide/neutral sphingomyelinase (nSMase) pathway initially described in oligodendrocytes, where ceramide generation via the nSMase pathway was necessary for PLP sorting into exosomes. It is unclear how this pathway is applicable outside highly specialized neuronal cells with complicated lipid biology such as oligodendrocytes, although sorting of other exosome cargoes has now been shown to be ceramide dependent (Singh, Pochampally, Watabe, Lu, & Mo, 2014). In melanoma cells, however, reducing nSMase did not affect MVB biogenesis or exosome secretion (van Niel et al., 2011). Instead, ESCRT-independent MVB biogenesis in these cells also seemed to follow a tetraspanin pathway reliant on the tetraspanin CD63.

In summary, exosome biogenesis encompasses both ESCRT-dependent and ESCRT-independent pathways. Many ESCRT-independent pathways appear to rely on tetraspanin-TeMs or lipid domains, such as ceramide. Specific cargoes may be sorted into exosomes via any one of the proposed pathways or via combinations of these pathways. It still remains to be seen if subsets of exosomes arise via distinct biogenic pathways or if these pathways interact to generate heterogeneous populations of exosomes (Fig. 1). It may be the case that further biogenesis pathways or networks exist because general models of membrane budding and abscission, as described above, would certainly allow for this.

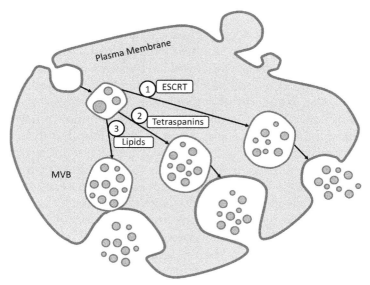

FIGURE 1 Proposed biogenesis pathways of cancer exosomes. Exosomes are often dependent on (1) ESCRT machineries, (2) tetraspanins, or (3) lipids such as ceramide to form. It is unclear to what extent different pathways are connected or independent from one another. It appears likely that these mechanisms act simultaneously on the same MVB, but to different extents in different cell lines. It is also likely that proportions of these machineries are used differently across subsets of formed MVBs and are thus implicated to various extents in the formation of exosome subpopulations.

3. Factors Affecting the Release of Exosomes From Cancer Cells

In several studies, cancer cells were shown to secrete significantly more exosomes than healthy cells. Logozzi and colleagues showed that melanoma-bearing SCID mice had five-fold higher levels of exosomes in the plasma compared with nontumor–bearing mice, and the number of circulating exosomes was proportional to tumor size. These exosomes were not only positive for CD63 but also contained high levels of caveolin. Similar results were found in melanoma patients with advanced disease (stage III–IV). Exosomes were isolated from healthy donors (n = 58) and patients with melanoma (n = 90) and quantified with the Exotest using CD63 and Cav1 as positive markers. The exosome concentration was significantly higher in cancer patients compared with healthy donors: 504 ± 315 CD63$^+$ exosomes in cancer patients vs. 223 ± 125 CD63$^+$ exosomes in healthy donors; 619 ± 310 Cav1$^+$ exosomes in melanoma patients vs. 228 ± 102 Cav1$^+$ exosomes in healthy donors (Logozzi et al., 2009).

Although cancer cells secrete more exosomes than healthy cells, several other factors including exposure to chemotherapeutics and other drugs can affect the exosomal secretion rate. For example, substantial release of EVs from cancer cells was observed only 1 h after exposure to phototherapy or doxorubicin. Prostate cancer cells (PC-3) treated with Foscan (temoprofin, a photosensitizer) and exposed to photodynamic therapy released 4×10^9 EV/ml within an hour of treatment (Aubertin et al., 2016), a ~400-fold increase over controls. Extracellular secretion increased as a function of Foscan concentration over lower non-lethal doses (0.02 μM–0.5 μM) with decreased vesicle release at toxic doses (0.5 μM–10 μM). Doxorubicin-treated PC-3 cells secreted 30 times more vesicles than untreated cells (2×10^9 EV/ml vesicles over a 24 h). Although the majority of the EVs were in the 300–400 μm range, the authors found using magnetic nanoprobes that a fraction of the EVs were of endosomal origin. Future work is needed to quantify what proportion of EVs was exosomal (Table 2).

Docosahexaenoic acid (DHA) (Hannafon et al., 2015), a natural compound with anti-cancer activity that inhibits angiogenesis has also been reported to increase exosome secretion from cancer cells. Changes in exosomal secretion were also associated with miRNA profile alterations. miRNAs (miR-23b, miR-27b, and miR-320b) involved in cellular migration and angiogenesis were significantly increased in exosomes derived from DHA-treated MCF-7 breast cancer cells. Exosomes isolated from these cells inhibited endothelial cell tube formation; and, by inhibiting the proteins responsible for exosome release such as Rab GTPase and Rab27a, the effects could be reversed (Table 2).

In addition to drugs, other factors such as hypoxia or cellular senescence can affect exosomal release from cancer cells. Hypoxia is an almost universal feature of malignant tumors as a result of an imbalance between oxygen supply and consumption by tumor cells. Hypoxia is often negatively correlated with therapeutic outcomes and can contribute to the aggressive phenotype (Vaupel & Harrison, 2004). King and colleagues found that breast cancer cells significantly increased exosome secretion when exposed to moderate (1% oxygen) and severe (0.1% oxygen) hypoxia. At moderate hypoxia, exosome secretion

Table 2 (A) Exosome Secretion From Healthy and Cancer Cells. (B) Effects of Drugs, Hypoxia, and Senescence on Exosome Secretion in Cancer Cells

Cell Type/Line	Ref.	Healthy Cells Versus Cancer Cells	Control Group	Fold Increase in Secretion	Type of Vesicles	Quantification Method
Human melanoma (Me501)	Logozzi et al. (2009)	Tumored SCID mice	Nontumored SCID mice	5	Exosomes	CD63 and Cav1 by ELISA
Humans		Melanoma patients	Healthy donors	2–3	Exosomes	CD63 and Cav1 by ELISA

Cell Type/Line	Ref.	Effects of Drugs, Hypoxia, and Senescence	Control Group	Fold Increase in Secretion	Type of Vesicles	Quantification Method
Human prostate cancer (PC-3)	Aubertin et al. (2016)	Photodynamic therapy: PDT+Foscan	No light exposure	~400	Extracellular vesicles	NTA+flow cytometry
		Doxorubicin	Nontreated cells	30	Extracellular vesicles	NTA+flow cytometry
Human breast cancer (MCF7)	Hannafon et al. (2015)	Docosahexaenoic acid	Nontreated cells	~1.1	Exosomes	CD63-GFP microscopy
Human breast cancer (MDA-MB-231)		Docosahexaenoic acid	Nontreated cells	~1.1	Exosomes	CD63-GFP microscopy
Human breast cancer (MCF7)	King, Michael, and Gleadle (2012)	Hypoxia (1% oxygen)	Normoxia	1.4	Exosomes	NTA
		Severe hypoxia (0.1% oxygen)	Normoxia	1.8	Exosomes	NTA
Human breast cancer (MDA-MB 231)		Hypoxia (1% oxygen)	Normoxia	1.3	Exosomes	NTA
		Severe hypoxia (0.1% oxygen)	Normoxia	2	Exosomes	NTA
Human breast cancer (SKBR3)		Hypoxia (1% oxygen)	Normoxia	1.2	Exosomes	NTA
		Severe hypoxia (0.1% oxygen)	Normoxia	1.9	Exosomes	NTA
Human prostate cancer (LNCaP)	Lehmann et al. (2008)	Irradiation-induced senescence (4 Gy)	Mock-irradiation	3	Exosome-like	DiL membrane staining
		Nutlin-3+irradiation (4 Gy)	Mock-irradiation	~8	Exosome-like	DiL membrane staining
Human lung cancer (A549)	Xiao et al. (2014)	Cisplatin	Nontreated cells	~4.5	Exosomes	BCA protein assay
Human pancreatic cancer (Panc1)	Mikamori et al. (2017)	Pre-miR-155	Control miRNA	~1.4	Exosomes	Bradford protein assay

from MCF-7, SKBR3, and MDA-MB-231 breast cancer cells increased by 1.24–1.4-fold. At 0.1% oxygen levels, exosome production increased 1.77–1.94-fold. Exosomes derived from cells exposed to hypoxia were also 2.77-fold more enriched in miR-210, an miRNA linked to the hypoxia pathway (King et al., 2012) and often upregulated in tumors (Devlin, Greco, Martelli, & Ivan, 2011).

Prostate cancer cells undergoing irradiation-associated senescence have been shown to secrete significantly higher numbers of exosomes or exosome-like microvesicles. TSG101 knockdown inhibited exosome secretion and showed that the vesicles were of endosomal origin. Nutilin-3-induced radiosensitization caused premature senescence and increased the release of exosome-like particles twofold. The authors also showed that senescence increased exosome production not only in cancer cells but also in normal human dermal fibroblasts (NHDFs). NHDFs that reached a state of natural senescence secreted ~15-fold more exosome-like vesicles than proliferating NHDFs cells at lower passage numbers (Lehmann et al., 2008).

Human lung adenocarcinoma cells (A549) treated with cisplatin showed ~4.5-fold increase in exosome secretion compared with nontreated cells. Treatment with cisplatin also led to a significant knockdown of miR-21 and miR-133b in A549 exosomes (Xiao et al., 2014). In addition to small molecules, noncoding RNAs can also affect exosome secretion. miR-155 induced exosome secretion in pancreatic cells (PDAC) by enhancing the formation of MVB (Mikamori et al., 2017).

4. Composition of Exosomes

4.1 Proteins in Cancer Exosomes

Proteins are integral to exosomal function, so significant efforts are underway to characterize the proteins present in exosomes of different origins. These efforts have generated a wealth of data that are being curated in the ExoCarta and Vesiclepedia online repositories, and researchers have used these proteomic data to identify diagnostic biomarkers and to understand the biological function of exosomal proteins. Proteins can be incorporated into the hydrophilic core or integrated into the lipid bilayer of exosomes. Depending on their state (whole or fragmented), they can exert their full function or only partial function. Proteins integrated into the lipid membrane can affect how exosomes interact with recipient cells and play an important role in cellular adhesion and uptake.

A cross-comparison of the exosomal protein expression profiles from four different cancer cell types (prostate cancer, colorectal cancer, squamous cancer, and neuroblastoma) showed that exosomes of different cellular origin displayed unique protein expression profiles and shared only 211 proteins. Common proteins were mostly related to localization and organization processes important for exosome biogenesis. Conversely, proteins that were unique to the different cancer types contained overrepresented classes of proteins with differing biological activities. Neuroblastoma cells had 453 unique proteins, with a significant overrepresentation of proteins involved in intracellular signal transduction;

squamous cell carcinoma had 235 unique proteins with fatty acid oxidation-related proteins being the most overrepresented class; prostate cancer cells had 297 unique proteins with proteolytic proteins being most overrepresented; and colorectal cancer cells had 545 unique proteins, a large number of which were related to RNA/mRNA splicing.

The gene ontology (GO) terms related to the molecular function of the cancer samples were obtained after comparing the proteins with all exosomal proteins in ExoCarta using Panther. This allowed for the assessment of overrepresented classes of proteins specific to different cancer types (Fig. 2). All cancer samples showed overrepresentation of cell adhesion molecules many of which were integrins. Differences between the cancer samples were quite striking: for example, neuroblastoma exosomes contained a high diversity of different small-molecule binding proteins, whereas squamous cancer exosomes were the only cancer type that contained overrepresented proteins related to receptor activity and signal transduction. These proteins seem to be present at the exclusion of a more diverse set of enzymatic proteins, which the other three cancers all contained with high diversity. Interestingly, similar classes of proteins were overrepresented across all four of these cancer types. For example, nucleic acid–binding proteins, especially RNA-binding proteins, were overrepresented to various levels of statistical significance, as were those proteins implicated in protein binding. Within this latter set, cadherin binding was in each case the most statistically overrepresented class of protein-binding proteins.

In general, integral membrane proteins are a common protein class found in exosomes from various cell sources, both healthy and cancerous, albeit to varying extents. Membrane proteins are involved in many of the biological effects of exosomes, not just cell adhesion and targeting. Common membrane proteins include tetraspanins, adhesion molecules such as integrins, thrombospondins, and intercellular adhesion molecules (ICAMs), and glycoproteins (Kooijmans, Vader, van Dommelen, van Solinge, & Schiffelers, 2012; Thery, Zitvogel, & Amigorena, 2002).

Tetraspanins are transmembrane proteins that are particularly promiscuous in their biological activities, including cellular motility, invasion, cellular adhesion, and membrane fusion. For instance, Yue, Mu, Erb, and Zoller (2015) showed that the metastatic potential of adenocarcinoma cell exosomes was mediated by the presence of the tetraspanins CD151 and Tspan8, with double knockdown reducing the cellular motility and invasive potential of the resulting exosomes. *In vivo*, CD151 or Tspan8 knockout animals had reduced lung metastases compared with controls, whereas four out of five animals were completely negative for lung metastases in double knockout animals. These results were partially attributed to the ability of the wild-type exosomes to induce expression of chemokine receptors, such as CXCR4 in recipient cells. Given the potential importance of tetraspanins in exosome biogenesis and incorporation of additional cargoes, it is unclear to what degree the observed effects were directly due to CD151 or Tspan8 versus other codependent protein cargoes.

Besides promotion of cellular migration and metastasis, tetraspanins also mediate additional cancer hallmarks such as immune suppression. Clayton, Al-Taei, Webber, Mason, and Tabi (2011) reported that 20% of total hydrolytic activity against soluble ATP in the pleural effusions of mesothelioma patients was directly attributable to exosomes, at

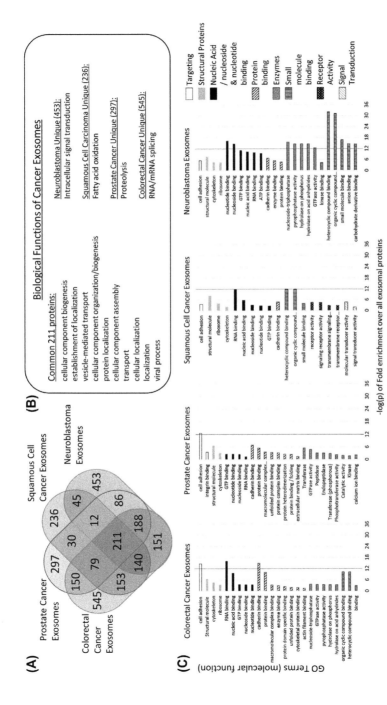

FIGURE 2 (A) Venn diagram of proteins identified in exosomes from colorectal cancer cells, prostate cancer cells, squamous cancer cells, and neuroblastoma. (B) Biological function of shared and unique exosomal proteins, (C) GO terms describing the molecular function of exosomal proteins derived from colorectal cancer, prostate cancer, squamous cancer, and neuroblastoma.

least partly through the tetraspanins CD39 and CD73. Adenosine triggers a cAMP response through the A2A receptor, and signaling in A2A receptor-positive T-cells inhibits their activation. Therefore, by hydrolyzing ATP within the extracellular space of the tumor, the cancer-released exosomes maintained a steady inhibitory effect over T-cells.

Apart from promoting cellular motility and metastasis, tetraspanins and tetraspanin-enriched domains have been implicated in mediating adhesion and uptake of exosomes into recipient cells. Although it is not fully understood how exosomes are exactly taken up into recipient cells, it has been suggested that integrins and ICAMs, which often found integrated into the tetraspanin-enriched domains, are responsible for binding and adhesion. For instance, ICAM-1 found on cancer exosomes was shown to bind to leukocytes, thereby blocking the leukocyte binding to endothelial cells (Lee et al., 2010).

In recent work from Hoshino et al. (2015), it was shown that the integrin constituents of cancer exosomes determine their organotropism and their binding to recipient cells. Exosomes with integrin alpha 6 (ITGα6) showed lung tropism, whereas exosomes with integrin beta 5 (ITGβ5) displayed liver tropism. Furthermore, they identified ITGβ3 as the main determinant of brain tropism of exosomes. Blocking these integrins with antibodies inhibited migration of those exosomes to the respective organs. This finding has offered an explanation as to why certain cancers preferentially metastasize to specific tissues.

EMT is one of the most studied cancer hallmarks and is often attributed to proteins found in cancer exosomes. Looking at the overrepresented proteins in cancer cells in ExoCarta, it is apparent that proteins involved in cadherin biology are often overrepresented in cancer exosomes; the GO term related to the molecular function of cadherin binding was overrepresented in prostate, colorectal, squamous and neuroblastoma cancers (Fig. 2). As cadherin is an important protein for cell adhesion and maintaining the epithelial morphology of cells, it seems likely that the protein repertoire of cancer exosomes may be especially important in EMT and the migratory potential of cancer cells.

Proteomic data corroborate an important role for exoproteins in facilitating EMT in recipient cells and in promoting the metastatic potential of cancer. In a study by Jeppesen et al. (2014), they focused on three human bladder carcinoma cell lines: a nonmetastatic control T24 and two lines with high metastatic potential SLT4 and FL3. Using quantitative iTRAQ proteomic analysis, they were able to identify proteins enriched in the exosomes from the metastatic cells compared with the nonmetastatic T24 cells, many of which are implicated in EMT. For example, vimentin, which promotes EMT through a protooncogene c-SRC modulation of E-cadherin/beta-catenin complexes, was upregulated over twofold in exosomes from both metastatic cell lines. Vimentin is also necessary for Slug- or H-Ras-V12g-mediated EMT. Additionally, casein kinase II (CK2) was upregulated in SLT4 and FL3 exosomes and imbalance in the alpha and beta subunits of CK2 is implicated in EMT in breast and colon cancers through effects on Snail and Twist. Additional EMT-related proteins such as hepatoma-derived growth factor, moesin, and the tetraspanin CD82 were also upregulated. The changes in protein content between the metastatic and nonmetastatic bladder cancer cells was poorly correlated with changes in mRNA expression, making it more likely that alterations in the biogenesis and protein sorting decisions in these cells are responsible for the observed differences.

Harris et al. (2015) examined the protein profiles of exosomes derived from MCF-7 breast cancer cells and the more metastatic breast cancer line MDA-MB-231. MDA-MB-231 but not MCF-7 exosomes stimulated cellular motility and vimentin expression in recipient MDA-MB-231 cells. These studies suggest that adoption of a more metastatic phenotype in cancer is related to the protein content of its released exosomes, especially in relation to EMT-promoting protein cargo.

The ability of cancer to metastasize efficiently may be highly related to the protein cargo of its released exosomes. Besides EMT, efficient movement of mesenchymal cells away from the tumor to promote metastasis is facilitated by many other factors mediated by cancer exosomes. One of the barriers to motile cells is the surrounding ECM. Many of the enzymatic proteins in cancer exosomes have activity against the ECM and have been shown to assist in ECM degradation and promotion of metastasis.

In a recent study by Thuma, Heiler, Schnolzer, and Zoller (2016), the authors demonstrated in a rat adenocarcinoma cell line that claudin-7 was important in promoting metastasis. Their study was especially complete in its ability to attribute specific phenotypic effects to specific molecular constituents of the exosomes as well as in providing an understanding into how these cargoes were incorporated into exosomes during formation. For example, they were able to show that claudin-7, but only when palmitoylated, partitioned into glycolipid-enriched microdomains (GEMs) where it associated with tetraspanin-8 and numerous integrins. Additionally, palmitoylated claudin-7 coimmunoprecipitated with matrix remodeling proteins, such as the matrix-metallopeptidases MMP-14 and MMP-9. GEM-located palmitoylated claudin-7 is also associated with the tetraspanin CD147, which is known to bind MMPs on neighboring cells. Such observations underscore the importance of protein interactions within discrete lipid locales for determining which cargoes are present in exosomes and highlight that cargo incorporation is highly codependent with lipid composition and protein posttranslational modifications.

Protein cargoes such as MMP-9 and MMP-14 are common within the sets of proteins reported for cancer exosomes. MMPs, or matrix metalloproteinases, show enzymatic activity against the ECM, degrading important protein networks, which typically serve to tether cells and lock them into an epithelial phenotype. Matrix remodeling is a highly explored function of cancer exosomes that also serves to promote a mesenchymal phenotype because cell attachments are lost as well as physically enabling cells to escape from the tumor bulk through a less intact architectural meshwork. It is likely that cancer exosomes contain numerous proteins with maintained enzymatic activity against the ECM. It has now been established for almost a decade that MMPs can be present in exosomes in the active form. Hakulinen, Sankkila, Sugiyama, Lehti, and Keski-Oja (2008) demonstrated that fibrosarcoma cells released active membrane-present MMP-14, with the exosomes showing an ability to degrade type 1 collagen and gelatin. MDA-MB-231-derived exosomes were also enriched in proteases, including serine proteases, urokinase-type plasminogen activator, and cathepsin D, which have been shown to promote cell motility and tumor invasion by degrading the ECM (Lu, Takai, Weaver, & Werb, 2011).

Tumor-derived exosomes can also promote tumor growth, either through increased cellular proliferation or through inhibition of apoptosis. Many tumor cells package inhibitors of apoptosis proteins (IAPs). The IAPs survivin, cIAP1, cIAP2, and XIAP were present in exosomes released by DLCL2, HeLa, MCF-7, Panc-1, and PC3 tumor cell lines (Valenzuela et al., 2015). Important to the promotion of cancer growth is the ability of cancer-released exosomes and EVs to use direct protein transfer. For instance, Al-Nedawi et al. (2008) reported that aggressive glioma cells bearing the truncated and mutant form of the EGF receptor (EGFRvIII) could package the protein in the membrane of released microvesicles and "share" it with glioma cells that did not express EGFRvIII. The mutant receptor was able to engage oncogenic signaling through the MAPK and Akt pathways in the recipient cells. Changes in expression of EGFRvIII-regulated genes such as VEGF, Bcl-x, and p27 were observed, leading to not only morphological transformation of recipient cells but also an increase in proliferative capacity of even nonadherent cells. Similar findings have been reported in colon cancer for mutagenic KRAS transfer to nontransformed cells (Demory Beckler et al., 2013), and in melanoma the MET tyrosine kinase receptor oncogene is transferred via exosomes to bone marrow progenitor cells educating them toward a metastatic phenotype (Peinado et al., 2012).

Although it has been shown that many functions of cancer exosomes are mediated directly by their protein cargoes, the most significant differences between healthy cells and cancer cells in terms of protein content are observed for RNA-binding and processing proteins. Looking at the proteins contained across many different cancer samples stored in the ExoCarta database revealed that RNA-binding proteins, in particular, were largely different between cancer and healthy cells. RNA packaging may be one of the most significant differences between healthy and cancerous settings, and many studies have examined the RNA cargo of exosomes, especially in relation to cancer communication.

4.2 Exosomal RNA in Cancer

Exosomes communicate via proteins, lipids, and nucleic acids. Not only are RNA molecules present in the exosomes released by cells but also there appear to be machineries within cells that sort specific RNAs into exosomes to exert phenotypic effects on recipient cells. This model of cellular communication is quite sophisticated and probably even more complicated than currently appreciated. In the case of cancer exosomes, there is a diverse population of RNA-binding proteins present in released vesicles as determined by proteomic studies. A general model of selective RNA loading into exosomes asserts that RNA-binding proteins recognizing RNAs with specific sequences serve as the so-called "active-sorting" machinery. An RNA-binding protein (e.g., hnRNPA2B1) may recognize "EXO-motif" sequences (e.g., GGAG) within cellular RNAs and bind to them; if this protein is then sorted into exosomes, bound RNAs are loaded as well (Villarroya-Beltri et al., 2013). These actively loaded RNA molecules include mRNAs, micro (mi)RNAs, transfer

RNAs (tRNAs), small nuclear (sn)RNAs, and other noncoding RNA classes and are sorted up to 1000 times or more in exosomes relative to their cellular levels (Baglio et al., 2015; Bolukbasi et al., 2012; Liu et al., 2015). Apart from active sorting, RNA molecules can be loaded into exosomes via passive diffusion as a function of mass action. Thus, generally speaking, highly expressed RNAs are found in exosomes at higher levels than RNAs expressed at low copy numbers.

miRNAs have received the most attention because of their known role in gene regulation and cancer biology. Impressively, cancer cells can deliver miRNAs to recipient cells at potent enough doses to initiate transcriptional reprogramming in favor of cancer dissemination and propagation. It was shown that exosomes from prostate cancer cells containing the oncogenic miRNAs miR-125b, miR-130b, and miR-155 were internalized by patient-derived adipose stem cells (ASCs), "subverting" these cells toward a neoplastic phenotype. When implanted in immunocompromised mice, the reprogrammed ASCs formed highly aggressive prostate-like tumors. When ASCs were primed with exosomes from a noncancer cell line and implanted, no tumors were formed (Abd Elmageed et al., 2014).

There is now a wealth of literature showing that specific miRNA cargoes mediate cancer hallmarks such as angiogenesis (Tadokoro et al., 2013; Ye et al., 2014; Zhuang et al., 2012a, 2012b; Umezu et al., 2013, 2014), EMT (Melo et al., 2014), promotion of migration, invasiveness, and metastasis (Ye et al., 2014; Singh et al., 2014; Yang et al., 2011; Kogure, Lin, Yan, Braconi, & Patel, 2011; Fabbri et al., 2012; Ostenfeld et al., 2014), transmission of chemoresistance (Chen et al., 2014), increased tumor cell proliferation (Kogure et al., 2011), immune system avoidance (Ye et al., 2014), and reprograming of cells at distant sites for initiation of a premetastatic niche (Fong et al., 2015; Zhou et al., 2014). The miRNA cargo often reflects the physiological status of the producing cell. For instance, the angiogenic miR-135b is released in response to hypoxia (Umezu et al., 2014). To exactly what extent miRNAs are responsible for these attributed effects is still unclear. It has been reported that additional cargoes including proteins such as Rab1a and 1b and Rab11b as well as H-ras and K-ras transcripts could mediate the transformation of the recipient ASCs (Abd Elmageed et al., 2014). Few studies address the pleiotropic effects arising from all of the molecular constituents present in exosomes. How this diverse range of cargoes work together to exert effects is often neglected because it is easier to examine one or a few cargoes in isolation.

Highlighting this point is the fact that other RNA cargoes besides miRNAs/mRNAs have received almost no research attention. Fig. 3 shows the profile of small RNAs of exosomes derived from MDA-MB-231 and MCF-7 breast cancer cells (Jenjaroenpun et al., 2013; Tosar et al., 2015). It is apparent that contrary to popular belief, miRNAs do not constitute the majority of the small RNAs present in those exosomes (Fiskaa et al., 2016). For instance, small nucleolar and small cajal body-specific RNAs (sno/scaRNA) make up ~74% of the small RNAs in MDA-MB-231 exosomes. scaRNAs represent a class of snoRNAs, themselves belonging to the "small nuclear" or snRNA class of ncRNAs. How exactly snoRNAs and scaRNAs affect recipient cells is not fully understood and requires further investigations.

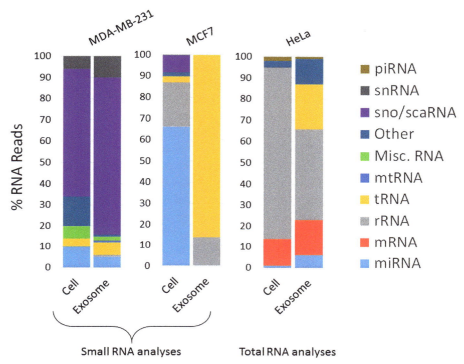

FIGURE 3 Small and total RNA analyses for MDA-MB-231, MCF-7, and HeLa exosomes and their parent cells. *Data adapted from Jenjaroenpun, P., Kremenska, Y., Nair, V. M., Kremenskoy, M., Joseph, B., & Kurochkin, I. V. (2013). Characterization of RNA in exosomes secreted by human breast cancer cell lines using next-generation sequencing, PeerJ, 1, e201, Tosar, J. P., Gambaro, F., Sanguinetti, J., Bonilla, B., Witwer, K. W., & Cayota, A. (2015). Assessment of small RNA sorting into different extracellular fractions revealed by high-throughput sequencing of breast cell lines, Nucleic Acids Research, 43(11), 5601–5616, and Schageman, J., Zeringer, E., Li, M., Barta, T., Lea, K., Gu, J., … Vlassov, A. V. (2013). The complete exosome workflow solution: From isolation to characterization of RNA cargo, BioMed Research International, 2013, 253957.*

snRNAs account for ~6–10% of the total RNA in MDA-MB-231 exosomes (Fig. 3). snRNAs are usually found in the cell nucleus and constitute parts of the spliceosomes that remove introns from primary mRNA transcripts. In a recent study, it was proposed that the formation of a premetastatic niche in the lung is established via the snRNA cargo carried in tumor-derived exosomes (Liu et al., 2016). In their model, using Lewis lung carcinoma cells or B16/F10 melanoma cells, they observed that nearly a quarter of all ncRNA reads from their exosome samples mapped to snRNAs. Some of these snRNAs were enriched 1000-fold in exosomes compared with parental tumor cells. They proposed a rather atypical model of RNA function in preparing the premetastatic niche preparation. The double-stranded structural snRNAs from tumor-derived EVs activated TLR3 receptors in lung endothelium, which then promoted chemokine release to attract neutrophils. This mechanism contributing to PMN formation also had clinical implications. Higher TLR3 levels and neutrophil infiltration were both correlated with poorer prognosis in patients.

Binning patients into TLR3 high and low groups or CD66b (a neutrophil marker) high and low groups both showed an approximately fourfold difference in overall survival at 5 years (~40% survival in TLR3 or CD66b low vs. ~10% survival in TLR3 or CD66b high; $P < .001$ for each Kaplan–Meier test).

Very surprisingly, the majority of small RNAs in MCF-7 exosomes consist of tRNA and ribosomal RNA (rRNA) (Fig. 3). Although miRNAs are present at ~66% of the small RNA reads in MCF-7 cells, they are virtually absent in exosomes as a proportion of small RNAs (less than 1%). If and how each of these RNA species contributes to the overall biological function is not known and requires further study. For HeLa cells, total RNA analyses were performed, thus mRNA is also captured in the exosomes (Schageman et al., 2013). Similar to the exosomes derived from MCF-7 cells and MDA-MB-231 cells, the miRNA content accounts for only a small percentage (~6%) of the total RNA. rRNA accounts for the largest percentage (~43%), followed by tRNA (~21%). Furthermore, the analyses showed that the RNA composition of the exosomes are not a simple representation of their parental cells but can be significantly different. It is clear that both the effects attributed to cancer exosomes as well, and the contribution of specific cargoes to these effects is highly dependent on the hypotheses of the researchers and what they choose to assay for. Many studies examining the role of cancer exosomes use experiments that can only uncover individual effects (e.g., invasion assays or angiogenesis assays) and perform either proteomic or miRNA profiling. It is likely that exosomes simultaneously promote cancer through various mechanisms, so how diverse subsets of exosomes contribute to particular effects should yield interesting insights in the near future.

miRNAs remain the most widely studied exo-RNA component, perhaps because they constitute ~90% of all ncRNAs (van Balkom, Eisele, Pegtel, Bervoets, & Verhaar, 2015). However, cancer cell exosomes often show completely different RNA profiles, and rarer and less studied noncoding RNAs may be of significant biological importance. Li et al. showed that the small RNA repertoire of exosomes extracted from glioma cells was strikingly different to the small RNA repertoire of normal glial cell exosomes (Li et al., 2013). Nearly 50% of the small ncRNAs identified through deep sequencing of the vesicles were unusual and of unknown function. Additionally, because the majority of these RNAs were not present in parental cells, it is possible that cancer cells produce novel RNA species intended for vesicular export. These vesicles underwent transcriptional reprogramming when they were seeded onto brain microvesicle endothelial cells even though the vesicles were depleted of miRNAs relative to the parental glioma cells. There is an expanding body of research implicating other ncRNAs besides miRNAs in transcriptional regulation.

Establishing the functions of these other ncRNAs will provide researchers in the exosome field with many more hypotheses about the different modes of communication afforded to cells via their exo-RNA cargo. Future studies should use unbiased sequencing approaches that reveal all of the RNA components of exosomes and not just profile for mRNAs or miRNAs composition.

Another interplay between exosomes and RNA is in RNA processing. Although it is generally assumed that the cargo packaged into exosomes remains unchanged over time, this may not always be the case. Dynamic processing of RNA has been observed within cancer-released exosomes (Melo et al., 2014). Over 48 h after exosomal collection, the amount of mature miRNA from MDA-MB-231, MCF-7, and 4T1 cells was shown to increase up to 32-fold for some species independent of cellular processing. This work suggests that pre-miRNAs may be sorted into exosomes and then processed to mature miRNAs via exosome-present Dicer after release. The temporal considerations of exosomal function are therefore called into question as well. Interestingly, this was observed for cancer cells, but accumulation of mature miRNAs in the exosomes of nontumorigenic mouse mammary epithelial cells was negligible. This serves as a good lesson that cancer exosomes especially may have hard-to-predict tricks up their sleeves.

4.3 Lipids in Exosomes

Several studies have shown that the lipid composition of exosomes is not an exact representation of that of the parental cells, with some lipids depleted and others enriched in exosomes. As exosomes are small, they naturally possess lipid bilayers with high curvature. Haraszti and colleagues showed that exosomes are generally enriched in inverted cone-shaped lipids, such as free fatty acids and lysophosphatidyl derivatives (Haraszti et al., 2016). Because these lipids promote the formation of positive curvatures, they are preferentially found in the outer lipid bilayer with the inner lipid bilayer usually consisting of lipids that promote negative curvature including cone-shaped lipids such as cardiolipin, which has four tails. Haraszti and colleagues found that lipids in the inner leaflet contained more diunsaturated and polyunsaturated fatty acids than lipids of the outer leaflet. This further supports the notion that lipids that adopt a cone shape and promote negative curvature are primarily located in the inner leaflet. Phosphatidylcholine, known to adopt a cylindrical shape, was significantly depleted in exosomes derived from mesenchymal stem cells (MSCs), U87 cancer cells, and Huh7 cells but was mostly enriched in cells. In the same study, phosphatidylinositol was depleted from exosomes derived from all the three cell lines, whereas phosphatidylethanolamine was depleted from MSC and Huh7 exosomes but not from U87 exosomes (Haraszti et al., 2016). Llorente and colleagues analyzed a panel of ~280 lipid species and showed that exosomes were enriched in lipids that contain saturated fatty acids and were relatively depleted in lipids with monounsaturated fatty acids compared with cells (Llorente et al., 2013).

   When comparing exosomes derived from the three different cancer cell lines SOJ6, PC-3, and U87 cells (Fig. 4), it is clear that although exosomes share a set of common lipids, their percentages differ (Haraszti et al., 2016; Llorente et al., 2013; Beloribi et al., 2012). Although cholesterol makes up to 45%–60% of total lipids in SOJ6 and PC-3 cells, it only accounts for ~3% in U87 cells. It is well established that the rigid structure of cholesterol significantly contributes to the mechanical stability of lipid membranes. How a low amount of cholesterol affects the vesicular stability of U87 exosomes compared with SOJ6 and PC-3 exosomes is currently not known. As cholesterol has also been linked to the formation of lipid rafts and contributes to protein anchoring to the membrane, it will be

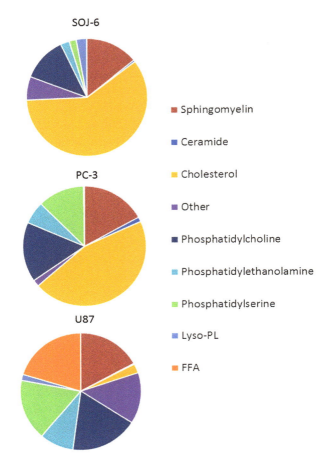

FIGURE 4 Lipid composition of exosomes derived from SOJ-6, U87 cells, and PC-3 cells. *Data adapted from Haraszti, R. A, Didiot, M. C., Sapp, E., Leszyk, J., Shaffer, S. A., Rockwell, H. E., …Khvorova, A. (2016). High-resolution proteomic and lipidomic analysis of exosomes and microvesicles from different cell sources. Journal of Extracellular Vesicles, 5, 32570; Llorente, A., Skotland, T., Sylvanne, T., Kauhanen, D., Rog, T., Orlowski, A., …Sandvig, K. (2013). Molecular lipidomics of exosomes released by PC-3 prostate cancer cells. Biochimica et Biophysica Acta, 1831(7), 1302–1309; Beloribi, S., Ristorcelli, E., Breuzard, G., Silvy, F., Bertrand-Michel, J., Beraud, E., …Lombardo, D. (2012). Exosomal lipids impact notch signaling and induce death of human pancreatic tumoral SOJ-6 cells. PLoS One, 7(10), e47480.*

important to understand how the total amount of cholesterol present in the lipid bilayer affects incorporation of proteins into the exosomal membrane (Simons & Ehehalt, 2002).

Phospholipids account for ~20–50% of all lipids in exosomes albeit with differing percentages of phosphatidylcholine, phosphatidylethanolamine, phosphatidylserine, and lysophospholipid. This shows that exosomes from different sources not only differ in their RNA and protein composition but also in their lipid composition. Lipid composition not only affects the stability of the exosomes but can also exert pharmacological and biological effects. For instance, sphingomyelin, which accounts for ~14–17% of exosome lipids in all the three cell types, is intrinsically biologically inactive. However, after exposure to anticancer agents or environmental stress, sphingomyelin breaks down into ceramide and sphingosine, both of which are implicated in inducing apoptosis by disrupting the PI3K-Akt pathway (Woodcock, 2006). It is not currently known if and how sphingomyelin-enriched exosomes affect recipient cells, but emerging studies indicate that the role of lipids is not solely confined to their function as structural components of cell membranes or biological vesicles but that they can also participate in physiological and pathophysiological processes. Arachidonic acid–derived eicosanoids, including prostaglandins, leukotrienes, and lipoxins are produced through enzymatic degradation from phospholipids and have been shown to interact with target cells via specific G-protein-coupled receptors (Harizi, Corcuff, & Gualde, 2008). As lipid-signaling molecules, these eicosanoids are involved in the development of many diseases including cancer. Eicosanoids and their derivatives promote tumor growth by triggering cell proliferation, angiogenesis, and by decreasing apoptosis. These examples show how important it is to also take the lipid composition of exosomes into consideration when assessing their biological and pharmacological effects.

5. Size and Charge of Cancer Exosomes

As exosomes can be secreted by a wide range of cells under a wide range of conditions, their reported size varies. Exosomes are commonly described as being 30–100 nm in diameter (Thery, Amigorena, Raposo, & Clayton, 2006). The size of exosomes can be determined using a variety of techniques including nanoparticle tracking analysis (NTA), dynamic light scattering (DLS), and various microscopy-based methods such as transmission electron microscopy (TEM) and atomic force microscopy (AFM) (Caponnetto et al., 2016). Depending on the methods used, the charge and size distribution of the exosomes can vary. NTA and DLS measure the hydrodynamic diameters of particles in their solvated state. Although measurements of monodisperse systems are straightforward, the measurements of heterogeneous or bimodal nanoparticle suspensions can be more problematic. Because the intensity of light scattered from large particles is stronger than that from small particles, the fraction of small particles is often underestimated. This also applies to measurements by NTA. In contrast, TEM and AFM measure exosome particle sizes in the dry, nonsolvated state. Furthermore, the field of view used for the analysis is relatively small, and a sufficiently high number of particles need to be analyzed to provide representative data. For example, different sizes were reported for exosomes derived from the same cell line. Su, Aldawsari, and Amiji (2016) reported pancreatic cancer (Panc-1) exosomes to be in the range of 30–40 nm, whereas Mikamori et al. (2017) reported Panc-1 exosomes to be larger in size (89±2.8 nm). This can result from different exosomal isolation methods or different size measurement techniques.

Knowing the size of nanoparticles is important for gaining a full understanding of their bioactivity (Shang, Nienhaus, & Nienhaus, 2014). Caponnetto et al. (2016) investigated the effect of different isolation methods on exosome size and their subsequent uptake by recipient cells. They found that when isolating exosomes using the precipitation reagent ExoQuick as opposed to ultracentrifugation, smaller exosome populations were isolated as measured by AFM, and the fraction of small exosomes was masked when analyzed by NTA. When comparing the smaller exosome samples isolated by ExoQuick to the larger ones isolated by ultracentrifugation, they found that there was significantly enhanced internalization of exosomes from the smaller isolated exosome fraction (Caponnetto et al., 2016).

Exosomes have been shown to interact with recipient cells, and these interactions can be mediated by a variety of factors including surface charge. The surface charge is most likely a function of exosomal lipid content and the proteins incorporated into the exosomal membrane. Table 3 provides an overview of the zeta potential of different exosomes derived from a wide range of cell types. Although a direct side-by-side comparison is difficult because of the use of different buffers and measurement methods, all exosomes displayed negative zeta potential values that ranged from −4 to −45 mV.

Akagi, Ichiki, and Ohshima (2016, pp. 469–473) measured the zeta potential of exosomes derived from noncancerous and cancerous cells and found that sialic acid was one of the main contributors to exosomal surface charge. They found that the exosomes from cancerous cells displayed a more negative zeta potential than exosomes isolated from healthy cells. For example, normal prostate epithelial cells displayed a zeta potential of −8.9 ± 1.9 mV, whereas the zeta potential of prostate cancer cells (PC-3) and PC-3 ML were reported to be more negatively charged: −12.8 ± 2.8 mV and −13.1 ± 2.2 mV, respectively. Furthermore, the authors measured the zeta potential of cells and their exosomes. They found that there was a strong correlation between the zeta potential of exosomes and that of their cells of origin. This strongly indicates that exosomes inherit many of the properties and compositions of the parental cells. The authors found that the negative zeta potential of cancer cells and cancer-derived exosomes was mainly controlled by the expression of sugar chains such as mannose, polylactosamine, α-2,6 sialic acid, and complex N-linked glycans. To assess the influence of sialic acid on the zeta potential of exosomes, the authors treated the exosomes with sialidase, after which the zeta potential of both the exosomes derived from PC-3 cancer cells and healthy PNT2 cells became less negatively charged to −4.1 ± 2.6 mV and −4.6 ± 3.3 mV. This strongly indicates that sialic acid is one of the key contributors to the surface charge of exosomes.

A study by Adachi and colleagues further underlined the notion that the properties of exosomes are often controlled by their parent cells. They found that when malignant melanoma cells (B16–F10) underwent a phenotypic change, a similar change could also be observed in their secreted exosome populations (Adachi et al., 2016). When cells with different expression levels of the Met oncoprotein, also known as the hepatocyte growth factor receptor, were compared, a notable change in the zeta potential and protein cargo in their exosomes was observed. For both low- and high-Met expressing melanoma cells, they measured the corresponding zeta potentials of the exosomes to be −35.1 and −31.1 mV, respectively. Because no standard deviations were included, it is unclear if the difference in the zeta potential is statistically significant. High-Met exosomes were also found to contain a larger quantity of not only Met but also tyrosinase-related protein 2 (TRP2) and heat shock protein 90 (HSP90), all of which have been found to be present at high concentrations in the advanced stages of melanoma (Peinado et al., 2012). The exosomes derived

Table 3 Particle Size and Zeta Potential of Exosomes Secreted From Cancer Cells

	Description	Cell Line	Size (nm)	Method (Size)	Zeta Potential (mV)	Condition (Zeta Potential)	Study
Murine exosomes	Melanoma	B16–F10	95±14	DLS	−11±5	In PBS	Hood, San, and Wickline (2011)
			110.3		−35.1	–	Adachi et al. (2016)
			112.9		−31.1	–	
		B16-BL6	66±11		−44.9±1.5	At 20°C	Takahashi et al. (2013)
			112±12	TRPS	−33±0.2	–	Morishita, Takahashi, Matsumoto, Nishikawa, and Takakura (2016)
	Myoblast	C2C12	100±4		−39.2±0.9	–	Charoenviriyakul et al. (2017)
	Fibroblast	NIH3T3	111±10		−38.9±1.1	–	
	Aortic endothelial cells	MAEC	106±2		−35.3±0.5	–	
			102±7		−41.6±1.5	–	
	Macrophage	RAW264.7	105±4		−38.8±0.6	–	
	Hepatic stellate cells	HSC	139±5	TEM	−26	–	Chen et al. (2014)
	Normal breast epithelium	NMuMG	80.5	DLS	−12.9	In ddH$_2$0	Shi, Ren, Zhen, and Qiu (2015)
	Breast cancer	4T1	89.6		−15.8		
Human exosomes	Embryonic kidney cells	HEK 293T	~120	NTA	−54±6	Exosomes were suspended in	Sokolova et al. (2011)
	Endothelial colony forming cells	ECFC	~110		−49±13	PBS and diluted 1:1000 with pure	
	Mesenchymal stem cells	MSC	~110		−52±4	water	

Continued

Table 3   Particle Size and Zeta Potential of Exosomes Secreted From Cancer Cells—cont'd

Description	Cell Line	Size (nm)	Method (Size)	Zeta Potential (mV)	Condition (Zeta Potential)	Study
Neuroblastoma	HTLA-230	68.05±6.58	DLS	−12.1±0.17	Exosomes were suspended 1:10 PBS:H$_2$O at 25°C	Marimpietri et al. (2013)
	IMR-32	72.54±8.9		−14.8±1.55		
	SH-SY5Y	83.66±9.22		−13.2±1.1		
	GI-LI-N	72.20±8.83		−12.0±0.15		
Prostate epithelium	CRL-2221	53.6±12.8		−13.8±0.67	Exosomes were suspended and analyzed in PBS	Dean et al. (2017)
	PNT2	–	–	−7.4±2.2		Akagi et al. (2016, pp. 469–473)
Prostate cancer	LNCaP	55.4±5.3	DLS	−11.2±0.62		Dean et al. (2017)
	PC-3	87.2±10.1		−11.0±0.69		
	PC-3Mluc-C6	–	–	−11.0±2.3		Akagi et al. (2016, pp. 469–473)
		–	–	−11.1±2.7		
Breast epithelium	MCF-10A	92.9±42.7	DLS	−12.6±1.85		Dean et al. (2017)
		–	–	−8.8±2.5		Akagi et al. (2016, pp. 469–473)
Breast cancer	SUM149	91.1±1.8	DLS	−11.7±0.48		Dean et al. (2017)
	MDA-MB-231	–	–	−10.2±2.4		Akagi et al. (2016, pp. 469–473)
	MDA-MB-231-luc-D3H2LN	–	–	−10.9±2.3		
Pancreatic cancer	Panc-1	39.03±2.28	DLS	−4.54±0.19		Su et al. (2016)
		89±2.8	NTA	–	–	Mikamori et al. (2017)
Glioblastoma	U87	141		−13	In H$_2$O	Didiot et al. (2016)
Lung epithelium	BEAS-2B	41.2±8.8	DLS	−13.0±0.91	In PBS	Dean et al. (2017)

DLS, dynamic light scattering (*ZetaSizer*); *NTA*, nanoparticle tracking analysis; *tRPS*, tunable resistive pulse sensing (qNano instrument); *TEM*, transmission electron microscopy.

from high Met–producing cells played a significant role in metastatic invasion and pre-metastatic niche formation in the lung (Adachi et al., 2016). Table 3 displays the size and zeta potential of exosomes derived from different cancer cells. Because of the different methods and buffers used, a direct comparison of the zeta potential values is not possible. Exosomes derived from different cancer cell lines seem to all display a negative zeta potential ranging from –4.5 mV to –45 mV.

6. Summary and Future Considerations

To fully understand the role of exosomes under physiological and pathophysiological conditions and as intercellular messengers between cells, elucidating their composition and physicochemical properties is of utmost importance (Ferguson et al., 2018). It has become evident that exosomes derived from different cells display distinct RNA, lipid, and protein profiles. How these exosomal RNAs, proteins, and lipids may function in concert to exert exosomal function is still not fully understood. Recent studies indicate that exosomes derived from the same batch of cells can consist of different subpopulations that differ in protein composition (Laulagnier et al., 2005; Oksvold et al., 2014; Smith et al., 2015). Although subsets of exosomes have been identified in terms of protein composition, it is unclear to what extent RNA and lipid composition is distributed across the heterogeneous exosome population. With recent studies showing that many naturally occurring miRNAs are present at very low copy numbers in exosomes (between 1.15×10^{-1}–6.97×10^{-1} miRNAs per exosome) (Chevillet et al., 2014), it is likely that subpopulations of exosomes will also exist with regard to their miRNA content. Most studies use experimental approaches that characterize exosomes in bulk. To fully understand the effects mediated by exosomes, a single particle approach with regard to protein, lipid, and RNA content will be needed. Despite the tremendous progress over the last few years, much work is still needed to fully characterize and understand exosomes, their composition, and their function. Questions remain as to how each cell type, healthy cells, cancerous cells, neuronal cells, and endothelial cells control exosomal cargo and composition and how much occurs through simple diffusion and random mass action.

Acknowledgment

JN acknowledges support by the NIH through the awards EB023262 and EB021454.

References

Abd Elmageed, Z. Y., Yang, Y., Thomas, R., Ranjan, M., Mondal, D., Moroz, K., … Abdel-Mageed, A. B. (2014). Neoplastic reprogramming of patient-derived adipose stem cells by prostate cancer cell-associated exosomes. *Stem Cells*, *32*(4), 983–997.

Abrami, L., Brandi, L., Moayeri, M., Brown, M. J., Krantz, B. A., Leppla, S. H., … van der Goot, F. G. (2013). Hijacking multivesicular bodies enables long-term and exosome-mediated long-distance action of anthrax toxin. *Cell Reports*, *5*(4), 986–996.

Adachi, E., Sakai, K., Nishiuchi, T., Imamura, R., Sato, H., & Matsumoto, K. (2016). Different growth and metastatic phenotypes associated with a cell-intrinsic change of Met in metastatic melanoma. *Oncotarget, 7*(43), 70779–70793.

Akagi, T., Ichiki, T., & Ohshima, H. (2016). *Evaluation of zeta-potential of individual exosomes secreted from biological cells using a Microcapillary Electrophoresis Chip, Encyclopedia of Biocolloid and Biointerface Science 2V set.* John Wiley & Sons, Inc.

Al-Nedawi, K., Meehan, B., Micallef, J., Lhotak, V., May, L., Guha, A., & Rak, J. (2008). Intercellular transfer of the oncogenic receptor EGFRvIII by microvesicles derived from tumour cells. *Nature Cell Biology, 10*(5), 619–624.

Andreu, Z., & Yanez-Mo, M. (2014). Tetraspanins in extracellular vesicle formation and function. *Frontiers in Immunology, 5*, 442.

Aubertin, K., Silva, A. K., Luciani, N., Espinosa, A., Djemat, A., Charue, D., … Wilhelm, C. (2016). Massive release of extracellular vesicles from cancer cells after photodynamic treatment or chemotherapy. *Scientific Reports, 6*, 35376.

Azmi, A. S., Bao, B., & Sarkar, F. H. (2013). Exosomes in cancer development, metastasis, and drug resistance: A comprehensive review. *Cancer and Metastasis Reviews, 32*(3–4), 623–642.

Baglio, S. R., Rooijers, K., Koppers-Lalic, D., Verweij, F. J., Perez Lanzon, M., Zini, N., … Pegtel, D. M. (2015). Human bone marrow- and adipose-mesenchymal stem cells secrete exosomes enriched in distinctive miRNA and tRNA species. *Stem Cell Research & Therapy, 6*, 127.

Baietti, M. F., Zhang, Z., Mortier, E., Melchior, A., Degeest, G., Geeraerts, A., … David, G. (2012). Syndecan-syntenin-ALIX regulates the biogenesis of exosomes. *Nature Cell Biology, 14*(7), 677–685.

Beloribi, S., Ristorcelli, E., Breuzard, G., Silvy, F., Bertrand-Michel, J., Beraud, E., … Lombardo, D. (2012). Exosomal lipids impact notch signaling and induce death of human pancreatic tumoral SOJ-6 cells. *PLoS One, 7*(10), e47480.

Bolukbasi, M. F., Mizrak, A., Ozdener, G. B., Madlener, S., Strobel, T., Erkan, E. P., … Saydam, O. (2012). miR-1289 and "Zipcode"-like sequence Enrich mRNAs in microvesicles, molecular therapy. *Nucleic acids, 1* e10.

Campanella, C., Bucchieri, F., Merendino, A. M., Fucarino, A., Burgio, G., Corona, D. F. V., … Cappello, F. (2012). The odyssey of Hsp60 from tumor cells to other destinations includes plasma membrane-associated stages and Golgi and exosomal protein-trafficking modalities. *PLoS One, 7*(7), e42008.

Caponnetto, F., Manini, I., Skrap, M., Palmai-Pallag, T., Di Loreto, C., Beltrami, A. P., … Ferrari, E. (2016). Size-dependent cellular uptake of exosomes. *Nanomedicine, 13*(3), 1011–1020.

Charoenviriyakul, C., Takahashi, Y., Morishita, M., Matsumoto, A., Nishikawa, M., & Takakura, Y. (2017). Cell type-specific and common characteristics of exosomes derived from mouse cell lines: Yield, physico-chemical properties, and pharmacokinetics. *European Journal of Pharmaceutical Sciences, 96*, 316–322.

Chen, L., Charrier, A., Zhou, Y., Chen, R., Yu, B., Agarwal, K., … Brigstock, D. R. (2014). Epigenetic regulation of connective tissue growth factor by MicroRNA-214 delivery in exosomes from mouse or human hepatic stellate cells. *Hepatology, 59*(3), 1118–1129.

Chen, W. X., Liu, X. M., Lv, M. M., Chen, L., Zhao, J. H., Zhong, S. L., … Tang, J. H. (2014). Exosomes from drug-resistant breast cancer cells transmit chemoresistance by a horizontal transfer of microRNAs. *PLoS One, 9*(4), e95240.

Chevillet, J. R., Kang, Q., Ruf, I. K., Briggs, H. A., Vojtech, L. N., Hughes, S. M., … Tewari, M. (2014). Quantitative and stoichiometric analysis of the microRNA content of exosomes. *Proceedings of the National Academy of Sciences of the United States of America, 111*(41), 14888–14893.

Clayton, A., Al-Taei, S., Webber, J., Mason, M. D., & Tabi, Z. (2011). Cancer exosomes express CD39 and CD73, which suppress T cells through adenosine production. *The Journal of Immunology, 187*(2), 676–683.

Colombo, M., Moita, C., van Niel, G., Kowal, J., Vigneron, J., Benaroch, P., … Raposo, G. (2013). Analysis of ESCRT functions in exosome biogenesis, composition and secretion highlights the heterogeneity of extracellular vesicles. *Journal of Cell Science, 126*(Pt 24), 5553–5565.

Dean, I., Dzinic, S. H., Bernardo, M. M., Zou, Y., Kimler, V., Li, X., … Sheng, S. (2017). The secretion and biological function of tumor suppressor maspin as an exosome cargo protein. *Oncotarget, 8*(5), 8043–8056.

Demory Beckler, M., Higginbotham, J. N., Franklin, J. L., Ham, A. J., Halvey, P. J., Imasuen, I. E., & Coffey, R. J. (2013). Proteomic analysis of exosomes from mutant KRAS colon cancer cells identifies intercellular transfer of mutant KRAS. *Molecular & Cellular Proteomics, 12*(2), 343–355.

Devlin, C., Greco, S., Martelli, F., & Ivan, M. (2011). miR-210: More than a silent player in hypoxia. *IUBMB Life, 63*(2), 94–100.

Didiot, M. C., Hall, L. M., Coles, A. H., Haraszti, R. A., Godinho, B. M., Chase, K., … Khvorova, A. (2016). Exosome-mediated delivery of Hydrophobically modified siRNA for Huntingtin mRNA silencing. *Molecular Therapy, 24*(10), 1836–1847.

Fabbri, M., Paone, A., Calore, F., Galli, R., Gaudio, E., Santhanam, R., … Croce, C. M. (2012). MicroRNAs bind to Toll-like receptors to induce prometastatic inflammatory response. *Proceedings of the National Academy of Sciences of the United States of America, 109*(31), E2110–E2116.

Fang, Y., Wu, N., Gan, X., Yan, W., Morrell, J. C., & Gould, S. J. (2007). Higher-order oligomerization targets plasma membrane proteins and HIV gag to exosomes. *PLoS Biology, 5*(6), e158.

Ferguson, S. W., & Nguyen, J. (2016). Exosomes as therapeutics: The implications of molecular composition and exosomal heterogeneity. *Journal of Controlled Release, 228*, 179–190.

Ferguson, S. W., Wang, J., Lee, C. J., Liu, M., Neelamegham, S., Canty, J. M., & Nguyen, J. (January 23, 2018). The microRNA regulatory landscape of MSC-derived exosomes: a systems view. *Scientific Reports, 8*(1), 1419. https://doi.org/10.1038/s41598-018-19581-x.

Fiskaa, T., Knutsen, E., Nikolaisen, M. A., Jorgensen, T. E., Johansen, S. D., Perander, M., & Seternes, O. M. (2016). Distinct small RNA signatures in extracellular vesicles derived from breast cancer cell lines. *PLoS One, 11*(8), e0161824.

Fong, M. Y., Zhou, W., Liu, L., Alontaga, A. Y., Chandra, M., Ashby, J., … Wang, S. E. (2015). Breast-cancer-secreted miR-122 reprograms glucose metabolism in premetastatic niche to promote metastasis. *Nature Cell Biology, 17*(2), 183–194.

Geminard, C., De Gassart, A., Blanc, L., & Vidal, M. (2004). Degradation of AP2 during reticulocyte maturation enhances binding of hsc70 and Alix to a common site on TFR for sorting into exosomes. *Traffic, 5*(3), 181–193.

Grant, B. D., & Donaldson, J. G. (2009). Pathways and mechanisms of endocytic recycling. *Nature Reviews. Molecular Cell Biology, 10*(9), 597–608.

Gross, J. C., Chaudhary, V., Bartscherer, K., & Boutros, M. (2012). Active Wnt proteins are secreted on exosomes. *Nature Cell Biology, 14*(10), 1036–1045.

Hakulinen, J., Sankkila, L., Sugiyama, N., Lehti, K., & Keski-Oja, J. (2008). Secretion of active membrane type 1 matrix metalloproteinase (MMP-14) into extracellular space in microvesicular exosomes. *Journal of Cellular Biochemistry, 105*(5), 1211–1218.

Hannafon, B. N., Carpenter, K. J., Berry, W. L., Janknecht, R., Dooley, W. C., & Ding, W.-Q. (2015). Exosome-mediated microRNA signaling from breast cancer cells is altered by the anti-angiogenesis agent docosahexaenoic acid (DHA). *Molecular Cancer, 14*(1), 133.

Haraszti, R. A., Didiot, M. C., Sapp, E., Leszyk, J., Shaffer, S. A., Rockwell, H. E., … Khvorova, A. (2016). High-resolution proteomic and lipidomic analysis of exosomes and microvesicles from different cell sources. *Journal of Extracellular Vesicles, 5*, 32570.

Harizi, H., Corcuff, J. B., & Gualde, N. (2008). Arachidonic-acid-derived eicosanoids: Roles in biology and immunopathology. *Trends in Molecular Medicine, 14*(10), 461–469.

Harris, D. A., Patel, S. H., Gucek, M., Hendrix, A., Westbroek, W., & Taraska, J. W. (2015). Exosomes released from breast cancer carcinomas stimulate cell movement. *PLoS One, 10*(3), e0117495.

Hood, J. L., San, R. S., & Wickline, S. A. (2011). Exosomes released by melanoma cells prepare sentinel lymph nodes for tumor metastasis. *Cancer Research, 71*(11), 3792–3801.

Hoshino, A., Costa-Silva, B., Shen, T. L., Rodrigues, G., Hashimoto, A., Tesic Mark, M., ... Lyden, D. (2015). Tumour exosome integrins determine organotropic metastasis. *Nature, 527*(7578), 329–335.

Hoshino, D., Kirkbride, K. C., Costello, K., Clark, E. S., Sinha, S., Grega-Larson, N., ... Weaver, A. M. (2013). Exosome secretion is enhanced by invadopodia and drives invasive behavior. *Cell Reports, 5*(5), 1159–1168.

Jenjaroenpun, P., Kremenska, Y., Nair, V. M., Kremenskoy, M., Joseph, B., & Kurochkin, I. V. (2013). Characterization of RNA in exosomes secreted by human breast cancer cell lines using next-generation sequencing. *PeerJ, 1*, e201.

Jeppesen, D. K., Nawrocki, A., Jensen, S. G., Thorsen, K., Whitehead, B., Howard, K. A., ... Ostenfeld, M. S. (2014). Quantitative proteomics of fractionated membrane and lumen exosome proteins from isogenic metastatic and nonmetastatic bladder cancer cells reveal differential expression of EMT factors. *Proteomics, 14*(6), 699–712.

King, H. W., Michael, M. Z., & Gleadle, J. M. (2012). Hypoxic enhancement of exosome release by breast cancer cells. *BMC Cancer, 12*, 421.

Kogure, T., Lin, W. L., Yan, I. K., Braconi, C., & Patel, T. (2011). Intercellular nanovesicle-mediated microRNA transfer: A mechanism of environmental modulation of hepatocellular cancer cell growth. *Hepatology, 54*(4), 1237–1248.

Kooijmans, S. A. A., Vader, P., van Dommelen, S. M., van Solinge, W. W., & Schiffelers, R. M. (2012). Exosome mimetics: A novel class of drug delivery systems. *International Journal of Nanomedicine, 7*, 1525–1541.

Kowal, J., Arras, G., Colombo, M., Jouve, M., Morath, J. P., Primdal-Bengtson, B., ... Thery, C. (2016). Proteomic comparison defines novel markers to characterize heterogeneous populations of extracellular vesicle subtypes. *Proceedings of the National Academy of Sciences of the United States of America, 113*(8), E968–E977.

Kwon, S. H., Oh, S., Nacke, M., Mostov, K. E., & Lipschutz, J. H. (2016). Adaptor protein CD2AP and L-type Lectin LMAN2 regulate exosome cargo protein trafficking through the Golgi complex. *Journal of Biological Chemistry, 291*(49), 25462–25475.

Laulagnier, K., Vincent-Schneider, H., Hamdi, S., Subra, C., Lankar, D., & Record, M. (2005). Characterization of exosome subpopulations from RBL-2H3 cells using fluorescent lipids. *Blood Cells, Molecules and Diseases, 35*(2), 116–121.

Lee, H. M., Choi, E. J., Kim, J. H., Kim, T. D., Kim, Y. K., Kang, C., & Gho, Y. S. (2010). A membranous form of ICAM-1 on exosomes efficiently blocks leukocyte adhesion to activated endothelial cells. *Biochemical and Biophysical Research Communications, 397*(2), 251–256.

Lehmann, B. D., Paine, M. S., Brooks, A. M., McCubrey, J. A., Renegar, R. H., Wang, R., & Terrian, D. M. (2008). Senescence-associated exosome release from human prostate cancer cells. *Cancer Research, 68*(19), 7864–7871.

Li, C. C., Eaton, S. A., Young, P. E., Lee, M., Shuttleworth, R., Humphreys, D. T., ... Suter, C. M. (2013). Glioma microvesicles carry selectively packaged coding and non-coding RNAs which alter gene expression in recipient cells. *RNA Biology, 10*(8), 1333–1344.

Liu, Y., Gu, Y., Han, Y., Zhang, Q., Jiang, Z., Zhang, X., ... Cao, X. (2016). Tumor exosomal RNAs promote lung pre-metastatic niche formation by activating alveolar epithelial TLR3 to recruit neutrophils. *Cancer Cell, 30*(2), 243–256.

Liu, Y., Li, D., Liu, Z., Zhou, Y., Chu, D., Li, X., ... Zhang, C. Y. (2015). Targeted exosome-mediated delivery of opioid receptor Mu siRNA for the treatment of morphine relapse. *Scientific Reports, 5*, 17543.

Llorente, A., Skotland, T., Sylvanne, T., Kauhanen, D., Rog, T., Orlowski, A., ... Sandvig, K. (2013). Molecular lipidomics of exosomes released by PC-3 prostate cancer cells. *Biochimica et Biophysica Acta, 1831*(7), 1302–1309.

Logozzi, M., De Milito, A., Lugini, L., Borghi, M., Calabro, L., Spada, M., … Fais, S. (2009). High levels of exosomes expressing CD63 and caveolin-1 in plasma of melanoma patients. *PLoS One, 4*(4), e5219.

Lu, P., Takai, K., Weaver, V. M., & Werb, Z. (2011). Extracellular matrix degradation and remodeling in development and disease. *Cold Spring Harbor Perspectives in Biology, 3*(12).

Luzio, J. P., Rous, B. A., Bright, N. A., Pryor, P. R., Mullock, B. M., & Piper, R. C. (2000). Lysosome-endosome fusion and lysosome biogenesis. *Journal of Cell Science, 113*(9), 1515–1524.

Marimpietri, D., Petretto, A., Raffaghello, L., Pezzolo, A., Gagliani, C., Tacchetti, C., … Pistoia, V. (2013). Proteome profiling of neuroblastoma-derived exosomes reveal the expression of proteins potentially involved in tumor progression. *PLoS One, 8*(9), e75054.

Mellman, I. (1996). Endocytosis and molecular sorting. *Annual Review of Cell and Developmental Biology, 12*, 575–625.

Melo, S. A., Luecke, L. B., Kahlert, C., Fernandez, A. F., Gammon, S. T., Kaye, J., … Kalluri, R. (2015). Glypican-1 identifies cancer exosomes and detects early pancreatic cancer. *Nature, 523*(7559), 177–182.

Melo, S. A., Sugimoto, H., O'Connell, J. T., Kato, N., Villanueva, A., Vidal, A., … Kalluri, R. (2014). Cancer exosomes perform cell-independent microRNA biogenesis and promote tumorigenesis. *Cancer Cell, 26*(5), 707–721.

Mikamori, M., Yamada, D., Eguchi, H., Hasegawa, S., Kishimoto, T., Tomimaru, Y., … Doki, Y. (2017). MicroRNA-155 controls exosome synthesis and promotes gemcitabine resistance in pancreatic ductal adenocarcinoma. *Scientific Reports, 7*, 42339.

Morishita, M., Takahashi, Y., Matsumoto, A., Nishikawa, M., & Takakura, Y. (2016). Exosome-based tumor antigens-adjuvant co-delivery utilizing genetically engineered tumor cell-derived exosomes with immunostimulatory CpG DNA. *Biomaterials, 111*, 55–65.

Mulcahy, L. A., Pink, R. C., & Carter, D. R. F. (2014). Routes and mechanisms of extracellular vesicle uptake. *Journal of Extracellular Vesicles, 3*. https://doi.org/10.3402/jev.v3.24641.

van Niel, G., Charrin, S., Simoes, S., Romao, M., Rochin, L., Saftig, P., … Raposo, G. (2011). The tetraspanin CD63 regulates ESCRT-independent and -dependent endosomal sorting during melanogenesis. *Developmental Cell, 21*(4), 708–721.

Oksvold, M. P., Kullmann, A., Forfang, L., Kierulf, B., Li, M., Brech, A., … Pedersen, K. W. (2014). Expression of B-cell surface antigens in subpopulations of exosomes released from B-cell lymphoma cells. *Clinical Therapeutics, 36*(6) 847–862 e1.

Ostenfeld, M. S., Jeppesen, D. K., Laurberg, J. R., Boysen, A. T., Bramsen, J. B., Primdal-Bengtson, B., … Orntoft, T. F. (2014). Cellular disposal of miR23b by RAB27-dependent exosome release is linked to acquisition of metastatic properties. *Cancer Research, 74*(20), 5758–5771.

Ostrowski, M., Carmo, N. B., Krumeich, S., Fanget, I., Raposo, G., Savina, A., … Thery, C. (2010). Rab27a and Rab27b control different steps of the exosome secretion pathway. *Nature Cell Biology, 12*(1), 19–30.

Peinado, H., Aleckovic, M., Lavotshkin, S., Matei, I., Costa-Silva, B., Moreno-Bueno, G., … Lyden, D. (2012). Melanoma exosomes educate bone marrow progenitor cells toward a pro-metastatic phenotype through MET. *Nat Med, 18*(6), 883–891.

Perez-Hernandez, D., Gutierrez-Vazquez, C., Jorge, I., Lopez-Martin, S., Ursa, A., Sanchez-Madrid, F., … Yanez-Mo, M. (2013). The intracellular interactome of tetraspanin-enriched microdomains reveals their function as sorting machineries toward exosomes. *Journal of Biological Chemistry, 288*(17), 11649–11661.

Rana, S., & Zoller, M. (2011). Exosome target cell selection and the importance of exosomal tetraspanins: A hypothesis. *Biochemical Society Transactions, 39*(2), 559–562.

Raposo, G., & Stoorvogel, W. (2013). Extracellular vesicles: Exosomes, microvesicles, and friends. *The Journal of Cell Biology, 200*(4), 373–383.

Schageman, J., Zeringer, E., Li, M., Barta, T., Lea, K., Gu, J., … Vlassov, A. V. (2013). The complete exosome workflow solution: From isolation to characterization of RNA cargo. *BioMed Research International, 2013,* 253957.

Schmidt, O., & Teis, D. (2012). The ESCRT machinery. *Current Biology, 22*(4), R116–R120.

Shang, L., Nienhaus, K., & Nienhaus, G. U. (2014). Engineered nanoparticles interacting with cells: Size matters. *Journal of Nanobiotechnology, 12,* 5.

Shi, J., Ren, Y., Zhen, L., & Qiu, X. (2015). Exosomes from breast cancer cells stimulate proliferation and inhibit apoptosis of CD133+ cancer cells in vitro. *Molecular Medicine Reports, 11*(1), 405–409.

Simons, K., & Ehehalt, R. (2002). Cholesterol, lipid rafts, and disease. *Journal of Clinical Investigation, 110*(5), 597–603.

Singh, R., Pochampally, R., Watabe, K., Lu, Z., & Mo, Y. Y. (2014). Exosome-mediated transfer of miR-10b promotes cell invasion in breast cancer. *Molecular Cancer, 13,* 256.

Smith, Z. J., Lee, C., Rojalin, T., Carney, R. P., Hazari, S., Knudson, A., … Wachsmann-Hogiu, S. (2015). Single exosome study reveals subpopulations distributed among cell lines with variability related to membrane content. *Journal of Extracellular Vesicles, 4,* 28533.

Sokolova, V., Ludwig, A. K., Hornung, S., Rotan, O., Horn, P. A., Epple, M., & Giebel, B. (2011). Characterisation of exosomes derived from human cells by nanoparticle tracking analysis and scanning electron microscopy. *Colloids and Surfaces B, Biointerfaces, 87*(1), 146–150.

Su, M. J., Aldawsari, H., & Amiji, M. (2016). Pancreatic cancer cell exosome-mediated macrophage reprogramming and the role of MicroRNAs 155 and 125b2 transfection using nanoparticle delivery systems. *Scientific Reports, 6,* 30110.

Tadokoro, H., Umezu, T., Ohyashiki, K., Hirano, T., & Ohyashiki, J. H. (2013). Exosomes derived from hypoxic leukemia cells enhance tube formation in endothelial cells. *The Journal of Biological Chemistry, 288*(48), 34343–34351.

Takahashi, Y., Nishikawa, M., Shinotsuka, H., Matsui, Y., Ohara, S., Imai, T., & Takakura, Y. (2013). Visualization and in vivo tracking of the exosomes of murine melanoma B16-BL6 cells in mice after intravenous injection. *Journal of Biotechnology, 165*(2), 77–84.

Tamai, K., Tanaka, N., Nakano, T., Kakazu, E., Kondo, Y., Inoue, J., … Sugamura, K. (2010). Exosome secretion of dendritic cells is regulated by Hrs, an ESCRT-0 protein. *Biochemical and Biophysical Research Communications, 399*(3), 384–390.

Thery, C., Amigorena, S., Raposo, G., & Clayton, A. (2006). Isolation and characterization of exosomes from cell culture supernatants and biological fluids. *Current Protocols in Cell Biology* Editorial board, Juan S. Bonifacino... [et al.] Chapter 3 Unit 3 22.

Thery, C., Boussac, M., Veron, P., Ricciardi-Castagnoli, P., Raposo, G., Garin, J., & Amigorena, S. (2001). Proteomic analysis of dendritic cell-derived exosomes: A secreted subcellular compartment distinct from apoptotic vesicles. *The Journal of Immunology, 166*(12), 7309–7318.

Thery, C., Zitvogel, L., & Amigorena, S. (2002). Exosomes: Composition, biogenesis and function. *Nature Reviews Immunology, 2*(8), 569–579.

Thuma, F., Heiler, S., Schnolzer, M., & Zoller, M. (2016). Palmitoylated claudin7 captured in glycolipid-enriched membrane microdomains promotes metastasis via associated transmembrane and cytosolic molecules. *Oncotarget, 7*(21), 30659–30677.

Tosar, J. P., Gambaro, F., Sanguinetti, J., Bonilla, B., Witwer, K. W., & Cayota, A. (2015). Assessment of small RNA sorting into different extracellular fractions revealed by high-throughput sequencing of breast cell lines. *Nucleic Acids Research, 43*(11), 5601–5616.

Umezu, T., Ohyashiki, K., Kuroda, M., & Ohyashiki, J. H. (2013). Leukemia cell to endothelial cell communication via exosomal miRNAs. *Oncogene, 32*(22), 2747–2755.

Umezu, T., Tadokoro, H., Azuma, K., Yoshizawa, S., Ohyashiki, K., & Ohyashiki, J. H. (2014). Exosomal miR-135b shed from hypoxic multiple myeloma cells enhances angiogenesis by targeting factor-inhibiting HIF-1. *Blood, 124*(25), 3748–3757.

Valenzuela, M. M., Ferguson Bennit, H. R., Gonda, A., Diaz Osterman, C. J., Hibma, A., Khan, S., & Wall, N. R. (2015). Exosomes secreted from human cancer cell lines contain inhibitors of apoptosis (IAP). *Cancer Microenviron, 8*(2), 65–73.

van Balkom, B. W., Eisele, A. S., Pegtel, D. M., Bervoets, S., & Verhaar, M. C. (2015). Quantitative and qualitative analysis of small RNAs in human endothelial cells and exosomes provides insights into localized RNA processing, degradation and sorting. *Journal of Extracellular Vesicles, 4*, 26760.

Vaupel, P., & Harrison, L. (2004). Tumor hypoxia: Causative factors, compensatory mechanisms, and cellular response. *The Oncologist, 9*(suppl 5), 4–9.

Vella, L. J. (2014). The emerging role of exosomes in epithelial-mesenchymal-transition in cancer. *Frontiers in Oncology, 4*, 361.

Villarroya-Beltri, C., Gutierrez-Vazquez, C., Sanchez-Cabo, F., Perez-Hernandez, D., Vazquez, J., Martin-Cofreces, N., … Sanchez-Madrid, F. (2013). Sumoylated hnRNPA2B1 controls the sorting of miRNAs into exosomes through binding to specific motifs. *Nature Communications, 4*, 2980.

White, I. J., Bailey, L. M., Aghakhani, M. R., Moss, S. E., & Futter, C. E. (2006). EGF stimulates annexin 1-dependent inward vesiculation in a multivesicular endosome subpopulation. *Embo Journal, 25*(1), 1–12.

Woodcock, J. (2006). Sphingosine and ceramide signalling in apoptosis. *IUBMB Life, 58*(8), 462–466.

Xiao, X., Yu, S., Li, S., Wu, J., Ma, R., Cao, H., … Feng, J. (2014). Exosomes: Decreased sensitivity of lung cancer A549 cells to cisplatin. *PLoS One, 9*(2), e89534.

Yang, M., Chen, J., Su, F., Yu, B., Su, F., Lin, L., … Song, E. (2011). Microvesicles secreted by macrophages shuttle invasion-potentiating microRNAs into breast cancer cells. *Molecular Cancer, 10*, 117.

Ye, S. B., Li, Z. L., Luo, D. H., Huang, B. J., Chen, Y. S., Zhang, X. S., … Li, J. (2014). Tumor-derived exosomes promote tumor progression and T-cell dysfunction through the regulation of enriched exosomal microRNAs in human nasopharyngeal carcinoma. *Oncotarget, 5*(14), 5439–5452.

Yue, S., Mu, W., Erb, U., & Zoller, M. (2015). The tetraspanins CD151 and Tspan8 are essential exosome components for the crosstalk between cancer initiating cells and their surrounding. *Oncotarget, 6*(4), 2366–2384.

Zhang, X., Yuan, X., Shi, H., Wu, L., Qian, H., & Xu, W. (2015). Exosomes in cancer: Small particle, big player. *Journal of Hematology & Oncology, 8*, 83.

Zhou, W., Fong, M. Y., Min, Y., Somlo, G., Liu, L., Palomares, M. R., … Wang, S. E. (2014). Cancer-secreted miR-105 destroys vascular endothelial barriers to promote metastasis. *Cancer Cell, 25*(4), 501–515.

Zhuang, G., Wu, X., Jiang, Z., Kasman, I., Yao, J., Guan, Y., … Ferrara, N. (2012a). Tumour-secreted miR-9 promotes endothelial cell migration and angiogenesis by activating the JAK-STAT pathway. *The EMBO Journal, 31.*

Zhuang, G., Wu, X., Jiang, Z., Kasman, I., Yao, J., Guan, Y., … Ferrara, N. (2012b). Tumour-secreted miR-9 promotes endothelial cell migration and angiogenesis by activating the JAK-STAT pathway. *The EMBO Journal, 31*(17), 3513–3523.

4

Heterogeneity of Tumor Exosomes – Role in Precision Medicine

Vincent Bernard, Jinjie Ling, Anirban Maitra

SHEIKH AHMED PANCREATIC CANCER RESEARCH CENTER, UT MD ANDERSON CANCER CENTER, HOUSTON, TX, UNITED STATES

CHAPTER OUTLINE

1. Introduction 59

2. Double-Stranded Genomic DNA in Circulating Exosomes 60

3. Detection of Mutational Signatures in Genomic DNA-Enriched Exosomes 61

4. Exosomes as Agents for Early Detection, Diagnosis, and Stratification 61

5. Genomic Molecular Profiling of Exosomal Cargo 63

6. Transcriptomic Characterization of Tumors Through Liquid Biopsies 65

7. Conclusions 65

Acknowledgments 66

References 66

1. Introduction

Exosomes are extracellular vesicles (EVs) exhibiting a diameter of 40–120 nm, conceived endogenously through the multivesicular endosome pathway and released to the extracellular space via fusion with the plasma membrane (Desrochers, Antonyak, & Cerione, 2016; Teis, Saksena, & Emr, 2009). Microvesicles (MVs) are a class of larger EVs with a diameter ranging from 0.2 to 1 µm and originate from the budding and fission at special "lipid raft" domains of the plasma membrane (Cocucci & Meldolesi, 2011). A methodology to reliably enrich for exosomes, but not MVs, currently does not exist, as there are significant overlap between size, shape, density and cell marker profiles (Lo Cicero, Stahl, & Raposo, 2015). Both are molecular vehicles reported to transfer a variety of biochemical cargo, including protein products, RNA transcripts, microRNAs, and fragmented DNA, but beyond their distinct biogenesis pathways and relative diameters, the two are difficult to delineate, and as a result, are frequently used interchangeably in literature (Chiba, Kimura, & Asari, 2012;

Cocucci & Meldolesi, 2015; Muralidharan-Chari et al., 2010; Skog et al., 2008). For the purpose of this chapter, we will use the term exosome to refer specifically to the population of small EVs produced through the multivesicular endosome biogenesis pathway.

2. Double-Stranded Genomic DNA in Circulating Exosomes

Circulating exosomes are known to facilitate intercellular communication through the exchange of numerous biochemical products such as proteins, lipids, mRNA transcripts, miRNA, and DNA of both chromosomal and mitochondrial origin (Ludwig & Giebel, 2012). The recent identification of double-stranded high molecular weight genomic DNA within circulating exosomes has proven to be an exciting discovery in the context of cancer liquid biopsies with translational implications for early detection, diagnosis, monitoring, and prognostic and therapeutic stratification of solid tumors, including deep seated visceral cancers. Specifically, exosome-derived DNA (exoDNA) may have a role in precision medicine, whereby molecular profiling of exoDNA may lead to the identification of effective therapeutic strategies based on the molecular makeup of a patient's underlying cancer from which the exosomes have been released into circulation. Similar in concept to the use of circulating tumor DNA (ctDNA), exoDNA allows for the profiling of an additional liquid biopsy compartment, whereby tumor profiling is possible through a minimally invasive approach compared to more invasive tumor biopsy procedures, thus allowing for repeated biopsy samples taken throughout disease treatment and progression.

The presence of chromosomal DNA cargo within exosomes was validated in exosomes isolated from healthy human plasma and from the culture supernatants of HEK293 human embryonic kidney cells and K562 human leukemia cells (Cai et al., 2013). In this study, isolated exosomes were treated with DNase to ascertain that the isolated genomic DNA presided in the interior, rather than the exterior of the exosomes, the latter of which would represent the circulating cell free DNA (cfDNA) fraction. Additionally, as opposed to cfDNA, which exists in the form of fragmented DNA molecules of ~170 bp, exoDNA consisted of high molecular weight DNA. Numerous groups have since reported the presence of exosomes enriched with long and/or fragmented genomic DNA of varying sizes from different sources, including plasma, urine, and pleural effusions, with whole-genome sequencing studies subsequently revealing that exoDNA covered the entire compendium of human chromosomes without a bias toward particular regions of the genome (Cai et al., 2013; Kahlert et al., 2014; San Lucas et al., 2016; Thakur et al., 2014).

The shielding of the genomic DNA by the exosome exterior appears to attenuate DNA degradation by extracellular DNAses, and enhances the stability of the exoDNA, an observation that raises the possibility of using tumor-secreted exosomes paired with next generation sequencing (NGS) as a liquid biopsy platform for comprehensive interrogation of the cancer genome (Jin et al., 2016). Additionally, having the ability to profile DNA from different sources in circulation may allow for characterization of differing biological underpinnings that occur during tumor progression. In other words, it is generally believed that cfDNA is released in circulation from cells undergoing active apoptosis or necrosis, versus exoDNA that may be derived from cells that undergoing rapid proliferation and active biogenesis of exosomes. A recent study hypothesizes a potential mechanism of DNA packaging within exosomes involving the enrichment of Histone H2B proteins within exosomes (Castillo et al., 2018). In other words, these proteins have a role in identification of foreign or aberrant cytosolic DNA molecules and have been shown to co-localize with exosomal proteins such as CD63, which is involved in exosome cargo trafficking. It is thus thought that exosomes may provide a mechanism for exporting of mutated DNA molecules out of the cells as a means of self-defense.

3. Detection of Mutational Signatures in Genomic DNA-Enriched Exosomes

Mutation detection within exoDNA of pancreatic cancer patient plasma initially demonstrated utility using polymerase chain reaction (PCR) and Sanger sequencing to determine trademark *KRAS* and *TP53* mutations, known common genetic drivers of pancreatic cancer (Kahlert et al., 2014). Subsequent reports confirmed that Sanger sequencing detection of mutational signatures can also be performed in exosomes secreted by prostate cancer cells (Lazaro-Ibanez et al., 2014). For prostate cancer patients, this methodology can be applied not only in excreted urine, but also in circulation, as high molecular weight exosomal DNA fragments were also identified in the plasma of prostate cancer patients (Lazaro-Ibanez et al., 2014; Ronquist et al., 2012). Interestingly, the presence of exosomes has been acknowledged in a range of biological fluids including blood, urine, milk, and saliva, creating several opportunities for applications that rely on the fluid context, such as the use of exosomes in the urine for urinary tract malignancies, exosomes in pleural fluids for lung/mesothelial cancers, or exosomes within the blood for visceral malignancies (Lasser et al., 2011).

4. Exosomes as Agents for Early Detection, Diagnosis, and Stratification

In recalcitrant cancers such as pancreatic ductal adenocarcinoma (PDAC), diagnosis often occurs at a late stage of the disease when the cancer becomes virtually uncurable. This is typically attributed to the late presentation of disease symptoms and an inability to discern low volume (early stage) disease. Currently, carbohydrate antigen 19-9 (CA 19-9) is the circulating tumor marker most commonly used for diagnosis in the clinic. Because CA 19-9 is not elevated in the early stages of PDAC, and is also present in many benign cases of pancreatitis and biliary obstruction, it has mostly been used as a prognostic tool to track tumor progression. As a result, new methodologies with the capacity to detect tumors at an early, treatable stage without the direct, invasive sampling of the cancer, are desperately needed to address these types of aggressive cancers. Noninvasive liquid biopsy strategies involving the isolation of circulating tumor cells (CTCs) and ctDNA from patient blood to determine the presence of an asymptomatic cancer have shown promise, but the diagnostic and early detection potential of circulating exosomal DNA (exoDNA) is just beginning to be understood (Bettegowda et al., 2014; Sergeant et al., 2012; Ting et al., 2014).

As the principal driver mutation, *KRAS* is near ubiquitous in PDAC, with an estimated ~95% of tumors exhibiting some *KRAS* mutations (Bailey et al., 2016; Bryant et al., 2014). This near ubiquitous presence of *KRAS* in PDAC tumors, and the fact that it represents one of the first mutations that is acquired during carcinogenesis, suggests that a strategy for its detection in the context of liquid biopsies may provide an avenue for early detection and treatment monitoring (Bernard, Fleming, & Maitra, 2016). Using an ultrasensitive mutation detection methodology known as droplet digital polymerase chain reaction (ddPCR), Allenson et al. demonstrated the feasibility of detecting *KRAS* mutations in exoDNA from PDAC patients, and determining the circulating mutant allele frequency (MAF) for the oncogenic allele amongst a sea of wild type DNA (Allenson et al., 2017). This study demonstrated the ability to detect mutant alleles in exoDNA obtained from all stages of PDAC, as well as allowing for stratification of patient survival outcomes based on the *KRAS* MAF. Notably, mutation detection of exosomal *KRAS* (exo*KRAS*) was seen in 7.4% of age matched healthy controls, 66.7% of localized disease, 80% of locally advanced disease, and 85% of metastatic PDAC, representing 75.4% sensitivity and 92.6% specificity for exo*KRAS* as a tumor biomarker for evaluating PDAC. Furthermore, a patient that tested positive for exo*KRAS* was 8.17 times more likely to have an early stage cancer rather than be tumor-free. Interestingly, exo*KRAS* MAF levels correspond with disease-free survival in patients with localized disease, where patients with an exo*KRAS* of >1.0% experiencing poorer disease free survival, a relationship that the prognostic biomarker CA 19-9 did not illustrate. This suggests that there may be a subpopulation of patients that may require more aggressive intervention and follow-up.

In the aforementioned study, it was also notable that exoDNA outperformed ctDNA in the detection of PDAC, and generated significantly higher detection rates of positivity across all stages of the disease, but most important in the early stages (resectable stages) of cancer. The ddPCR analysis of cfDNA revealed mutant cf*KRAS* detection in 14.8% of healthy controls, 45.5% of localized disease, 30.8% of locally advanced disease, and 57.9% of metastatic PDAC. A possible explanation for this discordance is that ctDNA is theorized to be released extensively into circulation only at the later phases of PDAC, where dying cells becomes more pervasive, and as a result, may be less effective at pinpointing early stage disease manifestations (Kidess & Jeffrey, 2013). Thus, exoDNA, a product of normal biogenesis pathways, may be a promising alternative to ctDNA for the earlier detection of PDAC. As a cautionary note, mutant *KRAS* was also found in healthy individuals (including two independent cohorts from the US and Europe), a phenomenon that appears to increase with age-related clonal hematopoiesis and/or the likely presence of *KRAS*-mutant precursor lesions within the pancreas, GI tract or lung. This finding serves to add an important caveat to the utility of the current methodology as an early diagnostic tool, and prevent the phenomenon of "overdiagnosis." It is thus important to consider limiting such screens to high-risk populations using current assay technologies, or develop methodologies that may increase specificity, such as detection of a panel of mutations that represents a higher probability for an underlying cancer (as opposed to a clinically insignificant precursor lesion).

5. Genomic Molecular Profiling of Exosomal Cargo

A key component of a precision medicine approach to cancer is the ability to profile the molecular characteristics of a patient's underlying cancer. This is particularly difficult for visceral cancers such as PDAC where attempts at surgical sampling of tumor tissue are inherently invasive and frequently limited by the obscure tumor location and risk accompanied with surgical procedure (Chantrill et al., 2015). Minimally invasive liquid biopsies have been attractive alternatives to direct tissue sampling. Investigators have previously used plasma-derived, cfDNA to identify key oncogenic "hotspot" drivers (i.e., *V-Raf murine sarcoma viral oncogene homolog B, KRAS, epidermal growth factor receptor (EGFR)*) via digital PCR, but the highly fragmented nature of cfDNA makes applications involving NGS platforms more challenging (El-Hefnawy et al., 2004; Mitchell et al., 2008; Mouliere et al., 2011). Though attempts at using cfDNA for targeted genomic profiling have been published (and companies such as Foundation Medicine, GRAIL, and Guardant Health are heavily investing in such cfDA "liquid biopsy" assays), the feasibility of circulating exosomes as means for tumor profiling and disease monitoring has only just begun to be described (Fig. 1).

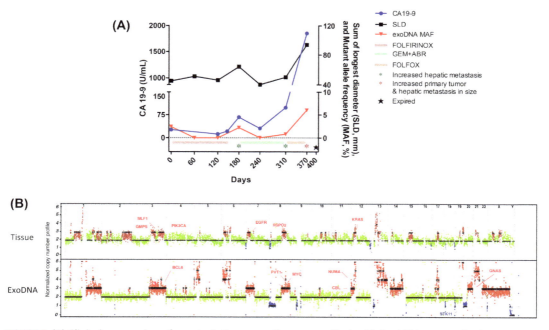

FIGURE 1 (A) Clinical progression of a metastatic pancreatic cancer patient with liquid biopsy monitoring. Mutational burden as measured by digital PCR *KRAS* mutant allelic frequency detection (*red*), is plotted with the standard clinical biomarker CA19-9 (*blue*), and sum of longest diameters (SLD) based on tumor lesion CT imaging (*black*). Liquid biopsies through exoDNA appear to closely correlate with clinical progression. (B) Normalized copy number profiles of pancreatic tumor tissue and exoDNA from whole genome sequencing. Putative cancer related driver copy number gains are annotated in red, and copy number losses in blue. Many events found in tissue are confirmed in exoDNA. ExoDNA appears to capture many more events not represented in tumor tissue, suggesting that liquid biopsies may capture heterogeneous events not profiled in single tissue biopsies.

A recent study sought to determine the efficacy of exosomes in visceral tumor genomic profiling (San Lucas et al., 2016). San Lucas et al. isolated circulating exosomes from various bodily fluid sources including peripheral whole blood and pleural effusions of metastatic PDAC patients. The exoDNA extracted from these exosomes contained genomic DNA of high molecular weight, which was representative of the entire human genome (65%–91% genomic representation on whole-genome sequencing). The exoDNA isolates further revealed high representation of tumor fraction ranging from 56% to 82%, suggesting that this liquid biopsy compartment may confer an enriched source of tumor derived material in circulation. This is further emphasized by the high cancer-derived DNA fraction found in exosomes obtained from a pleural effusion (82%) in the context of <1% malignant cells on cytospin in the same sample. Whole exome sequencing of exoDNA further revealed several potentially actionable mutations, including COSMIC (Catalog of Somatic Mutations in Cancer) alterations that could be used to monitor tumor genomic evolution over time, and COSMIC genes that could be addressed through a particular clinical trial or chemotherapy. Sequencing data of exoDNA also illustrated amplified copy numbers of major mutational signatures such as *KRAS, EGFR,* and *ERBB2*. In a particularly interesting case demonstrating the potential utility of exoDNA for therapeutic selection, the investigators detected an unexpected somatic *BRCA2* mutation, known to impair homologous recombination, a critical DNA repair mechanism in actively dividing cells. This patient subsequently achieved a striking response to a regimen comprising of cisplatin, a crosslinking agent that generates widespread DNA damage. Although retrospective and correlative in nature, this data suggests that further characterization of how mutation detection through exoDNA can impact therapeutic decision-making is further warranted.

In a separate study, Castillo et al. (2018) describe an enrichment methodology to specifically capture cancer-specific exosomes (CSEs) from the circulation, allowing for the ability to perform high resolution genomic characterization through the captured cargo (Castillo et al., 2018). The authors identified a panel of six proteins - CLDN4, EPCAM, CD151, LGALS3BP, HIST2H2BE and HIST2H2BF - that were specifically expressed on the surface of PDAC-derived CSEs ("surfaceome"), and could be exploited through an immunocapture approach for enriching CSEs. As opposed to ctDNA, which cannot be specifically captured from the total cfDNA compartment, exosomes have the added benefit of expressing tumor specific markers that can be used to separate tumor and normal tissue derived exosomes. This is particularly relevant in the context of those patients undergoing active therapy where circulating tumor burden can dramatically decrease to the point of making mutational events in circulation undetectable using current technologies, or in the context of early detection of an asymptomatic cancer where the volume of CSEs might be overwhelmed by the complement of normal exosomes. To overcome this limitation, Castillo et al. applied an antibody cocktail through an immunocapture technique that allowed for positive selection of CSEs, which can be subsequently used for mutation detection. Using this assay, they achieved an increase in mutation detection from 44% to 73% in non-captured versus captured exosomes. The authors also demonstrated the utility of this technique in being able to perform NGS on CSE-derived exoDNA through a molecular barcoding targeted sequencing approach. In an index case of a patient who initially responded to poly (ADP-ribose) polymerase inhibitor therapy secondary to a somatic *BRCA2* stop-gain mutation, and subsequently progressed, the authors were able to identify a putative mechanism of resistance through a second splice site mutation of the same gene, which allowed for reversion of the initial stop-gain ("*BRCA2* reversion mutation"). This further demonstrates the ability of exosomes to not only detect genomic vulnerabilities, but also to provide a means to identify mechanisms of resistance for real-time precision oncology decision making.

6. Transcriptomic Characterization of Tumors Through Liquid Biopsies

As a source a highly enriched tumor material, exosomes also contain a milieu of cargo that can be used for tumor characterization, such as mRNA. Whereas cfRNA is largely comprised of highly fragmented circulating RNA transcripts, limiting the molecular assays to those involved in microRNA detection, RNA within exosomes (exoRNA), provides a source of long mRNA transcripts that allow for more detailed characterization of tumors through liquid biopsies. In the study by San Lucas et al. (2016), exoRNA allowed for the orthogonal validation of gene amplifications, as in the case of overexpression of *ERBB2*. In addition, the benefits of this transcriptome-based approach may also extend toward the determination of novel cancer-specific fusion transcripts that may otherwise not be apparent from genomic sequencing only. Further, the identification of expressed cancer-derived neoantigens (both missense mutations and fusions) may facilitate emerging precision immunotherapies that rely on discovery of such neoantigens in a patient-specific manner. Ultimately, this may also allow for profiling of the dynamic changes in the neoantigen repertoire, which may occur from selective pressures and "antigen editing" that occurs during tumor progression. Through serial monitoring of tumor associated antigens and how these evolve over time, one can begin to suggest novel therapeutic approaches relating to ideal immunotherapeutic stratification. Specifically, quantitative estimates of neoantigen load through liquid biopsies may provide an early surrogate of response to immunotherapies such as vaccines or engineered T-cell receptors, of which there is currently no readily available biomarker.

7. Conclusions

NGS of circulating exosomes (including enriched CSEs) provides promising strategies for non-invasive tumor profiling and disease monitoring. Recent data from many laboratories suggests that exosomes are an important component of liquid biopsies, facilitating identification of actionable mutations critical to developing patient-tailored precision treatment regimens. In addition, the ability of exoRNA to profile the tumor transcriptome presents many new exciting opportunities, such as the discovery of novel neoantigens that may serve as the basis for emerging adoptive T-cell immunotherapies. This system also exhibits high clinical relevancy with abridged times from patient blood draw to exosome sequencing and data analysis. These promising data warrants the further development of exosomes as a complementary clinical tool in early disease detection, disease monitoring, and therapeutic stratification.

Acknowledgments

Khalifa Bin Zayed Foundation, MD Anderson Cancer Moonshot program, NCI CA196403, NCI CA224020.

References

Allenson, K., et al. (2017). High prevalence of mutant KRAS in circulating exosome-derived DNA from early-stage pancreatic cancer patients. *Annals of Oncology*, *28*(4), 741–747.

Bailey, P., et al. (2016). Genomic analyses identify molecular subtypes of pancreatic cancer. *Nature*, *531*(7592), 47–52.

Bernard, V., Fleming, J., & Maitra, A. (2016). Molecular and genetic basis of pancreatic carcinogenesis: Which concepts may be clinically relevant? *Surgical Oncology Clinics of North America*, *25*(2), 227–238.

Bettegowda, C., et al. (2014). Detection of circulating tumor DNA in early- and late-stage human malignancies. *Science Translational Medicine*, *6*(224), 224ra24.

Bryant, K. L., et al. (2014). KRAS: Feeding pancreatic cancer proliferation. *Trends in Biochemical Sciences*, *39*(2), 91–100.

Cai, J., et al. (2013). Extracellular vesicle-mediated transfer of donor genomic DNA to recipient cells is a novel mechanism for genetic influence between cells. *Journal of Molecular Cell Biology*, *5*(4), 227–238.

Castillo, J., et al. (2018). Surfaceome profiling enables isolation of cancer-specific exosomal cargo in liquid biopsies from pancreatic cancer patients. *Annals of Oncology* 2017, *29*(1), 223–229.

Chantrill, L. A., et al. (2015). Precision medicine for advanced pancreas cancer: The individualized molecular pancreatic cancer therapy (IMPaCT) trial. *Clinical Cancer Research*, *21*(9), 2029–2037.

Chiba, M., Kimura, M., & Asari, S. (2012). Exosomes secreted from human colorectal cancer cell lines contain mRNAs, microRNAs and natural antisense RNAs, that can transfer into the human hepatoma HepG2 and lung cancer A549 cell lines. *Oncology Reports*, *28*(5), 1551–1558.

Cocucci, E., & Meldolesi, J. (2011). *Ectosomes: Current Biology*, *21*(23), R940–R941.

Cocucci, E., & Meldolesi, J. (2015). Ectosomes and exosomes: Shedding the confusion between extracellular vesicles. *Trends in Cell Biology*, *25*(6), 364–372.

Desrochers, L. M., Antonyak, M. A., & Cerione, R. A. (2016). Extracellular vesicles: Satellites of information transfer in cancer and stem cell biology. *Developmental Cell*, *37*(4), 301–309.

El-Hefnawy, T., et al. (2004). Characterization of amplifiable, circulating RNA in plasma and its potential as a tool for cancer diagnostics. *Clinical Chemistry*, *50*(3), 564–573.

Jin, Y., et al. (2016). DNA in serum extracellular vesicles is stable under different storage conditions. *BMC Cancer*, *16*(1), 753.

Kahlert, C., et al. (2014). Identification of double-stranded genomic DNA spanning all chromosomes with mutated KRAS and p53 DNA in the serum exosomes of patients with pancreatic cancer. *Journal of Biological Chemistry*, *289*(7), 3869–3875.

Kidess, E., & Jeffrey, S. S. (2013). Circulating tumor cells versus tumor-derived cell-free DNA: Rivals or partners in cancer care in the era of single-cell analysis? *Genome Medicine, 5*(8), 70.

Lasser, C., et al. (2011). Human saliva, plasma and breast milk exosomes contain RNA: Uptake by macrophages. *Journal of Translational Medicine, 9,* 9.

Lazaro-Ibanez, E., et al. (2014). Different gDNA content in the subpopulations of prostate cancer extracellular vesicles: Apoptotic bodies, microvesicles, and exosomes. *The Prostate, 74*(14), 1379–1390.

Lo Cicero, A., Stahl, P. D., & Raposo, G. (2015). Extracellular vesicles shuffling intercellular messages: For good or for bad. *Current Opinion in Cell Biology, 35,* 69–77.

Ludwig, A. K., & Giebel, B. (2012). Exosomes: Small vesicles participating in intercellular communication. *The International Journal of Biochemistry & Cell Biology, 44*(1), 11–15.

Mitchell, P. S., et al. (2008). Circulating microRNAs as stable blood-based markers for cancer detection. *Proceedings of the National Academy of Sciences of the United States of America, 105*(30), 10513–10518.

Mouliere, F., et al. (2011). High fragmentation characterizes tumour-derived circulating DNA. *PLoS One, 6*(9), e23418.

Muralidharan-Chari, V., et al. (2010). Microvesicles: Mediators of extracellular communication during cancer progression. *Journal of Cell Science, 123*(Pt 10), 1603–1611.

Ronquist, G. K., et al. (2012). Prostasomes are heterogeneous regarding size and appearance but affiliated to one DNA-containing exosome family. *The Prostate, 72*(16), 1736–1745.

San Lucas, F. A., et al. (2016). Minimally invasive genomic and transcriptomic profiling of visceral cancers by next-generation sequencing of circulating exosomes. *Annals of Oncology, 27*(4), 635–641.

Sergeant, G., et al. (2012). Pancreatic cancer circulating tumour cells express a cell motility gene signature that predicts survival after surgery. *BMC Cancer, 12,* 527.

Skog, J., et al. (2008). Glioblastoma microvesicles transport RNA and proteins that promote tumour growth and provide diagnostic biomarkers. *Nature Cell Biology, 10*(12), 1470–1476.

Teis, D., Saksena, S., & Emr, S. D. (2009). SnapShot: The ESCRT machinery. *Cell, 137*(1), 182–182 e1.

Thakur, B. K., et al. (2014). Double-stranded DNA in exosomes: A novel biomarker in cancer detection. *Cell Research, 24*(6), 766–769.

Ting, D. T., et al. (2014). Single-cell RNA sequencing identifies extracellular matrix gene expression by pancreatic circulating tumor cells. *Cell Reports, 8*(6), 1905–1918.

5

Proteomic Profiling of Tumor Exosomes

Sara Yousuf, Alfred A. Simental, Salma Khan

LOMA LINDA UNIVERSITY, LOMA LINDA, CA, UNITED STATES

CHAPTER OUTLINE

1. Introduction **69**

2. Sources of Exosomes **71**
   2.1 Exosomes in Plasma 72
   2.2 Exosomes in Cerebrospinal Fluid 72
   2.3 Exosomes in Urine 72
   2.4 Exosomes in Breast Milk 73
   2.5 Exosomes in Saliva 73

3. Proteomic Content of Exosomes in Cancers **74**
   3.1 Breast Cancer 74
   3.2 Prostate Cancer 78
   3.3 Ovarian Cancer 78
   3.4 Head and Neck Cancer 79
   3.5 Leukemia 79
   3.6 Bladder Cancer 80
   3.7 Liver Cancer 80
   3.8 Pancreatic Cancer 81
   3.9 Brain Tumor 81
   3.10 Lung Cancer 82
   3.11 Colorectal Cancer 82

4. Conclusion **83**

Acknowledgments **83**

References **83**

1. Introduction

Exosomes are membrane vesicles of 40–100 nm in diameter that are released by most cell types on fusion of multivesicular bodies (MVBs) with the plasma membrane. Exosomes are secreted by various cell types. They are essentially cell messengers and are able to

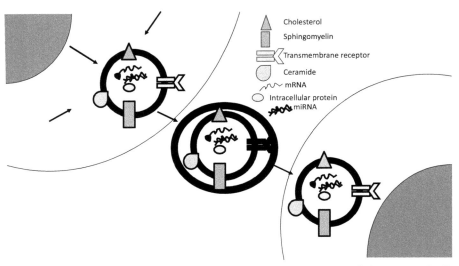

FIGURE 1 Schematic representation of exosomal content and secretion to the extracellular space and cell-to-cell communication in tumor.

communicate from cell to cell (Fontana, Saieva, Taverna, & Alessandro, 2013). Fig. 1 shows how an exosome works in a microenvironment in terms of cell-to-cell communication in cancer and was also shown previously by others (Thomas et al., 2013). Researchers have always been aware that exosomes existed, but using them for their potential diagnostic and therapeutic significance is a recent development. This knowledge has increased the research on exosomes. Exosomes represent a huge opportunity for biomarker discovery because they contain protein, messenger RNA, and microRNA (miRNA); all of which has been used to change the state of the disease (Street et al., 2012). Exosomes that are secreted from tumor cells contain antigens that can potentially be used for immunotherapeutic purposes (Bard et al., 2004). The fact that tumor cells actively shed exosomes shows that they play a role in tumor progression (Fontana et al., 2013). Exosomes are actively released, which shows the proteins and RNA come from an original cell. Douglas et al. discusse how miRNA has shown the type and origin of a tumor. More importantly, exosomes are able to remove specific drugs from cells, and cells releasing exosomes containing therapeutic drugs have been reported to be resistant to these drugs (Safaei et al., 2005).

Researchers characterize the proteome of tumor exosomes to differentiate with tumor-derived exosomes and normal cell exosomes. Proteomic profiling of exosomes can potentially help find biomarkers for various cancers and tumors such as breast cancer, prostate cancer, and lung cancer. One of the goals of clinical proteomics is to identify biomarkers in biological diseases to be able to diagnose early for diseases. Simpson, Lim, Moritz, and Mathivanan (2009) suggest that the presence of infectious cargo such as prions and retroviruses may be the entities for exosome purification strategies. Exosomes can be found in body fluids.

The first exosome proteomes were found in mesothelioma, which showed 38 proteins that were expressed (Bard et al., 2004; Hegmans et al., 2004). The function of exosomes can be reflected in their protein composition (Wubbolts et al., 2003). This could potentially be a powerful diagnostic role because physicians would be able to identify the expressed exosome helping with early diagnosis. To investigate the differences between tumor-derived exosomes and normal cell exosomes, researchers often characterize the proteome of tumor exosomes. Exosomes are representative of the cell from which they were derived, so it is reasonable to assume that proteome signatures would exhibit unique qualities dependent on the cellular origin. Indeed, one of the first exosomal proteomes to be characterized was from mesothelioma cells. This study identified 38 distinct proteins, many of which were previously shown to be expressed in exosomes (Hegmans et al., 2004). However, it is interesting to note that the authors also identified developmental endothelial locus-1 (DEL-1), a protein not previously identified in exosomes but suggested to play a role in angiogenesis (Ho et al., 2004).

Although cancer treatments and diagnostic tools have come very far, there are still two ever-critical problems: the design of more effective treatment regimens and the identification of biomarkers for patient diagnosis and prognosis. There is a need for specific and accurate diagnosis of disease state, and molecular monitoring of disease progression has become more important than ever (Suchorska & Lach, 2016; Wang et al., 2015; Yu, Cao, Shen, & Feng, 2015). Fortunately, the intensive search for cancer biomarkers, and potential novel therapeutic targets, has revealed a factor with significant potential, the exosome (Tickner, Urquhart, Stephenson, Richard, & O'Byrne, 2014). The tumor microenvironment is said to have a pivotal role in cancer development and progression (Iero et al., 2008). Tumor cells begin to shape the host environment, starting at early phases, to favor their survival and expansion. Although some host components, such as the immune system, may initially attempt to restrict disease progression, defenses are progressively blunted by the activation of suppressive pathways or even turned into tumor-promoting players (Rivoltini et al., 2005).

There is emerging evidence that cancer-derived exosomes contribute to the recruitment and reprogramming of constituents associated with tumor environment. Here, we discuss different mechanisms associated with biogenesis, payload, and transport of exosomes. We highlight the functional relevance of exosomes in cancer, as related to tumor microenvironment, tumor immunology, angiogenesis, and metastasis. Exosomes may exert an immunosuppressive function and trigger an antitumor response by presenting tumor antigens to dendritic cells. Exosomes may serve as cancer biomarkers and aid in the treatment of cancer (Kahlert & Kalluri, 2013).

2. Sources of Exosomes

These secreted vesicles can be detected in most bodily secreted fluids including blood, urine, ascites, milk, cerebrospinal fluid (CSF), and saliva.

2.1 Exosomes in Plasma

Plasma is one of the rich sources of exosomes compared with serum (Khan et al., 2012, 2014). Attempts have been made to optimize exosome isolation from human plasma and serum (Hao et al., 2017; Khan et al., 2012; Muller, Hong, Stolz, Watkins, & Whiteside, 2014; Witwer et al., 2013). The release of extracellular vesicles (EVs), which are found in all the bodily fluids, is enhanced in cancer, and a major focus of cancer proteomics is therefore targeted at EVs (Inal et al., 2013). The blood/plasma secretome is also a source of EVs, potentially diagnostic of infectious disease, whether from EVs released from infected cells or from the pathogens themselves. The release of EVs with specific molecular contents into urine and plasma may be useful biomarkers for kidney disease and cancers (Zhang, Peng et al., 2016; Zhang, Zhou et al., 2016).

2.2 Exosomes in Cerebrospinal Fluid

Studies have been conducted about proteomic profiling in CSF and brain metastatic (BM). Exosomes can transfer information between cells that cause pathology, which could be a biomarker discovery. Street et al. conducted a study to find out whether human CSF has exosomes and the variability of those exosomes between different people. This study discovered that exosomes were identified in human CSF. This could prove to be a reservoir for biomarkers in various neurological disorders such as Alzheimer's disease. They also found a significant amount of variability within different people meaning that the techniques used to concentrate exosomes need to be modified. Within this research, it was determined that the isolation of exosomes needs to be improved to get more accurate and relevant results. If the techniques were modified, and there was a standard protocol, then more practitioners and researchers could use it effectively on a wide variety of patients.

Camacho, Guerrero, and Marchetti (2013) and Camacho et al. (2004) investigated the differences in miRNA and protein profiles of BM and nonbrain metastatic (non-BM). The study researched the cargo proteins, analyzed the proteomic content, and showed that BM exosomes can be internalized between non-BM cells and transport their cargo into cells. According to Camacho et al. (2013), these findings help researchers move toward a better understanding of exosome roles in the brain and for the discovery and applications of biomarkers through proteomic profiling. As the proteins secreted by cells as a response to various stimuli are most likely secreted into blood/plasma, the identification and preselection of candidate protein biomarkers from cell secretomes with subsequent validation of their presence at higher levels in serum/plasma is a promising approach (Makridakis & Vlahou, 2010; Stastna & Van Eyk, 2012).

2.3 Exosomes in Urine

Human urine contains a large number of exosomes. Urine can be collected in large quantities in a noninvasive way, making it a more convenient option than blood. Exosome isolation in urine can help to find urine proteins that could help discover biomarkers.

Pisitkun, Johnstone, and Knepper (2006) and Pisitkun, Shen, and Knepper (2004) discovered that urinary vesicles have protein components of MVBs. This study was able to identify proteins that can be found within diseases having to do with blood pressure and kidney regulation. Gonzales et al. (2009) discovered a wider range of known proteome in human urinary exosomes. The study discovered that exosomes come from MVB and are delivered to the urine when the outer membranes fuse with the plasma membrane and the proteins that were found are recognized as components that come together for the formation of MVB (Gonzales et al., 2009).

A myriad of proteins and peptides can be identified in normal human urine. These are derived from a variety of sources including glomerular filtration of blood plasma, cell sloughing, apoptosis, proteolytic cleavage of cell surface glycosylphosphatidylinositol-linked proteins, and secretion of exosomes by epithelial cells. Mass spectrometry–based approaches to urinary protein and peptide profiling can, in principle, reveal changes in excretion rates of specific proteins/peptides that can have predictive value in the clinical arena, e.g., in the early diagnosis of disease, in the classification of disease with regard to likely therapeutic responses, in the assessment of prognosis, and in monitoring response to therapy. These approaches have potential value, not only for diseases of the kidney and urinary tract but also in systemic diseases that are associated with circulating small protein and peptide markers that can pass the glomerular filter (Pisitkun et al., 2006).

2.4 Exosomes in Breast Milk

In addition to urine, exosomes can also be found in breast milk. Breast milk is a complex liquid with immune-competent cells and soluble proteins that provide immunity to the infant and affect the maturation of the infant's immune system (Admyre et al., 2007). Exosomes carry immune relevant cells, which help direct the immune responses to different cancers and harmful agents. Human breast milk is known to be an important immunologic support system for the infant during the first months of life. Exosomes have been found to be involved in both immune stimulation and tolerization (Admyre et al., 2007).

2.5 Exosomes in Saliva

Previous investigations have revealed that tumors are often the primary source of circulating membrane vesicles and increased amount of tumor-derived protein, RNA, and DNA were found in the plasma/blood of cancer patients (Yang, Wei, Schafer, & Wong, 2014; Yang, Bucan, et al., 2015; Yang, Xing, et al., 2015). Exosomes are ubiquitous in most body fluids (breast milk, saliva, blood, urine, malignant ascites, amniotic, bronchoalveolar lavage, and synovial fluids) (Palanisamy et al., 2010), and this study showed differential protein expressions in oral keratinocytes. Exosomes secreted in human saliva contain proteins and nucleic acids that could be exploited for diagnostic purposes for many diseases such as breast, head and neck, oral, gastric, pancreatic, and many other cancers. Therefore, tissue-specific exosomes with their constituent tissue-specific biomarkers can serve as a biomarker source for the diagnosis, prognosis, and monitoring of disease.

3. Proteomic Content of Exosomes in Cancers

The expression and release of specific tumor-derived proteins into the peripheral circulation has served as the centerpiece of cancer screening and diagnosis. Tumors actively release exosomes, exhibiting proteins and RNAs derived from the originating cell, into the peripheral circulation and other biologic fluids. Currently, in over 75 investigations compiled in ExoCarta, over 2300 proteins and 270 miRNAs have been linked with exosomes derived from biologic fluids. Our previous work has indicated that these circulating exosomal proteins can serve as surrogates for the tumor cell–associated counterparts, extending their diagnostic potential to asymptomatic individuals (Khan et al., 2012, 2014; Turay et al., 2016). The exosomes derived from these approaches were assessed for quantity and quality of specific marker proteins. These results suggest that although each method purifies exosomal material, circulating exosomes isolated by ExoQuick precipitation produces exosomal RNA and protein with greater purity and quantity than chromatography, ultracentrifugation, and DynaBeads. Although this precipitation approach isolates exosomes in general and does not exhibit specificity for the originating cell, the increased quantity and quality of exosomal proteins and RNAs should enhance the sensitivity and accuracy of downstream pathway analyses, such as qRT-PCR profiling of miRNA and mass spectrometric and electrophoretic analyses of exosomal proteins. To date, the clinical utility of EVs has been hampered by issues with nomenclature and methods of isolation. The term "exosomes" was introduced in 1981 to denote any nanometer-sized vesicles released outside the cell and to differentiate them from intracellular vesicles. Based on this original definition, we use "exosomes" as synonymous with "EVs." Although our original studies used ultracentrifugation to isolate these vesicles (Khan et al., 2011), we immediately became aware of the significant impact of the isolation method on the number, type, content, and integrity of the vesicles isolated. Isolation can be done by ultracentrifugation (Taylor, Homesley, & Doellgast, 1983), density gradient centrifugation (Witwer et al., 2013), size exclusion chromatography (Taylor & Gercel-Taylor, 2005), filtration (Taylor & Shah, 2015), polymer-based precipitation (Taylor, Zacharias, & Gercel-Taylor, 2011), and immunoaffinity capture (Tauro et al., 2012). Summary of the exosomal proteins and resources in cancers are shown in Table 1.

3.1 Breast Cancer

Studies have shown findings in tumor markers, which help with the early detection of breast cancer with the help of proteomic profiling of exosomes. Andre et al. (2002) reported HSP, MHC-1, Her2/Neu, and Mart1 in peritoneal fluid of breast cancer. Later, another study also detected HER2 in condition media of breast cancer cells (Koga et al., 2005). Palazzolo et al. (2012) explored the concept that exosomes are released by neoplastic cells and their role in holding tumor antigens when secreted into an extracellular medium. The study compared the protein in exosome-like vesicles to the whole cell lysates proteome, which is very significant in cancer progression. The study by Palazzolo et al. contributed to the knowledge of vesiculation phenomenon by showing that vesicles

Table 1   Proteomic Contents of Tumor Exosomes in Different Cancer Types

Cancer Type	Methods	Source	Candidate Marker	References
Bladder	LC-MS/MS		EDIL-3/Del1	Beckham et al. (2014)
Bladder	SDS-PAGE/MALDI-TOF MS		Periostin	Silvers et al. (2016)
Bladder	LC-MALDI-TOF/TOF MS		Basigin, galectin-3, trophoblast glycoprotein	Welton et al. (2010)
Bladder	iTRAQ proteomics	Cell lines	Vimentin, hepatoma-derived growth factor casein kinase II α, and annexin A2	Jeppesen et al. (2014)
Bladder	LC-MS/MS	Urine	Tumor-associated calcium-signal transducer 2 (TACSTD2) CTNNA1, CTNNB1 VSAP, ITGA4, PAK1, DDR1, CDC42, RHOA, NRAS, RHO, PIK3AR1, MLC1, MMRN1, and CTTNBP2	Chen et al. (2012) and Kumari, Saxena, and Agrawal (2015)
Brain	ELISA	Condition media	VEGF	Al-Nedawi, Meehan, Kerbel, Allison, and Rak (2009)
Brain		Serum		Skog et al. (2008)
Brain	NMR spectroscopy	Condition media/plasma	EGFR, IDH1, PDGRα, HSP90, MHC-II, PDPN	Garnier, Jabado, and Rak (2013), Santiago-Dieppa et al. (2014), and Shao et al. (2015)
Brain	LC-MS/MS	Condition media	HNF4A, D283MED	Epple et al. (2012)
Brain	ELISPOT/ELISA	Cells/serum	EGFR, EGFRvIII, and TGF-beta	Graner et al. (2009)
Brain	Evolutionary relationships classification system	Cells	Caveolin1, merlin/NF2, tuberin	Camacho et al. (2004)
Breast	2DE/MALDI-TOF MS	Cell lines	GAS5	Koldemir, Ozgur, and Gezer (2017)
Breast	MALDI-ToF mass spectrometry		PRDX2, PRDX6, LAMC1, LG3BP, GBB2, ITA3, TPM4, UBIQ	Green, Alpaugh, Barsky, Rappa, and Lorico (2015) and Palazzolo et al. (2012)
Breast	LC-MS/MS	Serum/plasma	Survivin, splice variants	Khan et al. (2014)
Breast	Sandwich assay	Cell lines	CD24, CD63, EGFR, glypican-1	Etayash, McGee, Kaur, and Thundat (2016)
Breast	Tandem-mass-tag	Cell line	Syndecan-1, ALIX, CD9, CD 81	Clark et al. (2015)
Breast	Western blotting, immunoelectron microscopy	Peritoneal fluid	MHC-I, HSPs, Her2/Neu, Mart1, TRP, gp100	Andre et al. (2002)
Breast	Flow cytometry	Conditioned media	HER2	Koga et al. (2005)
Breast	Isotope labeling with amino acids in cell culture (SILAC)-based approach	Condition media	EGFR amd HIF-1a signaling	Thomas et al. (2013)
Breast	LC-MS/MS	Condition media	Periostin	Vardaki et al. (2016)
Breast	UPLC	Plasma	Fibronectin	Moon et al. (2016)
Breast	HPLC-MS/MS	Condition media	CYR 61	Sanchez-Bailon et al. (2015)
Breast	Atomic force microscopy	Conditioned media	Annexin II	Maji et al. (2017)
Breast	MALDI-ToF mass spectrometry		Flotillin-1, prohibitin, G-protein, annexin A2, HSP60, HSP70	Staubach, Razawi and Hanisch (2009)

Continued

Table 1 Proteomic Contents of Tumor Exosomes in Different Cancer Types—cont'd

Cancer Type	Methods	Source	Candidate Marker	References
Colorectal	Nano-LC-MS-MS, 2-D DIGE, cytokine array, western blotting, and MS		GPA 33, CEACAM5, EFNB1, annexins, ARFs, Rabs, ADAM10, CD44, NG2 ephrin-B1, MIF, beta-catenin, junction plakoglobin, galectin-4, RACK1, and tetraspanin-8, FASL, TRAI, FASLG, TNFSF10	Greening, Ji, Kapp, and Simpson (2013), Ji, Greening, Kapp, Moritz, and Simpson (2009), Mathivanan et al. (2010), Andreola et al. (2002), Abusamra et al. (2005b), Huber et al. (2008), Taylor and Gercel-Taylor (2005), Choi et al. (2017), Simpson et al. (2009), and Silva et al. (2012)
Colorectal	Western blotting	Plasma	HSP60	Campanella et al. (2015)
Colorectal	2DE/MALDI-TOF MS	Sera	RPL7, RPL10A, RPS15; mitochondria: MRPL3, ATP5H	Ragusa et al. (2014)
Colorectal	MALDI-MS/nano-HPLC/ESI-MS/MS	Sera	PGAM1, aldolase C, Vip36, Agrin	Klein-Scory et al. (2010)
Colorectal	ELISA/FACS, LC-MS/MS, Western blotting	Plasma, condition media	CD63, Alix, Rab5a, CD9, TSG101, Alix, and CD63 MET, S100A8, S100A9, TNC EFNB2, JAG1, SRC, TNIK CAV1, FLOT1, FLOT2, PROM1	Zarovni et al. (2015), Ji et al. (2013), and Demory Beckler et al. (2013)
Colorectal	Nano-LC-MS/MS	Ascites, serum	Alix, CD81, and Tsg101, 359 EV proteins, DKK-4	Choi et al. (2012), Choi et al. (2011), Jimenez, Knol, Meijer, and Fijneman (2010), and Lim et al. (2012)
Head and neck cancer	iTRAQ		MMP-13 containing proteins	You et al. (2015)
Head and neck cancer	LC-MS/MS	Saliva	808 proteins	Sivadasan et al. (2015)
Head and neck cancer	iTRAQ	Condition media	640 proteins up 51	Chan et al. (2015)
Head and neck cancer	Western blotting	Condition media	HIF-1a	Aga et al. (2014)
Head and neck cancer	LC-MS/MS	Condition media	217 proteins, Galectin 9, LMP-1	Jelonek et al. (2015) and Principe et al. (2013)
Leukemia	Western blotting	Condition media	BCL-w, BCL-xl, and survivin, TGFb	Raimondo et al. (2015)
Leukemia	ELISA/Western blot	Plasma	TGFβ-1l	Hong, Muller, Whiteside, and Boyiadzis (2014)
Leukemia		Condition media and plasma	VPS33B, GDI2, VPS16B, FLOT1	Gu et al. (2016)
Leukemia	ELISA	Plasma	TGF b	Szczepanski, Szajnik, Welsh, Whiteside, and Boyiadzis (2011)
Leukemia	LC-MS/MS	Plasma, condition media	TGF-β1, FasL, PD-1/PDL-1, MICA/MICB, CD39/CD73	Boyiadzis and Whiteside (2017)

Cancer	Method	Source	Markers/Proteins	Reference
Liver	Western blotting	Condition media	CD63 and HSP70	Zhu, Qu, Sun, Qian, and Zhao (2014)
Liver	Western blotting	Condition media	Integrins	Hoshino et al. (2015)
Lung	Nano-HPLC-chip-MS/MS	Urine	LRG1	Li, Zhang, Qiu, and Qiu (2011)
Lung	ELISA	Plasma	EGFR	Yamashita et al. (2013)
Lung	Extracellular vesicle (EV) array	Plasma	37 proteins	Jakobsen et al. (2015)
Lung	Triple SILAC	Condition media	EGFR, GRB2, and SRC	Clark, Fondrie, Yang, and Mao (2016)
Lung	EV array	Plasma	CD151, CD171, and tetraspanin 8 NY-ESO-1, EGFR, PLAP, EpCam, and Alix	Sandfeld-Paulsen, and Jakobsen (2016) and Sandfeld-Paulsen, and Aggerholm-Pedersen (2016)
Lung	LC-MS/MS	Saliva and serum	319 and 994 proteins	Sun et al. (2017)
Ovary	Western blotting	Ascites	ALCAM	Carbotti et al. (2013)
Ovary	Western blotting	Ascites	Alix, TSG101 Hsp84/90, Hsc70 MHC I and II, phosphate isomerase, peroxiredoxin, aldehyde reductase, fatty acid synthase	Dorayappan, Wallbillich, Cohn, and Selvendiran (2016) and Andre et al. (2002)
Ovary	Western blotting	Tumor and condition media	NKG2D and DNAM-1	Labani-Motlagh et al. (2016)
Ovary	Nano-LC–ESI-MS/MS	Condition media	TSG101 and Alix	Liang et al. (2013)
Ovary	ELISA, FCM	Ascites	TCR, CD20, HLA-DR, B7-2, HER2/neu, CA125 and histone H2A, FasL, and TRAIL	Peng, Yan, and Keng (2011)
Ovary	Western blotting and gelatin zymography	Condition media and ascites	CD24 and EpCAM	Runz et al. (2007)
Ovary	Nano-LC–ESI-MS/MS	Condition media	60 proteins	Sinha, Ignatchenko, Ignatchenko, Mejia-Guerrero, and Kislinger (2014)
Ovary	LC-MS/MS, western blotting	Plasma/ascites/ condition media	MRP2, A TGFβ-B1, TP7A, and ATP7B MAGE 3/6	Szajnik, Czystowska-Kuzmicz, Elishaev, and Whiteside (2016) and Szajnik et al. (2013)
Ovary	TEM, western blotting	Plasma	Claudin-4	Li et al. (2009)
Ovary	Western blotting	Ascites	CD44	Nakamura et al. (2017) and Nakamura et al. (2016)
Pancreas	LC-MS/MS	Condition media and plasma	Migration inhibitory factor	Costa-Silva et al. (2015)
Pancreas	iTRAQ	Condition media	4, 517 proteins	An et al. (2017)
Prostate	LC-MS/MS	Plasma/serum	Survivin	Khan et al. (2012)
Prostate	LC-MS/MS	Serum	DNA helicase homolog PIF1, four and a half LIM domain 3, glutathione S transferase omega 2, maternal embryonic leucine zipper kinase, iroquois homeobox protein 5, leucine-rich zipper containing 4, minichromosome maintenance complex component 5, mitochondrial tumor suppressor 1 isoform 4	Turay et al. (2016)

such as exosomes can convey signals to immune cells. Using SILAC-based approach, Thomas et al. (2013) showed epidermal growth factor receptor (EGFR) and HIF-1 signaling proteins in the exosomes. Khan et al. (2014) explored the exosomal Survivin and its splice variants, which could provide a biomarker that would help physicians with early diagnosis. Green et al. (2015) showed PRDX2, PRDx6, TPM4, and other proteins in the exosomes. In addition to this, it is crucial to find biomarkers that will be able to foreshadow the reoccurrence of cancer. Survivin is released from cancer cells that bind to the tumor via exosomes (Khan et al., 2011). Understanding this pathway through exosome would lead to more potential therapeutics for patients. Clark et al. found it challenging to "truly define the exosome proteome because of the challenge of discerning contaminant proteins that may be identified via mass spectrometry using various exosome enrichment strategies (Andre et al., 2002; Clark et al., 2015; Etayash et al., 2016; Green et al., 2015; Koga et al., 2005; Koldemir et al., 2017; Maji et al., 2017; Moon et al., 2016; Sanchez-Bailon et al., 2015; Staubach et al., 2009; Thomas et al., 2013; Vardaki et al., 2016). To better define the exosome proteome in breast cancer, they incorporated a combination of Tandem-mass-tag quantitative proteomics approach and support vector machine cluster analysis of three conditioned media-derived fractions (Clark et al., 2015).

3.2 Prostate Cancer

Proteomic profiling of exosomes has been important to prostate cancer to help identify biomarkers for early diagnosis and other therapies for patients. Khan et al. (2012) did a study where they found the plasma exosomal Survivin. The expression of this exosome was found in both new and old cases of prostate cancer. The discovery of this exosome will help for early detection of prostate cancer. Turay et al.'s (2016) study explores exosomes in prostate cancer with African-American and how proteomic exosomes can contain ethnically specifically biomarkers. This would be crucial with decreasing the mortality and incidence of prostate cancer through early detection among African-American males. The study "designed to profile the exosomal proteins from the plasma of ethnically diverse prostate cancer patients and control individuals with no diagnosis of prostate cancer and to compare these exosomal profiles between the different ethnic groups. We believe that in addition to diagnostic markers, prognostic, predictive, and therapeutic markers are needed to act as surrogate endpoints in forecasting disease severity, choosing appropriate treatment modalities, and monitoring responses to therapies (Mikolajczyk, Song, Wong, Matson, & Rittenhouse, 2004). The goal of Duijvesz et al.'s (2013) study was to determine the significance of exosomal proteins by comparing them with exosomes from noncancerous prostate cells. The study was able to verify differential expression between cancerous and noncancerous exosomes, which allowed researchers to see markers expressed in prostate cancer.

3.3 Ovarian Cancer

Studies showed that ovarian cancer–derived exosomes also carried tissue-specific proteins associated with tumorigenesis and metastasis, especially in ovarian carcinoma

(Andre et al., 2002; Carbotti et al., 2013; Dorayappan et al., 2016; Labani-Motlagh et al., 2016; Lea et al., 2017; Li et al., 2009; Liang et al., 2013; Peng et al., 2011; Runz et al., 2007; Sinha et al., 2014; Szajnik et al., 2016; Szajnik et al., 2013; Zhang, Peng, et al., 2016). Claudin-containing exosomes were detected in the plasma-derived exosomes in ovarian cancer (Li et al., 2009). Labani-Motlagh et al. (2016) showed that differential expression of ligands for NKG2D and DNAM-1 receptors by epithelial ovarian cancer–derived exosomes influenced NK cell cytotoxicity. It was shown that FasL and TRAIL on exosomes derived from ascites of ovarian cancer patients may partly account for the apoptosis of cells of the immune system (Peng et al., 2011). A very recent study showed that ovarian cancer–derived exosomes transfer CD44 to HPMCs, facilitating cancer invasion (Nakamura et al., 2016, 2017).

Based on the known roles of exosomes in cellular communication, these data indicate that exosomes released by ovarian cancer cells may play important roles in ovarian cancer progression and provide a potential source of blood-based protein biomarker.

3.4 Head and Neck Cancer

Human salivary proteome contains 3449 proteins; 808 of them have been reported as differentially expressed proteins in oral cancer tissues (Sivadasan et al., 2015). The current findings provide novel insight into the vital role of MMP13-containing exosomes in nasopharyngeal (NPC) progression, which might offer unique insights into potential therapeutic strategies for NPC progressions (You et al., 2015). Incubation of NPC cells with MSC-derived exosomes resulted in the uptake of exosomes by the cells, which promoted their proliferation, migration, and tumorigenesis (Shi et al., 2016); an iTRAQ-based quantitative proteomics was used to identify the differentially expressed proteins in C666-1 exosomes. Among the 640 identified proteins, 51 and 89 proteins were considered as up- and downregulated ($\geq$1.5-fold variations) in C666-1 exosomes compared with the normal counterparts (Chan et al., 2015). Radiation-induced 217 exosomal protein expression was reported (Jelonek et al., 2015). Principe et al. (2013) reported the expression of Galectin 9 and LMP-1 in exosome from the cells. Another study in NPC cell line exosome showed an increased expression of HIF-1a by western blotting (Aga et al., 2014). A recent study showed that tumor-derived exosomes regulate expression of immune function–related genes in human T-cell subsets (Muller, Mitsuhashi, Simms, Gooding, & Whiteside, 2016).

3.5 Leukemia

Chronic myeloid leukemia–derived exosomes promote, through an autocrine mechanism, the proliferation and survival of tumor cells (Boyiadzis & Whiteside, 2017; Raimondo et al., 2015; Szczepanski et al., 2011), both in vitro and in vivo, by activating antiapoptotic pathways (Raimondo et al., 2015). At diagnosis, protein and TGF-β1 levels were higher in acute myeloid leukemia than control exosomes (Hong et al., 2014). Gu et al. (2016) suggested that sorting protein VPS33 B regulates exosomal autocrine signaling to mediate hematopoiesis and leukemogenesis.

3.6 Bladder Cancer

Bladder cancer exosomes contain EDIL-3/Del1 and facilitate cancer progression (Beckham et al., 2014). Periostin was shown to get secreted in bladder cancer (Silvers et al., 2016). Concentrations of 24 proteins changed significantly between bladder cancer (n = 28) and hernia (n = 12), and tumor-associated calcium-signal transducer 2 (TACSTD2) was detected by ELISA and its potential value for the diagnosis of bladder cancer (Chen et al., 2012). Exosomes are rich sources of biological material (proteins and nucleic acids) secreted by both tumor and normal cells and found in the urine of the bladder cancer patients. Proteins in the nine top-ranked pathways included CTNNA1 (alpha-catenin), CTNNB1 (beta-catenin), VSAP, ITGA4, PAK1, DDR1, CDC42, RHOA, NRAS, RHO, PIK3AR1, MLC1, MMRN1, and CTTNBP2, and network analysis revealed 10 important hub proteins and identified inferred interactor NF2 (Kumari et al., 2015). To determine the potential of periostin as a bladder cancer indicator, it was shown that patient urinary EVs to have markedly higher levels of periostin than controls (Silvers et al., 2016). Welton et al. (2010) reported that 353 high-quality identifications with 72 proteins in bladder cancer exosomes, not previously identified by other human exosome proteomics studies. Several proteins linked to epithelial–mesenchymal transition, including increased abundance of vimentin and hepatoma-derived growth factor in the membrane, and casein kinase II α and annexin A2 in the lumen of exosomes, respectively, from metastatic cells (Jeppesen et al., 2014). Another study investigated the exosomal proteome of a bladder cancer cell line. The use of exosomes in bladder cancer studies is particularly interesting because exosomes can be isolated from urine. The concentration of tumor-derived exosomes would theoretically be greater in a urine sample from a bladder cancer patient (in comparison with exosomes isolated from sera of the same patient), thus providing a potentially powerful diagnostic tool. The study by Welton et al. identified 353 proteins contained within HT1376-derived exosomes, 72 of which were not previously associated with exosomes (Beckham et al., 2014; Chen et al., 2012; Jeppesen et al., 2014; Nawaz et al., 2014; Pisitkun et al., 2004; Welton et al., 2010; Wood, Knowles, Thompson, Selby, & Banks, 2013). It will be interesting to see if any of these proteins are also found in urinary exosomes of bladder cancer patients. To this point, a Urinary Exosome Protein Database has been compiled based on two separate studies by the same group (Gonzales et al., 2009; Pisitkun et al., 2004, 2006). However, in each case, fewer than 10 patients were enrolled in the study, and all patients were healthy at the time of the study (Gonzales et al., 2009; Pisitkun et al., 2004, 2006). This database currently contains 1160 identified proteins and will undoubtedly become essential in the identification of urinary biomarkers of cancer patients (Henderson & Azorsa, 2012).

3.7 Liver Cancer

Typical exosome proteins, such as the transmembrane protein CD63 and heat shock protein 70, were confirmed in the exosomes of hepatocellular carcinoma cells. Two potential hepatoma-associated proteins were also identified. TGM2 was first found to exist in the exosomes of human liver cancer cells, but annexin A2 was not secreted into exosomes (Zhu et al., 2014). Exosomal proteomics revealed distinct integrin expression patterns, in which the exosomal integrins α6β4 and α6β1 were associated with lung metastasis, although exosomal integrin αvβ5 was linked to liver metastasis (Hoshino et al., 2015).

3.8 Pancreatic Cancer

Costa-Silva et al. (2015) found that macrophage migration inhibitory factor was highly expressed in pancreatic ductal adenocarcinomas–-derived exosomes, and its blockade prevented liver premetastatic niche formation and metastasis. Approximately, 700–800 exosomal proteins per sample were detected from serum exosomes from pancreatic patients undergoing chemotherapy, several of which have been implicated in metastasis and treatment resistance. The differential loading of exosomes during a course of therapy suggests that exosomes may provide novel insights into the development of treatment resistance and metastasis (An et al., 2017). It was identified 4517 proteins in exosomes from Panc02 and Panc02-H7 cells via iTRAQ quantitative proteomic analyses, 79 of which were differentially expressed between the two cell lines. A recent study showed an exosome-derived wnt5 secretion in Panc1 cell line (Harada et al., 2017). Bioinformatics analyses showed that most of the differentially expressed proteins were involved in pancreatic cancer growth, invasion, and metastasis, and metabolism-related signaling pathways were involved in exosome-mediated intracellular communication. However, further studies are needed to confirm whether these proteins are potential pancreatic cancer diagnostic/prognostic markers or novel therapeutic targets.

3.9 Brain Tumor

The intercellular exchange of proteins and genetic material via exosomes is a potentially effective approach for cell-to-cell communication, and it may perform multiple functions aiding to tumor survival and metastasis. Certain proteins were found from both exosome and condition media (Camacho et al., 2004). ELISA study showed higher VEGF, EGFR, TGF b in exosomes in the brain (Al-Nedawi et al., 2009; Epple et al., 2012; Graner et al., 2009). Proteomic profiling of BM versus non-BM cell–derived exosomes showed high expression of proteins implicated in cell communication, cell cycle, and in key cancer invasion and metastasis pathways such as Annexin IV, Chk1, TFR, BCL.xL (Camacho et al., 2013). With NMR spectroscopy, Garnier et al. (2013) showed several proteins. Proteomic analysis of hypoxic Glioma secretome showed a panel of proteins: a total of 239 proteins were identified from the exosome and soluble fractions. Vascular endothelial growth factor, stanniocalcin 1 (STC1) and stanniocalcin 2, and insulin-like growth factor–binding protein 3 and 6, enriched in the soluble fraction, and lysyl oxidase homolog 2 enriched in the exosomal fraction were identified as upregulated proteins by hypoxia based on a label-free quantitative analysis and EGFRVIII expression was reported by ELISA from serum (Skog et al., 2008; Yoon et al., 2014).

3.10 Lung Cancer

The leucine-rich α-2-glycoprotein (LRG1) was found to be expressed at higher levels in urinary exosomes and lung tissue of non-small cell lung cancer (NSCLC) patients (Li et al., 2011). Yamashita et al. (2013) reported a significantly higher exosomal EGFR expression levels by ELISA in lung cancer compared with control. The extracellular vesicle array was used to phenotype exosomes directly from the plasma samples. The array contained 37 antibodies targeting lung cancer–related proteins and was used to capture exosomes, which were visualized with a cocktail of biotin-conjugated CD9, CD63, and CD81 antibodies (Jakobsen et al., 2015). The study revealed proteins associated with cell adhesion, the extracellular matrix, and a variety of signaling molecules were enriched in NSCLC exosomes. Data reveal a protein profile associated with NSCLC exosomes that suggest a role these vesicles have in the progression of lung carcinogenesis, as well as identifies several promising candidates that could be used as a multimarker protein panel in a diagnostic platform for NSCLC (Clark et al., 2016). Another group demonstrated exosome protein profiling as a promising diagnostic tool in lung cancer independently of stage and histological subtype: CD151, CD171, and tetraspanin 8 were the strongest separators of patients with cancer of all histological subtypes versus patients without cancer (Jakobsen et al., 2015; Sandfeld-Paulsen, Jakobsen, et al., 2016). The 49 proteins attached to the exosomal membrane were evaluated. NY-ESO-1, EGFR, PLAP, EpCam, and Alix had a significant concentration-dependent impact on inferior overall survival (Sandfeld-Paulsen, Aggerholm-Pedersen, et al., 2016). Recently, label-free quantification was applied to systematically compare the protein profiling in saliva and serum exosomes. 319 and 994 exosomal proteins were identified from saliva and serum by LC-MS/MS, respectively (Sun et al., 2017).

3.11 Colorectal Cancer

To identify the best molecular targets for total exosome capture from diverse biological sources and for selective enrichment in populations of interest (e.g., tumor-derived exosomes), several exosomes displayed proteins and respective antibodies have been evaluated for plate and bead functionalization. Moreover, we have optimized and directly implemented downstream steps allowing online quantification and characterization of bound exosome markers, namely proteins and RNAs. Thus, assembled assays enabled rapid overall quantification and validation of specific exosome-associated targets in/on plasma exosomes, with multifold increased yield and enrichment ratio over benchmarking technologies. Studies reported protein derived from exosomes in colorectal cancer using different methods (Andre et al., 2002; Campanella et al., 2015; Choi et al., 2011, 2012; Demory Beckler et al., 2013; Harada et al., 2017; Ji et al., 2013; Jimenez et al., 2010; Klein-Scory et al., 2010; Lim et al., 2012; Ragusa et al., 2014; Silva et al., 2012; Simpson et al., 2009; Zarovni et al., 2015)

4. Conclusion

Developing blood-based tests is appealing for noninvasive disease diagnosis, especially when biopsy is difficult, costly, and sometimes not even an option. Tumor-derived exosomes have attracted increasing interest in noninvasive cancer diagnosis and monitoring of treatment response. However, the biology and clinical value of exosomes remains largely unknown due in part to current technical challenges in rapid isolation, molecular classification, and comprehensive analysis of exosomes. The microfluidic exosome analysis platform and NanoSight technology (to characterize exosome), both NanoSight and qNano are excellent complimentary tools to study and characterize exosomes, will form the basis for critically needed infrastructures for advancing the biology and clinical utilization of exosomes (He, Crow, Roth, Zeng, & Godwin, 2014).

Cancer cells release exosomes into the extracellular environment before metastasis. Tetraspanin is a type of four times transmembrane proteins. It may be involved in cell motility, adhesion, morphogenesis, as well as cell and vesicular membrane fusion. The exosomal tetraspanin network is a molecular scaffold connecting various proteins for signal transduction. The complex of tetraspanin–integrin determines the recruiting cancer exosomes to premetastatic sites. Tetraspanin is a key element for the exosomes uptake that may lead to the reprogramming of target cells. Reprogrammed target cells assist premetastatic niche formation. We and others have described the biogenesis, secretion, and intercellular interaction of exosomes in various tumors. We hope that exosome-based drug deliveries will be a great success to overcome therapeutic resistance, which is a most challenging issue in cancer therapy.

Acknowledgments

Authors would like to acknowledge the Department of Biochemistry, Center for Health Disparities and Molecular Medicine, and Division of Head and Neck Surgery for financial support. Authors also like to acknowledge the laboratory of Dr. Nathan Wall, where all the exosome-related papers were published.

References

Abusamra, A. J., Zhong, Z., Zheng, X., Li, M., Ichim, T. E., Chin, J. L., Min, W. P. (2005). Tumor exosomes expressing Fas ligand mediate CD8⁺ T-cell apoptosis. *Blood Cells, Molecules and Diseases, 35*, 169–173. https://www.sciencedirect.com/science/article/pii/S1079979605001105?via%3Dihub.

Admyre, C., Johansson, S. M., Qazi, K. R., Filen, J. J., Lahesmaa, R., Norman, M., … Gabrielsson, S. (2007). Exosomes with immune modulatory features are present in human breast milk. *The Journal of Immunology, 179*(3), 1969–1978.

Aga, M., Bentz, G. L., Raffa, S., Torrisi, M. R., Kondo, S., Wakisaka, N., … Shackelford, J. (2014). Exosomal HIF1alpha supports invasive potential of nasopharyngeal carcinoma-associated LMP1-positive exosomes. *Oncogene, 33*(37), 4613–4622. https://doi.org/10.1038/onc.2014.66.

Al-Nedawi, K., Meehan, B., Kerbel, R. S., Allison, A. C., & Rak, J. (2009). Endothelial expression of autocrine VEGF upon the uptake of tumor-derived microvesicles containing oncogenic EGFR. *Proceedings of the National Academy of Sciences of the United States of America, 106*(10), 3794–3799. https://doi.org/10.1073/pnas.0804543106.

Andre, F., Schartz, N. E., Movassagh, M., Flament, C., Pautier, P., Morice, P., ... Zitvogel, L. (2002). Malignant effusions and immunogenic tumour-derived exosomes. *Lancet, 360*(9329), 295–305. https://doi.org/10.1016/s0140-6736(02)09552-1.

Andreola, G., Rivoltini, L., Castelli, C., Huber, V., Perego, P., Deho, P., ... Fais, S. (2002). Induction of lymphocyte apoptosis by tumor cell secretion of FasL-bearing microvesicles. *The Journal of Experimental Medicine, 195*, 1303–1316. https://www.ncbi.nlm.nih.gov/pmc/articles/PMC2193755/pdf/011624.pdf.

An, M., Lohse, I., Tan, Z., Zhu, J., Wu, J., Kurapati, H., ... Lubman, D. M. (2017). Quantitative proteomic analysis of serum exosomes from patients with locally advanced pancreatic cancer undergoing chemoradiotherapy. *Journal of Proteome Research, 16*(4), 1763–1772. https://doi.org/10.1021/acs.jproteome.7b00024.

Bard, M. P., Hegmans, J. P., Hemmes, A., Luider, T. M., Willemsen, R., Severijnen, L. A., ... Lambrecht, B. N. (2004). Proteomic analysis of exosomes isolated from human malignant pleural effusions. *American Journal of Respiratory Cell and Molecular Biology, 31*(1), 114–121. https://doi.org/10.1165/rcmb.2003-0238OC.

Beckham, C. J., Olsen, J., Yin, P. N., Wu, C. H., Ting, H. J., Hagen, F. K., ... Lee, Y. F. (2014). Bladder cancer exosomes contain EDIL-3/Del1 and facilitate cancer progression. *The Journal of Urology, 192*(2), 583–592. https://doi.org/10.1016/j.juro.2014.02.035.

Boyiadzis, M., & Whiteside, T. L. (2017). The emerging roles of tumor-derived exosomes in hematological malignancies. *Leukemia, 31*(6), 1259–1268. https://doi.org/10.1038/leu.2017.91.

Camacho, L., Guerrero, P., & Marchetti, D. (2013). MicroRNA and protein profiling of brain metastasis competent cell-derived exosomes. *PLoS One, 8*(9), e73790. https://doi.org/10.1371/journal.pone.0073790.

Camacho, M. E., Leon, J., Entrena, A., Velasco, G., Carrion, M. D., Escames, G., ... Espinosa, A. (2004). 4,5-dihydro-1H-pyrazole derivatives with inhibitory nNOS activity in rat brain: Synthesis and structure-activity relationships. *Journal of Medicinal Chemistry, 47*(23), 5641–5650. https://doi.org/10.1021/jm0407714.

Campanella, C., Rappa, F., Sciume, C., Marino Gammazza, A., Barone, R., Bucchieri, F., ... Cappello, F. (2015). Heat shock protein 60 levels in tissue and circulating exosomes in human large bowel cancer before and after ablative surgery. *Cancer, 121*(18), 3230–3239. https://doi.org/10.1002/cncr.29499.

Carbotti, G., Orengo, A. M., Mezzanzanica, D., Bagnoli, M., Brizzolara, A., Emionite, L., ... Fabbi, M. (2013). Activated leukocyte cell adhesion molecule soluble form: A potential biomarker of epithelial ovarian cancer is increased in type II tumors. *International Journal of Cancer, 132*(11), 2597–2605. https://doi.org/10.1002/ijc.27948.

Chan, Y. K., Zhang, H., Liu, P., Tsao, S. W., Lung, M. L., Mak, N. K., ... Ying-Kit Yue, P. (2015). Proteomic analysis of exosomes from nasopharyngeal carcinoma cell identifies intercellular transfer of angiogenic proteins. *International Journal of Cancer, 137*(8), 1830–1841. https://doi.org/10.1002/ijc.29562.

Chen, C. L., Lai, Y. F., Tang, P., Chien, K. Y., Yu, J. S., Tsai, C. H., ... Chen, Y. T. (2012). Comparative and targeted proteomic analyses of urinary microparticles from bladder cancer and hernia patients. *Journal of Proteome Research, 11*(12), 5611–5629. https://doi.org/10.1021/pr3008732.

Choi, D., Lee, T. H., Spinelli, C., Chennakrishnaiah, S., D'Asti, E., & Rak, J. (2017). Extracellular vesicle communication pathways as regulatory targets of oncogenic transformation. *Seminars in Cell & Developmental Biology, 67*, 11–22. https://www.sciencedirect.com/science/article/pii/S1084952117300046?via%3Dihub.

Choi, D. S., Choi, D. Y., Hong, B. S., Jang, S. C., Kim, D. K., Lee, J., ... Gho, Y. S. (2012). Quantitative proteomics of extracellular vesicles derived from human primary and metastatic colorectal cancer cells. *Journal of Extracellular Vesicles, 1*. https://doi.org/10.3402/jev.v1i0.18704.

Choi, D. S., Park, J. O., Jang, S. C., Yoon, Y. J., Jung, J. W., Choi, D. Y., … Gho, Y. S. (2011). Proteomic analysis of microvesicles derived from human colorectal cancer ascites. *Proteomics, 11*(13), 2745–2751. https://doi.org/10.1002/pmic.201100022.

Clark, D. J., Fondrie, W. E., Liao, Z., Hanson, P. I., Fulton, A., Mao, L., & Yang, A. J. (2015). Redefining the breast cancer exosome proteome by Tandem mass Tag quantitative proteomics and multivariate cluster analysis. *Analytical Chemistry, 87*(20), 10462–10469. https://doi.org/10.1021/acs.analchem.5b02586.

Clark, D. J., Fondrie, W. E., Yang, A., & Mao, L. (2016). Triple SILAC quantitative proteomic analysis reveals differential abundance of cell signaling proteins between normal and lung cancer-derived exosomes. *Journal of Proteomics, 133,* 161–169. https://doi.org/10.1016/j.jprot.2015.12.023.

Costa-Silva, B., Aiello, N. M., Ocean, A. J., Singh, S., Zhang, H., Thakur, B. K., … Lyden, D. (2015). Pancreatic cancer exosomes initiate pre-metastatic niche formation in the liver. *Nature Cell Biology, 17*(6), 816–826. https://doi.org/10.1038/ncb3169.

Demory Beckler, M., Higginbotham, J. N., Franklin, J. L., Ham, A. J., Halvey, P. J., Imasuen, I. E., … Coffey, R. J. (2013). Proteomic analysis of exosomes from mutant KRAS colon cancer cells identifies intercellular transfer of mutant KRAS. *Molecular & Cellular Proteomics, 12*(2), 343–355. https://doi.org/10.1074/mcp.M112.022806.

Dorayappan, K. D., Wallbillich, J. J., Cohn, D. E., & Selvendiran, K. (2016). The biological significance and clinical applications of exosomes in ovarian cancer. *Gynecologic Oncology, 142*(1), 199–205. https://doi.org/10.1016/j.ygyno.2016.03.036.

Duijvesz, D., Burnum-Johnson, K. E., Gritsenko, M. A., Hoogland, A. M., Vredenbregt-van den Berg, M. S., Willemsen, R., … Jenster, G. (2013). Proteomic profiling of exosomes leads to the identification of novel biomarkers for prostate cancer. *PLoS One, 8*(12), e82589. https://doi.org/10.1371/journal.pone.0082589.

Epple, L. M., Griffiths, S. G., Dechkovskaia, A. M., Dusto, N. L., White, J., Ouellette, R. J., … Graner, M. W. (2012). Medulloblastoma exosome proteomics yield functional roles for extracellular vesicles. *PLoS One, 7*(7), e42064. https://doi.org/10.1371/journal.pone.0042064.

Etayash, H., McGee, A. R., Kaur, K., & Thundat, T. (2016). Nanomechanical sandwich assay for multiple cancer biomarkers in breast cancer cell-derived exosomes. *Nanoscale, 8*(33), 15137–15141. https://doi.org/10.1039/c6nr03478k.

Fontana, S., Saieva, L., Taverna, S., & Alessandro, R. (2013). Contribution of proteomics to understanding the role of tumor-derived exosomes in cancer progression: State of the art and new perspectives. *Proteomics, 13*(10–11), 1581–1594. https://doi.org/10.1002/pmic.201200398.

Garnier, D., Jabado, N., & Rak, J. (2013). Extracellular vesicles as prospective carriers of oncogenic protein signatures in adult and paediatric brain tumours. *Proteomics, 13*(10–11), 1595–1607. https://doi.org/10.1002/pmic.201200360.

Gonzales, P. A., Pisitkun, T., Hoffert, J. D., Tchapyjnikov, D., Star, R. A., Kleta, R., … Knepper, M. A. (2009). Large-scale proteomics and phosphoproteomics of urinary exosomes. *Journal of the American Society of Nephrology, 20*(2), 363–379. https://doi.org/10.1681/asn.2008040406.

Graner, M. W., Alzate, O., Dechkovskaia, A. M., Keene, J. D., Sampson, J. H., Mitchell, D. A., & Bigner, D. D. (2009). Proteomic and immunologic analyses of brain tumor exosomes. *The FASEB Journal, 23*(5), 1541–1557. https://doi.org/10.1096/fj.08-122184.

Green, T. M., Alpaugh, M. L., Barsky, S. H., Rappa, G., & Lorico, A. (2015). Breast cancer-derived extracellular Vesicles: Characterization and contribution to the metastatic phenotype. *BioMed Research International, 2015,* 634865. https://doi.org/10.1155/2015/634865.

Greening, D. W., Ji, H., Kapp, E. A., & Simpson, R. J. (2013). Sulindac modulates secreted protein expression from LIM1215 colon carcinoma cells prior to apoptosis. *Biochimica et Biophysica Acta, 1834,* 2293–2307. https://www.sciencedirect.com/science/article/pii/S1570963913002793?via%3Dihub.

Gu, H., Chen, C., Hao, X., Wang, C., Zhang, X., Li, Z., … Zheng, J. (2016). Sorting protein VPS33B regulates exosomal autocrine signaling to mediate hematopoiesis and leukemogenesis. *Journal of Clinical Investigation, 126*(12), 4537–4553. https://doi.org/10.1172/jci87105.

Hao, Y. X., Li, Y. M., Ye, M., Guo, Y. Y., Li, Q. W., Peng, X. M., … Xiao, W. H. (2017). KRAS and BRAF mutations in serum exosomes from patients with colorectal cancer in a Chinese population. *Oncology Letters*, *13*(5), 3608–3616. https://doi.org/10.3892/ol.2017.5889.

Harada, T., Yamamoto, H., Kishida, S., Kishida, M., Awada, C., Takao, T., & Kikuchi, A. (2017). Wnt5b-associated exosomes promote cancer cell migration and proliferation. *Cancer Science*, *108*(1), 42–52. https://doi.org/10.1111/cas.13109.

He, M., Crow, J., Roth, M., Zeng, Y., & Godwin, A. K. (2014). Integrated immunoisolation and protein analysis of circulating exosomes using microfluidic technology. *Lab on a Chip*, *14*(19), 3773–3780. https://doi.org/10.1039/c4lc00662c.

Hegmans, J. P., Bard, M. P., Hemmes, A., Luider, T. M., Kleijmeer, M. J., Prins, J. B., … Lambrecht, B. N. (2004). Proteomic analysis of exosomes secreted by human mesothelioma cells. *The American Journal of Pathology*, *164*(5), 1807–1815. https://doi.org/10.1016/s0002-9440(10)63739-x.

Henderson, M. C., & Azorsa, D. O. (2012). The genomic and proteomic content of cancer cell-derived exosomes. *Frontiers in Oncology*, *2*, 38. https://doi.org/10.3389/fonc.2012.00038.

Ho, H. K., Jang, J. J., Kaji, S., Spektor, G., Fong, A., Yang, P., … Cooke, J. P. (2004). Developmental endothelial locus-1 (Del-1), a novel angiogenic protein: Its role in ischemia. *Circulation*, *109*(10), 1314–1319. https://doi.org/10.1161/01.cir.0000118465.36018.2d.

Hong, C. S., Muller, L., Whiteside, T. L., & Boyiadzis, M. (2014). Plasma exosomes as markers of therapeutic response in patients with acute myeloid leukemia. *Frontiers in Immunology*, *5*, 160. https://doi.org/10.3389/fimmu.2014.00160.

Hoshino, A., Costa-Silva, B., Shen, T. L., Rodrigues, G., Hashimoto, A., Tesic Mark, M., … Lyden, D. (2015). Tumour exosome integrins determine organotropic metastasis. *Nature*, *527*(7578), 329–335. https://doi.org/10.1038/nature15756.

Huber, V., Filipazzi, P., Iero, M., Fais, S., & Rivoltini, L. (2008). More insights into the immunosuppressive potential of tumor exosomes. *Journal of Translational Medicine*, *6*, 63–67. https://www.ncbi.nlm.nih.gov/pmc/articles/PMC2590595/pdf/1479-5876-6-63.pdf.

Iero, M., Valenti, R., Huber, V., Filipazzi, P., Parmiani, G., Fais, S., & Rivoltini, L. (2008). Tumour-released exosomes and their implications in cancer immunity. *Cell Death and Differentiation*, *15*(1), 80–88. https://doi.org/10.1038/sj.cdd.4402237.

Inal, J. M., Kosgodage, U., Azam, S., Stratton, D., Antwi-Baffour, S., & Lange, S. (2013). Blood/plasma secretome and microvesicles. *Biochimica et Biophysica Acta*, *1834*(11), 2317–2325. https://doi.org/10.1016/j.bbapap.2013.04.005.

Jakobsen, K. R., Paulsen, B. S., Baek, R., Varming, K., Sorensen, B. S., & Jorgensen, M. M. (2015). Exosomal proteins as potential diagnostic markers in advanced non-small cell lung carcinoma. *Journal of Extracellular Vesicles*, *4*, 26659. https://doi.org/10.3402/jev.v4.26659.

Jelonek, K., Wojakowska, A., Marczak, L., Muer, A., Tinhofer-Keilholz, I., Lysek-Gladysinska, M., … Pietrowska, M. (2015). Ionizing radiation affects protein composition of exosomes secreted in vitro from head and neck squamous cell carcinoma. *Acta Biochimica Polonica*, *62*(2), 265–272. https://doi.org/10.18388/abp.2015_970.

Jeppesen, D. K., Nawrocki, A., Jensen, S. G., Thorsen, K., Whitehead, B., Howard, K. A., … Ostenfeld, M. S. (2014). Quantitative proteomics of fractionated membrane and lumen exosome proteins from isogenic metastatic and nonmetastatic bladder cancer cells reveal differential expression of EMT factors. *Proteomics*, *14*(6), 699–712. https://doi.org/10.1002/pmic.201300452.

Ji, H., Greening, D. W., Barnes, T. W., Lim, J. W., Tauro, B. J., Rai, A., … Simpson, R. J. (2013). Proteome profiling of exosomes derived from human primary and metastatic colorectal cancer cells reveal differential expression of key metastatic factors and signal transduction components. *Proteomics*, *13*(10–11), 1672–1686. https://doi.org/10.1002/pmic.201200562.

Ji, H., Greening, D. W., Kapp, E. A., Moritz, R. L., & Simpson, R. J. (2009). Secretome-based proteomics reveals sulindac-modulated proteins released from colon cancer cells. *PROTEOMICS - Clinical Applications*, *3*, 433–451. http://onlinelibrary.wiley.com/doi/10.1002/prca.200800077/epdf?r3_referer=wol&tracking_action=preview_click&show_checkout=1&purchase_referrer=onlinelibrary.wiley.com&purchase_site_license=LICENSE_DENIED_NO_CUSTOMER.

Jimenez, C. R., Knol, J. C., Meijer, G. A., & Fijneman, R. J. (2010). Proteomics of colorectal cancer: Overview of discovery studies and identification of commonly identified cancer-associated proteins and candidate CRC serum markers. *Journal of Proteomics*, *73*(10), 1873–1895. https://doi.org/10.1016/j.jprot.2010.06.004.

Kahlert, C., & Kalluri, R. (2013). Exosomes in tumor microenvironment influence cancer progression and metastasis. *Journal of Molecular Medicine (Berlin)*, *91*(4), 431–437. https://doi.org/10.1007/s00109-013-1020-6.

Khan, S., Bennit, H. F., Turay, D., Perez, M., Mirshahidi, S., Yuan, Y., & Wall, N. R. (2014). Early diagnostic value of survivin and its alternative splice variants in breast cancer. *BMC Cancer*, *14*, 176. https://doi.org/10.1186/1471-2407-14-176.

Khan, S., Jutzy, J. M., Aspe, J. R., McGregor, D. W., Neidigh, J. W., & Wall, N. R. (2011). Survivin is released from cancer cells via exosomes. *Apoptosis: An International Journal on Programmed Cell Death*, *16*(1), 1–12. https://doi.org/10.1007/s10495-010-0534-4.

Khan, S., Jutzy, J. M., Valenzuela, M. M., Turay, D., Aspe, J. R., Ashok, A., … Wall, N. R. (2012). Plasma-derived exosomal survivin, a plausible biomarker for early detection of prostate cancer. *PLoS One*, *7*(10), e46737. https://doi.org/10.1371/journal.pone.0046737.

Klein-Scory, S., Kübler, S., Diehl, H., Eilert-Micus, C., Reinacher-Schick, A., Stühler, K., … Schwarte-Waldhoff, I. (2010). Immunoscreening of the extracellular proteome of colorectal cancer cells. *BMC Cancer*, *10*, 70. https://doi.org/10.1186/1471-2407-10-70.

Koga, K., Matsumoto, K., Akiyoshi, T., Kubo, M., Yamanaka, N., Tasaki, A., … Katano, M. (2005). Purification, characterization and biological significance of tumor-derived exosomes. *Anticancer Research*, *25*(6a), 3703–3707.

Koldemir, O., Ozgur, E., & Gezer, U. (2017). Accumulation of GAS5 in exosomes is a marker of apoptosis induction. *Biomedical Research*, *6*(3), 358–362. https://doi.org/10.3892/br.2017.848.

Kumari, N., Saxena, S., & Agrawal, U. (2015). Exosomal protein interactors as emerging therapeutic targets in urothelial bladder cancer. *Journal of the Egyptian National Cancer Institute*, *27*(2), 51–58. https://doi.org/10.1016/j.jnci.2015.02.002.

Labani-Motlagh, A., Israelsson, P., Ottander, U., Lundin, E., Nagaev, I., Nagaeva, O., … Mincheva-Nilsson, L. (2016). Differential expression of ligands for NKG2D and DNAM-1 receptors by epithelial ovarian cancer-derived exosomes and its influence on NK cell cytotoxicity. *Tumor Biologyogy*, *37*(4), 5455–5466. https://doi.org/10.1007/s13277-015-4313-2.

Lea, J., Sharma, R., Yang, F., Zhu, H., Ward, E. S., & Schroit, A. J. (2017). Detection of phosphatidylserine-positive exosomes as a diagnostic marker for ovarian malignancies: A proof of concept study. *Oncotarget*, *8*(9), 14395–14407. https://doi.org/10.18632/oncotarget.14795.

Liang, B., Peng, P., Chen, S., Li, L., Zhang, M., Cao, D., … Shen, K. (2013). Characterization and proteomic analysis of ovarian cancer-derived exosomes. *Journal of Proteomics*, *80*, 171–182. https://doi.org/10.1016/j.jprot.2012.12.029.

Lim, J. W., Mathias, R. A., Kapp, E. A., Layton, M. J., Faux, M. C., Burgess, A. W., … Simpson, R. J. (2012). Restoration of full-length APC protein in SW480 colon cancer cells induces exosome-mediated secretion of DKK-4. *Electrophoresis*, *33*(12), 1873–1880. https://doi.org/10.1002/elps.201100687.

Li, J., Sherman-Baust, C. A., Tsai-Turton, M., Bristow, R. E., Roden, R. B., & Morin, P. J. (2009). Claudin-containing exosomes in the peripheral circulation of women with ovarian cancer. *BMC Cancer*, *9*, 244. https://doi.org/10.1186/1471-2407-9-244.

Li, Y., Zhang, Y., Qiu, F., & Qiu, Z. (2011). Proteomic identification of exosomal LRG1: A potential urinary biomarker for detecting NSCLC. *Electrophoresis*, *32*(15), 1976–1983. https://doi.org/10.1002/elps.201000598.

Maji, S., Chaudhary, P., Akopova, I., Nguyen, P. M., Hare, R. J., Gryczynski, I., & Vishwanatha, J. K. (2017). Exosomal Annexin II promotes angiogenesis and breast cancer metastasis. *Molecular Cancer Research*, *15*(1), 93–105. https://doi.org/10.1158/1541-7786.mcr-16-0163.

Makridakis, M., & Vlahou, A. (2010). Secretome proteomics for discovery of cancer biomarkers. *Journal of Proteomics*, *73*(12), 2291–2305. https://doi.org/10.1016/j.jprot.2010.07.001.

Mathivanan, S., Ji, H., & Simpson, R. J. (2010). Exosomes: Extracellular organelles important in intercellular communication. *Journal of Proteomics*, *73*, 1907–1920. https://www.sciencedirect.com/science/article/pii/S1874391910001843?via%3Dihub.

Mikolajczyk, S. D., Song, Y., Wong, J. R., Matson, R. S., & Rittenhouse, H. G. (2004). Are multiple markers the future of prostate cancer diagnostics? *Clinical Biochemistry*, *37*(7), 519–528. https://doi.org/10.1016/j.clinbiochem.2004.05.016.

Moon, P.-G., Lee, J.-E., Cho, Y.-E., Lee, S. J., Chae, Y. S., Jung, J. H., … Baek, M.-C. (2016). Fibronectin on circulating extracellular vesicles as a liquid biopsy to detect breast cancer. *Oncotarget*, *7*(26), 40189–40199. https://doi.org/10.18632/oncotarget.9561.

Muller, L., Hong, C. S., Stolz, D. B., Watkins, S. C., & Whiteside, T. L. (2014). Isolation of biologically-active exosomes from human plasma. *Journal of Immunological Methods*, *411*, 55–65. https://doi.org/10.1016/j.jim.2014.06.007.

Muller, L., Mitsuhashi, M., Simms, P., Gooding, W. E., & Whiteside, T. L. (2016). Tumor-derived exosomes regulate expression of immune function-related genes in human T cell subsets. *Scientific Reports*, *6*, 20254. https://doi.org/10.1038/srep20254.

Nakamura, K., Sawada, K., Kinose, Y., Yoshimura, A., Toda, A., Nakatsuka, E., … Kimura, T. (2017). Exosomes promote ovarian cancer cell invasion through transfer of CD44 to peritoneal mesothelial cells. *Molecular Cancer Research*, *15*(1), 78–92. https://doi.org/10.1158/1541-7786.mcr-16-0191.

Nakamura, K., Sawada, K., Yoshimura, A., Kinose, Y., Nakatsuka, E., & Kimura, T. (2016). Clinical relevance of circulating cell-free microRNAs in ovarian cancer. *Molecular Cancer*, *15*(1), 48. https://doi.org/10.1186/s12943-016-0536-0.

Nawaz, M., Camussi, G., Valadi, H., Nazarenko, I., Ekstrom, K., Wang, X., … Kislinger, T. (2014). The emerging role of extracellular vesicles as biomarkers for urogenital cancers. *Nature Reviews. Urology*, *11*(12), 688–701. https://doi.org/10.1038/nrurol.2014.301.

Palanisamy, V., Sharma, S., Deshpande, A., Zhou, H., Gimzewski, J., & Wong, D. T. (2010). Nanostructural and transcriptomic analyses of human saliva derived exosomes. *PLoS One*, *5*(1), e8577. https://doi.org/10.1371/journal.pone.0008577.

Palazzolo, G., Albanese, N. N., DI Cara, G., Gygax, D., Vittorelli, M. L., & Pucci-Minafra, I. (2012). Proteomic analysis of exosome-like vesicles derived from breast cancer cells. *Anticancer Research*, *32*(3), 847–860.

Peng, P., Yan, Y., & Keng, S. (2011). Exosomes in the ascites of ovarian cancer patients: Origin and effects on anti-tumor immunity. *Oncology Reports*, *25*(3), 749–762. https://doi.org/10.3892/or.2010.1119.

Pisitkun, T., Johnstone, R., & Knepper, M. A. (2006). Discovery of urinary biomarkers. *Molecular & Cellular Proteomics*, *5*(10), 1760–1771. https://doi.org/10.1074/mcp.R600004-MCP200.

Pisitkun, T., Shen, R. F., & Knepper, M. A. (2004). Identification and proteomic profiling of exosomes in human urine. *Proceedings of the National Academy of Sciences of the United States of America*, *101*(36), 13368–13373. https://doi.org/10.1073/pnas.0403453101.

Principe, S., Hui, A. B., Bruce, J., Sinha, A., Liu, F. F., & Kislinger, T. (2013). Tumor-derived exosomes and microvesicles in head and neck cancer: Implications for tumor biology and biomarker discovery. *Proteomics*, *13*(10–11), 1608–1623. https://doi.org/10.1002/pmic.201200533.

Ragusa, M., Statello, L., Maugeri, M., Barbagallo, C., Passanisi, R., Alhamdani, M. S., … Purrello, M. (2014). Highly skewed distribution of miRNAs and proteins between colorectal cancer cells and their exosomes following cetuximab treatment: Biomolecular, genetic and translational implications. *Oncoscience*, *1*(2), 132–157.

Raimondo, S., Saieva, L., Corrado, C., Fontana, S., Flugy, A., Rizzo, A., … Alessandro, R. (2015). Chronic myeloid leukemia-derived exosomes promote tumor growth through an autocrine mechanism. *Cell Communication and Signaling*, *13*, 8. https://doi.org/10.1186/s12964-015-0086-x.

Rivoltini, L., Canese, P., Huber, V., Iero, M., Pilla, L., Valenti, R., … Parmiani, G. (2005). Escape strategies and reasons for failure in the interaction between tumour cells and the immune system: How can we tilt the balance towards immune-mediated cancer control? *Expert Opinion on Biological Therapy*, *5*(4), 463–476. https://doi.org/10.1517/14712598.5.4.463.

Runz, S., Keller, S., Rupp, C., Stoeck, A., Issa, Y., Koensgen, D., … Altevogt, P. (2007). Malignant ascites-derived exosomes of ovarian carcinoma patients contain CD24 and EpCAM. *Gynecologic Oncology*, *107*(3), 563–571. https://doi.org/10.1016/j.ygyno.2007.08.064.

Safaei, R., Larson, B. J., Cheng, T. C., Gibson, M. A., Otani, S., Naerdemann, W., & Howell, S. B. (2005). Abnormal lysosomal trafficking and enhanced exosomal export of cisplatin in drug-resistant human ovarian carcinoma cells. *Molecular Cancer Therapeutics*, *4*(10), 1595–1604. https://doi.org/10.1158/1535-7163.mct-05-0102.

Sanchez-Bailon, M. P., Calcabrini, A., Mayoral-Varo, V., Molinari, A., Wagner, K. U., Losada, J. P., … Martin-Perez, J. (2015). Cyr61 as mediator of Src signaling in triple negative breast cancer cells. *Oncotarget*, *6*(15), 13520–13538. https://doi.org/10.18632/oncotarget.3760.

Sandfeld-Paulsen, B., Aggerholm-Pedersen, N., Baek, R., Jakobsen, K. R., Meldgaard, P., Folkersen, B. H., … Sorensen, B. S. (2016). Exosomal proteins as prognostic biomarkers in non-small cell lung cancer. *Molecular Oncology*, *10*(10), 1595–1602. https://doi.org/10.1016/j.molonc.2016.10.003.

Sandfeld-Paulsen, B., Jakobsen, K. R., Baek, R., Folkersen, B. H., Rasmussen, T. R., Meldgaard, P., … Sorensen, B. S. (2016). Exosomal proteins as diagnostic biomarkers in lung cancer. *Journal of Thoracic Oncology*, *11*(10), 1701–1710. https://doi.org/10.1016/j.jtho.2016.05.034.

Santiago-Dieppa, D. R., Steinberg, J., Gonda, D., Cheung, V. J., CArter, B. S., & Chen, C. C. (2014). Extracellular vesicles as a platform for 'liquid biopsy' in glioblastoma patients. *Expert Review of Molecular Diagnostics*, *14*, 819–825. https://www.ncbi.nlm.nih.gov/pmc/articles/PMC4436244/pdf/nihms688228.pdf.

Shao, H., Chung, J., Lee, K., Balaj, L., Min, C., Carter, B. S., … Weissleder, R. (2015). Chip-based analysis of exosomal mRNA mediating drug resistance in glioblastoma. *Nature Communications*, *6*, 6999. https://www.ncbi.nlm.nih.gov/pmc/articles/PMC4430127/pdf/ncomms7999.pdf.

Shi, S., Zhang, Q., Xia, Y., You, B., Shan, Y., Bao, L., … Gu, Z. (2016). Mesenchymal stem cell-derived exosomes facilitate nasopharyngeal carcinoma progression. *American Journal of Cancer Research*, *6*(2), 459–472.

Silva, J., Garcia, V., Rodriguez, M., Compte, M., Cisneros, E., Veguillas, P., … Bonilla, F. (2012). Analysis of exosome release and its prognostic value in human colorectal cancer. *Genes Chromosomes & Cancer*, *51*(4), 409–418.

Silvers, C. R., Liu, Y. R., Wu, C. H., Miyamoto, H., Messing, E. M., & Lee, Y. F. (2016). Identification of extracellular vesicle-borne periostin as a feature of muscle-invasive bladder cancer. *Oncotarget*, *7*(17), 23335–23345. https://doi.org/10.18632/oncotarget.8024.

Simpson, R. J., Lim, J. W., Moritz, R. L., & Mathivanan, S. (2009). Exosomes: Proteomic insights and diagnostic potential. *Expert Review of Proteomics*, *6*(3), 267–283. https://doi.org/10.1586/epr.09.17.

Sinha, A., Ignatchenko, V., Ignatchenko, A., Mejia-Guerrero, S., & Kislinger, T. (2014). In-depth proteomic analyses of ovarian cancer cell line exosomes reveals differential enrichment of functional categories compared to the NCI 60 proteome. *Biochemical and Biophysical Research Communications*, *445*(4), 694–701. https://doi.org/10.1016/j.bbrc.2013.12.070.

Sivadasan, P., Gupta, M. K., Sathe, G. J., Balakrishnan, L., Palit, P., Gowda, H., … Sirdeshmukh, R. (2015). Human salivary proteome–a resource of potential biomarkers for oral cancer. *Journal of Proteomics, 127*(Pt A), 89–95. https://doi.org/10.1016/j.jprot.2015.05.039.

Skog, J., Wurdinger, T., van Rijn, S., Meijer, D. H., Gainche, L., Sena-Esteves, M., … Breakefield, X. O. (2008). Glioblastoma microvesicles transport RNA and proteins that promote tumour growth and provide diagnostic biomarkers. *Nature Cell Biology, 10*(12), 1470–1476. https://doi.org/10.1038/ncb1800.

Stastna, M., & Van Eyk, J. E. (2012). Secreted proteins as a fundamental source for biomarker discovery. *Proteomics, 12*(4–5), 722–735. https://doi.org/10.1002/pmic.201100346.

Staubach, S., Razawi, H., & Hanisch, F. G. (2009). Proteomics of MUC1-containing lipid rafts from plasma membranes and exosomes of human breast carcinoma cells MCF-7. *Proteomics, 9*(10), 2820–2835. https://doi.org/10.1002/pmic.200800793.

Street, J. M., Barran, P. E., Mackay, C. L., Weidt, S., Balmforth, C., Walsh, T. S., … Dear, J. W. (2012). Identification and proteomic profiling of exosomes in human cerebrospinal fluid. *Journal of Translational Medicine, 10*, 5. https://doi.org/10.1186/1479-5876-10-5.

Suchorska, W. M., & Lach, M. S. (2016). The role of exosomes in tumor progression and metastasis (Review). *Oncology Reports, 35*(3), 1237–1244. https://doi.org/10.3892/or.2015.4507.

Sun, Y., Liu, S., Qiao, Z., Shang, Z., Xia, Z., Niu, X., … Xiao, H. (2017). Systematic comparison of exosomal proteomes from human saliva and serum for the detection of lung cancer. *Analytica Chimica Acta, 982*, 84–95. https://doi.org/10.1016/j.aca.2017.06.005.

Szajnik, M., Czystowska-Kuzmicz, M., Elishaev, E., & Whiteside, T. L. (2016). Biological markers of prognosis, response to therapy and outcome in ovarian carcinoma. *Expert Review of Molecular Diagnostics, 16*(8), 811–826. https://doi.org/10.1080/14737159.2016.1194758.

Szajnik, M., Derbis, M., Lach, M., Patalas, P., Michalak, M., Drzewiecka, H., … Whiteside, T. L. (2013). Exosomes in plasma of patients with ovarian carcinoma: Potential biomarkers of tumor progression and response to therapy. *Gynecol Obstet (Sunnyvale)* (Suppl. 4), 3. https://doi.org/10.4172/2161-0932.s4-003.

Szczepanski, M. J., Szajnik, M., Welsh, A., Whiteside, T. L., & Boyiadzis, M. (2011). Blast-derived microvesicles in sera from patients with acute myeloid leukemia suppress natural killer cell function via membrane-associated transforming growth factor-β1. *Haematologica, 96*(9), 1302–1309. https://doi.org/10.3324/haematol.2010.039743.

Tauro, B. J., Greening, D. W., Mathias, R. A., Ji, H., Mathivanan, S., Scott, A. M., & Simpson, R. J. (2012). Comparison of ultracentrifugation, density gradient separation, and immunoaffinity capture methods for isolating human colon cancer cell line LIM1863-derived exosomes. *Methods, 56*(2), 293–304. https://doi.org/10.1016/j.ymeth.2012.01.002.

Taylor, D. D., & Gercel-Taylor, C. (2005). Tumour-derived exosomes and their role in cancer-associated T-cell signalling defects. *British Journal of Cancer, 92*(2), 305–311. https://doi.org/10.1038/sj.bjc.6602316.

Taylor, D. D., Homesley, H. D., & Doellgast, G. J. (1983). "Membrane-associated" immunoglobulins in cyst and ascites fluids of ovarian cancer patients. *American Journal of Reproductive Immunology, 3*(1), 7–11.

Taylor, D. D., & Shah, S. (2015). Methods of isolating extracellular vesicles impact down-stream analyses of their cargoes. *Methods, 87*, 3–10. https://doi.org/10.1016/j.ymeth.2015.02.019.

Taylor, D. D., Zacharias, W., & Gercel-Taylor, C. (2011). Exosome isolation for proteomic analyses and RNA profiling. *Methods in Molecular Biology, 728*, 235–246. https://doi.org/10.1007/978-1-61779-068-3_15.

Thomas, S. N., Liao, Z., Clark, D., Chen, Y., Samadani, R., Mao, L., … Yang, A. J. (2013). Exosomal proteome profiling: A potential multi-marker cellular phenotyping tool to characterize hypoxia-induced radiation resistance in breast cancer. *Proteomes, 1*(2), 87–108. https://doi.org/10.3390/proteomes1020087.

Tickner, J. A., Urquhart, A. J., Stephenson, S. A., Richard, D. J., & O'Byrne, K. J. (2014). Functions and therapeutic roles of exosomes in cancer. *Frontiers in Oncology, 4*, 127. https://doi.org/10.3389/fonc.2014.00127.

Turay, D., Khan, S., Diaz Osterman, C. J., Curtis, M. P., Khaira, B., Neidigh, J. W., … Wall, N. R. (2016). Proteomic profiling of serum-derived exosomes from ethnically diverse prostate cancer patients. *Cancer Investigation, 34*(1), 1–11. https://doi.org/10.3109/07357907.2015.1081921.

Vardaki, I., Ceder, S., Rutishauser, D., Baltatzis, G., Foukakis, T., & Panaretakis, T. (2016). Periostin is identified as a putative metastatic marker in breast cancer-derived exosomes. *Oncotarget, 7*(46), 74966–74978. https://doi.org/10.18632/oncotarget.11663.

Wang, X., Ding, X., Nan, L., Wang, Y., Wang, J., Yan, Z., … Yu, L. (2015). Investigation of the roles of exosomes in colorectal cancer liver metastasis. *Oncology Reports, 33*(5), 2445–2453. https://doi.org/10.3892/or.2015.3843.

Welton, J. L., Khanna, S., Giles, P. J., Brennan, P., Brewis, I. A., Staffurth, J., … Clayton, A. (2010). Proteomics analysis of bladder cancer exosomes. *Molecular & Cellular Proteomics, 9*(6), 1324–1338. https://doi.org/10.1074/mcp.M000063-MCP201.

Witwer, K. W., Buzas, E. I., Bemis, L. T., Bora, A., Lasser, C., Lotvall, J., … Hochberg, F. (2013). Standardization of sample collection, isolation and analysis methods in extracellular vesicle research. *Journal of Extracellular Vesicles, 2*. https://doi.org/10.3402/jev.v2i0.20360.

Wood, S. L., Knowles, M. A., Thompson, D., Selby, P. J., & Banks, R. E. (2013). Proteomic studies of urinary biomarkers for prostate, bladder and kidney cancers. *Nature Reviews. Urology, 10*(4), 206–218. https://doi.org/10.1038/nrurol.2013.24.

Wubbolts, R., Leckie, R. S., Veenhuizen, P. T., Schwarzmann, G., Mobius, W., Hoernschemeyer, J., … Stoorvogel, W. (2003). Proteomic and biochemical analyses of human B cell-derived exosomes. Potential implications for their function and multivesicular body formation. *Journal of Biological Chemistry, 278*(13), 10963–10972. https://doi.org/10.1074/jbc.M207550200.

Yamashita, T., Kamada, H., Kanasaki, S., Maeda, Y., Nagano, K., Abe, Y., … Tsunoda, S. (2013). Epidermal growth factor receptor localized to exosome membranes as a possible biomarker for lung cancer diagnosis. *Die Pharmazie, 68*(12), 969–973.

Yang, Y., Bucan, V., Baehre, H., von der Ohe, J., Otte, A., & Hass, R. (2015). Acquisition of new tumor cell properties by MSC-derived exosomes. *International Journal of Oncology, 47*(1), 244–252. https://doi.org/10.3892/ijo.2015.3001.

Yang, J., Wei, F., Schafer, C., & Wong, D. T. (2014). Detection of tumor cell-specific mRNA and protein in exosome-like microvesicles from blood and saliva. *PLoS One, 9*(11), e110641. https://doi.org/10.1371/journal.pone.0110641.

Yang, Y., Xing, Y., Liang, C., Hu, L., Xu, F., & Chen, Y. (2015). Crucial microRNAs and genes of human primary breast cancer explored by microRNA-mRNA integrated analysis. *Tumor Biologyogy, 36*(7), 5571–5579. https://doi.org/10.1007/s13277-015-3227-3.

Yoon, J. H., Kim, J., Kim, K. L., Kim, D. H., Jung, S. J., Lee, H., … Lee, T. G. (2014). Proteomic analysis of hypoxia-induced U373MG glioma secretome reveals novel hypoxia-dependent migration factors. *Proteomics, 14*(12), 1494–1502. https://doi.org/10.1002/pmic.201300554.

You, Y., Shan, Y., Chen, J., Yue, H., You, B., Shi, S., … Cao, X. (2015). Matrix metalloproteinase 13-containing exosomes promote nasopharyngeal carcinoma metastasis. *Cancer Science, 106*(12), 1669–1677. https://doi.org/10.1111/cas.12818.

Yu, S., Cao, H., Shen, B., & Feng, J. (2015). Tumor-derived exosomes in cancer progression and treatment failure. *Oncotarget, 6*(35), 37151–37168. https://doi.org/10.18632/oncotarget.6022.

Zarovni, N., Corrado, A., Guazzi, P., Zocco, D., Lari, E., Radano, G., … Chiesi, A. (2015). Integrated isolation and quantitative analysis of exosome shuttled proteins and nucleic acids using immunocapture approaches. *Methods, 87*, 46–58. https://doi.org/10.1016/j.ymeth.2015.05.028.

Zhang, W., Peng, P., Kuang, Y., Yang, J., Cao, D., You, Y., & Shen, K. (2016). Characterization of exosomes derived from ovarian cancer cells and normal ovarian epithelial cells by nanoparticle tracking analysis. *Tumor Biologyogy, 37*(3), 4213–4221. https://doi.org/10.1007/s13277-015-4105-8.

Zhang, W., Zhou, X., Zhang, H., Yao, Q., Liu, Y., & Dong, Z. (2016). Extracellular vesicles in diagnosis and therapy of kidney diseases. *American Journal of Physiology Renal Physiology, 311*(5), F844–f851. https://doi.org/10.1152/ajprenal.00429.2016.

Zhu, L., Qu, X. H., Sun, Y. L., Qian, Y. M., & Zhao, X. H. (2014). Novel method for extracting exosomes of hepatocellular carcinoma cells. *World Journal of Gastroenterology, 20*(21), 6651–6657. https://doi.org/10.3748/wjg.v20.i21.6651.

6

Nucleic Acid Profiling in Tumor Exosomes

Malav S. Trivedi, Maria Abreu

NOVA SOUTHEASTERN UNIVERSITY, FORT LAUDERDALE, FL, UNITED STATES

CHAPTER OUTLINE

1. Introduction **94**
   1.1 Introduction to Exosomes 94

2. Composition of Exosomes **95**

3. Nucleic Acid Loading Into Exosomes **95**
   3.1 miRNA Loading Into Exosomes 95
   3.2 Endosomal Vesicle Pathways 96

4. Isolation Techniques for RNA/DNA From Exosomes **97**
   4.1 Source of Extracellular Vesicles 97
      4.1.1 Cell Culture Supernatants 97
      4.1.2 Body Fluids 98
   4.2 Isolation of Exosomes and RNA/DNA 98
   4.3 Isolation of microRNAs/Noncoding RNAs Specifically 99

5. Profiling of Nucleic Acid—RNA **100**
   5.1 Analysis of the Quality, Quantity, and Diversity of evRNA 100
      5.1.1 Assessing evRNA Quality 100
      5.1.2 Evaluating Quantity of evRNA 101
   5.2 Characterizing evRNA 103
      5.2.1 Deep Sequencing Platforms 103
      5.2.2 Library Preparations From evRNA Samples 104
      5.2.3 Validation of Deep Sequencing Data 106
      5.2.4 Validation by qRT-PCR 106
      5.2.5 Microarray Analysis of RNA From Exosome 107
      5.2.6 NanoString 107

6. Single EV Analysis **108**

7. Statistical and Bioinformatic Software for evRNA Characterization ... **109**

Diagnostic and Therapeutic Applications of Exosomes in Cancer. https://doi.org/10.1016/B978-0-12-812774-2.00006-7

8. Experimental Artifacts and Contaminants Affecting evRNA Analysis **109**
   8.1 Non-EV-Associated RNA and Other Lab-Derived Contaminations Can Amplify During the
       Library Preparation Step **109**
   8.2 Presence of DNA? **110**

9. Conclusions **110**

References **111**

1. Introduction

1.1 Introduction to Exosomes

Exosomes are generally about 30–100 nm and are also known as extracellular membrane vesicles with endocytic origin (Minciacchi, Freeman, & Di Vizio, 2015). These microvesicles are generally released into the extracellular environment after merging with multivesicular bodies with the plasma membrane (Mulcahy, Pink, & Carter, 2014). It is also known that such extracellular microvesicles can be taken up by neighboring cells or travel in the bodily fluids and can be taken up by other cells in the body (Budnik, Ruiz-Cañada, & Wendler, 2016; Gutiérrez-Vázquez, Villarroya-Beltri, Mittelbrunn, & Sánchez-Madrid, 2013; Mulcahy et al., 2014). However, the complete mechanism for the interaction of exosomes and the recipient cells is unclear (Lee, EL Andaloussi, & Wood, 2012; Zaborowski, Balaj, Breakefield, & Lai, 2015). Current hypothesis with some experimental data suggests receptor–ligand interaction, fusion with the plasma membrane, or internalization of the exosomes by the recipient cells by endocytosis to be some of the probable mechanisms (Bastos, Ruivo, da Silva, & Melo, 2017; van Dongen, Masoumi, Witwer, & Pegtel, 2016; Mulcahy et al., 2014). Most cells, such as mast cells, dendritic cells, tumor cells, B cells, epithelial cells, and T cells, are found to release exosomes (Hong, Schouest, & Xu, 2017; Robbins & Morelli, 2014). Moreover, many biological fluids including bronchoalveolar lavage fluid, saliva, plasma, urine, and breast milk also have microvesicles (Keller, Ridinger, Rupp, Janssen, & Altevogt, 2011; Pisitkun, Shen, & Knepper, 2004; Yáñez-Mó et al., 2015). Regardless of the cellular origin, several common proteins are found in exosomes, including chaperones and cytoskeletal proteins (Conde-Vancells et al., 2008). Moreover, others and we have also showed that large quantities of RNA are also being transferred from one cell to another (Quesenberry, Aliotta, Deregibus, & Camussi, 2015; Su, Aldawsari, & Amiji, 2016; Tkach & Théry, 2016; Trivedi, Talekar, Shah, Ouyang, & Amiji, 2016). Although the functions of such exosomal protein or RNA content is not yet fully understood, some of the known functions include antigen presentation, induction of drug tolerance, as well as the transfer of genetic material (De Toro, Herschlik, Waldner, & Mongini, 2015; Yáñez-Mó et al., 2015). Interestingly, recent studies have also demonstrated the spread of oncogenes by exosomes and microvesicles secreted by tumor cells (Trivedi et al., 2016; Whiteside, 2016). Exosomes are also known to play a crucial role in spreading genetic material and pathogens such as prions and viruses from one cell to another (Schwab et al., 2015).

The RNA component of exosomes (evRNA) is of particular diagnostic interest as naked RNAs (outside of exosomes/ microvesicles) in bodily fluids are not stable due to the presence and exposure to RNases. In the last decade, there is an exponential growth in interest toward exosomes (Théry, 2011). This is mainly directed toward understanding their function in the body to more practical applications, such as use in diagnostics and therapeutics development. Critical to such mechanistic understanding and practical applications is the development of reagents, tools and protocols for their isolation, as well as characterization and analysis of their RNA and protein contents. In this chapter, we will focus on the different technologies used to profile the nucleic acid content in exosomes. We will also discuss the effects of different methods of exosome isolation on the resulting RNA content and quality. Lastly, we will also discuss the specific applications of different technologies for use of nucleic acids from exosomes for diagnostic purposes.

2. Composition of Exosomes

In mammalian cells, exosomes are characterized by specific sets of lipids, proteins, and RNAs. Several reports have been published to date, using quantitative reverse transcription-polymerase chain reaction (qRT-PCR) and next-generation sequencing for initial characterization of the RNA content of exosomes. Exosomes carry a unique repertoire of mRNAs and their fragments, rRNAs, long noncoding RNAs, microRNAs (miRNAs), other small noncoding RNAs (piRNA, snRNA, snoRNA, scaRNA, Y RNA), natural antisense RNAs, tRNAs and their fragments (Li et al., 2014). Often the RNA profile of exosomes does not reflect one of its cellular origin, which suggests that RNA packaging into exosomes could be a selective process (Ragusa et al., 2017). Although a common mechanism for loading or RNA content in cells is unknown, studies show that exosomal loading of miRNA might be selected by proteins, target mRNAs, or posttranscriptional modifications (Janas, Janas, Sapoń, & Janas, 2015; Kajdos, Janas, Kolasa-Zwierzchowska, Wilczyński, & Stetkiewicz, 2015).

3. Nucleic Acid Loading Into Exosomes

As mentioned earlier and will be further discussed, cellular processes that lead to specific RNA loading into exosomes are of great interest for diagnostic and therapeutic purposes. Hence, it is important to understand the mechanism for loading of RNA content into exosomes.

3.1 miRNA Loading Into Exosomes

Almost all miRNAs are encoded in the intronic genomic region and are transcribed as long primary miRNAs (pri-miRNAs) by RNA polymerase II (Olena & Patton, 2010; Wahid, Shehzad, Khan, & Kim, 2010). They undergo sequential processing by RNase III enzymes Drosha and Dicer to produce 19–24 nucleotide mature miRNA duplexes (Ha & Kim, 2014). Dicer then transfers the duplex to one of four Argonaute (Ago) proteins that allow the RNA binding.

Alternative Drosha-independent miRNA processing pathways have also been described, including mirtrons, and snoRNA- and tRNA-derived miRNAs. Although primarily the Ago protein loads one miRNA strand for mRNA binding for subsequent repression, the other miRNA strand could possibly be transported by RNA-binding proteins (RBPs) toward microvesicle bodies as well for subsequent secretion in exosomes. Such strands are also called passenger strands and studies have shown abundance of such passenger stands into microvesicles. For example, miRNA content of cardiac fibroblast-derived exosomes has a relatively high abundance of many miRNA passenger strands (Bang et al., 2014). However, because Ago2 is generally known to be absent from exosomes, it is suggested that exosomal miRNAs are protected from degradation and/or sorted by other RBPs (Zhang et al., 2015). miRNAs are known to be loaded into exosomes based on the certain disease conditions, different cells and tissue types which offers a specificity that can be exploited for diagnostic and therapeutic purposes (Bertoli, Cava, & Castiglioni, 2015; Iorio & Croce, 2012). Moreover, as discussed in this book, the miRNA signature for tumor and tumor microenvironment plays a significant role in tumor development and progression. Identifying the processes that support such characteristic loading of RNA into microvesicles will play an important role in chemotherapeutic applications.

3.2 Endosomal Vesicle Pathways

Solutes, nutrients, ligands, and components of the plasma membrane are all transported in the mammalian cells via early endosomes (Bissig & Gruenberg, 2013). When the endosome matures into late endosomes, the inward budding from the limiting membrane of endosome leads to the formation of MVBs. During MVB formation, cytosolic RNAs are taken up and undergo inward budding from the limiting membrane of endosomes. This budding-in process also supports the RNA loading into exosomes. As discussed below, it is suggested that there are specific lipid-mediated mechanisms for specific loading of RNA into exosomes.

There are specific exosome-sorting RNA motifs (the EXOmotif) for both miRNA and mRNA (Janas et al., 2015; Villarroya-Beltri et al., 2013; Zhang et al., 2015). As mentioned earlier, RBPs other than the Ago complex can deliver miRNAs to the raft-like region on membrane; for example, miRNA association with heterogeneous nuclear ribonucleoproteins A2/B1 (hnRNPA2B1) promotes their exosomal release (Villarroya-Beltri et al., 2013). Such RBP can also bind to raft-like regions on the limiting membrane of the MVB without the miRNA and can be found in exosomes independent of the miRNA. Such protection by the raft-like region could explain in part the relatively high abundance of many miRNA passenger strands in exosomes (Janas et al., 2015).

Similar to the raft-like regions and the RBPs, the hydrophobic modification of RNA can also affect its interaction with the membrane-bound vesicles. For example, methylation of some miRNAs or isopentenylation of tRNAs can increase affinity to the raft-like region of the membrane (Janas et al., 2015; Janas, Janas, & Yarus, 2012). It is also important to note that in the same manner that there is specific sorting of miRNAs and mRNAs in the vesicles, the tRNAs are also specifically selected for loading into exosomes. For example, a recent report indicates that breast cancer–specific miRNA signature unique to extracellular vesicles is accompanied by a unique breast cancer–specific tRNA fragments (Fiskaa et al., 2016; Guzman et al., 2015). Hence, there are several different factors that affect the miRNA loading into the exosomes, and these factors could be general for all exosome secretion or specific for distinctive cells, tissue types, or diseases. Identification of such factors could help in modifying the exosome secretion and altering the miRNA/mRNA content of the exosomes in certain specialized cells or tissues leading to new paths of therapeutic development.

4. Isolation Techniques for RNA/DNA From Exosomes

4.1 Source of Extracellular Vesicles

Cell culture supernatants and biological fluids such as cerebrospinal fluid, urine, breast milk, plasma, and serum can be used to isolate vesicles. Such diverse starting materials require unique isolation techniques. Different parameters, depending on the source, affect the quality and quantity of exosomes and exosome-derived nucleic acids. These factors, discussed below, need to be considered when designing experiments to isolate exosomes and EV nucleic acids from different sources including cell culture supernatant and different bodily fluids.

4.1.1 Cell Culture Supernatants

As indicated by the seminal paper published from the International Society of Extracellular Vesicles, studies that use cell culture supernatants as a source of exosomes for analysis should consider several different parameters for evRNA analysis and report (Hill et al., 2013; Witwer et al., 2013). Some small factors are inclusion of a detailed description in the manuscript regarding the number of cells, media volume, centrifugation parameters (including rotor type, centrifugal force, and time), and the downstream storage conditions before RNA extraction. Other important factors to consider are as follows: (1) The health of the cell cultures before evRNA analysis should be included via confluency, and viability of the cell culture. (2) Because culture systems can be contaminated with mycoplasma, cell cultures should also be checked for the presence of these contaminants. (3) Ideal controls for apoptosis should also be included because the evRNA population can be contaminated by release of RNA-containing apoptotic bodies or nucleoprotein complexes released from apoptotic or necrotic cells in cell cultures. Such contaminants can be separated using fractionation of pelleted vesicles using density gradient media (such as sucrose or iodixanol) (Cantin, Diou, Bélanger, Tremblay, & Gilbert, 2008). (4) It is also important to mention whether cell culture additives such as fetal calf serum or fetal bovine serum were depleted of exosomes before use. Postisolation characterization should be done to ensure the quality and purity of the vesicles. For example, techniques such as transmission electron microscopy, nanoparticle tracking analysis (NTA; Nanosight), and tunable resistive pulse sensing (qTRS; qNano) provide approximate vesicle size postisolation (Filipe, Hawe, & Jiskoot, 2010; Vogel et al., 2016). Additionally, flow cytometry, NTA, and dynamic light scattering have also been used to undertake quantitative analysis for exosomes (Vogel et al., 2016).

Lastly, characterizing the protein markers for commonly found in exosome, such as CD63 further demonstrates the purity of vesicles isolated (Gallart-Palau et al., 2015).

4.1.2 Body Fluids

As discussed in this chapter, the methods used for isolation, collection, or purification of exosome from body fluids and its subsequent storage procedures can impact the quality and quantity of evRNA content (Gardiner et al., 2016). Such method selection mostly depends on the type, volume, and viscosity of the bodily fluids used for isolation of EV (Hill et al., 2013; Szatanek, Baran, Siedlar, & Baj-Krzyworzeka, 2015). The parameters discussed earlier related to description, purification, and necessary controls for cell cultures also apply to the exosomes isolated from bodily fluids. However, temperature and the duration of delay in processing also alter the yield of exosome and subsequently the extraction of RNA from these exosomes (Bæk, Søndergaard, Varming, & Jørgensen, 2016; Hill et al., 2013). Witwer et al. provide a detailed checklist for sample choice and sample collection/processing (Witwer et al., 2013). One of the most important considerations is the presence of RNA species that may be bound to other molecules such as proteins and lipid complexes other than evRNA (Turchinovich, Weiz, Langheinz, & Burwinkel, 2011; Vickers, Palmisano, Shoucri, Shamburek, & Remaley, 2011). Such contamination can be avoided by using nucleases such as RNase that can cleave the extracellular RNA, not evRNA, since the EV lipid bilayer protects the evRNA. Importantly, nuclease levels in body fluids can change because of pathological conditions. In cancer patients, for example, increased levels of serum nucleases have been observed (Hill et al., 2013; Kottel, Hoch, Parsons, & Hoch, 1978). Such variations in nuclease levels should be taken into account when analyzing evRNA present in clinical samples. Although RNaseA, which is specific for single-stranded RNA, has been mostly used in evRNA analysis, a potential risk is that high concentrations of RNases are difficult to inhibit, and residual activity may affect the yields of RNA isolated from the sample. Hence, RNase treatment during sample preparation depends on the research question being addressed. For instance, if the purpose of the experiment is to examine how RNAs are selectively incorporated into EV biogenesis, treatment with RNase may be essential. Besides the presence of extracellular RNA, some biological fluids are rich in DNA (for instance, plasma). Moreover, none of the available RNA extraction methods excludes coisolation of DNA entirely, and such contaminating DNA may interfere in RNA Bioanalyzer profiling and deep sequencing. It is therefore advisable to treat samples suspected of DNA contamination with DNase before evRNA isolation.

4.2 Isolation of Exosomes and RNA/DNA

Efficient EV purification before lysis and extraction is paramount to RNA quality. Overall, the differential ultracentrifugation process is considered as a "gold standard" for isolating exosome and the enclosed RNA (P. Li, Kaslan, Lee, Yao, & Gao, 2017) because it permits a good yield and quality. However, it is an extremely time-consuming process, in addition to the need of a well-optimized protocol and noteworthy capital investment.

But unfortunately, in the absence of a consistent universal standardized protocol, different investigators use various preprocessing protocols including insufficient generation of platelet, or inadequate separation from cellular debris as well as different protocols for ultracentrifugation itself (Greening, Xu, Ji, Tauro, & Simpson, 2015; Hill et al., 2013; Szatanek et al., 2015). Hence, as a result of these variables and lack of standardized methods, it is extremely difficult to compare results between studies. Although isolation of exosomes is still not standardized, the isolation methods of nucleic acid from exosomes are much developed. This is mainly using a robust method that yields RNA in quality and quantity similar to ultracentrifugation. Several researchers have actually developed such a method by incorporating a variety of approaches. These include the use of size-based filters (e.g., ExoMir, Bioscientific), antibody-based capture (e.g., Immunobeads, HansaBioMed), and polymer-based precipitation reagents (e.g., Life Technologies, System Biosciences Inc.) for specific applications (Hill et al., 2013). For example, the kits using specific size-based filters generally lack specificity for the EV fraction because any particle that matches the size of the filter will be retained by it, including cellular debris, protein complexes, or even lymphocytes (Hill, 2017; Hill et al., 2013). Similarly, precipitation using polymers followed by lysis using chaotropic salts also lacks specificity and generally coisolate undesired protein-bound extracellular RNA (Deun et al., 2014; Enderle et al., 2015; Momen-Heravi et al., 2013). Although the antibody-based purification of exosome has higher specificity, it requires an estimation/information of EV protein content; which is in contrast with the rising and unknown field of exosomes (Zaborowski et al., 2015). Recently, researchers have developed spin column–based method to isolate exosome and extract the RNA contents from plasma and serum in an easy and reproducible workflow, and RNA integrity and size distribution, purity, and yield are in comparison with an optimized ultracentrifugation procedure (Enderle et al., 2015). This new procedure captures nearly 100% of mRNA from plasma samples and is equal to or better than ultracentrifugation in mRNA yield. It also allows isolating and collecting intact vesicles that can be examined by electron microscopy.

4.3 Isolation of microRNAs/Noncoding RNAs Specifically

miRNA-based biomarkers for diagnostics and therapeutic purposes are of great interest in the current medical and scientific communities (Price & Chen, 2014). This includes the understanding of tumor biology, tumor microenvironment, chemoresistance, and tumor metastasis. However, there are several technical issues that need to be considered before employing exosomal miRNAs for clinical practice (Moldovan et al., 2014). miRNA recovery from different sample types is largely influenced by the RNA isolation method used (El-Khoury, Pierson, Kaoma, Bernardin, & Berchem, 2016). For example, the miRCURY kit allows isolating highly pure and better quality RNA in comparison with miRNeasy and Trizol LS resulting in an optimal RT-qPCR efficiency, but the recovery of miRNAs is highly dependent on the quantity of starting material. In fact, studies suggest that miRCURY columns tend to be saturated by large RNA species when the starting number of cells increases, thus impeding the optimal recovery of miRNAs. Moreover, studies also support the column saturation hypothesis for miRCURY kit, such that better results were obtained for miRNA detection, when the starting cell number with an optimum cell number; for example, 200,000 rather than 800,000 (El-Khoury et al., 2016; Liu et al., 2011). In such case, the Trizol LS–extracted RNA was better suited when using high input material as compared with miRCURY kit as evident from the RT-qPCR results. Importantly, when using as few as 100 or 1000 cells, miRNeasy kit permitted a better miRNA detection when compared with miRCURY kit (Hill et al., 2013; Ramón-Núñez et al., 2017). Such effect does not necessarily reflect a clogged column issue but could indicate a lower adsorption efficiency of miRCURY column with relatively low amounts of input material (El-Khoury et al., 2016; Moldovan et al., 2014). However, it is noteworthy to mention that the reduced recovery of miRNAs by using the miRCURY kit is not an indicator of a reduced ability of the kit for the isolation of all small-RNA species; for example, the detection of U6 snRNA from miRCURY-isolated RNA is shown to be better than its detection when isolated by other methods (El-Khoury et al., 2016; Popov, Szabo, & Mandys, 2015). These findings suggest that miRNAs might be the most affected by the suboptimal capacity of the columns.

5.   Profiling of Nucleic Acid—RNA

5.1   Analysis of the Quality, Quantity, and Diversity of evRNA

Several different types of RNAs including long and short, coding and noncoding are reported in exosome. However, such characterization is challenging because the quality and quantity of isolated evRNA is affected by the isolated exosomes as discussed earlier. There are several different methods to characterize and investigate these different parameters.

5.1.1   Assessing evRNA Quality

Isolation of intact (nondegraded) RNA or RNA integrity is of great importance in quantitative gene expression profiling experiments (Opitz et al., 2010). However, it should be noted that RIN provided by the Bioanalyzer software is not an indicator of small evRNA integrity because RIN provides values that are based on major 18S and 28S ribosomal RNA peaks, abundantly present in cells but not so much in EV (Mateescu et al., 2017). As discussed above, RNA can be degraded in many ways: by enzymes, namely ribonucleases (RNases), which are both ubiquitous and extremely stable; by mechanical stress introduced by freezing, thawing, or centrifugation; by base-catalyzed hydrolysis; by heat, especially in the presence of divalent cations; and by UV damage. Any such exposure can cause RNA damage and resulting in altered downstream quantitative applications (Pucci et al., 2016). Such risk is more important when working with small quantities of RNA, especially because such smaller quantities of RNA are more prone to be fragmented over the course of the isolation process (Tan & Yiap, 2009). However, it is noteworthy to mention that processed fragments of RNA, especially longer RNA molecules are already present in the exosome,

Table 1   Methods for Determining evRNA Purity and Integrity

Method	Use	Pros	Cons
Agilent Bioanalyzer chips	Integrity	• Small volume required • Highly sensitive • Total length profile of RNA	• Not suited for assessing small RNA integrity • Assessment based on intact 18S/28S rRNAs generally depleted from extracellular vesicles(EVs) • Sensitive to contaminants such as DNA
Next-generation sequencing	Integrity and purity	• Detects fragmentation, for example, as 3′ bias in mRNA reads after poly-A selection • Detects presence of foreign genetic material (e.g., derived from fetal bovine serum)	• Erroneous assessment of fragments in the case of highly modified RNA types • Long reads (i.e., PacBio) most useful but require lots of material
RT-PCR and derivatives (i.e., 5′/3′ RACE)	Integrity	• Robust and sensitive, can map exact sites of fragmentation	• Analysis of single transcripts only
Proteinase-nuclease protection assay	Purity	• Rigorously determine that RNA is present in EV lumen	• Leftover nucleases may still be active at point of vesicle lysis
PicoGreen	Purity	• Test for presence of dsDNA	• Not DNA-specific in samples with RNA concentrations over 130 ng/mL

Modified from Mateescu, B., Kowal, E. J. K., van Balkom, B. W. M., Bartel, S., Bhattacharyya, S. N., Buzás, E. I., … Nolte-'t Hoen, E. N. M. (2017). Obstacles and opportunities in the functional analysis of extracellular vesicle RNA – an ISEV position paper. *Journal of Extracellular Vesicles, 6*(1). https://doi.org/10.1080/20013078.2017.1286095.

and it could be biologically relevant. Another important quality measure is the purity of RNA, and some of the methods that can determine the RNA integrity and purity are listed and compared in Table 1.

5.1.2 Evaluating Quantity of evRNA

Similar to the problems associated with the characterization of cellular RNA, there are issues associated with the study of evRNA. One of the major issues is the limited quantity of RNA obtained from EV as compared with RNA derived directly from cells (Fritz et al., 2016). Moreover, EV samples are mostly devoid of intact large and small ribosomal RNA subunits. As a result, the required RNA quantity for specific analysis methods (e.g., sequencing, microarrays, or quantitative reverse transcription-polymerase chain reaction (RT-qPCR)) does not necessarily match respective recommendations for cellular RNA samples (Deun et al., 2014; Rekker et al., 2014).

NanoDrop spectrophotometer family (NanoDrop 1000, 2000, or 2000c; Thermo Fisher Scientific, Wilmington, USA) is generally used to quantify the cellular RNA; but it cannot be used to measure the evRNA. This is mainly because the NanoDrop quantifies the RNA in microliter volumes of RNA based on UV-absorbance, which is accurate in the range of 2–3 μg/μL to 1–2 ng/μL. However, the evRNA is typically present in less than 2 ng/μL concentration. Similarly, the Qubit RNA HS (high sensitivity) assay (Thermo Fisher Scientific) is also specific for RNA but has a limit of >0.2 ng/μL even while using the maximum volume for the kit (20 μL of sample) (Li, Ben-Dov, Mauro, & Williams, 2015). Hence, unless we use a large volume of the sample, this method cannot be used to measure the evRNA concentration. However, it is noteworthy to mention that this is a good method to isolate RNA in presence of contaminating DNA because it has been shown that spike-in RNA (to bring sample RNA concentration above the minimum) allows the Qubit RNA HS assay to quantify small amounts of RNA with high specificity.

Bioanalyzer Pico chip (Agilent Technologies, Foster city, USA) is one of the most sensitive RNA quantification methods currently available with a lower detection limit of 50 pg/μL. More importantly, the technique requires only 1 μL of sample. The Bioanalyzer also provides electrophoresis-like length profiles, which are useful for estimating the size distribution of RNA in EV samples. Hence, volume of sample required and quantity of RNA in the samples can still be detected with accuracy even at lower values. However, one of the major limitations is that the method is designed to assess quality based on the large ribosomal subunits that are generally absent in the exosome as described previously. Despite these challenges, Bioanalyzer is currently the most preferred technique to measure evRNA quantity. In comparison, the small-RNA chip (Agilent Technologies) provides a fast and sensitive analysis to resolve small nucleic acids in the size range of 6–150 nucleotides such a miRNAs. It has a high resolution such that miRNA and tRNA can be segregated. Furthermore, it requires minimal sample consumption because it uses as little as 50 pg of purified miRNA or 10 ng of total RNA for analysis in 1 μL volume. Although these qualities make it preferable for use for measuring EV-associated RNA content and length, it is prone to errors since the RNA concentration is determined relative to a supplied RNA ladder and internal marker peaks, which can result in variability between different measurements (Li et al., 2015). In addition, aggregates of RNA dye can also result in their separate peaks on the electrophoresis profiles contributing to errors in measurement (Becker, Hammerle-Fickinger, Riedmaier, & Pfaffl, 2010; Mateescu et al., 2017). The nanochip is less sensitive to salt, whereas the Pico assay is highly sensitive to differences in salt concentration. In addition, Bioanalyzer 2100 chip will also detect contaminating DNA in evRNA isolates because the method uses a non-RNA-specific dye. In lack of specific evRNA quantitation equipment, some researchers have also used the Quant-iT RiboGreen RNA Assay Kit (Thermo Fisher Scientific), which is based on a nucleic acid–specific fluorescent dye that can be used to quantify RNA with a linear detection range of 1–200 ng using any standard fluorescence microplate reader. Although such measurement is sensitive to DNA contamination, it is less sensitive to protein and phenol chloroform. This method is particularly useful because it uses a standard curve that can be adjusted for low-input RNA samples, e.g., evRNA. Hence, sensitive techniques such as Agilent Bioanalyzer pico chip and the Quant-iT RiboGreen RNA Assay are far more suitable for evRNA quantification than the NanoDrop. Detection of the levels of particular transcripts by highly sensitive RT-qPCR as discussed later may be used as a proxy for total RNA quantity in samples containing a very low amount of RNA. Most techniques, with the exception of the Qubit RNA HS Assay, are also sensitive to DNA contamination. We therefore recommend pretreatment of samples with DNase for accurate RNA quantitation. All these methods are further compared in their detection limit in Table 2.

Table 2   Suitability of RNA Detection Methods for Quantification of evRNA

Method	Lower Detection Limit	Remarks
Nanodrop spectrophotometer family (NanoDrop, Thermo Fisher Scientific)	3 µg/µL to 2 ng/µL (µL volumes of RNA)	Not generally suited for measuring evRNA because of high lower limit of detection
Qubit RNA HS (high sensitivity) assay (Thermo Fisher Scientific)	>0.2 ng/µL	Not generally suited for measuring evRNA because of high lower limit of detection
Bioanalyzer Pico chip (Agilent Technologies)	50 pg/µL	Most sensitive quantification method for total RNA, but prone to errors. Most relevant for assessing total RNA content and length distribution
Bioanalyzer small-RNA chip (Agilent Technologies)	50 pg/µL of purified miRNA or 10 ng/µL of total (cell) RNA in size range of 6–150 nucleotides	Similar properties as Pico chip. Useful for resolving miRNA from tRNA and other small-RNA species.
Quant-iT RiboGreen RNA Assay kit (Thermo Fisher Scientific)	Detection range of 1–200 ng (sample diluted to 1 mL)	Less sensitive to contaminants, such as protein and phenol chloroform
Quantitative reverse transcription-polymerase chain reaction (RT-qPCR)	1 fg (~2500 copies for mRNA) of a particular transcript	Most sensitive quantification method overall but does not analyze total RNA, must select primers specific for targets and validate to check for off-target amplification

Modified from Mateescu, B., Kowal, E. J. K., van Balkom, B. W. M., Bartel, S., Bhattacharyya, S. N., Buzás, E. I., … Nolte-'t Hoen, E. N. M. (2017). Obstacles and opportunities in the functional analysis of extracellular vesicle RNA – an ISEV position paper. Journal of Extracellular Vesicles, 6(1). https://doi.org/10.1080/20013078.2017.1286095.

5.2 Characterizing evRNA

Although there are several different methods for measuring the quality and quantity of evRNA, the ultimate goal of the evRNA isolation and characterization is to identify the specific RNA transcripts that are secreted in the exosomes (Tang et al., 2017). Because such identification will ultimately provide the specific biomarkers for diagnostic or therapeutic purposes, the advent of next-generation sequencing and microarray platform has definitely propelled the field of evRNA analysis in parallel to cytosolic RNA analysis. Although there are numerous books and review articles discussing the specifics of RNA sequencing, deep sequencing or microarray technologies pertaining to evRNA analysis, we highlight some of important aspects of these technologies, their comparison against each other in performance related to evRNA analysis, specific applications of each of these tools as well as pros and cons as it pertains to evRNA analysis.

5.2.1 Deep Sequencing Platforms
The major factor in selecting a specific sequencing platform or selecting between the personal or service-based order is the depth of coverage obtained with the specific equipment (Conesa et al., 2016; Kukurba & Montgomery, 2015). This is important because deeper coverage permits analysis of rare sequences and provides relatively more confident results toward the number of reads per transcript. The depth is highly critical because transcripts that are low in abundance might be missed by a lower depth analysis; hence, such different

parameters need to be considered for analyzing the evRNA (Trapnell et al., 2013). Several technologies have been developed for deep sequencing over the last couple of years. Many of these technology platforms rely on proprietary kits and equipment for undertaking deep sequencing. The major systems currently used are Illumina HiSeq, Roche 454 pyrosequencing, and the SOLiD system from Applied Biosystems. However, most of these platforms are used for high-coverage sequencing and mainly operated via service providers. Hence, there are also smaller deep sequencing platforms mainly geared toward individual laboratories for sequencing. These mainly include the MiSeq (Illumina), Ion Torrent Personal Genome Machine (Life Technologies), and the GS Junior (Roche). Hence, based on the cost and time constraints as well as sample type and source as discussed earlier, a specific sequencing platform should be considered for evRNA analysis.

5.2.2 Library Preparations From evRNA Samples

Although exosomes mainly contain small RNAs (20–200 nucleotides), recent evidence also indicates that larger RNAs, such as mRNAs and low levels of 18/28S ribosomal RNA might also be present in the exosome (Bellingham, Coleman, & Hill, 2012; Nolte-'t Hoen et al., 2012; Valadi et al., 2007). Such differential length of RNAs is an important factor to consider while sequencing because small versus large RNAs require differential preparation for sequencing including library preparation protocol. Table 3 is indicative of some comparison between

Table 3   Small-RNA-Sequencing Library Preparation: Most Widely Used Library Construction Methods

Method	Use	Pros	Cons	
Hybridization method (e.g., SREK for SOLiD)	Two double-stranded adapters that contain degenerate 5′- or 3′-end overhangs	Ligation adapters with T4 RNA ligase	Reverse transcription into cDNA	Further amplification by PCR using primers that anneal to both adapter sequences
Polyadenylation based (A)	Addition 3′-poly A tail using Poly (A) polymerase	Ligation RNA adapter to the 5′-end using T4 RNA ligase	Reverse transcription into cDNA	Further amplification by PCR
Polyadenylation based (B)	Addition 3′-poly A tail using Poly (A) polymerase	Reverse transcription into cDNA	Ligation adapter to the 5′-end using T4 RNA ligase 1	Further amplification by PCR
Sequential adapter ligation (e.g., Illumina) (A)	Sequential ligation of 3′- and 5′-adapter oligonucleotides directly to small RNA	Ligation RNA adapter to the 5′-end using T4 RNA ligase	Reverse transcription into cDNA	Further amplification by PCR
Sequential adapter ligation (e.g., Illumina) (B)	Sequential ligation of 3′- and 5′-adapter oligonucleotides directly to small RNA	Reverse transcription into cDNA	Ligation adapter to the 5′-end using T4 RNA ligase 1	Further amplification by PCR

Modified from Hill, A. F., Pegtel, D. M., Lambertz, U., Leonardi, T., O'Driscoll, L., Pluchino, S., … Nolte-'t Hoen, E. N. M. (2013). ISEV position paper: Extracellular vesicle RNA analysis and bioinformatics. *Journal of Extracellular Vesicles, 2*. https://doi.org/10.3402/jev.v2i0.22859.

different methods commonly used for library preparation. Moreover, another important factor to consider is the possibility of cleavage products of RNAs or the small evRNAs themselves can overlap with larger noncoding RNAs in these exosomes (Rinn & Chang, 2012). Such fragments of RNA can be a result of RNA isolation process before library preparation and can lead to a nonspecific interpretation and erroneous small-evRNA profile. Hence, as discussed in detail earlier in this chapter, the EV isolation process, RNA isolation procedure, as well as library preparation methods should be carefully considered to avoid any confounders and ensure a valid robust analysis and interpretation of sequencing results as listed in Table 4. Normal qualitative and quantitative parameters should definitely be characterized before beginning library preparation and noted in the publication for reference.

In brief, library preparation for sequencing involves ligation of adapter to both ends of the RNAs followed by reverse RT into cDNA and finally amplifying via adapter-region specific primers. Several different confounders can be introduced during the library preparation resulting skewed sequencing analysis. For example, (1) differential efficiency of adapter ligation to certain RNAs due to the T4 ligase; (2) modifications in the barcode sequences (in case of multiplex sequencing); or (3) owing to differential amplification during polymerase chain reaction (PCR). Different techniques in addition to sequencing are generally recommended to validate and confirm relative expression differences between samples. The cDNA ligated to the adapter can be run on a gel, and gel-bands can be excised to exclude adapters without inserts. Specific size of cDNA can be selected for sequencing and can improve sequencing robustness; however, an increased depth of sequencing is usually sufficient to detect less abundant evRNAs. In addition, enzymes such as phosphatases are sometimes added to the pool of library preparation to enrich/eliminate specific classes of RNAs during sequencing (Munafó & Robb, 2010). Although external spike-in RNA controls can be useful to evaluate the sensitivity, accuracy, and comparability of RNA-sequencing experiments, one of the most important parameters to consider is consistency of protocols for library preparation to compare and validate data for reproducibility and robustness (Jiang et al., 2011). The most widely used library construction methods are listed in Table 3.

Table 4   Common Sources of Bias in RNA Isolation and Sequencing Methods

Source	Example	Cons
Size selection	Size selection Underrepresentation of midsize RNAs in RNA sequencing experiments	Tailor size selectivity of RNA purification technique to size of RNA of interest. If analyzing total RNA, perform multiple extractions for differently sized populations.
Library preparation kit or protocol	Adapter ligation bias	Use newly developed strategies to control for ligation bias, i.e., 4N adapter-based kits.
Sequencing platform	Different biases in different sequencing platforms	Use of identical platforms for experiments to be directly compared. Corroborate important conclusions with a second technique.

Modified from Hill, A. F., Pegtel, D. M., Lambertz, U., Leonardi, T., O'Driscoll, L., Pluchino, S., ... Nolte-'t Hoen, E. N. M. (2013). ISEV position paper: Extracellular vesicle RNA analysis and bioinformatics. *Journal of Extracellular Vesicles, 2*. https://doi.org/10.3402/jev.v2i0.22859.

Technical and/or biological replicates and the quantity of such replicates is a major factor to consider while undertaking RNA deep sequencing, especially because this can be cost-prohibitory. Biological variability should be taken into consideration to detect significant differences in RNA abundance, especially if the deep sequencing of evRNA is applied to patient samples for diagnostics purposes (Auer & Doerge, 2010). Although small-RNA sequences may have lower genetic variability (Parts et al., 2012), it is not clear if such genetic variability has downstream effects on the differential expression of evRNAs in different individuals. Hence, it is important to consider the choice to either pool patient samples or perform RNA sequencing on individual samples, which is indirectly dependent on amount of RNA obtained and the availability of financial resources. As mentioned earlier, in this case, source-specific effects on such evRNA, appropriate control groups should be included (Witwer et al., 2013).

The library preparation step can introduce some variation depending on the protocol and platform as discussed above (Head et al., 2014). Different laboratories and service providers have proprietary protocols for addition of adapters as part of the library preparation and generally vary between the sequencing platforms employed. Such differential addition of adapter sequences varies widely and needs to be assessed for confounding effects (Conesa et al., 2016). Hence, as suggested earlier a secondary method for characterization is recommended for validation as discussed later. However, the sequencing itself is highly reproducible, and repeated sequencing of the same library is not required, unless needed to increase the sequence depth coverage allowing detection of rare transcripts. Furthermore, such sequence coverage can also identify sequences of other bacterial and viral origins (Houldcroft, Beale, & Breuer, 2017). Hence there are different sources of biological and technical variability, and such parameters need to be considered while designing the experiments for nucleic acid profiling in exosomes.

5.2.3 Validation of Deep Sequencing Data
Validating deep sequencing data arising from EV samples with a different technique is important for a number of reasons. Firstly, the preparation of sequencing libraries and platforms differs between providers. Secondly, validating data sets is important for standardization between different laboratories.

5.2.4 Validation by qRT-PCR
The most commonly employed validation technique is quantitative real-time PCR (qPCR). Several researcher use RT-qPCR to quantify transcript abundances in evRNA preparations (Enderle et al., 2015; Zeringer et al., 2013). As performed with cellular RNA, RT-qPCR can quantify levels of a particular nucleic acid transcript in a sample by measuring its increase in concentration over time (using fluorescent nucleotides or a fluorescent probe) when subjected to exponential amplification by PCR. However, this method does not directly measure total RNAs, but it is extremely sensitive and is able to detect 1 fg or ~2500 copies of a given transcript in an optimized system (Levesque-Sergerie, Duquette, Thibault, Delbecchi, & Bissonnette, 2007). This is highly applicable to the often observed lower

quantity of evRNA. Especially, quantifying the levels of specific transcript or a panel of individual transcripts by RT-qPCR not only serves as a proxy for total RNA content but can also allow to characterize the content of evRNA (Mateescu et al., 2017). Although RT-qPCR is sensitive to DNA contamination, this can be minimized by additional experimental validation, for example, running a gel to verify a single amplicon of the expected size or designing primers over exon–exon junctions in the case of mRNA.

Different biological replicates can be sequenced without the need for deep sequencing, if sequencing is coupled with validation via qPCR (Zeng & Mortazavi, 2012). Furthermore, it is also important to select a specific control for normalization when analyzing changes in expression levels between replicates and treatment conditions. This is mainly because the housekeeping genes that are generally used as normalization controls for the cellular RNA analysis cannot be used for evRNA analysis as they might be expressed at different levels in different exosomes or not expressed at all. Hence, the appropriate use of normalization controls should be first validated by measuring the expression of such evRNA and showing steady expression in controls and experimental conditions. In fact, because of such reasons, it is highly advised to use multiple reference genes for qPCR data normalization. Digital PCR is highly used to identify experiment-specific targets for normalization; however, a recently adopted approach is to use a global normalization strategy to quantitative data (Kuroda et al., 2015; Sanders, Mason, Foy, & Huggett, 2013). This includes calculating the mean expression values of all expressed small RNAs and using it for normalization (Mestdagh et al., 2009).

5.2.5 Microarray Analysis of RNA From Exosome

Although deep sequencing is the gold standard for evRNA analysis, microarray technology can also be used for this purpose and is a well-established, relatively easier and cost-effective way for gene-expression measurements of known fragments of mRNA, miRNA, long noncoding RNA (lncRNA) species. Microarray technology has been used in studies of both exosomes from cell culture systems and those isolated from bodily fluids (Ismail et al., 2013; Noerholm et al., 2012). As mentioned later, the major advantage over the deep sequencing technique is that the microarray platform does not use specialized sequencing equipment and/or complex bioinformatics approaches. In contrast, the major limitation is the lack of ability to identify any novel sequences in exosome because microarrays use only known sequences as targets. Another major limitation is the need to update the microarrays when any updates are introduced to the databases such as miRBase. Hence, although microarrays are a useful tool for profiling evRNAs and used for several studies, they have their own limitations but in combination with sequencing can provide robust results (Nolte-'t Hoen et al., 2012).

5.2.6 NanoString

With excellent specificity, low false-positive rates, and a direct digital readout, the expression panels for NanoString's miRNA and mRNA measurements are unique and powerful biomarker discovery tools (Chatterjee et al., 2015). The major advantage of the assay is that it uses NanoString's nCounter platform that provides direct digital counts of each miRNA without the use of RT making it highly specific. As noted above, avoiding such amplification also prevents any nonspecific off-target amplification and any contaminating moieties, promoting its use and application for biomarker discovery and signature development tool.

nCounter NanoString technology is based on a novel method of direct molecular barcoding and detection of target molecules digitally through the use of color-coded probe pairs (Kulkarni, 2011). Sample Preparation Kit provides reagents for ligating unique oligonucleotide tags onto the 3′ end of target miRNAs/mRNAs so that short RNA targets can be detected by nCounter probes. Sample preparation involves multiplexed ligation of the specific tags to their target miRNA and an enzymatic purification to remove nonligated tags. Sequence specificity of the ligation reaction is ensured by the use of melting temperature-optimized bridging oligos that are complementary to a portion of both the target miRNA/mRNA and miRNA/mRNA-specific tag along with careful, stepwise control of annealing and ligation temperatures. One of the benefits of the direct digital counting methodology afforded by the NanoString nCounter platform is that it is perfectly positioned for translational research needs, enabling the utility of very poor or low-quality RNA, such as the evRNA. We have previously used such NanoString technologies for characterizing the miRNA profiles in lung cancer. Several studies have used such NanoString panels for similar applications (Das et al., 2016; Hyeon et al., 2017). It is noteworthy to mention the higher specificity, low input quantity, absence of amplification with less contamination steps have allowed NanoString analysis to be much more relevant and important to evRNA analysis and for validation of specific set of genes identified by deep sequencing.

6. Single EV Analysis

Owing to the heterogeneity of exosome origin and content, it is becoming quite clear that to fully comprehend the functional implications and purpose of RNA incorporation into exosomes, we cannot rely on bulk exosome analysis but will have to undertake analysis at the EV subpopulation or individual EV level. However, as imaginable, this is not at all a small feat and faces several challenges (Groot Kormelink et al., 2016; Im et al., 2014; Wunsch et al., 2016). Detecting and enumerating RNA molecules in individual exosomes face unmet challenges and none of the currently used single EV analysis cytometry-based methods is capable of detecting a single fluorescein molecule. Fluorescent RNA-tracking dyes (e.g., acridine orange, Syto84, or Syto-RNA Select) have been applied for high-resolution flow cytometry analysis of the evRNA with poor signal to noise ratio, limited probe specificity as well as limited fluorescence per EV (Mateescu et al., 2017).

Most RNA dyes are intercalating agents, and the structure of a small miRNA molecular can only bind few molecules of intercalating dyes and does not support intercalation of hundreds of dye molecules limiting its application for RNA analysis in single EV. Hence, markers with higher quantum yield (and better signal: noise ratio), such as quantum nanocrystal-tagged oligonucleotides, are much more suitable for detection of molecular components in single exosome. Other innovative methods indirectly detect RNA in EV by microscopy, notably, by using RNA-binding fluorescent protein probes (Lai et al., 2015). Use of microfluidics might enhance the signal to noise ratio and allow detection of RNA within single EV probably using rolling circle amplification technique; however, such system is still being developed (Zhao, Yang, Zeng, & He, 2016). Thus, although it is highly pertinent for diagnostic purposes to characterize the single EV for RNA content, it will take us some time to have technologies that can perform such detection.

7. Statistical and Bioinformatic Software for evRNA Characterization

EVpedia is an integrated database of high-throughput data sets from exosome derived from prokaryotes and eukaryotes and updates every 6 months (Kim et al., 2013). It is a free Web-based database and a useful resource to elucidate functional roles of exosome, but there is a dire need for high-quality EV data sets with a unified criteria and standards for high-throughput identification and characterization. Although the detailed procedures can vary between studies, most of these studies are fundamentally common and use combinations of filtration, differential centrifugation, and density gradient centrifugation methods to purify EV followed by isolation of evRNA and a specific type of technique to characterize evRNAs. Further to the need of uniformity and standards for the isolation and characterization methodology, there is also an important need for coordinated standards for high-throughput data production, analysis as well as interpretation and correlation of high-throughput data, such as mass spectrometry–based proteomics, microarray- or sequencing-based transcriptomics. Hence, it is crucial to identify and collect segregated data sets and compile a compendium that designates a clear and detailed guideline for the preparation of exosome, evRNAs, and the resultant high-throughput data.

ExoCarta provides information on proteins/RNA identified in exosomes freely to the users through the Web. As noted above, there is an exponential demand of such software because of the ever-growing interest on the identification of exosome-based biomarkers in different disorders including tumor diagnostics. Owing to such data explosion, ExoCarta is also updated regularly with biophysical data, including protein–protein interactions, posttranslational modifications, and lipid composition. With the addition of more data, the compendium provides useful insights into the molecules that are present in exosomes and cannot act as a primary resource for future exosome-based studies.

8. Experimental Artifacts and Contaminants Affecting evRNA Analysis

8.1 Non-EV-Associated RNA and Other Lab-Derived Contaminations Can Amplify During the Library Preparation Step

Several RNA-containing structures such as ribonucleoprotein complexes (RNPs), viral particles, and lipoproteins (HDL and LDL) from the EV or non-EV source, for example, the fetal bovine serum used in cell culture media as discussed above can act as a major source of RNA in EV samples (Shelke, Lässer, Gho, & Lötvall, 2014; Wei, Batagov, Carter, & Krichevsky, 2016). Whereas any precautionary step incorporated to increase rigor of the isolation protocol, for example, washing and repelleting exosome after centrifugation can decrease efficiency as well as lead to inclusion of contaminating particles present in the first pellet that may repellet together with exosome. Higher centrifugal force can also force the RNPs to nonspecifically associate with the exosome surface. As noted earlier, it is important to distinguish between RNA encapsulated within exosome from RNA outside-exosome, and this can be achieved via treatment of exosome with proteinase and RNase. Low-starting RNA quantities will significantly amplify the contaminating RNA moieties and lead to skewed sequencing results, and hence a blank run of a buffer sample can be performed that will amplify these contaminants can control for these possibilities. In addition, such limitations need to be taken into consideration while interpreting and reporting the results especially in patient's samples for diagnostic purposes.

8.2 Presence of DNA?

Although extracellular DNA is known to be present in various biological fluids (e.g., plasma and urine) and in culture medium because of necrosis/apoptosis or active cellular secretion processes (Kondratova, Serd'uk, Shelepov, & Lichtenstein, 2005), it is not clear if it is also present inside exosome. Hence, to demonstrate protection or lack thereof, it is highly recommended to characterize the presence of DNA by treating with proteinase and DNase before exosome lysis and DNA readout (Thakur et al., 2014). Owing to nonspecific binding to exosome or similarity in size and molecular weight, bigger DNA fragments (>3 kb) can bind to exosome fractions during centrifugation; however, the source of such DNA is still debated.

There are several other artifacts that can be considered relevant to the isolation and profiling of evRNAs; for example, (1) does the subcellular localization of the RNA/miRNA transcript affect the transport into the exosome?; (2) how can we explain the differential correlation between the RNA/miRNA expression in cellular fraction versus the exosomes?; (3) what are the other RBPs that affect the RNA/miRNA transport into the exosomes? These are just few listed and fundamental to our understanding and more so the application of evRNA for tumor diagnostic and chemotherapy.

9. Conclusions

Valuable information pertaining to the messages in form of evRNAs exchanged between cells can be obtained by in-depth analysis of evRNA using tools such as advanced sequencing. However, one of the major challenges is the lack of consistent and defined protocol, different starting sources of exosome, effects of techniques and equipment on quality and quantity of evRNA can limit the robust characterization and profiling of such evRNAs. Hence, although the field has progressed tremendously in the last 10 years with the advent in sequencing technology, the application of such evRNA for diagnostic and further therapeutic purposes indirectly is restricted by fundamental challenging parameters. Hence, it is recommended to undertake a well-considered experimental setup allowing for interlaboratory comparisons and avoiding any confounders that influence experimental outcomes and include appropriate controls. Furthermore, including a detailed description of such techniques in scientific publications allows other researchers to follow a consistent protocol as well and propelling the evRNA field forward. And the raw data also need to be shared on different platforms for other researchers to assess and conduct independent analysis/validation of the data. This chapter provides a small window into status of such analysis and important parameters to be considered while setting up evRNA analysis. Although we have done our best to include all parameters, the current chapter is not all comprehensive, and the field is constantly evolving in parallel to the sequencing and Nanostring field. However, we hope that the chapter helps scientists to think critically for planning and executing their experiments for profiling of evRNAs.

References

Auer, P. L., & Doerge, R. W. (2010). Statistical design and analysis of RNA sequencing data. *Genetics, 185*(2), 405–416. https://doi.org/10.1534/genetics.110.114983.

Bang, C., Batkai, S., Dangwal, S., Gupta, S. K., Foinquinos, A., Holzmann, A., … Thum, T. (2014). Cardiac fibroblast-derived microRNA passenger strand-enriched exosomes mediate cardiomyocyte hypertrophy. *The Journal of Clinical Investigation, 124*(5), 2136–2146. https://doi.org/10.1172/JCI70577.

Bastos, N., Ruivo, C. F., da Silva, S., & Melo, S. A. (2017). Exosomes in cancer: Use them or target them? *Seminars in Cell & Developmental Biology.* https://doi.org/10.1016/j.semcdb.2017.08.009.

Becker, C., Hammerle-Fickinger, A., Riedmaier, I., & Pfaffl, M. W. (2010). mRNA and microRNA quality control for RT-qPCR analysis. *Methods (San Diego, Calif.), 50*(4), 237–243. https://doi.org/10.1016/j.ymeth.2010.01.010.

Bellingham, S. A., Coleman, B. M., & Hill, A. F. (2012). Small RNA deep sequencing reveals a distinct miRNA signature released in exosomes from prion-infected neuronal cells. *Nucleic Acids Research, 40*(21), 10937–10949. https://doi.org/10.1093/nar/gks832.

Bertoli, G., Cava, C., & Castiglioni, I. (2015). MicroRNAs: New biomarkers for diagnosis, prognosis, therapy prediction and therapeutic tools for breast cancer. *Theranostics, 5*(10), 1122–1143. https://doi.org/10.7150/thno.11543.

Bissig, C., & Gruenberg, J. (2013). Lipid sorting and multivesicular endosome biogenesis. *Cold Spring Harbor Perspectives in Biology, 5*(10). https://doi.org/10.1101/cshperspect.a016816.

Budnik, V., Ruiz-Cañada, C., & Wendler, F. (2016). Extracellular vesicles round off communication in the nervous system. *Nature Reviews Neuroscience, 17*(3), 160–172. https://doi.org/10.1038/nrn.2015.29.

Bæk, R., Søndergaard, E. K. L., Varming, K., & Jørgensen, M. M. (2016). The impact of various preanalytical treatments on the phenotype of small extracellular vesicles in blood analyzed by protein microarray. *Journal of Immunological Methods, 438*(Suppl. C), 11–20. https://doi.org/10.1016/j.jim.2016.08.007.

Cantin, R., Diou, J., Bélanger, D., Tremblay, A. M., & Gilbert, C. (2008). Discrimination between exosomes and HIV-1: Purification of both vesicles from cell-free supernatants. *Journal of Immunological Methods, 338*(1–2), 21–30. https://doi.org/10.1016/j.jim.2008.07.007.

Chatterjee, A., Leichter, A. L., Fan, V., Tsai, P., Purcell, R. V., Sullivan, M. J., & Eccles, M. R. (2015). A cross comparison of technologies for the detection of microRNAs in clinical FFPE samples of hepatoblastoma patients. *Scientific Reports, 5*(10438). https://doi.org/10.1038/srep10438.

Conde-Vancells, J., Rodriguez-Suarez, E., Embade, N., Gil, D., Matthiesen, R., Valle, M., … Falcon-Perez, J. M. (2008). Characterization and comprehensive proteome profiling of exosomes secreted by hepatocytes. *Journal of Proteome Research*, *7*(12), 5157–5166.

Conesa, A., Madrigal, P., Tarazona, S., Gomez-Cabrero, D., Cervera, A., McPherson, A., … Mortazavi, A. (2016). A survey of best practices for RNA-seq data analysis. *Genome Biology*, *17*. https://doi.org/10.1186/s13059-016-0881-8.

Das, K., Chan, X. B., Epstein, D., Teh, B. T., Kim, K.-M., Kim, S. T., … Tan, P. (2016). NanoString expression profiling identifies candidate biomarkers of RAD001 response in metastatic gastric cancer. *ESMO Open*, *1*(1). https://doi.org/10.1136/esmoopen-2015-000009.

De Toro, J., Herschlik, L., Waldner, C., & Mongini, C. (2015). Emerging roles of exosomes in normal and pathological conditions: New insights for diagnosis and therapeutic applications. *Frontiers in Immunology*, *6*. https://doi.org/10.3389/fimmu.2015.00203.

Deun, J. V., Mestdagh, P., Sormunen, R., Cocquyt, V., Vermaelen, K., Vandesompele, J., … Hendrix, A. (2014). The impact of disparate isolation methods for extracellular vesicles on downstream RNA profiling. *Journal of Extracellular Vesicles*, *3*(1), 24858. https://doi.org/10.3402/jev.v3.24858.

El-Khoury, V., Pierson, S., Kaoma, T., Bernardin, F., & Berchem, G. (2016). Assessing cellular and circulating miRNA recovery: The impact of the RNA isolation method and the quantity of input material. *Scientific Reports*, *6*. https://doi.org/10.1038/srep19529.

Enderle, D., Spiel, A., Coticchia, C. M., Berghoff, E., Mueller, R., Schlumpberger, M., … Noerholm, M. (2015). Characterization of RNA from exosomes and other extracellular vesicles isolated by a novel spin column-based method. *Plos One*, *10*(8), e0136133. https://doi.org/10.1371/journal.pone.0136133.

Filipe, V., Hawe, A., & Jiskoot, W. (2010). Critical evaluation of Nanoparticle Tracking Analysis (NTA) by NanoSight for the measurement of nanoparticles and protein aggregates. *Pharmaceutical Research*, *27*(5), 796–810. https://doi.org/10.1007/s11095-010-0073-2.

Fiskaa, T., Knutsen, E., Nikolaisen, M. A., Jørgensen, T. E., Johansen, S. D., Perander, M., et al. (2016). Distinct small RNA signatures in extracellular vesicles derived from breast cancer cell lines. *PLoS One*, *11*(8), e0161824. https://doi.org/10.1371/journal.pone.0161824.

Fritz, J. V., Heintz-Buschart, A., Ghosal, A., Wampach, L., Etheridge, A., Galas, D., et al. (2016). Sources and functions of extracellular small RNAs in human circulation. *Annual Review of Nutrition*, *36*, 301–336. https://doi.org/10.1146/annurev-nutr-071715-050711.

Gallart-Palau, X., Serra, A., Wong, A. S. W., Sandin, S., Lai, M. K. P., Chen, C. P., … Sze, S. K. (2015). Extracellular vesicles are rapidly purified from human plasma by PRotein Organic Solvent PRecipitation (PROSPR). *Scientific Reports*, *5*. https://doi.org/10.1038/srep14664.

Gardiner, C., Vizio, D. D., Sahoo, S., Théry, C., Witwer, K. W., Wauben, M., et al. (2016). Techniques used for the isolation and characterization of extracellular vesicles: Results of a worldwide survey. *Journal of Extracellular Vesicles*, *5*. https://doi.org/10.3402/jev.v5.32945.

Greening, D. W., Xu, R., Ji, H., Tauro, B. J., & Simpson, R. J. (2015). A protocol for exosome isolation and characterization: Evaluation of ultracentrifugation, density-gradient separation, and immunoaffinity capture methods. *Methods in Molecular Biology (Clifton, N.J.)*, *1295*, 179–209. https://doi.org/10.1007/978-1-4939-2550-6_15.

Groot Kormelink, T., Arkesteijn, G. J. A., Nauwelaers, F. A., van den Engh, G., Nolte-'t Hoen, E. N. M., & Wauben, M. H. M. (2016). Prerequisites for the analysis and sorting of extracellular vesicle subpopulations by high-resolution flow cytometry. *Cytometry. Part A: The Journal of the International Society for Analytical Cytology*, *89*(2), 135–147. https://doi.org/10.1002/cyto.a.22644.

Gutiérrez-Vázquez, C., Villarroya-Beltri, C., Mittelbrunn, M., & Sánchez-Madrid, F. (2013). Transfer of extracellular vesicles during immune cell-cell interactions. *Immunological Reviews*, *251*(1), 125–142. https://doi.org/10.1111/imr.12013.

Guzman, N., Agarwal, K., Asthagiri, D., Yu, L., Saji, M., Ringel, M. D., & Paulaitis, M. E. (2015). Breast cancer-specific miR signature unique to extracellular vesicles includes "microRNA-like" tRNA fragments. *Molecular Cancer Research: MCR*, *13*(5), 891–901. https://doi.org/10.1158/1541-7786.MCR-14-0533.

Ha, M., & Kim, V. N. (2014). Regulation of microRNA biogenesis. *Nature Reviews Molecular Cell Biology*, *15*(8), 509. https://doi.org/10.1038/nrm3838.

Head, S. R., Komori, H. K., LaMere, S. A., Whisenant, T., Van Nieuwerburgh, F., Salomon, D. R., & Ordoukhanian, P. (2014). Library construction for next-generation sequencing: Overviews and challenges. *Biotechniques*, *56*(2). https://doi.org/10.2144/000114133. 61–passim.

Hill, A. F. (Ed.). (2017). *Exosomes and microvesicles* (Vol. 1545). New York, NY: Springer. https://doi.org/10.1007/978-1-4939-6728-5.

Hill, A. F., Pegtel, D. M., Lambertz, U., Leonardi, T., O'Driscoll, L., Pluchino, S., … Nolte-'t Hoen, E. N. M. (2013). ISEV position paper: Extracellular vesicle RNA analysis and bioinformatics. *Journal of Extracellular Vesicles*, *2*. https://doi.org/10.3402/jev.v2i0.22859.

Hong, X., Schouest, B., & Xu, H. (2017). Effects of exosome on the activation of CD4+ T cells in rhesus macaques: A potential application for HIV latency reactivation. *Scientific Reports*, *7*. https://doi.org/10.1038/s41598-017-15961-x.

Houldcroft, C. J., Beale, M. A., & Breuer, J. (2017). Clinical and biological insights from viral genome sequencing. *Nature Reviews Microbiology*, *15*(3), 183. https://doi.org/10.1038/nrmicro.2016.182.

Hyeon, J., Cho, S. Y., Hong, M. E., Kang, S. Y., Do, I., Im, Y. H., & Cho, E. Y. (2017). NanoString nCounter® approach in breast cancer: A comparative analysis with quantitative real-time polymerase chain reaction, in situ hybridization, and immunohistochemistry. *Journal of Breast Cancer*, *20*(3), 286–296. https://doi.org/10.4048/jbc.2017.20.3.286.

Im, H., Shao, H., Park, Y. I., Peterson, V. M., Castro, C. M., Weissleder, R., & Lee, H. (2014). Label-free detection and molecular profiling of exosomes with a nano-plasmonic sensor. *Nature Biotechnology*, *32*(5), 490–495. https://doi.org/10.1038/nbt.2886.

Iorio, M. V., & Croce, C. M. (2012). MicroRNA dysregulation in cancer: Diagnostics, monitoring and therapeutics. A comprehensive review. *EMBO Molecular Medicine*, *4*(3), 143–159. https://doi.org/10.1002/emmm.201100209.

Ismail, N., Wang, Y., Dakhlallah, D., Moldovan, L., Agarwal, K., Batte, K., … Marsh, C. B. (2013). Macrophage microvesicles induce macrophage differentiation and miR-223 transfer. *Blood*, *121*(6), 984–995. https://doi.org/10.1182/blood-2011-08-374793.

Janas, T., Janas, M. M., Sapoń, K., & Janas, T. (2015). Mechanisms of RNA loading into exosomes. *FEBS Letters*, *589*(13), 1391–1398. https://doi.org/10.1016/j.febslet.2015.04.036.

Janas, T., Janas, T., & Yarus, M. (2012). Human tRNASec associates with HeLa membranes, cell lipid liposomes, and synthetic lipid bilayers. *RNA*, *18*(12), 2260–2268. https://doi.org/10.1261/rna.035352.112.

Jiang, L., Schlesinger, F., Davis, C. A., Zhang, Y., Li, R., Salit, M., … Oliver, B. (2011). Synthetic spike-in standards for RNA-seq experiments. *Genome Research*, *21*(9), 1543–1551. https://doi.org/10.1101/gr.121095.111.

Kajdos, M., Janas, Ł., Kolasa-Zwierzchowska, D., Wilczyński, J. R., & Stetkiewicz, T. (2015). Microvesicles as a potential biomarker of neoplastic diseases and their role in development and progression of neoplasm. *Przegląd Menopauzalny = Menopause Review*, *14*(4), 283–291. https://doi.org/10.5114/pm.2015.56540.

Keller, S., Ridinger, J., Rupp, A.-K., Janssen, J. W. G., & Altevogt, P. (2011). Body fluid derived exosomes as a novel template for clinical diagnostics. *Journal of Translational Medicine*, *9*(86). https://doi.org/10.1186/1479-5876-9-86.

Kim, D.-K., Kang, B., Kim, O. Y., Choi, D., Lee, J., Kim, S. R., … Gho, Y. S. (2013). EVpedia: An integrated database of high-throughput data for systemic analyses of extracellular vesicles. *Journal of Extracellular Vesicles*, *2*. https://doi.org/10.3402/jev.v2i0.20384.

Kondratova, V. N., Serd'uk, O. I., Shelepov, V. P., & Lichtenstein, A. V. (2005). Concentration and isolation of DNA from biological fluids by agarose gel isotachophoresis. *Biotechniques*, *39*(5), 695–699.

Kottel, R. H., Hoch, S. O., Parsons, R. G., & Hoch, J. A. (1978). Serum ribonuclease activity in cancer patients. *British Journal of Cancer*, *38*(2), 280–286.

Kukurba, K. R., & Montgomery, S. B. (2015). RNA sequencing and analysis. *Cold Spring Harbour Protocols*, *2015*(11), 951–969. https://doi.org/10.1101/pdb.top084970.

Kulkarni, M. M. (2011). Digital multiplexed gene expression analysis using the NanoString nCounter system. *Current Protocols in Molecular Biology*. https://doi.org/10.1002/0471142727.mb25b10s94. Chapter 25, Unit25B.10.

Kuroda, T., Yasuda, S., Matsuyama, S., Tano, K., Kusakawa, S., Sawa, Y., … Sato, Y. (2015). Highly sensitive droplet digital PCR method for detection of residual undifferentiated cells in cardiomyocytes derived from human pluripotent stem cells. *Regenerative Therapy*, *2*(Suppl. C), 17–23. https://doi.org/10.1016/j.reth.2015.08.001.

Lai, C. P., Kim, E. Y., Badr, C. E., Weissleder, R., Mempel, T. R., Tannous, B. A., et al. (2015). Visualization and tracking of tumour extracellular vesicle delivery and RNA translation using multiplexed reporters. *Nature Communications*, *6*, 7029. https://doi.org/10.1038/ncomms8029.

Lee, Y., EL Andaloussi, S., & Wood, M. J. A. (2012). Exosomes and microvesicles: Extracellular vesicles for genetic information transfer and gene therapy. *Human Molecular Genetics*, *21*(R1), R125–R134. https://doi.org/10.1093/hmg/dds317.

Levesque-Sergerie, J.-P., Duquette, M., Thibault, C., Delbecchi, L., & Bissonnette, N. (2007). Detection limits of several commercial reverse transcriptase enzymes: Impact on the low- and high-abundance transcript levels assessed by quantitative RT-PCR. *BMC Molecular Biology*, *8*, 93. https://doi.org/10.1186/1471-2199-8-93.

Li, X., Ben-Dov, I. Z., Mauro, M., & Williams, Z. (2015). Lowering the quantification limit of the QubitTM RNA HS assay using RNA spike-in. *BMC Molecular Biology*, *16*, 9. https://doi.org/10.1186/s12867-015-0039-3.

Li, P., Kaslan, M., Lee, S. H., Yao, J., & Gao, Z. (2017). Progress in exosome isolation techniques. *Theranostics*, *7*(3), 789–804. https://doi.org/10.7150/thno.18133.

Liu, W., Tao, K., You, N., Liu, Z., Zhang, H., & Dou, K. (2011). Differences in the properties and mirna expression profiles between side populations from hepatic cancer cells and normal liver cells. *Plos One*, *6*(8), e23311. https://doi.org/10.1371/journal.pone.0023311.

Li, M., Zeringer, E., Barta, T., Schageman, J., Cheng, A., & Vlassov, A. V. (2014). Analysis of the RNA content of the exosomes derived from blood serum and urine and its potential as biomarkers. *Philosophical Transactions of the Royal Society B: Biological Sciences*, *369*(1652). https://doi.org/10.1098/rstb.2013.0502.

Mateescu, B., Kowal, E. J. K., van Balkom, B. W. M., Bartel, S., Bhattacharyya, S. N., Buzás, E. I., … Nolte-'t Hoen, E. N. M. (2017). Obstacles and opportunities in the functional analysis of extracellular vesicle RNA – an ISEV position paper. *Journal of Extracellular Vesicles*, *6*(1). https://doi.org/10.1080/20013078.2017.1286095.

Mestdagh, P., Vlierberghe, P. V., Weer, A. D., Muth, D., Westermann, F., Speleman, F., et al. (2009). A novel and universal method for microRNA RT-qPCR data normalization. *Genome Biology*, *10*(6), R64. https://doi.org/10.1186/gb-2009-10-6-r64.

Minciacchi, V. R., Freeman, M. R., & Di Vizio, D. (2015). Extracellular vesicles in cancer: Exosomes, microvesicles and the emerging role of large oncosomes. *Seminars in Cell & Developmental Biology*, *40*, 41–51. https://doi.org/10.1016/j.semcdb.2015.02.010.

Moldovan, L., Batte, K. E., Trgovcich, J., Wisler, J., Marsh, C. B., & Piper, M. (2014). Methodological challenges in utilizing miRNAs as circulating biomarkers. *Journal of Cellular and Molecular Medicine*, *18*(3), 371–390. https://doi.org/10.1111/jcmm.12236.

Momen-Heravi, F., Balaj, L., Alian, S., Mantel, P.-Y., Halleck, A. E., Trachtenberg, A. J., … Kuo, W. P. (2013). Current methods for the isolation of extracellular vesicles. *Biological Chemistry*, *394*(10), 1253–1262. https://doi.org/10.1515/hsz-2013-0141.

Mulcahy, L. A., Pink, R. C., & Carter, D. R. F. (2014). Routes and mechanisms of extracellular vesicle uptake. *Journal of Extracellular Vesicles*, *3*. https://doi.org/10.3402/jev.v3.24641.

Munafó, D. B., & Robb, G. B. (2010). Optimization of enzymatic reaction conditions for generating representative pools of cDNA from small RNA. *RNA*, *16*(12), 2537–2552. https://doi.org/10.1261/rna.2242610.

Noerholm, M., Balaj, L., Limperg, T., Salehi, A., Zhu, L. D., Hochberg, F. H., … Skog, J. (2012). RNA expression patterns in serum microvesicles from patients with glioblastoma multiforme and controls. *BMC Cancer*, *12*(22). https://doi.org/10.1186/1471-2407-12-22.

Nolte-'t Hoen, E. N. M., Buermans, H. P. J., Waasdorp, M., Stoorvogel, W., Wauben, M. H. M., & 't Hoen, P. A. C. (2012). Deep sequencing of RNA from immune cell-derived vesicles uncovers the selective incorporation of small non-coding RNA biotypes with potential regulatory functions. *Nucleic Acids Research*, *40*(18), 9272–9285. https://doi.org/10.1093/nar/gks658.

Olena, A. F., & Patton, J. G. (2010). Genomic organization of microRNAs. *Journal of Cellular Physiology*, *222*(3), 540–545. https://doi.org/10.1002/jcp.21993.

Opitz, L., Salinas-Riester, G., Grade, M., Jung, K., Jo, P., Emons, G., … Gaedcke, J. (2010). Impact of RNA degradation on gene expression profiling. *BMC Medical Genomics*, *3*(36). https://doi.org/10.1186/1755-8794-3-36.

Parts, L., Hedman, Å. K., Keildson, S., Knights, A. J., Abreu-Goodger, C., van de Bunt, M., … Lindgren, C. M. (2012). Extent, causes, and consequences of small RNA expression variation in human adipose tissue. *PLoS Genetics*, *8*(5), e1002704. https://doi.org/10.1371/journal.pgen.1002704.

Pisitkun, T., Shen, R.-F., & Knepper, M. A. (2004). Identification and proteomic profiling of exosomes in human urine. *Proceedings of the National Academy of Sciences of the United States of America*, *101*(36), 13368–13373. https://doi.org/10.1073/pnas.0403453101.

Popov, A., Szabo, A., & Mandys, V. (2015). Small nucleolar RNA U91 is a new internal control for accurate microRNAs quantification in pancreatic cancer. *BMC Cancer*, *15*. https://doi.org/10.1186/s12885-015-1785-9.

Price, C., & Chen, J. (2014). MicroRNAs in cancer biology and therapy: Current status and perspectives. *Genes & Diseases*, *1*(1), 53–63. https://doi.org/10.1016/j.gendis.2014.06.004.

Pucci, F., Garris, C., Lai, C. P., Newton, A., Pfirschke, C., Engblom, C., … Pittet, M. J. (2016). SCS macrophages suppress melanoma by restricting tumor-derived vesicle-B cell interactions. *Science (New York, N.Y.)*, *352*(6282), 242–246. https://doi.org/10.1126/science.aaf1328.

Quesenberry, P. J., Aliotta, J., Deregibus, M. C., & Camussi, G. (2015). Role of extracellular RNA-carrying vesicles in cell differentiation and reprogramming. *Stem Cell Research & Therapy*, *6*. https://doi.org/10.1186/s13287-015-0150-x.

Ragusa, M., Barbagallo, C., Cirnigliaro, M., Battaglia, R., Brex, D., Caponnetto, A., … Purrello, M. (2017). Asymmetric RNA distribution among cells and their secreted Exosomes: Biomedical meaning and considerations on diagnostic applications. *Frontiers in Molecular Biology*, *4*. https://doi.org/10.3389/fmolb.2017.00066.

Ramón-Núñez, L. A., Martos, L., Fernández-Pardo, Á., Oto, J., Medina, P., España, F., et al. (2017). Comparison of protocols and RNA carriers for plasma miRNA isolation. Unraveling RNA carrier influence on miRNA isolation. *PLoS One*, *12*(10), e0187005. https://doi.org/10.1371/journal.pone.0187005.

Rekker, K., Saare, M., Roost, A. M., Kubo, A.-L., Zarovni, N., Chiesi, A., … Peters, M. (2014). Comparison of serum exosome isolation methods for microRNA profiling. *Clinical Biochemistry*, *47*(1–2), 135–138. https://doi.org/10.1016/j.clinbiochem.2013.10.020.

Rinn, J. L., & Chang, H. Y. (2012). Genome regulation by long noncoding RNAs. *Annual Review of Biochemistry*, *81*. https://doi.org/10.1146/annurev-biochem-051410-092902.

Robbins, P. D., & Morelli, A. E. (2014). Regulation of immune responses by extracellular vesicles. *Nature Reviews Immunology*, *14*(3), 195–208. https://doi.org/10.1038/nri3622.

Sanders, R., Mason, D. J., Foy, C. A., & Huggett, J. F. (2013). Evaluation of digital PCR for absolute RNA quantification. *PLoS One*, *8*(9). https://doi.org/10.1371/journal.pone.0075296.

Schwab, A., Meyering, S. S., Lepene, B., Iordanskiy, S., van Hoek, M. L., Hakami, R. M., et al. (2015). Extracellular vesicles from infected cells: Potential for direct pathogenesis. *Frontiers in Microbiology*, *6*. https://doi.org/10.3389/fmicb.2015.01132.

Shelke, G. V., Lässer, C., Gho, Y. S., & Lötvall, J. (2014). Importance of exosome depletion protocols to eliminate functional and RNA-containing extracellular vesicles from fetal bovine serum. *Journal of Extracellular Vesicles, 3.* https://doi.org/10.3402/jev.v3.24783.

Su, M.-J., Aldawsari, H., & Amiji, M. (2016). Pancreatic cancer cell exosome-mediated macrophage reprogramming and the role of MicroRNAs 155 and 125b2 transfection using nanoparticle delivery systems. *Scientific Reports, 6,* srep30110. https://doi.org/10.1038/srep30110.

Szatanek, R., Baran, J., Siedlar, M., & Baj-Krzyworzeka, M. (2015). Isolation of extracellular vesicles: Determining the correct approach (review). *International Journal of Molecular Medicine, 36*(1), 11–17.

Tang, Y.-T., Huang, Y.-Y., Zheng, L., Qin, S.-H., Xu, X.-P., An, T.-X., … Wang, Q. (2017). Comparison of isolation methods of exosomes and exosomal RNA from cell culture medium and serum. *International Journal of Molecular Medicine, 40*(3), 834–844. https://doi.org/10.3892/ijmm.2017.3080.

Tan, S. C., & Yiap, B. C. (2009). DNA, RNA, and protein extraction: The past and the present. *Journal of Biomedicine and Biotechnology.* https://doi.org/10.1155/2009/574398.

Thakur, B. K., Zhang, H., Becker, A., Matei, I., Huang, Y., Costa-Silva, B., … Lyden, D. (2014). Double-stranded DNA in exosomes: A novel biomarker in cancer detection. *Cell Research, 24*(6), 766–769. https://doi.org/10.1038/cr.2014.44.

Théry, C. (2011). Exosomes: Secreted vesicles and intercellular communications. *F1000 Biology Reports, 3.* https://doi.org/10.3410/B3-15.

Tkach, M., & Théry, C. (2016). Communication by extracellular vesicles: where we are and where we need to go. *Cell, 164*(6), 1226–1232. https://doi.org/10.1016/j.cell.2016.01.043.

Trapnell, C., Hendrickson, D. G., Sauvageau, M., Goff, L., Rinn, J. L., & Pachter, L. (2013). Differential analysis of gene regulation at transcript resolution with RNA-seq. *Nature Biotechnology, 31*(1). https://doi.org/10.1038/nbt.2450.

Trivedi, M., Talekar, M., Shah, P., Ouyang, Q., & Amiji, M. (2016). Modification of tumor cell exosome content by transfection with wt-p53 and microRNA-125b expressing plasmid DNA and its effect on macrophage polarization. *Oncogenesis, 5*(8), e250. https://doi.org/10.1038/oncsis.2016.52.

Turchinovich, A., Weiz, L., Langheinz, A., & Burwinkel, B. (2011). Characterization of extracellular circulating microRNA. *Nucleic Acids Research, 39*(16), 7223–7233. https://doi.org/10.1093/nar/gkr254.

Valadi, H., Ekström, K., Bossios, A., Sjöstrand, M., Lee, J. J., & Lötvall, J. O. (2007). Exosome-mediated transfer of mRNAs and microRNAs is a novel mechanism of genetic exchange between cells. *Nature Cell Biology, 9*(6), 654–659. https://doi.org/10.1038/ncb1596.

van Dongen, H. M., Masoumi, N., Witwer, K. W., & Pegtel, D. M. (2016). Extracellular vesicles exploit viral entry routes for cargo delivery. *Microbiology and Molecular Biology Reviews, 80*(2), 369–386. https://doi.org/10.1128/MMBR.00063-15.

Vickers, K. C., Palmisano, B. T., Shoucri, B. M., Shamburek, R. D., & Remaley, A. T. (2011). MicroRNAs are transported in plasma and delivered to recipient cells by high-density lipoproteins. *Nature Cell Biology, 13*(4), 423–433. https://doi.org/10.1038/ncb2210.

Villarroya-Beltri, C., Gutiérrez-Vázquez, C., Sánchez-Cabo, F., Pérez-Hernández, D., Vázquez, J., Martin-Cofreces, N., … Sánchez-Madrid, F. (2013). Sumoylated hnRNPA2B1 controls the sorting of miRNAs into exosomes through binding to specific motifs. *Nature Communications, 4*(2980). https://doi.org/10.1038/ncomms3980.

Vogel, R., Coumans, F. A. W., Maltesen, R. G., Böing, A. N., Bonnington, K. E., Broekman, M. L., … Pedersen, S. (2016). A standardized method to determine the concentration of extracellular vesicles using tunable resistive pulse sensing. *Journal of Extracellular Vesicles, 5.* https://doi.org/10.3402/jev.v5.31242.

Wahid, F., Shehzad, A., Khan, T., & Kim, Y. Y. (2010). MicroRNAs: Synthesis, mechanism, function, and recent clinical trials. *Biochimica et Biophysica Acta (BBA) - Molecular Cell Research, 1803*(11), 1231–1243. https://doi.org/10.1016/j.bbamcr.2010.06.013.

Wei, Z., Batagov, A. O., Carter, D. R. F., & Krichevsky, A. M. (2016). Fetal bovine serum RNA interferes with the cell culture derived extracellular RNA. *Scientific Reports, 6*(31175). https://doi.org/10.1038/srep31175.

Whiteside, T. L. (2016). Tumor-derived exosomes and their role in cancer progression. *Advances in Clinical Chemistry, 74*, 103–141. https://doi.org/10.1016/bs.acc.2015.12.005.

Witwer, K. W., Buzás, E. I., Bemis, L. T., Bora, A., Lässer, C., Lötvall, J., … Hochberg, F. (2013). Standardization of sample collection, isolation and analysis methods in extracellular vesicle research. *Journal of Extracellular Vesicles, 2.* https://doi.org/10.3402/jev.v2i0.20360.

Wunsch, B. H., Smith, J. T., Gifford, S. M., Wang, C., Brink, M., Bruce, R. L., … Astier, Y. (2016). Nanoscale lateral displacement arrays for the separation of exosomes and colloids down to 20 nm. *Nature Nanotechnology, 11*(11), 936–940. https://doi.org/10.1038/nnano.2016.134.

Yáñez-Mó, M., Siljander, P. R. -M., Andreu, Z., Zavec, A. B., Borràs, F. E., Buzas, E. I., … De Wever, O. (2015). Biological properties of extracellular vesicles and their physiological functions. *Journal of Extracellular Vesicles, 4.* https://doi.org/10.3402/jev.v4.27066.

Zaborowski, M. P., Balaj, L., Breakefield, X. O., & Lai, C. P. (2015). Extracellular vesicles: Composition, biological relevance, and methods of study. *BioScience, 65*(8), 783–797. https://doi.org/10.1093/biosci/biv084.

Zeng, W., & Mortazavi, A. (2012). Technical considerations for functional sequencing assays. *Nature Immunology, 13*(9), 802–807. https://doi.org/10.1038/ni.2407.

Zeringer, E., Li, M., Barta, T., Schageman, J., Pedersen, K. W., Neurauter, A., … Vlassov, A. V. (2013). Methods for the extraction and RNA profiling of exosomes. *World Journal of Methodology, 3*(1), 11–18. https://doi.org/10.5662/wjm.v3.i1.11.

Zhang, J., Li, S., Li, L., Li, M., Guo, C., Yao, J., & Mi, S. (2015). Exosome and exosomal MicroRNA: Trafficking, sorting, and function. *Genomics, Proteomics & Bioinformatics, 13*(1), 17–24. https://doi.org/10.1016/j.gpb.2015.02.001.

Zhao, Z., Yang, Y., Zeng, Y., & He, M. (2016). A microfluidic ExoSearch chip for multiplexed exosome detection towards blood-based ovarian cancer diagnosis. *Lab on a Chip, 16*(3), 489–496. https://doi.org/10.1039/c5lc01117e.

7

Nanotechnology Platforms for Cancer Exosome Analyses

Hyungsoon Im, Katherine S. Yang, Hakho Lee, Cesar M. Castro

MASSACHUSETTS GENERAL HOSPITAL, BOSTON, MA, UNITED STATES

CHAPTER OUTLINE

1. Introduction... 119

2. Nanoplasmonic Sensing.. 120

3. Electrochemical Sensing 122

4. Immunomagnetic Exosome RNA Analysis.. 124

5. Conclusions... 126

References.. 126

1. Introduction

A common theme in early cancer diagnosis, personalized cancer treatment, and treatment response is the need for reliable and specific bioassay (Basik et al., 2013; Bidard, Weigelt, & Reis-Filho, 2013; Pantel & Alix-Panabieres, 2013). The current gold standard is tissue biopsy, which is invasive and inherently difficult to repeat. These factors often limit tissue biopsy from capturing the heterogeneity and temporal evolution of a tumor (Gerlinger et al., 2012). Conventional imaging techniques offer noninvasive alternatives, but they are costly when serially used and insensitive to detect subtle invasion, micrometastases, and early stages of cancer formation (Brindle, 2008; Condeelis & Weissleder, 2010; Weissleder & Pittet, 2008). Liquid biopsies, based on novel biomarkers in circulation derived from exosomes (or more broadly extracellular vesicles, EVs), represent a wealth of information about the primary, as well as metastatic tumor sites (Théry, 2015). EVs carry cell-specific cargo (lipids, proteins, and genetic material), which can be exploited as a minimally invasive means to probe the molecular status of tumors (Alderton, 2015; Im et al., 2014; Rahbari, Rahbari, Reissfelder, Weitz & Kahlert, 2016; Shao et al., 2015). Importantly, the tumor-derived vesicle concentrations in biological fluids and their molecular profiles could indicate tumor burden and treatment response (Im et al., 2014; Shao et al., 2015).

Diagnostic and Therapeutic Applications of Exosomes in Cancer. https://doi.org/10.1016/B978-0-12-812774-2.00007-9

Routine EV analyses, however, have not been achieved yet mainly because of the lack of tools that can analyze EVs with high sensitivity and accuracy. Current challenges include lengthy and extensive processing for EV isolation and low sensitivity with conventional analytical methods (e.g., Western blotting, ELISA) that often hinder throughput, multiplexed analyses. Most prior EV studies thus focused on RNA analysis to harness the power of polymerase chain reaction (PCR) amplification. Conversely, EV proteomic analyses through conventional approaches, where no amplification safety net exists, have been facing technical hurdles. Several recent studies have demonstrated EV protein screening using flow cytometry (Arraud, Gounou, Turpin, & Brisson, 2016; Nolte et al., 2012; Van Der Vlist, Nolte, Stoorvogel, Arkesteijn, & Wauben, 2012) but have involved specialized high-end equipments to handle EVs' small, but dispersed size (Pospichalova et al., 2015). More recent studies have applied novel nanomaterials [e.g., graphene oxide (Zhang, He, & Zeng, 2016) nanorod particles (Liang et al., 2017)] or microfluidic analytical systems (Son et al., 2016; Su, Goldberg, & Stoltz, 2016; Zhao, Yang, Zeng, & He, 2016) to detect EVs and identify their protein contents. Development and advancement of such diverse, ultrasensitive detection technologies could offer additional insight into understanding the heterogeneity and production dynamics of EVs.

From the translational research perspective, EVs represent novel diagnostic biomarkers poised for further exploitation. Furthermore, their integral roles in cell to cell communication (Pucci et al., 2016; Tkach & Théry, 2016), creation of the premetastatic niche (Costa-Silva et al., 2015), and high potential as drug delivery carriers (Andaloussi, Mäger, Breakefield, & Wood, 2013) offer interdisciplinary opportunities to generate novel platforms that align with patient preferences (e.g., liquid biopsies) and biorepository needs (e.g., precious specimen amounts).

In this chapter, we will describe three platforms recently developed for the analysis of EV protein and RNA biomarkers directly from clinical specimens and discuss factors that could facilitate their translation into routine clinic use.

2. Nanoplasmonic Sensing

Surface plasmon resonance (SPR) sensors are sensitive to local refractive index changes near the sensor surface. In biosensing applications, binding of target molecules to affinity ligands immobilized on the sensor surface changes the local refractive index, which can be detected by a spectral shift of optical resonance. SPR sensing does not require a secondary label for detection; this allows rapid, label-free sensing with minimal sample processing. With the label-free sensing capability, SPR sensors have been widely used to characterize molecular interactions between antibodies, antigens, proteins, nucleic acids, and small molecules (Cooper, 2002; Homola, 2008).

Recently, several reports showed that SPR is an attractive EV detection technology for rapid label-free analyses (Grasso et al., 2015; Im et al., 2014; Rupert et al., 2014; Sina et al., 2016; Zhu et al., 2014). Among the various SPR systems, nanohole-based SPR sensor, named nano-plasmonic exosome (nPLEX) system, is of particular interest because of high detection sensitivity for EVs directly from clinical samples in a high-throughput manner.

The first generation nPLEX sensor comprises subwavelength nanohole arrays made in an opaque gold film. Compared with conventional SPR systems, nPLEX is more uniquely suited for EV analysis, as the sensing depth can be readily matched to target EV size to maximize the detection sensitivity. Moreover, it operates in a colinear transmission mode on nanohole structures, rather than in angular reflectance mode commonly used in conventional SPR platforms, which allows for system miniaturization and the construction of densely packed sensing arrays for high-throughput analyses. We have shown that nPLEX analyses are: (1) exquisitely sensitive, (2) allow for profiling of dozens of proteins in EVs to identify the cell of origin, and (3) facilitate assessment of tumor burden over time.

More recently, we developed the second generation of the system. The system, called new nanoplasmonic sensor (NPS), is further optimized for robust, high-throughput analyses of EVs in clinical samples (Fig. 7.1) (Yang et al., 2017). High-throughput chip fabrication has been implemented to meet the patient volume demands of clinical use. Interference lithography was used to fabricate periodic nanostructures in a wafer-scale batch process (Menezes, Barea, Chillcce, Frateschi, & Cescato, 2012; Yanik et al., 2011). Combined with

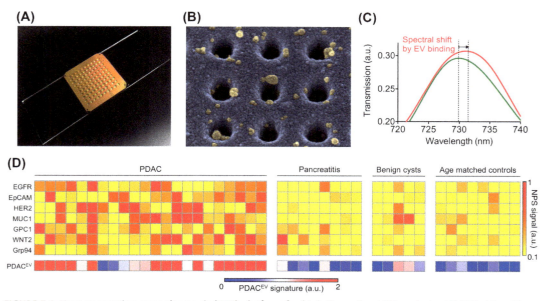

FIGURE 7.1 Next-generation nanoplasmonic (NPS) platform for high-throughput EV analysis. (A) NPS chip with 100 sensing arrays. Solutions printed on a gold film with a microarray spotter. (B) Scanning electron micrograph of the nanohole sensor surface with bound EVs. (C) EV binding on the nanohole surface red-shifts an optical resonance peak. (D) Heatmap analysis of EV markers. The PDAC[EV] signature, a combined marker panel of EPCAM, EGFR, MUC1, GPC1, and WNT2, differentiates PDAC patients from pancreatitis and control groups. *Reproduced from Yang, K. S., et al. (2017). Multiparametric plasma EV profiling facilitates diagnosis of pancreatic malignancy.* Science Translational Medicine, 9 *(391), eaal3226.*

conventional semiconductor fabrication processes, a large number of chips can be fabricated at decreased cost as production rates scale up (Im et al., 2011, 2014). This process allows integration of a vast number of sensing arrays similar to protein microarray sensors and was used to detect pancreatic cancer on chips containing 100 measurement sites. This improved sensor yielded data for 25 antibodies or markers in quadruplicate per patient sample. Further improvements to the system included the use of a molecular printer to automatically dispense antibodies and EVs onto the sensor surface. Finally, a piezoelectric microscope stage was incorporated into the system for automated scanning of the sensor array and collection of the transmission data.

The NPS system demonstrated the utility of the device for rapid and sensitive measurement of 7 biomarkers in 100 clinical populations. Through the analysis of tumor-derived EVs from more than 100 patients with pancreatic diseases and cancers, we identified a marker combination that showed higher accuracy for detecting pancreatic cancer than any single biomarker. NPS analysis correctly classified benign pancreatic disease versus pancreatic adenocarcinoma with a sensitivity of 86% (CI, 65%–95%), specificity of 81% (CI, 58%–95%), and accuracy of 84% (CI, 69%–93%), using a five biomarker PDAC^{EV} signature consisting of antibodies against EpCAM, EGFR, MUC1, GPC1, WNT-2. This type of analysis will be expanded into larger clinical trials of pancreatic and other cancers.

3. Electrochemical Sensing

Electrochemical sensing approaches that provide rapid assays and affordable readout systems could be an effective detection modality. Electrochemical sensors operate by reacting with the target substance of interest and producing an electrical signal proportional to the target material. Miniaturized electrochemical sensors have been used for decades to routinely detect toxic gases and chemicals, as well as biomolecules. They can prove highly sensitive when combined with certain enzyme reporters for signal amplification.

In EV analysis, a hindrance to the clinical translation of new technologies is due in part to the additional procedures and equipment needed for EV isolation and the technical complexities and high costs associated with chip fabrication and analytical instruments. A recently developed integrated magnetic-electrochemical exosome (iMEX) platform is a miniaturized analytical system that can rapidly isolate and detect EVs in clinical specimens (Fig. 7.2) (Jeong et al., 2016). It combines magnetic isolation and electrochemical sensing in an integrated approach. The iMEX can isolate cell-specific EVs directly from plasma samples using magnetic microbeads. It also achieves high detection sensitivity in a parallel measurement format. A portable iMEX system was able to detect <10^5 vesicles from 10 μL of samples in 1 h.

The first iMEX prototype contains eight sensing elements for parallel measurements. Each sensor is equipped with a potentiostat and reads signals instantly through metal electrodes. A magnet holder is placed underneath the electrode to concentrate magnetic beads on the electrode surface for improved detection sensitivity. The iMEX system

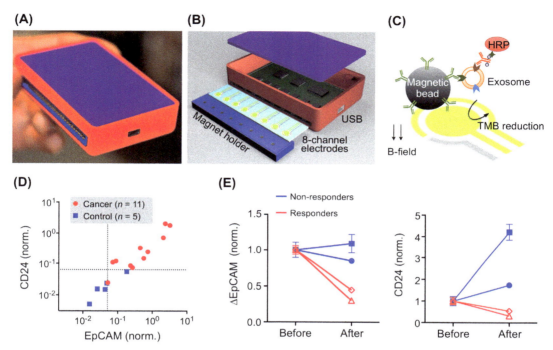

FIGURE 7.2 Integrated magnetic-electrochemical exosome (iMEX) platform. (A) The packaged iMEX device has a small form factor ($9 \times 6 \times 2\,cm^3$). (B) Integrated circuits to simultaneously measure signals from eight electrodes. A small cylindrical magnet is located below each electrode to concentrate immunomagnetically captured EVs. (C) EVs are captured on magnetic beads and labeled with HRP enzyme for electrochemical detection. Magnetic beads are coated with capture antibodies (e.g., CD63 antibody). (D) Plasma samples from ovarian cancer patients and healthy controls were analyzed by the iMEX assay. (E) EpCAM and CD24 levels in EVs were analyzed before and after treatment. All measurements were in duplicate. *Reproduced from Jeong, S., et al. (2016). Integrated magneto–electrochemical sensor for exosome analysis.* ACS Nano, 10*(2), 1802–1809.*

showed comparable performance to commercially available electrochemical systems but is much smaller and cost-effective.

For molecular profiling, EVs are first captured onto immunomagnetic beads coated with antibodies against CD63, one of the markers enriched in exosomes. The captured EVs are then labeled by biotinylated antibodies against specific cancer biomarkers followed by incubation with streptavidin-conjugated horseradish peroxidase (HRP) enzymes. The magnetic bead-Ev conjugates are mixed with a 3,3′,5,5′-tetramethylbenzidine solution and loaded on the top of electrodes for electrical current measurements.

In clinical applications, the iMEX assay isolated and characterized EVs directly from plasma in a rapid, high-throughput manner. Clinical plasma samples were aliquoted into 10 μL volumes; each aliquot was assessed for a target protein marker. As proof of concept, the iMEX assay was tested for detecting cancer EVs in plasma samples. In a test with plasma samples from ovarian cancer patients ($n=11$) and healthy controls ($n=5$), the iMEX showed elevated express levels of EpCAM and CD24 in EVs from ovarian cancer patients, in concordance with previous studies using nPLEX (Im et al., 2014). In addition, serial testing from four ovarian cancer patients undergoing drug treatment showed that EV expression levels of EpCAM and CD24 increased in nonresponding patients, whereas both markers decreased in responding patients.

The low cost, portability, and integrated assay render the iMEX system highly attractive for clinical analyses of EVs. A unique advantage of iMEX is the capability of EV isolation and detection in a single platform. The magnetic isolation simplifies assay procedures and improves detection sensitivities. The electrochemical sensing facilitates high-throughput screening using a miniaturized device. With the unique features, the iMEX system is ready for testing large cohort clinical samples.

4. Immunomagnetic Exosome RNA Analysis

Recently studies have showed that EVs contain RNAs (including mRNA and miRNA) and DNAs (Balaj et al., 2011; Valadi et al., 2007) inside of vesicles. In addition to EV proteins, these nucleic acids are regarded as markers that highly reflect the underlying disease. Several studies have identified specific mRNA and miRNA markers in EVs for different cancer subtypes (Bryant et al., 2012; Silva et al., 2011; Skog et al., 2008).

A microfluidic platform, immunomagnetic exosome RNA (iMER), has been developed for detection of nucleic acid markers in EVs (Fig. 7.3). The iMER system integrates three features on the single microfluidic chip: immunomagnetic tumor EV enrichment, RNA purification, and RNA marker detection using real-time RT-qPCR. The iMER platform can capture EVs directly from 100 μL of serum with >93% capture efficiency. A single chip achieves an integrated work flow, from EV isolation to RNA analysis, in 2 h (Shao et al., 2015).

The iMER chip uses magnetic microbeads (3 μm in diameter) to enrich tumor-specific EVs. For glioblastoma multiforme (GBM) samples, the magnetic microbes are coated with antibodies against EGFR and EGFRvIII to capture and enrich GBM-specific EVs in serum samples. The enriched tumor EVs are lysed in the chip and introduced to a glass bead filter in which mRNAs are captured onto glass beads by electrostatic interaction between mRNAs and glass beads. The purified mRNAs are then reverse transcribed, amplified, and measured by qPCR. The platform uses torque-activated valves to control buffer flow and operation in each step (Shao et al., 2015).

Analyses of EVs have examined their relationship with parental cell origin and their potential application as cancer diagnostic markers. As described earlier, EV proteins have been examined with the different sensing platforms and showed good correlation of protein profiles between EVs and their parental cells (Im et al., 2014; Jeong et al., 2016; Shao et al., 2012). Similarly, the iMER assay showed that mRNA profiles between cells and EVs showed high correlation ($R^2 > 0.9$). In addition, iMER platform leveraged mRNA as predictive markers for drug resistance. Because the detection of EV proteins related to drug resistance poses technical challenges due to their scant amount, the iMER platform uses mRNA as the counterpart to proteins. In a pilot clinical study (Shao et al., 2015), the

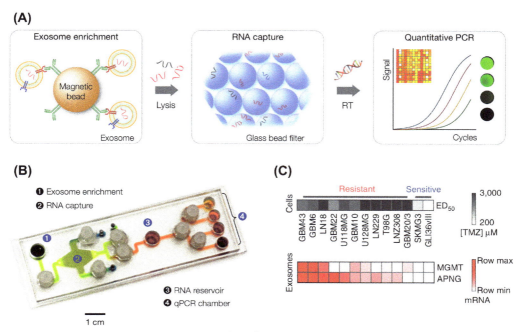

FIGURE 7.3 Immunomagnetic exosome RNA (iMER) platform. (A) Schematic of iMER assay. Cancer EVs were first isolated by immunomagnetic microbeads. After EV lysis, RNA were captured by glass beads and analyzed through quantitative on-chip PCR. (B) The microfluidic iMER prototype. Torque-activated valves were used to control the flow of solutions. (C) iMER analysis showed that the levels of MGMT, APNG, or both were elevated in EVs from resistant cell lines. *Reproduced from Shao, H., et al. (2015). Chip-based analysis of exosomal mRNA mediating drug resistance in glioblastoma.* Nature Communications, 6, *6999.*

iMER assay measured mRNA profiles of GBM-derived EVs and their dynamic changes on treatment initiation. The study identified key EV mRNA markers, O^6-methylguanine DNA methyltransferase (MGMT,) and alkylpurine-DNA-N-glycosylase (APNG) potentially predictive of temozolomide resistance and showed the capacity of EV-RNA for probing the epigenetic status of primary tumors. The platform was also applied to profile EVs in blood from GBM patients and healthy controls and showed (1) elevated levels of markers in GBM patients; (2) methylation status through MGMT mRNA in EVs; and (3) longitudinal assessment of MGMT and APNG mRNA in EVs for potential use of the markers as predictor variables.

The integrated and miniaturized on-chip RNA processing capacity of the iMER platform affords various advantages, including simplicity, rapid sample analysis (~2 h), minimal sample volume requirement (~100 μL for multiplexed analysis), and high sensitivity. The platform demonstrated the potential use of the technology for both predictive and diagnostic applications. Combining iMER's mRNA detection modalities with EV's protein detection can provide comprehensive snapshots of tumor. Such potential advances will facilitate both clinical applications and biological studies of EVs.

5. Conclusions

Conventional analytical methods (e.g., Western blot and ELISA) are often not amenable to comprehensive EV profiling, given their requirements of extensive purification, large sample volumes, or different assay schemes per detection targets. Beside the platforms reviewed here, many EV biosensors have been developed to facilitate EV preparation and analyses (Ko et al., 2016; Lee, Kim, Jeong, & Rhee, 2016; Lee, Shao, Weissleder & Lee, 2015; Liang et al., 2017; Zhao et al., 2016; Zhu et al., 2014; Zong et al., 2016). Such technology development will help deepening our understanding of EV biology, as well as assessing clinical values of EVs.

Widespread adoption of EV diagnostics into practice would require more than proof-of-concept systems, and integration of other technical components such as assay automation, minimal hands-on processing, and data analytics should be addressed. For clinical translation and commercialization, a host of other key issues should be considered in prototype systems. These include cost and protocols that align with clinical workflow. In a cost perspective, developing robust, modular platforms that require little maintenance and used by end users of various skill sets could help justify purchasing. Assays based on highly specific preanalytical handling from unique specimen collection methods to sample processing could also limit widespread clinical adoption. This is especially important to EV analysis, as the efficient isolation and purification of EVs often becomes technical bottlenecks due to nanoscale sizes. For rigorous clinical studies, it is also important to establish the standardization of methods, including sample collection and EV isolation protocols, statistical consideration, and randomized clinical trials in large patient cohorts. Practically, multiinstitutional or international efforts would be required to accelerate clinical translation of new nanotechnology. It is thus encouraging to witness the convergence in multidisciplines ranging from engineering and basic sciences to new technology development for EV research.

References

Alderton, G. K. (2015). Diagnosis: Fishing for exosomes. *Nature Reviews Cancer, 15,* 453.

Andaloussi, S. E. L., Mäger, I., Breakefield, X. O., & Wood, M. J. A. (2013). Extracellular vesicles: Biology and emerging therapeutic opportunities. *Nature Reviews Drug Discovery, 12,* 347–357.

Arraud, N., Gounou, C., Turpin, D., & Brisson, A. R. (2016). Fluorescence triggering: A general strategy for enumerating and phenotyping extracellular vesicles by flow cytometry. *Cytometry, Part A, 89,* 184–195.

Balaj, L., et al. (2011). Tumour microvesicles contain retrotransposon elements and amplified oncogene sequences. *Nature Communications, 2,* 180.

Basik, M., et al. (2013). Biopsies: Next-generation biospecimens for tailoring therapy. *Nature Reviews Clinical Oncology, 10,* 437–450.

Bidard, F. C., Weigelt, B., & Reis-Filho, J. S. (2013). Going with the flow: From circulating tumor cells to DNA. *Science Translational Medicine, 5,* 207ps14.

Brindle, K. (2008). New approaches for imaging tumour responses to treatment. *Nature Reviews Cancer, 8,* 94–107.

Bryant, R. J., et al. (2012). Changes in circulating microRNA levels associated with prostate cancer. *British Journal of Cancer, 106*, 768–774.

Condeelis, J., & Weissleder, R. (2010). In vivo imaging in cancer. *Cold Spring Harbor Perspectives in Biology, 2*, a003848.

Cooper, M. A. (2002). Optical biosensors in drug discovery. *Nature Reviews Drug Discovery, 1*, 515–528.

Costa-Silva, B., et al. (2015). Pancreatic cancer exosomes initiate pre-metastatic niche formation in the liver. *Nature Cell Biology, 17*, 816–826.

Gerlinger, M., et al. (2012). Intratumor heterogeneity and branched evolution revealed by multiregion sequencing. *The New England Journal of Medicine, 366*, 883–892.

Grasso, L., et al. (2015). Molecular screening of cancer-derived exosomes by surface plasmon resonance spectroscopy. *Analytical and Bioanalytical Chemistry, 407*, 5425–5432.

Homola, J. (2008). Surface plasmon resonance sensors for detection of chemical and biological species. *Chemical Reviews, 108*, 462–493.

Im, H., et al. (2011). Template-stripped smooth Ag nanohole arrays with silica shells for surface plasmon resonance biosensing. *ACS Nano, 5*, 6244–6253.

Im, H., et al. (2014). Label-free detection and molecular profiling of exosomes with a nano-plasmonic sensor. *Nature Biotechnology, 32*, 490–495.

Jeong, S., et al. (2016). Integrated magneto–electrochemical sensor for exosome analysis. *ACS Nano, 10*, 1802–1809.

Ko, J., et al. (2016). Smartphone-enabled optofluidic exosome diagnostic for concussion recovery. *Scientific Reports, 6*, 31215.

Lee, J. H., Kim, J. A., Jeong, S., & Rhee, W. J. (2016). Simultaneous and multiplexed detection of exosome microRNAs using molecular beacons. *Biosensors & Bioelectronics, 86*, 202–210.

Lee, K., Shao, H., Weissleder, R., & Lee, H. (2015). Acoustic purification of extracellular microvesicles. *ACS Nano, 9*, 2321–2327.

Liang, K., et al. (2017). Nanoplasmonic quantification of tumour-derived extracellular vesicles in plasma microsamples for diagnosis and treatment monitoring. *Nature Biomedical Engineering, 1*, 0021.

Menezes, J. W., Barea, L. A. M., Chillcce, E. F., Frateschi, N., & Cescato, L. (2012). Comparison of plasmonic arrays of holes recorded by interference lithography and focused ion beam. *IEEE Xplore: IEEE Photonics Journal, 4*, 544–551.

Nolte, E. N. M., et al. (2012). Quantitative and qualitative flow cytometric analysis of nanosized cell-derived membrane vesicles. *Nanomedicine, 8*, 712–720.

Pantel, K., & Alix-Panabieres, C. (2013). Real-time liquid biopsy in cancer patients: Fact or fiction? *Cancer Research, 73*, 6384–6388.

Pospichalova, V., et al. (2015). Simplified protocol for flow cytometry analysis of fluorescently labeled exosomes and microvesicles using dedicated flow cytometer. *Journal of Extracellular Vesicles, 4*, 25530.

Pucci, F., et al. (2016). SCS macrophages suppress melanoma by restricting tumor-derived vesicle–B cell interactions. *Science, 352*, 242–246.

Rahbari, M., Rahbari, N., Reissfelder, C., Weitz, J., & Kahlert, C. (2016). Exosomes: Novel implications in diagnosis and treatment of gastrointestinal cancer. *Langenbeck's Archives of Surgery, 401*, 1097–1110.

Rupert, D. L. M., et al. (2014). Determination of exosome concentration in solution using surface plasmon resonance spectroscopy. *Analytical Chemistry, 86*, 5929–5936.

Shao, H., et al. (2012). Protein typing of circulating microvesicles allows real-time monitoring of glioblastoma therapy. *Nature Medicine, 18*, 1835–1840.

Shao, H., et al. (2015). Chip-based analysis of exosomal mRNA mediating drug resistance in glioblastoma. *Nature Communications, 6*, 6999.

Silva, J., et al. (2011). Vesicle-related microRNAs in plasma of nonsmall cell lung cancer patients and correlation with survival. *The European Respiratory Journal, 37,* 617–623.

Sina, A. A. I., et al. (2016). Real time and label free profiling of clinically relevant exosomes. *Scientific Reports, 6,* 30460.

Skog, J., et al. (2008). Glioblastoma microvesicles transport RNA and proteins that promote tumour growth and provide diagnostic biomarkers. *Nature Cell Biology, 10,* 1470–1476.

Son, K. J., et al. (2016). Microfluidic compartments with sensing microbeads for dynamic monitoring of cytokine and exosome release from single cells. *Analyst, 141,* 679–688.

Su, J., Goldberg, A. F. G., & Stoltz, B. M. (2016). Label-free detection of single nanoparticles and biological molecules using microtoroid optical resonators. *Light Science & Applications, 5,* e16001.

Théry, C. (2015). Cancer: Diagnosis by extracellular vesicles. *Nature, 523,* 161–162.

Tkach, M., & Théry, C. (2016). Communication by extracellular vesicles: Where we are and where we need to go. *Cell, 164,* 1226–1232.

Valadi, H., et al. (2007). Exosome-mediated transfer of mRNAs and microRNAs is a novel mechanism of genetic exchange between cells. *Nature Cell Biology, 9,* 654–659.

Van Der Vlist, E. J., Nolte, E. N. M., Stoorvogel, W., Arkesteijn, G. J. A., & Wauben, M. H. M. (2012). Fluorescent labeling of nano-sized vesicles released by cells and subsequent quantitative and qualitative analysis by high-resolution flow cytometry. *Nature Protocols, 7,* 1311–1326.

Weissleder, R., & Pittet, M. J. (2008). Imaging in the era of molecular oncology. *Nature, 452,* 580–589.

Yang, K. S., et al. (2017). Multiparametric plasma EV profiling facilitates diagnosis of pancreatic malignancy. *Science Translational Medicine, 9,* eaal3226.

Yanik, A. A., et al. (2011). Seeing protein monolayers with naked eye through plasmonic Fano resonances. *Proceedings of the National Academy of Sciences of the United States of America, 108,* 11784–11789.

Zhang, P., He, M., & Zeng, Y. (2016). Ultrasensitive microfluidic analysis of circulating exosomes using a nanostructured graphene oxide/polydopamine coating. *Lab on a Chip, 16,* 3033–3042.

Zhao, Z., Yang, Y., Zeng, Y., & He, M. (2016). A microfluidic ExoSearch chip for multiplexed exosome detection towards blood-based ovarian cancer diagnosis. *Lab on a Chip, 16,* 489–496.

Zhu, L., et al. (2014). Label-free quantitative detection of tumor-derived exosomes through surface plasmon resonance imaging. *Analytical Chemistry, 86,* 8857–8864.

Zong, S., et al. (2016). Facile detection of tumor-derived exosomes using magnetic nanobeads and SERS nanoprobes. *Analytical Methods, 8,* 5001–5008.

8

Exosome RNAs as Biomarkers and Targets for Cancer Therapy

Akhil Srivastava, Narsireddy Amreddy, Rebaz Ahmed, Mohammed A. Razaq, Katherine Moxley, Rheal Towner, Yan D. Zhao, Allison Gillaspy, Ali S. Khan, Anupama Munshi, Rajagopal Ramesh

UNIVERSITY OF OKLAHOMA HEALTH SCIENCES CENTER, OKLAHOMA CITY, OK, UNITED STATES

CHAPTER OUTLINE

1. Introduction 130
2. History of RNA in Exosomes 130
3. RNA Species Present in Exosomes 132
   3.1 Small Noncoding RNAs 132
      3.1.1 Therapeutic Application of miRNA 138
      3.1.2 Exosome-Based Delivery of Therapeutic miRNAs 140
      3.1.3 Exosomes for Imaging and Chemotherapeutics 142
   3.2 Long Noncoding RNA in Exosomes 142
4. Sources of Exosomes 144
5. Technology for the Study of Exosomal miRNA 145
   5.1 Microarray and Sequencing for Exosomal Content Profiling 145
   5.2 In Silico Analysis 150
      5.2.1 The Database for Annotation, Visualization, and Integrated Discovery 151
      5.2.2 Ingenuity Pathway Analysis 151
6. Challenges and Perspective 151
7. Conclusion 152
Acknowledgments 152
Conflict of Interest 153
References 153
Further Reading 159

Diagnostic and Therapeutic Applications of Exosomes in Cancer. https://doi.org/10.1016/B978-0-12-812774-2.00008-0

1. Introduction

In a multicellular system, cellular communication plays an intrinsic and essential role in survival, growth, and normal cellular physiology. Signaling molecules, such as peptides, hormones, growth factors, and nucleic acids, are the key entities of cellular communication. However, how these molecules are involved in the cellular communication network and the mechanism through which cell signaling occurs remains a key question. The endocrine, paracrine, juxtacrine, and synaptic pathways are thought to be involved in cellular communication. In the past decade, research performed with small cellular vesicles, specifically exosomes, revealed a new aspect of intercellular communication and has convincingly established that exosomes carry bioactive molecules as cellular cargo that is deeply involved in communication between cells and with the extracellular environment (Sun, Zhuang, et al., 2013; Valadi et al., 2007). Studies have shown the presence of small peptides, nucleic acids (DNAs and various species of RNAs), and lipids packaged into the lumen of exosomes that are released from the cells and then shuttle intracellularly and between the extracellular milieu, delivering biomolecules (Kahlert & Kalluri, 2013). Interestingly, the molecular profile of the exosomes contents not only mirrors the cell of their origin but also reflects the physiological conditions of the cell at the time of release (Zhang et al., 2015). Furthermore, it has been observed that the release of exosomes considerably depends on the state of the cell and is increased when cells experience some type of physical or biological stress (Melo et al., 2014). The content of exosomes was found to remarkably differ when the exosomes are released from the cells of a diseased organ or tissue compared with neighboring normal cells from the same individual organism (Frydrychowicz, Kolecka-Bednarczyk, Madejczyk, Yasar, & Dworacki, 2015; Yang & Robbins, 2011).

It has been widely hypothesized that this difference in content can serve as unique molecular signature that can be exploited for the development of biomarkers for emerging modalities for early detection of disease and/or understanding the therapeutic responses of the treatment currently being administered to the patient and will aid physicians in offering personalized medicine to their patients (Taylor & Gercel-Taylor, 2008). After researchers realized that exosomes can carry various molecules as cargo, nucleic acids emerged as one of the major components of the cargo. RNA and its various species were one of the first molecules that were recognized to be present in exosomes, followed by the recent discovery of DNAs that were packaged into exosomes and subsequently transferred from the cell (Théry, Zitvogel, & Amigorena, 2002; Thakur et al., 2014). In the present chapter, we will critically discuss nucleic acids, especially RNAs, as cargo molecules in the exosomes, and the utility of exosomal cargo as a biomarker for the condition of cells and the individual organism of origin.

2. History of RNA in Exosomes

In 1984, Rose Johnstone, Adam, and Pan originally discovered exosomes in maturing reticulocytes. These vesicles were initially considered the trash bag of the cell, used to expunge toxic and harmful molecules from the cells. It took some time for researchers to grasp the potential of exosomes in cellular communication and cell signaling (Lakkaraju & Rodriguez-Boulan, 2008), although the exact mechanism of how exosomes were involved in cell-to-cell communication remained elusive. Exosomes were observed to bind to recipient cells through receptor–ligand interactions or through the cell adhesion molecules on their surface. The exosomes are subsequently internalized by recipient cells through endocytosis. Inside the recipient cells, exosomes release their content and ultimately induce biological effects/activities, such as the development of cellular tolerance and antigen presentation on T cells (Kalluri, 2016).

Later, deeper analysis and utilization of advanced techniques, including liquid chromatography–mass spectrometry (LC-MS), led to the identification of various proteins and lipid molecules inside the exosome lumen. RNAs were among the first nucleic acids identified in exosomes. Valadi et al. (2007) reported the presence of various species of RNAs in exosomes derived from MC/9 mouse cells and HMC-1 human mast cells, and termed these RNA "exosomal shuttle RNA", or esRNA. The researchers further identified functionally active mRNAs in sufficient amounts capable of coding for approximately 1300 genes present in exosomes. Ribosomal RNAs (18S and 28S) were not detected, but small RNAs, especially microRNA (miRNAs) were detected within the exosomes. Their observation that RNA from mast cell–derived exosomes can be transferred to other mouse mast cells, as well as to human mast cells, and translated into the functionally active proteins in both cell types was a major discovery supporting the hypothesis that exosomes are involved in cellular communications.

Since then, numerous RNA species have been identified in exosomes isolated from different sources. These species include long noncoding RNA (lncRNA), small nuclear RNA (snRNA), small nucleolar RNA (snoRNA), long intergenic noncoding RNA (lincRNA), piwi-interacting RNA (piRNA), and ribosomal and transfer RNA (tRNA) (Lambertz et al., 2015; M. Li et al., 2014; Silva & Melo, 2015) (Fig. 1). Among these, miRNAs and lncRNAs have been studied extensively for their role in the regulation of various cellular processes and signaling pathways involved in maintaining cellular homeostasis and disease. In the following section, the putative functional significance of each of miRNA and lncRNA and their utility as biomarkers to assess cell and body condition will be discussed in detail.

The ideal biomarker must be accessible using noninvasive protocols, inexpensive to quantify, specific to the disease of interest, translatable from model systems to humans, and capable of providing reliable early indications of disease before clinical symptoms appear. Biomarkers that can be used to stratify disease and assess response to therapeutics are medically valuable. Although most current biomarkers are protein based, challenges for developing new protein-based biomarkers include the complexity of protein composition in most biological samples (especially blood), the assorted posttranslational modifications of proteins, and the low abundance of many proteins of interest in serum and plasma. Nucleic acid (miRNA)–based biomarkers on the other hand would be useful, as they are more stable in blood and other biofluids, especially when enclosed in vesicles such as exosomes.

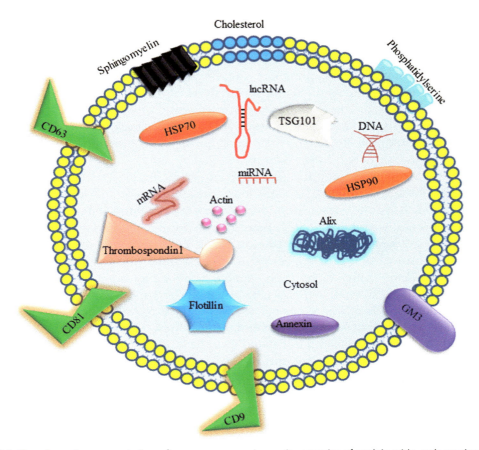

FIGURE 1 The schematic representation of an exosome carrying various species of nucleic acids, such as microRNA (miRNA), long noncoding RNAs (lncRNAs), and fragments of mRNA and DNA.

3. RNA Species Present in Exosomes

3.1 Small Noncoding RNAs

Small noncoding RNAs species are present in exosomes dominated by miRNAs. miRNAs are ~18–22 nucleotide (nt)-long, small, noncoding RNA molecules that are involved in the posttranscriptional negative regulation of mRNA through targeted mRNA cleavage, translational repression, and mRNA deadenylation. A single miRNA can target and regulate multiple mRNAs; conversely, a single mRNA can be targeted by multiple miRNAs indicating complex and intense involvement of miRNAs in regulation of gene expression. miRNA biogenesis is a two-step process that begins inside the nucleus and ends in the cytosol. The first step is initiated via transcription by RNA polymerase II enzyme, generating primary miRNA transcripts known as pri-miRs. pri-miRs are processed by ribonuclease III

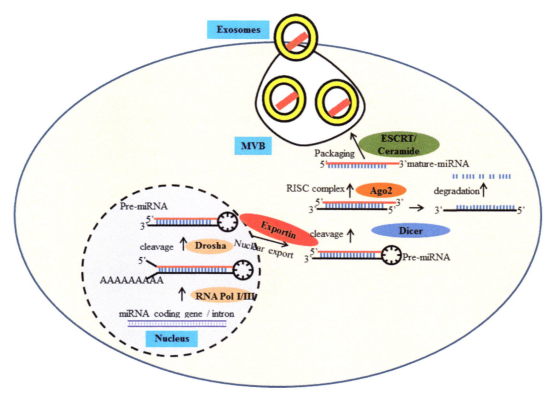

FIGURE 2 The schematic representation of synthesis of miRNA in the nucleus from pri-miRNA, its export to the cytoplasm, and its packaging into the exosomal lumen. The red and green colors represent low and high enrichment of miRNAs in the exosomal lumen.

Drosha and its cofactor to generate ~65 nt-long, hairpin-shaped precursor miRNAs known as pre-miRs. In the second stage, the pre-miRs are exported by the nuclear transport factor complex exportin-5 (Exp5) to the cytoplasm, where pre-miRs are modified by the Dicer enzyme to double-stranded 20–22 nucleotide, mature miRNAs. The antisense strand of the mature miRNA is selected by argonaute (Ago) to join the RNA-induced silencing complex (RISC), whereas the other strand is degraded. The miRNA strand and RISC complex then binds to the 8-mer seed region of the target mRNA at its three prime untranslated region (3′ UTR), also known as miRNA recognition elements, and causes it to degrade. Mature miRNAs silence gene expression by promoting translational repression and/or mRNA degradation (Fig. 2) (Bartel, 2004; Lages et al., 2012).

Calin (2009) was the first to report the role of aberrant expression of miRNA in cancer progression. Studies involving miRNA expression profiling in tumor cells and normal tissues have reported unique miRNA signatures correlated with cancer status. With recent evidence of miRNA in exosomal cargo, exosomal miRNAs have been actively investigated for their involvement in various cancer-related processes, including cell differentiation, cell cycle, proliferation, apoptosis, and angiogenesis (Di Leva, Garofalo, & Croce, 2014; Farazi et al., 2013). The miRNA content in exosomes has been found to be dysregulated in cancer cells or cells under pathological conditions when compared with exosomal miRNAs released from normal cells from the same individuals. This differential expression of miRNA profile in exosomes isolated from cancer patients has prompted interest in exosomes as potential diagnostic biomarkers to assess the status of disease, to record response to treatment, and/or for therapeutic targets.

Biomarkers are defined as measurable entities that represent the status of the body and response to therapeutic interventions. Various researchers have established that miRNAs are involved in many pathological conditions, and the expression profile of miRNAs can be correlated with the body's condition (Hannafon & Ding, 2013; Lan, Lu, Wang, & Jin, 2015; Xiao et al., 2012; Ye et al., 2014). The convenience of isolating exosomes from bodily fluids and the stability of miRNA inside exosomes encouraged many researchers to screen and identify unique miRNA profiles in exosomes that can be used as biomarkers and therapeutic indicators in different cancer types. Such an approach may lead to the development of a noninvasive method of diagnosis that bypasses the need for tumor mass biopsies for detection and monitoring of recurrence.

In an early study investigating exosomal miRNA status in cancer, Taylor and Gercel-Taylor (2008) found a high degree of correlation between miRNA from tumor cells and from the corresponding exosomes (0.71–0.90). Furthermore, they identified miRNA candidates who were closely correlated in primary tumor tissue and circulating exosomes from women with ovarian cancer; these miRNAs were significantly overexpressed compared with levels in healthy controls. This outcome strongly suggested that the expression signatures of exosomal miRNAs can be exploited as diagnostic biomarkers in ovarian cancer.

A study conducted in our laboratory showed enrichment of hsa-miRNAs-200c-3p, 23b-3p, and 100-5p in exosomes from patients with endometrial cancer compared with exosomes isolated from symptomatic individuals without cancer (Srivastava et al., unpublished data). The ability to successfully differentiate between symptomatic and asymptomatic patients based on exosomal miRNA profiles substantiates the potential of exosomal miRNA as a diagnostic marker.

Furthermore, the cancer genome atlas (TCGA) analysis revealed that hsa-miR-200c-3p expression was altered in 2.1% of patients with endometrial cancer, and patients with hsa-miR-200c-3p alterations had poor survival. In silico mining by starBase 2.0 and Pan-Cancer Project Analysis showed approximately a two-fold increase in hsa-miR-200c-3p miRNA expression in patients with endometrial cancer over healthy individuals. In addition, mRNA target analysis using MirTarBase (www.mirTarbase.mbc.ntu.edutw) showed Zeb1, Zeb2, and BMI-1 as prominent molecular targets of hsa-miR200c-3p (Chou et al., 2016). All of these findings support hsa-miR-200c-3p as a promising candidate as a biomarker and therapeutic target in endometrial cancer.

In lung cancer, the poor 5-year survival rate is allegedly because of late detection of the disease, which leads to insufficient treatment outcomes. In one of the earliest studies on lung cancer, the mean exosome concentration and subsequent miRNA concentration of exosomal miRNA were measured in a cohort of 27 patients with lung adenocarcinoma and 9 healthy controls. The comparison of disease with normal samples showed a significant elevation of both exosome number and exosomal miRNA concentration, revealing a possible correlation between exosomal miRNA levels and stage of disease. The exosomes in this study were isolated from serum (Rabinowits, Gerçel-Taylor, Day, Taylor, & Kloecker, 2009). Other studies have shown that let-7, miRNA-10, miRNA-15, miRNA-30, miRNA-21, miRNA-23, and miRNA-181 are enriched in exosomes derived from patients with lung cancer (Jin et al., 2017; Liu et al., 2017; Zhang et al., 2017).

In a study to assess the utility of exosomes as prognostic markers, we used massive parallel sequencing to screen the miRNAs isolated from the urine of patients with lung cancer ($n=5$), before and after chemotherapy. The study is ongoing, but initial results obtained from a small cohort of patients ($n=5$) matched pre- and posttreatment revealed a distinguishing pattern of miRNA enrichment. In general, a depletion in miRNA levels was observed in posttreated samples compared to matched pretreatment samples (Srivastava et al. unpublished data).

Munagala, Aqil, and Gupta (2016) examined serum-derived exosomes from nude mice with subcutaneous primary and recurrent xenograft lung tumors. The findings revealed significant upregulation of miRNA-21 and miRNA-155 in animals with recurrent tumors, compared with animals without tumors or with primary tumors, suggesting that miRNA profiling of exosomes can be a promising strategy for noninvasive diagnosis. Another group of researchers profiled cellular vesicles for potential development of a noninvasive diagnostic tool for lung cancer detection. There was a difference in vesicle-related miRNAs in plasma from nonsmall cell lung cancer patients and healthy controls, as shown by analysis using a Taqman low density array. The differential levels of let-7f and miR-30e-3p showed a strong correlation with poor prognosis and survival (Silva et al., 2011).

For exosomal miRNA–based methods of lung cancer diagnosis to be clinically relevant, large-scale miRNA profiling should be refined. To that end, Cazzoli et al. (2013) narrowed the potential pool of 746 miRNAs to 14 miRNAs. The researchers finally selected miR-200b-5p as a potential biomarker to assess the recurrence of small-cell lung tumors after surgical resection. The group also suggested that a distinction between normal and lung adenocarcinoma samples could be achieved based on the expression of the following exosomal miRNAs: miRNA-378a, miRNA-379, miRNA-139-5p, and miRNA-200-5p. In 2014, Rodríguez et al. demonstrated the potential use of exosomal miRNAs that were enriched in the plasma, but not the bronchoalveolar lavage (BAL) fluid, of patients with lung cancer as biomarkers (Rodríguez et al., 2014).

Taylor and Gercel-Taylor (2008) identified levels of eight miRNAs (miRNA-21, miRNA-141, miRNA-200a, miRNA-200c, miRNA-200b, miRNA-203, miRNA-205, and miRNA-214) in exosomes isolated from serum of women with benign disease and various stages of ovarian cancer. Through miRNA expression profiling analysis of 55 advanced ovarian tumors, it was shown that miRNA-200a, miRNA-200b, and miRNA-429 in the miR-200b-429 cluster are significantly associated with cancer recurrence and overall survival

(Taylor & Gercel-Taylor, 2008). miRNA-200 family members seem to be differentially expressed in exosomes derived from ovarian cancer cell lines.

A separate study examined ovarian cancer effusion supernatants isolated from the peritoneum or pleura to study the clinical relevance and biological role of exosomal miRNA in cancer progression. The researchers identified 11 miRNAs with significant associations with effusion sites and stage of disease. In survival and prognosis analysis, high levels of miRNA-21, 23b, and 29a were associated with poor progression-free survival ($P=.01$, $P=.015$, and $P=.009$, respectively), whereas high expression of miRNA-21 was correlated with poor overall survival ($P=0.017$; Vaksman, Tropé, Davidson, & Reich, 2014).

The levels of exosomes are excessively elevated in epithelial ovarian cancer (EOC) patients. Such exosomes are expected to have different miRNA levels from their normal counterparts as a unique signature of the disease. In a cohort of 163 patient serum samples, the levels of miRNA-373 ($P=0.004$), miRNA-200a ($P=0.0001$), miRNA-200b ($P=0.0001$), and miRNA-200c ($P=0.008$) were higher in comparison with levels in healthy individuals. The levels also correlated with the International Federation of Gynecology and Obstetrics (FIGO) defined stages of EOC. Thus, the levels of miRNA-200 family members and miRNA-21 are higher in exosomes from patients with advanced ovarian cancer, and these miRNAs can be explored as promising candidates for biomarkers or therapeutic targets (Meng et al., 2016). In an interesting study, miRNA let-7 and miRNA-200 were compared in the SKOV3 and OVCAR3 cell lines, which differ in invasiveness, and in the exosomes derived from these cell lines. Findings revealed that the let-7 family miRNAs positively correlated with invasiveness compared with miRNA-200 family miRNAs, which are only present in OVCAR-3 cells and exosomes. This result suggests that miRNA-200 is not involved in invasiveness (Kobayashi et al., 2014).

A noninvasive (or minimally invasive) circulating miRNA-based diagnostic test for early detection of colorectal cancer (CRC) is highly warranted, in light of the currently available painful and cumbersome colonoscopy screening. Deep RNA sequencing–based analysis has revealed dysregulated miRNAs in serum from patients with CRC. Ogata-Kawata et al. (2014) found that patients with colon cancer, even in the early stages, exhibited significantly higher levels of seven exosomal miRNAs (let-7a, miRNA-1229, miRNA-1246, miRNA-150, miRNA-21, miRNA-223, and miRNA-23a), compared with healthy controls. Ng et al. (2009) reported that miRNA-17-3p and miRNA-92 had 89% sensitivity and 70% specificity in discriminating CRC from normal controls. Pu et al. (2010) reported elevated levels of miRNAs, including miRNA-21, miRNA-221, and miRNA-222, in plasma samples from patients with CRC. Furthermore, they correlated the expression levels with p53 status, which suggests that amplification of plasma miRNA-221 could be a biomarker for CRC. Another group investigated 12 selected miRNAs using 157 plasma samples from advanced adenomas and carcinomas of colorectal neoplasia, as well as 59 healthy control samples. The findings showed that miRNA-29a and miRNA-92a have significant cancer diagnostic values, with 83% sensitivity and 84.7% specificity for assigning the disease condition (Fesler, Jiang, Zhai, & Ju, 2014; Huang et al., 2010).

Another study reported that both miRNA-29a and miRNA-18a are upregulated in the serum of patients with stage III CRC compared with controls, as determined by microarray analysis followed by qRT-PCR validation and normalization to a panel of three endogenous controls (Yang, Gu, Zhou, Xiang, & Chen, 2013). Recent studies show consistent results for miRNA-92a as a diagnostic and prognostic biomarker, based on 200 serum samples of CRC and 50 advanced adenomas. In addition, serum miRNA-21 has potential diagnostic value, as reported in several of the abovementioned studies. These two miRNAs could be helpful in distinguishing patients with precancerous adenoma from controls and exhibit high sensitivity and specificity.

Since early detection of CRC is key for successful treatment, the identification of biomarkers for precancerous adenoma is advantageous. Kanaan et al. (2013) determined that a panel of eight plasma miRNAs, including miRNA-532-3p, miRNA-331, miRNA-195, miRNA-17, miRNA-142-3p, miRNA-15b, miRNA-532, and miRNA-652, could identify adenoma in patients, with a sensitivity of 88% and a specificity of 64%.

In another study involving miRNA profiling on pooled plasma samples from 10 patients with CRC (5 stage II and 5 stage III) and 10 normal controls, followed by validation via qRT-PCR on 191 independent samples with 90 CRC patients, 43 advanced adenoma patients, and 58 healthy controls, miRNA-601 and miRNA-760 were decreased in CRC and in advanced adenomas compared with controls. Combined, these miRNAs showed 72.1% sensitivity and 62.1% specificity in distinguishing patients with advanced adenoma from controls. In discriminating CRC from control, they demonstrated 83.3% specificity and 93.1% specificity (Cheng et al., 2011). Consistent with other findings, this study also revealed that miRNA-29a and miRNA-92a were upregulated in CRC.

In a study that attempted to focus on differentially expressed plasma miRNAs that correlated with miRNAs that were differentially expressed at the tissue level, Yong, Law, and Wang (2013) found that a combination of miRNA-193a-3p, miRNA-23a, and miRNA-338-5p demonstrated 80.0% sensitivity and 84.4% specificity for disease determination. These three miRNAs demonstrated increased expression in patients with CRC, in tissue and in the circulation, and have potential as therapeutic targets or biomarker candidates.

In gastric cancer, miRNA-21 and miRNA-1225-5p appear to be dysregulated after curative gastric cancer surgery and can serve as biomarkers, thus providing a novel approach to early diagnosis of gastric cancer. Exosomal miRNA signatures in bodily fluids in gastric cancer patients show decreased levels of miRNAs, including miRNA-34b/c, miRNA-218 and miRNA-10b, whereas the miRNAs including miRNA-21, miRNA-103, and miRNA-223 showed significant increased levels.

Shi et al. (2015) reported that increases in the level of exosomal miR-21 in cerebrospinal fluid correlate with poor prognosis and cancer recurrence in patients with glioma. miRNAs have also been detected in the urine of patients with prostate and bladder cancer (Hessvik, Sandvig, & Llorente, 2013; Long et al., 2015). Bryzgunova et al. (2016) reported that the level of miRNA-19b, normalized to that of miRNA-16, in urine-derived exosomes was significantly lower in patients with prostate cancer than in healthy donors.

let-7 is one of the earliest candidate miRNAs to be identified in humans. Members of the let-7 miRNA family have been isolated from exosomes derived from highly metastatic gastric cancer cells (Ohshima et al., 2010). miRNA-21, miRNA-221, and miRNA-222 are overexpressed approximately 40-fold in exosomes from patients with CRC compared with normal control subjects (Pu et al., 2010). PCR array–based screening also identified miRNA-320 and miRNA-574-3p as potential exosomal markers of glioblastoma multiforme diagnosis (ROC curve analysis: miRNA-320 AUC = 0.719, P = .0067; miRNA-574-3p AUC = 0.738, P = 0.0055). However, in an independent validation cohort, no significant differences in expression were noted between patients and matched normal controls (Manterola et al., 2014).

It is not only in serum or plasma that exosomes exert their potential as cancer biomarkers. Recently, a miRNA panel extracted from bile-derived exosomes from patients with cholangiocarcinoma was proposed to be relevant for disease diagnosis. Results from the work of Liu, Sun, et al. (2014) showed high expression levels of miRNA-21 and miRNA-146a in exosomes derived from cervicovaginal lavages compared with those from HPV-positive and HPV-negative (normal) samples.

The key requirements for early cancer detection biomarkers are high sensitivity and specificity. The miRNA-200, miRNA-21, and let-7 families are frequently enriched in exosomes belonging to different cancers mentioned above and strongly suggest the utility of these miRNA families as biomarkers. Furthermore, there may be immense potential in using exosome-derived miRNAs in clinics for early diagnosis of cancer and for prognostic purposes.

Melo et al. (2014) recently reported that exosomal miRNA may play a role in the regulation of cancer-related activities in the cell. Conventionally, exosomes are considered biologically inert entities that can ferry biological material, such as miRNA, but cannot perform any biological function on their own. The study by Melo et al. (2014) has broken this conventional assumption by showing the presence of pre-miRNA along with all other components, including Dicer, Ago2, TRBP, and RISC loading complex, that are required for processing pre-miRNA to mature miRNA, independent from cells. This discovery further illustrates the need to analyze the role of exosomal miRNAs in physiological functions and disease.

3.1.1 Therapeutic Application of miRNA

Exosome-delivered miRNAs can reprogram the gene expression profiles in the recipient cells through targeted suppression of mRNA by posttranscription modifications. In cancer, exosome-delivered miRNAs have been identified as both oncomirs (i.e., inhibit tumor suppressor mRNAs) and tumor suppressor miRNAs (i.e., suppress oncogenic mRNAs) (Berindan-Neagoe, Monroig, Pasculli, & Calin, 2014). Because of their regulatory role, the miRNAs and/or their target mRNAs are often considered important targets in cancer therapeutics (Table 1). Many miRNAs are naturally packaged in exosomes and are reported to aid in metastasis by creating a favorable premetastatic niche for the establishment of

Table 1 List of Oncogenic and Tumor Suppressor Exosome miRNAs in Different Cancers

Cancer	Exosome miRNA	Target Genes	Status	Reference(s)
Gastric	let-7 family	KRAS, HRAS, HMGA2	Tumor suppressor	Motoyama et al. (2008)
Breast, prostate, bladder	miRNA-34a	BCL2, MYC, CD4, CDK6	Tumor suppressor/ Oncomir	Grammatikakis, Gorospe, and Abdelmohsen (2013)
Gastric	miRNA-181b	Cytochrome c	Oncomir	Liu, Sun, et al. (2014)
Hepatocellular carcinoma	miRNA-500, miRNA-18a, miRNA-221/222		Oncomir	Ali et al. (2017)
Hepatocellular carcinoma	miRNA-101, 106b, 122, 195		Tumor suppressor	Ali et al. (2017)
Ovarian	miRNA-21, 92, 93, 126, 29a		Oncomir	Kinose, Sawada, Nakamura, and Kimura (2014)
Ovarian	miRNA-200 family, let-7 family		Oncomir	Li, Zhang, Wang, and Wan (2010)
Lung	miRNA-21, 96	PTEM/Akt	Oncomir	Inamura (2017)
Lung	miRNA-221/222, 15b, 110, 125b/217		Tumor suppressor	Inamura (2017)
Leukemia	miRNA-155, 128-3p		Oncomir	Mets et al. (2014) and Schotte, Pieters, and Den Boer (2012)
Prostate	miRNA-34a	SIRT1, E2F3, E2F1, Bcl-1	Oncomir	Pang, Young, and Yuan (2010)
Breast	miRNA-155, 34a, 27a	Bcl-6, NOTCH1, CYP1B1	Oncomir	Jiang et al. (2010) and Mertens-Talcott, Chintharlapalli, Li, and Safe (2007)
Breast	miRNA-206, 21, 214, 28	PTEN, BRCA1	Tumor suppressor	Zhu et al. (2008)
Liver	miRNA-125b	LIN28B, CCND1, SOX2, MYC, CDK6	Oncomir	Sun et al. (2013)

secondary tumor sites. Exosomal miRNA-494 and miRNA-542-3p released from highly metastatic, poorly adherent Bsp73ASML rat adenocarcinoma cells show modulation of proteases, adhesion molecules, cell cycle components, angiogenesis-promoting genes, and genes engaged in oxidative response to prepare the premetastatic niche for seeding and proliferation of primary tumor cells (Rana, Malinowska, & Zöller, 2013). Another example is miR-105, present in metastatic breast cancer cell–derived exosomes, which is known to target tight junction protein ZO-1 to create a metastatic niche in endothelial cells. In lung cancer, miRNAs from tumor-derived exosomes activate toll-like receptor 3 (TLR3), which results in the formation of a favorable metastatic environment (Zhou, Fong, et al., 2014).

Researchers have recently shown that highly metastatic, cell-derived exosomes transfer miRNA-200 to weakly metastatic cells and induce the metastatic potential of tumors.

miRNA 200 family members regulate the epithelial to mesenchymal transition and potentiate tumor growth and metastasis (Pencheva & Tavazoie, 2013).

The basic role of the exosome in cell physiology is to flush out unwanted substances from the intracellular environment to reduce the cell toxicity. Consistent with this functional role, exosomes from cancer cells have been found to remove chemotherapeutic drugs from the cells, resulting in the induction of drug resistance (Zhao, Liu, Xiao, & Cao, 2015). Recent efforts have been made to understand the contribution of exosomal miRNAs to the development of drug resistance (Azmi, Bao, & Sarkar, 2013).

Exosomal miRNA also confers resistance by perturbing the cell or tumor microenvironment and protecting the tumor cells from therapeutic effects. miRNA-21 present in exosomes derived from M2 macrophages confers drug resistance on gastric cancer cells that are treated with cisplatin-based chemotherapeutics. The M2 exosomes show enriched miRNA-21 compared with inactivated macrophages; they shuttle miRNA-21 to gastric cancer cells and promote cisplatin resistance by deregulating the PTEN pathway and regulating antiapoptosis protein Bcl-2 (Zheng et al., 2017).

Interestingly, certain miRNAs are known to regulate exosome production, which in turn facilitates drug resistance in recipient cells. Pancreatic ductal adenocarcinoma (PDAC) is commonly treated with gemcitabine (GEM), but with continued exposure to GEM, drug resistance that is immediately transferred to all PDAC tissue in the patient's body has been observed. In a study to elucidate the mechanism of the development of drug resistance, Mikamori et al. (2017) found that long-term exposure to GEM increased the expression of miR-155 in PDAC cells. The increased miR-155 further increased production of exosomes, which facilitated the induction and spread of chemoresistance in other PDAC cells. A similar phenomenon of exosomal miRNA–mediated reduced sensitivity in cisplatin-treated A549 cancer cells was reported by Xiao et al. (2014). Table 2 lists several studies that explored the implications of miRNA from exosomes for chemoresistance.

3.1.2 Exosome-Based Delivery of Therapeutic miRNAs

The various studies mentioned above suggest the importance of exosomal miRNAs in its various capacities in the pathophysiology of cancer, such as oncomirs, tumor suppressors, or regulators of drug sensitivity (Thind & Wilson, 2016). A single miRNA can target several mRNAs (genes) of the downstream signaling pathway that is critical for processes involved in cancer progression. Thus, miRNAs have been actively investigated for the development of novel targeted anticancer therapies.

mRNAs can play pivotal role in the development of precise, personalized, effective, and successful treatment regimens. Exosomes can naturally carry miRNAs of therapeutic importance. Specific miRNA of interest can be loaded into exosomes to produce a therapeutic response by negatively regulating molecules involved in disease progression or poor prognosis. Regulatory small RNAs (siRNA or miRNA) cannot be administered to cells directly and are often administered through synthetic drug carriers that are usually based on polymers or are metallic nanoparticles. As they are artificial or synthetic

Table 2   Exosome miRNAs Involved in Chemosensitivity in Different Cancers

Cancer	Exosome miRNA	Drug Sensitivity	Target Molecules	Reference(s)
Small cell lung	miRNA-34	Chemoresistance	PTEN, CDH17, MAL	Zhou, Fong, et al. (2014)
Melanoma, glioblastoma	miRNA-9	Chemoresistance	MYC, VEGFA, MMP9, CDH1	Munoz, Rodriguez-Cruz, and Rameshwar (2015)
Breast, pancreatic, prostate, ovarian, melanoma	miRNA-221/222	Chemoresistance (Tamoxifen)	KIT, PTEN, CDKs, KRAS, MMP	Magee, Shi, and Garofalo (2015)
Melanoma, glioblastoma, neuroblastoma	miRNA-9	Temozolomide resistance	MYC, MYCN, VEGFA, MMP9, CDH1	Munoz et al. (2015)
Breast	miRNA-134	Doxorubicin, cisplatin resistance	STAT5b	Liu et al. (2015) and Pogribny et al. (2010)
Ovarian	miRNA-21-5p	Cisplatin resistance	NAV3	Pink et al. (2015)
Glioma	miRNA-221	Temozolomide resistance	DNM3	Yang et al. (2017)
Lung	miRNA-96	Cisplatin resistance	LMO7	Wu et al. (2016)
Lung	miRNA-100-5p	Cisplatin resistance		Qin et al. (2017)
Lung	miRNA-217	Cisplatin resistance		Guo, Feng, Huang, Wang, and Lu (2014)
Lung	miR-196a	Cisplatin resistance	Ddp	Li et al. (2016)
Pancreatic	miRNA-155	Cisplatin resistance		Shi et al. (2016)
Gastric	miRNA-21	Paclitaxel resistance	p-gp apoptic	Jin, Liu, and Wang (2015)
Gastric	miRNA-21	Cisplatin resistance	PTEN	Yang, Gu, et al. (2013)
Gastric	miRNA-19, 181b, 34a	Adriamycin (Doxorubicin) vincristine, gemcitabine, taxol		An, Sarmiento, Tan, and Zhu (2017)
Hepatocellular carcinoma	miRNA-221, 21	Sorafenib	Caspase 3 PTEN	Fornari et al. (2017)
Ovarian	miRNA-130a	Paclitaxel, cisplatin		Li et al. (2015)
Lung adenocarcinoma	miRNA-21	Gefitinib	PTEN/Akt	Shen et al. (2014)
Breast	miRNA-127, 34a, 27b, 206, 21, 214, 28, 451	Doxorubicin	Bcl-6, NOTCH1, CYP1B1, PTEN, BRCA1	Kutanzi, Yurchenko, Beland, Checkhun, and Pogribny (2011)

in nature, these particles often have drawbacks such as low enhanced permeability and retention effect, immunogenicity and off-target cytotoxicity. Because of their ability to ferry macromolecules, exosomes have been explored as a method to transfer miRNAs of therapeutic importance into recipient cells. Alvarez-Erviti et al. (2011) first reported the potential of exosomes as drug delivery vehicles when they demonstrated targeted delivery siRNA into the brains of mice, suggesting that exosomes can not only deliver therapeutic molecules in the body but can also reach organs that are isolated from rest of the body through stringent barriers, such as the blood–brain barrier. In this context, the miRNAs already packaged in exosomes can cause cancer initiation, progression, and therapeutic interventions.

The intriguing involvement of miRNAs in the regulation of cancer cell physiology makes miRNAs an ideal tool for cancer therapy. Although several researchers have explored the ways exosomes can be used as drug delivery vehicles, progress is slow and the work is still in the preliminary stage. More research is needed for the technology to move into the clinic for direct patient benefits. Nevertheless, the use of exosomes as delivery vehicles for cancer therapeutics may become a major avenue for precision and personalized medicine in cancer therapy.

3.1.3 Exosomes for Imaging and Chemotherapeutics

Exosomes can also be used as chemotherapeutic delivery vehicles. In our previous study, we established a delivery vehicle consisting of exosomes (Exo) loaded with gold nanoparticles (GNPs) conjugated through a pH cleavable linker to an anticancer drug "doxorubicin (Dox)" and termed this complex of Exo-GNP-Dox as "nanosomes." The results from the study showed that nanosomes were successfully able to deliver Dox to lung cancer cell lines, H1299 and A549 and also showed an enhanced therapeutic effect compared with when drug was delivered either in a complex with GNPs (GNP-Dox) or with exosomes (Exo-Dox). In addition to this, the unique combination of Exo-GNP-Dox in nanosomes circumvented the Dox-mediated cardiotoxicity effect, a major drawback of Dox that hinders its application in the treatment of many cancers (Srivastava et al., 2016). Inspired by nanosomes, we developed another drug carrier system for theranostic applications such that simultaneous delivery of drug and imaging of cancer site (diagnosis) can be achieved. To this end, exosomes were loaded with superparamagnetic iron oxide (MNP) nanoparticles that were conjugated with doxorubicin (Exo-MNP-Dox) and termed as "fexosomes." Because MNPs are good contrasting agents they can be exploited for magnetic resonance imaging (MRI). We envision that these MNP-loaded exosomes can be used as a theranostic with the dual functionality of delivering an anticancer drug and enabling molecular imaging of cancer cells. Initial experiments in our laboratory using A549 lung cancer cells showed increased uptake of Exo-MNP-Dox with enhanced contrast by MR imaging (Fig. 3). In future, these developed exosome-based carriers will be exploited for delivery of therapeutic miRNAs and siRNAs in combination with MNP in the treatment and monitoring of various cancers.

3.2 Long Noncoding RNA in Exosomes

Recent advances in RNA sequencing technologies have made it possible to survey transcriptomes in greater detail. These advances have revealed a new species of noncoding RNA known as lncRNA. Typically, lncRNA codes for transcripts longer than a minimum of 200 nucleotides up to 100 kb, and does not code for any proteins, although it modulates

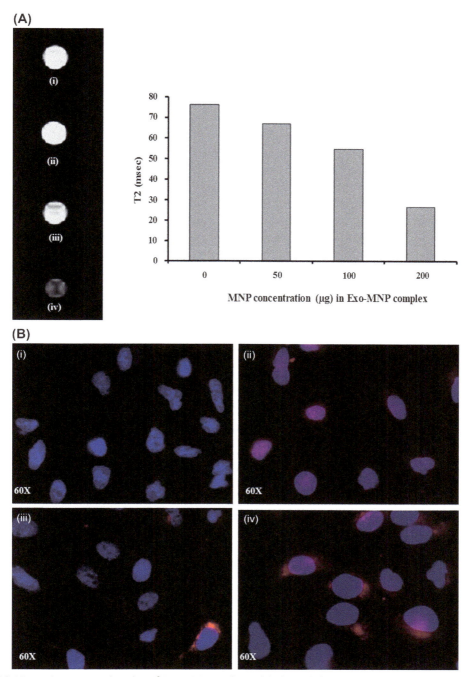

FIGURE 3 Magnetic resonance imaging of Exo-MNP complexes. (A) The T2 (relaxation time) MRI phantom images were captured at different concentrations of magnetic nanoparticles (MNPs) [(i) 0, (ii) 50, (iii) 100, and (iv) 200 μg] in Exo-MNP complex after treating A549 cells for 24 h. The images and the adjacent graph shows that with increasing concentration of MNPs [from (i) to (iv)] the T2 values are decreasing and dark contrast is increasing suggesting MR imaging capability of Exo-MNP complex in the cell. (B) The Exo-MNP-complexed with fluorescence drug doxorubicin (Exo-MNP-Dox) exhibited more cell uptake in A549 cells than by MNP-Dox and free Dox as observed by fluorescence microscopy after 24 h treatment; (i) control, (ii) free Dox, (iii) MNP-Dox, and (iv) Exo-MNP-Dox.

DNA transcription epigenetically (Fang & Fullwood, 2016; Prensner & Chinnaiyan, 2011). Exosomes are known to carry lncRNA, although they are less explored in studies in comparison with miRNA. ZFAS1 is a newly identified lncRNA present in the circulating exosomes of the serum from patients with gastric cancer. ZFAS1 levels are elevated in gastric cancer samples and in cell proliferation and migration of gastric cancer cells (Pan et al., 2017).

Furthermore, lncRNA-RoR is enriched in exosomes derived from HEPG2 hepatocellular carcinoma cells (HCC). Exosomal transfer of lncRNA-RoR between HepG2 HCC cells conferred a degree of chemoresistance in response to increasing doses of sorafenib, doxorubicin, and camptothecin, and siRNA-mediated knockdown of lncRNA-RoR was associated with a significant reduction in cell viability (Liu et al., 2015). Although this analysis did not conduct mechanistic studies, for the first time, an exosomal lncRNA, which may be targetable with siRNA has been implicated in the regulation of tumor chemoresistance.

HOTAIR is another type of lncRNA that is widely studied in a number of malignancies. HOTAIR interacts with the polycomb repressive complex to suppress target gene expression. It has been implicated in the pathogenesis of numerous malignancies (Hajjari & Salavaty, 2015). Alone, exosomal HOTAIR expression was capable of distinguishing benign from malignant disease with a sensitivity of 92.3% and specificity of 7.2%; when combined with miRNA-21, the sensitivity and specificity were improved to 94.2% and 73.5%, respectively (Loewen, Jayawickramarajah, Zhuo, & Shan, 2014). These data suggest that the exploitation of exosomes as a natural mechanism of ncRNA transport may provide a promising therapeutic mechanism to target tumor tissue, modulate the tumor microenvironment, or sensitize tumor cells to conventional anticancer drugs.

4. Sources of Exosomes

The idea of using exosome RNA as a biomarker and for subsequent cancer prevention is based on the premise of minimally invasive liquid biopsy and personalized medicine. Exosomes appear ideal for this purpose, as these vesicles carry a plethora of RNA species that are representative of disease condition and therapeutic response, and thus may be promising biomarkers specific to the genetic makeup of the individual and the disease type. Exosomes have been isolated from almost all bodily fluids, including blood serum/plasma, breast milk, ascites, urine, saliva, semen, ocular effluent and aqueous humor, saliva, cerebrospinal fluids, BAL fluids, and bile (Srivastava et al., 2015).

The ubiquitous presence of exosomes in biofluids makes them further suitable for use in liquid biopsy analysis for different cancers. Exosomes isolated from blood plasma are one of the most widely investigated exosomes for studying numerous cancers, including lung, breast, ovarian, and pancreatic cancer (Lässer, 2012). Blood is readily available and can be obtained with minimal invasion and discomfort. Plasma obtained from 1 mL of blood is sufficient for the isolation of the amount of exosomes required for miRNA purification. However, plasma as a source of exosomes is associated with challenges. The environment of plasma is complex, as it consists of many nucleoprotein complex aggregates, other vesicles are present, and there is ambiguity in identifying the origin of cells of specific exosomes, as plasma exosomes are usually derived from a mixed population of cells (Whiteside, 2015). However, studies have revealed modified methods to isolate specific exosomes from the mixed pool of exosomes present in the plasma. These methods largely use the techniques of size exclusion and affinity chromatography, followed by density gradient centrifugation and have considerably improved the purity of isolation of specific exosomes from plasma (Muller, Hong, Stolz, Watkins, & Whiteside, 2014).

Urine is the second most widely used biofluid for studying exosomal miRNAs involved in different cancers. Urine is an excellent source of exosomes, as these vesicles are naturally excreted and can be readily collected without any invasive procedure. In our studies identifying miRNA-based biomarkers in lung, ovarian, and endometrial cancer, we used urine as the primary material to isolate exosomes and study the miRNA profiles. We used the ultracentrifugation method of exosome isolation. Fig. 4 shows the transmission electron microscopy (TEM) and atomic force microscopy images and the nanoparticle tracking analysis profile of exosomes isolated from urine (Srivastava et al. unpublished data).

Prostate cancer is often diagnosed by prostate-specific antigen (PSA) testing. Because PSA testing is nonspecific in nature, patients often end up receiving biopsies, which are invasive and painful (Motamedinia et al., 2016). Similarly, colonoscopy for CRC diagnosis and cystoscopy for gall bladder cancer diagnosis are accompanied by significant physical discomfort; hence, development of diagnosed tools based on urinary exosomes will be helpful in diagnosing these cancers. Promising results from research have shown identification of miRNA signatures from the urinary exosomes from pancreatic, prostate, gall bladder, and hepatocellular cancer (Øverbye et al., 2015; Riches et al., 2016). Exosomes derived from cerebrospinal fluid and human breast milk have been used to diagnose glioma and breast cancer, respectively (Qin et al., 2016).

5. Technology for the Study of Exosomal miRNA

5.1 Microarray and Sequencing for Exosomal Content Profiling

The work flow involved in the identification of unique miRNA signatures that can be used as potential biomarkers is described in Fig. 5. With the advent of newer technologies and availability of powerful computational resources, selection of optimal technologies is essential for analyzing RNA isolated from the exosomes to ensure that the data are relevant and biologically meaningful. The commonly used methods for profiling miRNA are (1) quantitative (q) PCR, (2) high-throughput microarray methods, and (3) next-generation whole RNA sequencing (Table 3).

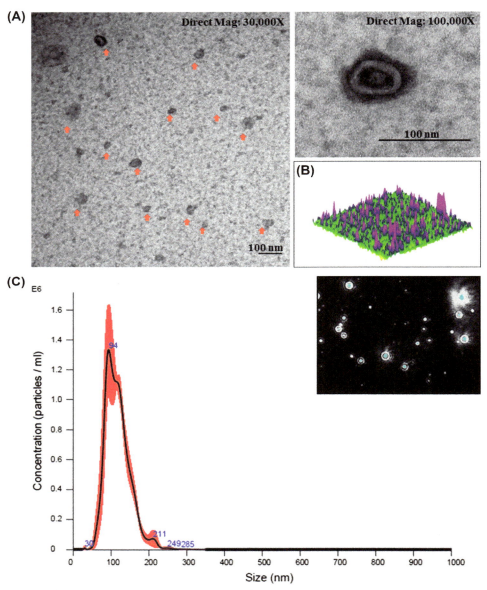

FIGURE 4 Transmission electron microscopy (TEM), atomic force microscopy (AFM), and nanotracker analysis (NTA) are tools that are frequently used to characterize exosomes. (A) TEM of exosomes at two magnifications. The right panel is a magnified image that shows the typical bilayer structure of exosomes. (B) AFM image of exosomes. Each crest in the figure represents the exosome. (C) NTA shows the size (diameter) and number of exosomes present in the isolations. Inset shows a screen capture from NTA with images of exosomes used for tracking and quantifying exosomes.

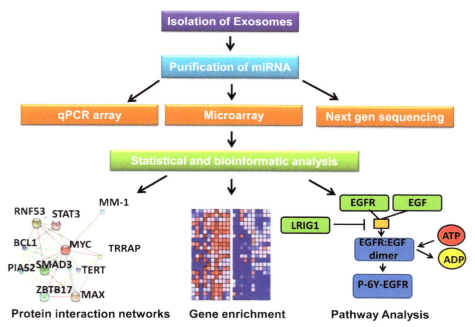

FIGURE 5 Work flow involved in an exosome miRNA study of novel miRNA markers and elucidation of their biological significance. The first step is to isolate exosomes from body fluids, followed by purification of total or miRNA from the isolated exosomes. In the next step, depending on the experimental design, miRNA is profiled with different platforms. The data then generated are analyzed using bioinformatics suits compatible to the platform used for profiling. To ascertain the biological meaning, the results are visualized using a variety of gene enrichment, networking, and pathway analysis tools described in the text.

The qPCR method of profiling miRNA is done on presynthesized PCR plates carrying probes for individual miRNAs with appropriate controls (Table 4). This effective, quick method examines a limited number of miRNAs that are most frequently altered in different cancers. The upregulated and downregulated miRNAs can be quantified with qPCR. Customized plates for specific cancer or pathways can also be used. The drawback of this method is that only a limited number of miRNAs can be explored simultaneously, and discovery of novel miRNAs is not possible. To overcome these limitations, one can use high-throughput microarray, in which expression patterns of 700–1600 miRNAs can be studied in a single slide. This method enhances the range of miRNA exploration and is widely used in many cancer studies (Kadota, Yoshioka, Fujita, Kuwano, & Ochiya, 2017) but is also limited by the number of miRNAs spotted on the slide.

In addition to RNA-based study, protein array analysis can be used to examine the proteomic profile present in exosomes. Fig. 6 shows the miRNA bioanalyzer assay results from urine exosomes obtained from a patient with lung cancer and a protein profiling study by a protein microarray system, such as those provided by Full Moon Biosystems. Next-generation massive sequencing platforms are currently the most popular methods, as this approach provides the freedom to study both known and unknown miRNAs, giving the

Table 3 Technologies Available and Widely Used for miRNA Analysis

Method	Name of Assay	Properties
qPCR	RT^2 PCR profiler	Quick and robust assay, but a limited number of miRNA can be studied
Micorarray	Microarray plates from several vendors	Large-scale miRNA profiles can be visualized. Time-consuming, expensive, and repeatability are the challenges
Massive parallel next-gen sequencing	Illumina or Solexa platforms	Robust, fast, cheap, and permits discovery of new miRNA or miRNAs with very low expression levels

Table 4 List of miRNAs Analyzed Using the SB Biosciences PCR Array System

miRBase or NCBI Accession No.	Mature miRNA ID or Gene Symbol
MIMAT0000433	hsa-miR-142-5p
MIMAT0000441	hsa-miR-9-5p
MIMAT0000451	hsa-miR-150-5p
MIMAT0000419	hsa-miR-27b-3p
MIMAT0000099	hsa-miR-101-3p
MIMAT0000065	hsa-let-7d-5p
MIMAT0000101	hsa-miR-103a-3p
MIMAT0000069	hsa-miR-16-5p
MIMAT0000082	hsa-miR-26a-5p
MIMAT0000090	hsa-miR-32-5p
MIMAT0000083	hsa-miR-26b-5p
MIMAT0000414	hsa-let-7g-5p
MIMAT0000244	hsa-miR-30c-5p
MIMAT0000095	hsa-miR-96-5p
MIMAT0000455	hsa-miR-185-5p
MIMAT0000434	hsa-miR-142-3p
MIMAT0000080	hsa-miR-24-3p
MIMAT0000646	hsa-miR-155-5p
MIMAT0000449	hsa-miR-146a-5p
MIMAT0003393	hsa-miR-425-5p
MIMAT0000257	hsa-miR-181b-5p
MIMAT0000715	hsa-miR-302b-3p
MIMAT0000420	hsa-miR-30b-5p
MIMAT0000076	hsa-miR-21-5p
MIMAT0000692	hsa-miR-30e-5p
MIMAT0000617	hsa-miR-200c-3p
MIMAT0000417	hsa-miR-15b-5p
MIMAT0000280	hsa-miR-223-3p
MIMAT0000460	hsa-miR-194-5p
MIMAT0000267	hsa-miR-210-3p
MIMAT0000068	hsa-miR-15a-5p
MIMAT0000256	hsa-miR-181a-5p
MIMAT0000423	hsa-miR-125b-5p
MIMAT0000097	hsa-miR-99a-5p
MIMAT0000085	hsa-miR-28-5p

Table 4 List of miRNAs Analyzed Using the SB Biosciences PCR Array System—cont'd

miRBase or NCBI Accession No.	Mature miRNA ID or Gene Symbol
MIMAT0000510	hsa-miR-320a
MIMAT0000443	hsa-miR-125a-5p
MIMAT0000100	hsa-miR-29b-3p
MIMAT0000086	hsa-miR-29a-3p
MIMAT0000432	hsa-miR-141-3p
MIMAT0000073	hsa-miR-19a-3p
MIMAT0000072	hsa-miR-18a-5p
MIMAT0000727	hsa-miR-374a-5p
MIMAT0004748	hsa-miR-423-5p
MIMAT0000062	hsa-let-7a-5p
MIMAT0000422	hsa-miR-124-3p
MIMAT0000092	hsa-miR-92a-3p
MIMAT0000078	hsa-miR-23a-3p
MIMAT0000081	hsa-miR-25-3p
MIMAT0000066	hsa-let-7e-5p
MIMAT0000720	hsa-miR-376c-3p
MIMAT0000445	hsa-miR-126-3p
MIMAT0000436	hsa-miR-144-3p
MIMAT0001341	hsa-miR-424-5p
MIMAT0000087	hsa-miR-30a-5p
MIMAT0000418	hsa-miR-23b-3p
MIMAT0004697	hsa-miR-151a-5p
MIMAT0000461	hsa-miR-195-5p
MIMAT0000435	hsa-miR-143-3p
MIMAT0000245	hsa-miR-30d-5p
MIMAT0000440	hsa-miR-191-5p
MIMAT0000415	hsa-let-7i-5p
MIMAT0000684	hsa-miR-302a-3p
MIMAT0000279	hsa-miR-222-3p
MIMAT0000063	hsa-let-7b-5p
MIMAT0000074	hsa-miR-19b-3p
MIMAT0000070	hsa-miR-17-5p
MIMAT0000093	hsa-miR-93-5p
MIMAT0000456	hsa-miR-186-5p
MIMAT0001080	hsa-miR-196b-5p
MIMAT0000084	hsa-miR-27a-3p
MIMAT0000077	hsa-miR-22-3p
MIMAT0000425	hsa-miR-130a-3p
MIMAT0000064	hsa-let-7c-5p
MIMAT0000681	hsa-miR-29c-3p
MIMAT0004597	hsa-miR-140-3p
MIMAT0000424	hsa-miR-128-3p
MIMAT0000067	hsa-let-7f-5p
MIMAT0000421	hsa-miR-122-5p
MIMAT0000075	hsa-miR-20a-5p
MIMAT0000680	hsa-miR-106b-5p
MIMAT0000252	hsa-miR-7-5p
MIMAT0000098	hsa-miR-100-5p
MIMAT0000717	hsa-miR-302c-3p

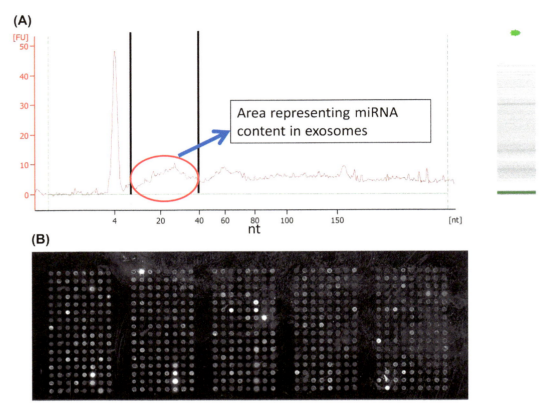

FIGURE 6 (A) The bioanalyzer assay showing the presence of miRNA in exosomes. (B) Protein microarray panel used for high-throughput study of differential expression of proteins in exosomes.

opportunity to discover unique miRNAs that can be exclusively associated with the cancer under investigation (Markopoulos et al., 2017).

5.2 In Silico Analysis

To understand the data generated after using high-throughput miRNA analysis platforms, various in silico tools are available that will produce biologically meaningful results. ExoCarta (www.exocarta.org) and Vesiclepedia (www.microvesicles.org) are two valuable resources in which the information related to the presence of various biomolecules in exosomes are carefully curated (Table 5).

These databases can serve as excellent reservoirs with which to analyze the discovery of miRNAs. miRTarBase (www.mirtarbase.mbc.ntu.edutw) and TargetScan (www.targetscan. org) are additional valuable resources with which the targets of miRNA can be assessed. The miRTarBase curates putative target mRNA information based on published results, whereas TargetScan has algorithms to predict the putative mRNA targets with various stringencies. Currently, several free-to-use and commercial standalone software packages are available for analyzing the EV-based OMICS data sets. The tools adopt a common

Table 5   The Contents of Extracellular Vesicles (EVs), and Exosomes, as Described in 538 (in EVs) and 286 (in Exosomes) Independent Studies From Two Databases

Molecules	Vesiclepedia Database[a]	Exocarta Database[a]
Proteins	92,897	9769
mRNA	27,642	3408
miRNA	4934	2838
Lipids	584	1116

[a]Accessed on May 24, 2017.

strategy to map the provided list of genes/proteins systematically to custom databases and provide a list of biological annotations that are statistically enriched in the query data set. In addition, these tools aid in identifying key pathways and processes regulating the biological function through enrichment analysis. Although there are several tools that can perform global analysis, some of the tools that are most commonly used will be discussed briefly.

5.2.1 The Database for Annotation, Visualization, and Integrated Discovery

The Database for Annotation, Visualization, and Integrated Discovery (DAVID) is a largely used Web-based enrichment analysis tool provided by the National Institute of Allergy and Infectious Diseases (NIAID), NIH. This tool can be used for functional annotation and enrichment of gene ontology terms. The user can cluster the redundant gene groups and visualize the obtained gene sets in BioCarta and KEGG pathway maps. One of the attractive features of this tool is that the user can convert the list of genes into various gene identifiers for cross-database accession identifier conversion.

5.2.2 Ingenuity Pathway Analysis

The ingenuity pathway analysis (IPA) tool is another Web-based omics analysis tool that maps the most relevant signaling and metabolic pathways based on data from high-throughput analysis. This tool also provides ways to identify molecular networks and biological functions for genes, predicts upstream and downstream transcription factors, and compares affected pathways and phenotypes across multiple testing conditions.

6. Challenges and Perspective

The ability of exosomes to encapsulate biomolecules, such as nucleic acids, and how to store these exosomes stably in the outside environment has garnered attention from the scientific community. Isolation of miRNA and lncRNA that represent the state of the cell of origin has shown promising results. This early work suggests that exosomes could be used as biomarkers for early detection of many cancers, especially those cancers for which survival is poor because of the lack of current diagnostic tools.

Because the study of exosomes is new, some challenges must be overcome before exosome-based applications can be used in clinical settings. The most prominent challenge is the isolation and purification of exosomes. Although many protocols have been developed for the isolation of exosomes, the method varies according to sample, and none of the presently available methods can be used universally. Isolation of pure, well-characterized populations of exosomes is essential to facilitate their development into biomarkers and therapeutic agents. Because exosomes exist alongside several other micro- and nanosized vesicular structures, it is essential to avoid any contamination from other structures during exosomal preparation.

The second challenge is that there is no universal definition of exosomes, which may often be grouped with other microvesicles. Although the International Society for Extracellular Vesicles (ISEV) has published an action paper for laying down the criteria to define exosomes, its strict implementation remains to be seen. In addition, there is currently no direct method available to quantify and characterize exosomes, even if proper isolations are performed. Methods including electron microscopy (TEM and SEM) and nanoparticle tracking systems have been used for characterization, but these methods are not suitable for routine clinical purposes.

Apart from these scientific issues, other basic issues, such as methods for sample collections, development of protocols, processing, and data analysis, must be resolved, standardized, and made consistent with good clinical and laboratory practices. The development of exosome-based applications for clinical purposes, such as methods for sample collection and handling, should be standardized before an exosome-based application can brought into clinics.

7. Conclusion

The exosome science is in nascent era exploration. In the last decade, researchers have discovered that exosomes stably carry active genetic material, such as mRNA, miRNA, and DNA, and can thus be used for various purposes in the treatment and diagnosis of diseases, including cancer. Exosomes isolated from body fluids are likely to reflect the state of the organ of exosome origin; thus, exosomes could play a decisive role in identifying the state of disease and responses to treatment. Research is underway to harness the potential of exosomes as diagnostic and prognostic markers. Some success has been achieved. However, more information is needed regarding the biology of exosome generation and packaging of molecules, and rapid research and development of newer and better technologies is warranted. The future of exosome science is exciting and has the capability to revolutionize medicine.

Acknowledgments

The study was supported in part by a grant received from the National Institutes of Health R01 CA167516 (RR), an Institutional Development Award (IDeA) from the National Institute of General Medical Sciences (P20 GM103639) of the National Institutes of Health (AM & RR), and by funds received from the Presbyterian Health Foundation Seed Grant (RR), Presbyterian Health Foundation Bridge Grant (RR, AM),

PHF Equipment Grant (RR), Stephenson Cancer Center Seed Grant (RR), and Jim and Christy Everest Endowed Chair in Cancer Developmental Therapeutics (RR), the University of Oklahoma Health Sciences Center. We thank Ms. Kathy Kyler at the office of Vice President of Research, OUHSC, for editorial assistance. Rajagopal Ramesh is an Oklahoma TSET Research Scholar and holds the Jim and Christy Everest Endowed Chair in Cancer Developmental Therapeutics.

Conflict of Interest

The authors declare no competing financial interests.

References

Ali, H. E. A., Hameed, R. A., Effat, H., Ahmed, E. K., Atef, A. A., Sharawi, S. K., & Wahab, A. H. A. (2017). Circulating microRNAs panel as a diagnostic tool for discrimination of HCV-associated hepatocellular carcinoma. *Clinics and Research in Hepatology & Gastroenterology*. https://doi.org/10.1016/j.clinre.2017.06.004.

Alvarez-Erviti, L., Seow, Y., Yin, H., Betts, C., Lakhal, S., & Wood, M. J. (2011). Delivery of siRNA to the mouse brain by systemic injection of targeted exosomes. *Nature Biotechnology, 29*(4), 341–345.

An, X., Sarmiento, C., Tan, T., & Zhu, H. (2017). Regulation of multidrug resistance by microRNAs in anticancer therapy. *Acta Pharmaceutica Sinica B, 7*(1), 38–51.

Azmi, A. S., Bao, B., & Sarkar, F. H. (2013). Exosomes in cancer development, metastasis, and drug resistance: A comprehensive review. *Cancer & Metastasis Reviews, 32*(3–4), 623–642.

Bartel, D. P. (2004). MicroRNAs: Genomics, biogenesis, mechanism, and function. *Cell, 116*(2), 281–297.

Berindan-Neagoe, I., Monroig, P., Pasculli, B., & Calin, G. A. (2014). MicroRNAome genome: A treasure for cancer diagnosis and therapy. *CA: A Cancer Journal for Clinicians, 64*(5), 311–336.

Bryzgunova, O. E., Zaripov, M. M., Skvortsova, T. E., Lekchnov, E. A., Grigor'eva, A. E., Zaporozhchenko, I. A., & Laktionov, P. P. (2016). Comparative study of extracellular vesicles from the urine of healthy individuals and prostate cancer patients. *PLoS One, 11*(6), e0157566. https://doi.org/10.1371/journal.pone.0157566.

Calin, G. A. (2009). MicroRNAs and cancer: What we know and what we still have to learn. *Genome Medicine, 1*(8), 78.

Cazzoli, R., Buttitta, F., Di Nicola, M., Malatesta, S., Marchetti, A., & Pass, H. I. (2013). MicroRNAs derived from circulating exosomes as non-invasive biomarkers for screening and diagnose lung cancer. *Journal of Thoracic Oncology, 8*(9), 1156–1162.

Cheng, H., Zhang, L., Cogdell, D. E., Zheng, H., Schetter, A. J., Nykter, M., & Zhang, W. (2011). Circulating plasma MiR-141 is a novel biomarker for metastatic colon cancer and predicts poor prognosis. *PLoS One, 6*(3), e17745.

Chou, C.-H., Chang, N.-W., Shrestha, S., Hsu, S.-D., Lin, Y.-L., Lee, W.-H., … Huang, H.-D. (2016). miRTar-Base 2016: Updates to the experimentally validated miRNA-target interactions database. *Nucleic Acids Research, 44*(Database issue), D239–D247.

Di Leva, G., Garofalo, M., & Croce, C. M. (2014). microRNAs in cancer. *Annual Review of Pathology, 9*, 287–314.

Fang, Y., & Fullwood, M. J. (2016). Roles, functions, and mechanisms of long non-coding RNAs in cancer. *Genomics, Proteomics & Bioinformatics, 14*(1), 42–54.

Farazi, T. A., Hoell, J. I., Morozov, P., & Tuschl, T. (2013). microRNAs in human cancer. *Advances in Experimental Medicine and Biology, 774*, 1–20.

Fesler, A., Jiang, J., Zhai, H., & Ju, J. (2014). Circulating microRNA testing for the early diagnosis and follow-up of colorectal cancer patients. *Molecular Diagnosis and Therapy, 18*(3), 303–308.

Fornari, F., Pollutri, D., Patrizi, C., La Bella, T., Marinelli, S., Gardini, A. C., & Callegari, E. (2017). In hepato-cellular carcinoma miR-221 modulates Sorafenib resistance through inhibition of caspase-3 mediated apoptosis. *Clinical Cancer Research, 23*(14), 3953–3965.

Frydrychowicz, M., Kolecka-Bednarczyk, A., Madejczyk, M., Yasar, S., & Dworacki, G. (2015). Exosomes–structure, biogenesis and biological role in non-small-cell lung cancer. *Scandinavian Journal of Immunology, 81*(1), 2–10.

Grammatikakis, I., Gorospe, M., & Abdelmohsen, K. (2013). Modulation of cancer traits by tumor suppressor microRNAs. *International Journal of Molecular Sciences, 14*(1), 1822–1842.

Guo, J., Feng, Z., Huang, Z., Wang, H., & Lu, W. (2014). MicroRNA-217 functions as a tumour suppressor gene and correlates with cell resistance to cisplatin in lung cancer. *Molecules and Cells, 37*(9), 664–671.

Hajjari, M., & Salavaty, A. (2015). HOTAIR: An oncogenic long non-coding RNA in different cancers. *Cancer Biology & Medicine, 12*(1), 1–9.

Hannafon, B. N., & Ding, W.-Q. (2013). Intercellular communication by exosome-derived microRNAs in cancer. *International Journal of Molecular Sciences, 14*(7), 14240–14269.

Hessvik, N. P., Sandvig, K., & Llorente, A. (2013). Exosomal miRNAs as biomarkers for prostate cancer. *Frontiers in Genetics, 4*, 36. https://doi.org/10.3389/fgene.2013.00036.

Huang, Z., Huang, D., Ni, S., Peng, Z., Sheng, W., & Du, X. (2010). Plasma microRNAs are promising novel biomarkers for early detection of colorectal cancer. *International Journal of Cancer, 127*(1), 118–126.

Inamura, K. (2017). Major tumor suppressor and oncogenic non-coding RNAs: Clinical relevance in lung cancer. *Cells, 6*(2), 12. https://doi.org/10.3390/cells6020012.

Jiang, S., Zhang, H. W., Lu, M. H., He, X. H., Li, Y., Gu, H., & Wang, E. D. (2010). MicroRNA-155 functions as an OncomiR in breast cancer by targeting the suppressor of cytokine signaling 1 gene. *Cancer Research, 70*(8), 3119–3127.

Jin, B., Liu, Y., & Wang, H. (2015). Antagonism of miRNA-21 sensitizes human gastric cancer cells to Paclitaxel. *Cell Biochemistry and Biophysics, 72*(1), 275–282.

Jin, X., Chen, Y., Chen, H., Fei, S., Chen, D., Cai, X., & Su, M. (2017). Evaluation of tumor-derived exosomal miRNA as potential diagnostic biomarkers for early-stage non-small-cell lung cancer using next-generation sequencing. *Clinical Cancer Research.* https://doi.org/10.1158/1078-0432.CCR-17-0577.

Johnstone, R. M., Adam, M., & Pan, B. T. (1984). The fate of the transferrin receptor during maturation of sheep reticulocytes in vitro. *Canadian Journal of Biochemistry and Cell Biology, 62*(11), 1246–1254.

Kadota, T., Yoshioka, Y., Fujita, Y., Kuwano, K., & Ochiya, T. (March 2017). Extracellular vesicles in lung cancer—from bench to bedside. In *Seminars in cell & developmental biology*. Academic Press.

Kahlert, C., & Kalluri, R. (2013). Exosomes in tumor microenvironment influence cancer progression and metastasis. *Journal of Molecular Medicine, 91*(4), 431–437.

Kalluri, R. (2016). The biology and function of exosomes in cancer. *The Journal of Clinical Investigation, 126*(4), 1208–1215.

Kanaan, Z., Roberts, H., Eichenberger, M. R., Billeter, A., Ocheretner, G., Pan, J., … Galandiuk, S. (2013). A plasma microRNA panel for detection of colorectal adenomas: A step toward more precise screening for colorectal cancer. *Annals of Surgery, 258*(3), 400–408.

Kinose, Y., Sawada, K., Nakamura, K., & Kimura, T. (2014). The role of microRNAs in ovarian cancer. *Biomed Research International, 2014*. https://doi.org/10.1155/2014/249393.

Kobayashi, M., Salomon, C., Tapia, J., Illanes, S. E., Mitchell, M. D., & Rice, G. E. (2014). Ovarian cancer cell invasiveness is associated with discordant exosomal sequestration of Let-7 miRNA and miR-200. *Journal of Translational Medicine, 12*, 4. https://doi.org/10.1186/1479-5876-12-4.

Kutanzi, K. R., Yurchenko, O. V., Beland, F. A., Checkhun, V. F., & Pogribny, I. P. (2011). MicroRNA-mediated drug resistance in breast cancer. *Clinical Epigenetics, 2*(2), 171–185.

Lages, E., Ipas, H., Guttin, A., Nesr, H., Berger, F., & Issartel, J.-P. (2012). MicroRNAs: Molecular features and role in cancer. *Frontiers in Bioscience, 17*, 2508–2540.

Lakkaraju, A., & Rodriguez-Boulan, E. (2008). Itinerant exosomes: Emerging roles in cell and tissue polarity. *Trends in Cell Biology, 18*(5), 199–209.

Lambertz, U., Oviedo Ovando, M. E., Vasconcelos, E. J., Unrau, P. J., Myler, P. J., & Reiner, N. E. (2015). Small RNAs derived from tRNAs and rRNAs are highly enriched in exosomes from both old and new world Leishmania providing evidence for conserved exosomal RNA packaging. *BMC Genomics, 16*(1), 151. https://doi.org/10.1186/s12864-015-1260-7.

Lan, H., Lu, H., Wang, X., & Jin, H. (2015). MicroRNAs as potential biomarkers in cancer: Opportunities and challenges. *BioMed Research International, 2015*. https://doi.org/10.1155/2015/125094.

Lässer, C. (2012). Exosomal RNA as biomarkers and the therapeutic potential of exosome vectors. *Expert Opinion on Biological Therapy, 12*(Suppl. 1), S189–S197.

Li, J. H., Luo, N., Zhong, M. Z., Xiao, Z. Q., Wang, J. X., Yao, X. Y., & Cao, J. (2016). Inhibition of microRNA-196a might reverse cisplatin resistance of A549/DDP non-small-cell lung cancer cell line. *Tumour Biology, 37*(2), 2387–2394.

Li, M., Zeringer, E., Barta, T., Schageman, J., Cheng, A., & Vlassov, A. V. (2014). Analysis of the RNA content of the exosomes derived from blood serum and urine and its potential as biomarkers. *Philosophical Transactions of the Royal Society B: Biological Sciences, 369*(1652). https://doi.org/10.1098/rstb.2013.0502.

Li, N., Yang, L., Wang, H., Yi, T., Jia, X., Chen, C., & Xu, P. (2015). MiR-130a and MiR-374a function as novel regulators of cisplatin resistance in human ovarian cancer A2780 cells. *PLoS One, 10*(6), e0128886. https://doi.org/10.1371/journal.pone.0128886.

Li, S. D., Zhang, J. R., Wang, Y. Q., & Wan, X. P. (2010). The role of microRNAs in ovarian cancer initiation and progression. *Journal of Cellular and Molecular Medicine, 14*(9), 2240–2249.

Liu, G., Zheng, X., Xu, Y., Lu, J., Chen, J., & Huang, X. (2015). Long Non-coding RNAs expression profile in HepG2 cells reveals the potential role of long non-coding RNAs in the cholesterol metabolism. *Chinese Medical Journal, 128*(1), 91–97.

Liu, J., Sun, H., Wang, X., Yu, Q., Li, S., Yu, X., & Gong, W. (2014). Increased exosomal microRNA-21 and microRNA-146a levels in the cervicovaginal lavage specimens of patients with cervical cancer. *International Journal of Molecular Sciences, 15*(1), 758–773.

Liu, Q., Yu, Z., Yuan, S., Xie, W., Li, C., Hu, Z., & Li, Y. (2017). Circulating exosomal microRNAs as prognostic biomarkers for non-small-cell lung cancer. *Oncotarget, 8*(8), 13048–13058.

Loewen, G., Jayawickramarajah, J., Zhuo, Y., & Shan, B. (2014). Functions of lncRNA HOTAIR in lung cancer. *Journal of Hematology & Oncology, 7*, 90. https://doi.org/10.1186/s13045-014-0090-4.

Long, J. D., Sullivan, T. B., Humphrey, J., Logvinenko, T., Summerhayes, K. A., Kozinn, S., & Rieger-Christ, K. M. (2015). A non-invasive miRNA based assay to detect bladder cancer in cell-free urine. *American Journal of Translational Research, 7*(11), 2500–2509.

Magee, P., Shi, L., & Garofalo, M. (2015). Role of microRNAs in chemoresistance. *Annals of Translational Medicine, 3*(21), 332. https://doi.org/10.3978/j.issn.2305-5839.2015.11.32.

Manterola, L., Guruceaga, E., Pérez-Larraya, J. G., González-Huarriz, M., Jauregui, P., Tejada, S., & Alonso, M. M. (2014). A small noncoding RNA signature found in exosomes of GBM patient serum as a diagnostic tool. *Neuro-Oncology, 16*(4), 520–527.

Markopoulos, G. S., Roupakia, E., Tokamani, M., Chavdoula, E., Hatziapostolou, M., Polytarchou, C., & Kolettas, E. (2017). A step-by-step microRNA guide to cancer development and metastasis. *Cellular Oncology, 40*(4), 303–339.

Melo, S. A., Sugimoto, H., O'Connell, J. T., Kato, N., Villanueva, A., Vidal, A., … Kalluri, R. (2014). Cancer exosomes perform cell-independent microRNA biogenesis and promote tumorigenesis. *Cancer Cell, 26*(5), 707–721.

Meng, X., Müller, V., Milde-Langosch, K., Trillsch, F., Pantel, K., & Schwarzenbach, H. (2016). Diagnostic and prognostic relevance of circulating exosomal miR-373, miR-200a, miR-200b and miR-200c in patients with epithelial ovarian cancer. *Oncotarget*, *7*(13), 16923–16935.

Mertens-Talcott, S. U., Chintharlapalli, S., Li, X., & Safe, S. (2007). The oncogenic microRNA-27a targets genes that regulate specificity protein transcription factors and the G2-M checkpoint in MDA-MB-231 breast cancer cells. *Cancer Research*, *67*(22), 11001–11011.

Mets, E., Van Peer, G., Van der Meulen, J., Boice, M., Taghon, T., Goossens, S., … Rondou, P. (2014). MicroRNA-128-3p is a novel oncomiR targeting PHF6 in T-cell acute lymphoblastic leukemia. *Haematologica*, *99*(8), 1326–1333.

Mikamori, M., Yamada, D., Eguchi, H., Hasegawa, S., Kishimoto, T., Tomimaru, Y., & Doki, Y. (2017). MicroRNA-155 controls exosome synthesis and promotes gemcitabine resistance in pancreatic ductal adenocarcinoma. *Scientific Reports*, *7*. https://doi.org/10.1038/srep42339.

Motamedinia, P., Scott, A. N., Bate, K. L., Sadeghi, N., Salazar, G., Shapiro, E., & Russo, L. M. (2016). Urine exosomes for non-invasive assessment of gene expression and mutations of prostate cancer. *PLoS One*, *11*(5), e0154507. https://doi.org/10.1371/journal.pone.0154507.

Motoyama, K., Inoue, H., Nakamura, Y., Uetake, H., Sugihara, K., & Mori, M. (2008). Clinical significance of high mobility group A2 in human gastric cancer and its relationship to let-7 microRNA family. *Clinical Cancer Research*, *14*(8), 2334–2340.

Muller, L., Hong, C.-S., Stolz, D. B., Watkins, S. C., & Whiteside, T. L. (2014). Isolation of biologically-active exosomes from human plasma. *Journal of Immunological Methods*, *411*, 55–65.

Munagala, R., Aqil, F., & Gupta, R. C. (2016). Exosomal miRNAs as biomarkers of recurrent lung cancer. *Tumour Biology*, *37*(8), 10703–10714.

Munoz, J. L., Rodriguez-Cruz, V., & Rameshwar, P. (2015). High expression of miR-9 in CD133+ glioblastoma cells in chemoresistance to temozolomide. *Journal of Cancer Stem Cell Research*, *3*, e100. https://doi.org/10.14343/JCSCR.2015.3e1003.

Ng, E. K., Chong, W. W., Lam, E. K., Shin, V. Y., Yu, J., Poon, T. C., & Sung, J. J. (2009). Differential expression of microRNAs in plasma of colorectal cancer patients: A potential marker for colorectal cancer screening. *Gut*. https://doi.org/10.1136/gut.2008.167817.

Ogata-Kawata, H., Izumiya, M., Kurioka, D., Honma, Y., Yamada, Y., Furuta, K., & Tsuchiya, N. (2014). Circulating exosomal microRNAs as biomarkers of colon cancer. *PLoS One*, *9*(4), e92921. https://doi.org/10.1371/journal.pone.0092921.

Ohshima, K., Inoue, K., Fujiwara, A., Hatakeyama, K., Kanto, K., Watanabe, Y., … Mochizuki, T. (2010). Let-7 microRNA family is selectively secreted into the extracellular environment via exosomes in a metastatic gastric cancer cell line. *PLoS One*, *5*(10), e13247. https://doi.org/10.1371/journal.pone.0013247.

Øverbye, A., Skotland, T., Koehler, C. J., Thiede, B., Seierstad, T., Berge, V., & Llorente, A. (2015). Identification of prostate cancer biomarkers in urinary exosomes. *Oncotarget*, *6*(30), 30357–30376.

Pan, L., Liang, W., Fu, M., Huang, Z. H., Li, X., Zhang, W., & Zhang, X. (2017). Exosomes-mediated transfer of long noncoding RNA ZFAS1 promotes gastric cancer progression. *Journal of Cancer Research and Clinical Oncology*, *143*(6), 991–1004.

Pang, Y., Young, C. Y., & Yuan, H. (2010). MicroRNAs and prostate cancer. *Acta Biochim Biophys Sin*, *42*(6), 363–369.

Pencheva, N., & Tavazoie, S. F. (2013). Control of metastatic progression by microRNA regulatory networks. *Nature Cell Biology*, *15*(6), 546–554.

Pink, R. C., Samuel, P., Massa, D., Caley, D. P., Brooks, S. A., & Carter, D. R. F. (2015). The passenger strand, miR-21-3p, plays a role in mediating cisplatin resistance in ovarian cancer cells. *Gynecologic Oncology*, *137*(1), 143–151.

Pogribny, I. P., Filkowski, J. N., Tryndyak, V. P., Golubov, A., Shpyleva, S. I., & Kovalchuk, O. (2010). Alterations of microRNAs and their targets are associated with acquired resistance of MCF-7 breast cancer cells to cisplatin. *International Journal of Cancer, 127*(8), 1785–1794.

Prensner, J. R., & Chinnaiyan, A. M. (2011). The emergence of lncRNAs in cancer biology. *Cancer Discovery, 1*(5), 391–407.

Pu, X. X., Huang, G. L., Guo, H. Q., Guo, C. C., Li, H., Ye, S., & Lin, T. Y. (2010). Circulating miR-221 directly amplified from plasma is a potential diagnostic and prognostic marker of colorectal cancer and is correlated with p53 expression. *Journal of Gastroenterology and Hepatology, 25*(10), 1674–1680.

Qin, W., Tsukasaki, Y., Dasgupta, S., Mukhopadhyay, N., Ikebe, M., & Sauter, E. R. (2016). Exosomes in human breast milk promote EMT. *Clinical Cancer Research, 22*(17), 4517–4524.

Qin, X., Yu, S., Zhou, L., Shi, M., Hu, Y., Xu, X., & Feng, J. (2017). Cisplatin-resistant lung cancer cell–derived exosomes increase cisplatin resistance of recipient cells in exosomal miR-100–5p-dependent manner. *International Journal of Nanomedicine, 12*, 3721–3733.

Rabinowits, G., Gerçel-Taylor, C., Day, J. M., Taylor, D. D., & Kloecker, G. H. (2009). Exosomal microRNA: A diagnostic marker for lung cancer. *Clinical Lung Cancer, 10*(1), 42–46.

Rana, S., Malinowska, K., & Zöller, M. (2013). Exosomal tumor microRNA modulates premetastatic organ cells. *Neoplasia, 15*(3), 281–295.

Riches, A., Powis, S., Mullen, P., Harrison, D., Hacker, C., Lucocq, J., & Chinn, D. J. (2016). Human urinary exosomes in bladder cancer patients: Properties, concentrations and possible clinical application. *Bladder, 3*(1), e19. https://doi.org/10.14440/bladder.2016.63.

Rodríguez, M., Silva, J., López-Alfonso, A., López-Muñiz, M. B., Peña, C., Domínguez, G., & García, V. (2014). Different exosome cargo from plasma/bronchoalveolar lavage in non-small-cell lung cancer. *Genes, Chromosomes & Cancer, 53*(9), 713–724.

Schotte, D., Pieters, R., & Den Boer, M. L. (2012). MicroRNAs in acute leukemia: From biological players to clinical contributors. *Leukemia, 26*(1), 1–12.

Shen, H., Zhu, F., Liu, J., Xu, T., Pei, D., Wang, R., & Shu, Y. (2014). Alteration in Mir-21/PTEN expression modulates gefitinib resistance in non-small cell lung cancer. *PLoS One, 9*(7), e103305. https://doi.org/10.1371/journal.pone.0013247.

Shi, J. (2016). Considering exosomal miR-21 as a biomarker for cancer. *Journal of Clinical Medicine, 5*(4), 42. https://doi.org/10.3390/jcm5040042.

Shi, R., Wang, P.-Y., Li, X.-Y., Chen, J.-X., Li, Y., Zhang, X.-Z., & Cheng, S.-J. (2015). Exosomal levels of miRNA-21 from cerebrospinal fluids associated with poor prognosis and tumor recurrence of glioma patients. *Oncotarget, 6*(29), 26971–26981.

Silva, J., García, V., Zaballos, A., Provencio, M., Lombardía, L., Almonacid, L., & Herrera, M. (2011). Vesicle-related microRNAs in plasma of non-small cell lung cancer patients and correlation with survival. *European Respiratory Journal, 37*(3), 617–623.

Silva, M., & Melo, S. A. (2015). Non-coding RNAs in exosomes: New players in cancer biology. *Current Genomics, 16*(5), 295–303.

Srivastava, A., Amreddy, N., Babu, A., Panneerselvam, J., Mehta, M., Muralidharan, R., & Ramesh, R. (2016). Nanosomes carrying doxorubicin exhibit potent anticancer activity against human lung cancer cells. *Scientific Reports, 6*. https://doi.org/10.1038/srep38541.

Srivastava, A., Filant, J., Moxley, K. M., Sood, A., McMeekin, S., & Ramesh, R. (2015). Exosomes: A role for naturally occurring nanovesicles in cancer growth, diagnosis and treatment. *Current Gene Therapy, 15*(2), 182–192.

Sun, D., Zhuang, X., Zhang, S., Deng, Z. B., Grizzle, W., Miller, D., & Zhang, H. G. (2013). Exosomes are endogenous nanoparticles that can deliver biological information between cells. *Advanced Drug Delivery Reviews, 65*(3), 342–347.

Taylor, D. D., & Gercel-Taylor, C. (2008). MicroRNA signatures of tumor-derived exosomes as diagnostic biomarkers of ovarian cancer. *Gynecologic Oncology, 110*(1), 13–21.

Thakur, B. K., Zhang, H., Becker, A., Matei, I., Huang, Y., Costa-Silva, B., & Williams, C. (2014). Double-stranded DNA in exosomes: A novel biomarker in cancer detection. *Cell Research, 24*(6), 766. https://doi.org/10.1038/cr.2014.44.

Théry, C., Zitvogel, L., & Amigorena, S. (2002). Exosomes: Composition, biogenesis and function. *Nature Reviews Immunology, 2*(8), 569–579.

Thind, A., & Wilson, C. (2016). Exosomal miRNAs as cancer biomarkers and therapeutic targets. *Journal of Extracellular Vesicles, 5*. https://doi.org/10.3402/jev.v5.31292.

Vaksman, O., Tropé, C., Davidson, B., & Reich, R. (2014). Exosome-derived miRNAs and ovarian carcinoma progression. *Carcinogenesis, 35*(9), 2113–2120.

Valadi, H., Ekström, K., Bossios, A., Sjöstrand, M., Lee, J. J., & Lötvall, J. O. (2007). Exosome-mediated transfer of mRNAs and microRNAs is a novel mechanism of genetic exchange between cells. *Nature Cell Biology, 9*(6), 654–659.

Whiteside, T. L. (2015). The potential of tumor-derived exosomes for noninvasive cancer monitoring. *Expert Review of Molecular Diagnostics, 15*(10), 1293–1310.

Wu, L., Pu, X., Wang, Q., Cao, J., Xu, F., Xu, L. I., & Li, K. (2016). miR-96 induces cisplatin chemoresistance in non-small cell lung cancer cells by downregulating SAMD9. *Oncology Letters, 11*(2), 945–952.

Xiao, D., Ohlendorf, J., Chen, Y., Taylor, D. D., Rai, S. N., Waigel, S., & McMasters, K. M. (2012). Identifying mRNA, microRNA and protein profiles of melanoma exosomes. *PLoS One, 7*(10), e46874. https://doi.org/10.1371/journal.pone.0046874.

Xiao, X., Yu, S., Li, S., Wu, J., Ma, R., Cao, H., & Feng, J. (2014). Exosomes: Decreased sensitivity of lung cancer A549 cells to cisplatin. *PLoS One, 9*(2), e89534. https://doi.org/10.1371/journal.pone.0089534.

Yang, C., & Robbins, P. D. (2011). The roles of tumor-derived exosomes in cancer pathogenesis. *Clinical and Developmental Immunology*. https://doi.org/10.1155/2011/842849.

Yang, J. K., Yang, J. P., Tong, J., Jing, S. Y., Fan, B., Wang, F., & Jiao, B. H. (2017). Exosomal miR-221 targets DNM3 to induce tumor progression and temozolomide resistance in glioma. *Journal of Neuro-Oncology, 131*(2), 255–265.

Yang, Y., Gu, X., Zhou, M., Xiang, J., & Chen, Z. (2013). Serum microRNAs: A new diagnostic method for colorectal cancer. *Biomedical Reports, 1*(4), 495–498.

Ye, S., Li, Z.-L., Luo, D., Huang, B., Chen, Y.-S., Zhang, X., & Li, J. (2014). Tumor-derived exosomes promote tumor progression and T-cell dysfunction through the regulation of enriched exosomal microRNAs in human nasopharyngeal carcinoma. *Oncotarget, 5*(14), 5439–5452.

Yong, F. L., Law, C. W., & Wang, C. W. (2013). Potentiality of a triple microRNA classifier: miR-193a-3p, miR-23a and miR-338-5p for early detection of colorectal cancer. *BMC Cancer, 13*, 280. https://doi.org/10.1186/1471-2407-13-280.

Zhang, X., Yuan, X., Shi, H., Wu, L., Qian, H., & Xu, W. (2015). Exosomes in cancer: Small particle, big player. *Journal of Hematology & Oncology, 8*(1), 83. https://doi.org/10.1186/s13045-015-0181-x.

Zhang, Y., Sui, J., Shen, X., Li, C., Yao, W., Hong, W., & Liang, G. (2017). Differential expression profiles of microRNAs as potential biomarkers for the early diagnosis of lung cancer. *Oncology Reports, 37*(6), 3543–3553.

Zhao, L., Liu, W., Xiao, J., & Cao, B. (2015). The role of exosomes and "exosomal shuttle microRNA" in tumorigenesis and drug resistance. *Cancer Letters, 356*(2), 339–346.

Zheng, P., Chen, L., Yuan, X., Luo, Q., Liu, Y., Xie, G., & Shen, L. (2017). Exosomal transfer of tumor-associated macrophage-derived miR-21 confers cisplatin resistance in gastric cancer cells. *Journal of Experimental & Clinical Cancer Research, 36*(1), 53. https://doi.org/10.1186/s13046-017-0528-y.

Zhou, W., Fong, M. Y., Min, Y., Somlo, G., Liu, L., Palomares, M. R., … Wang, S. E. (2014). Cancer-secreted miR-105 destroys vascular endothelial barriers to promote metastasis. *Cancer Cell, 25*(4), 501–515.

Zhu, S., Wu, H., Wu, F., Nie, D., Sheng, S., & Mo, Y. Y. (2008). MicroRNA-21 targets tumor suppressor genes in invasion and metastasis. *Cell Research, 18*(3), 350–359.

Further Reading

Lu, L., Ju, F., Zhao, H., & Ma, X. (2015). MicroRNA-134 modulates resistance to doxorubicin in human breast cancer cells by downregulating ABCC1. *Biotechnology Letters, 37*(12), 2387–2394.

9

Diagnostic Potential of Tumor Exosomes

Philip Hochendoner, Zheng Zhao, Mei He
KANSAS STATE UNIVERSITY, MANHATTAN, KS, UNITED STATES

CHAPTER OUTLINE

1. Introduction 161
2. Methods and Tools for Exosome Isolation 163
   2.1 Ultracentrifugation 163
   2.2 ExoQuick 163
   2.3 Filtration 163
   2.4 Microfluidic 164
3. Tumor Exosomes for Cancer Diagnosis 164
4. Additional Diagnostic Potential of Exosomes 170
5. Conclusion/Discussion 170
References 171

1. Introduction

Cancer is a difficult disease to diagnose, treat, and live with. The pervasive nature of tumors and metastasis is one of great importance for us to mitigate as life expectancy continues to rise. The early detection and use of diagnostic tools have always been key for treating cancers in the past. Traditional methods of cancer diagnosis mainly rely on the imaging or surgical techniques to find tumors (e.g., MRI and tumor biopsy), which are costly, invasive, and associated with health risks that patients may not want to undergo (Zhou et al., 2017). In contrast, blood-based diagnosis is simple and less invasive, which only requires very little body fluids, and can be done without additional risks. The blood-based diagnosis also holds the promise for detecting early stage of cancer before the solid tumor forms (Carter et al., 2017; Esposito et al., 2017). However, several blood-based biomarker tests have been developed for decades with results that are still arguable, including prostate-specific antigen screening for prostate cancer, carcinoembryonic antigen for colorectal cancer, CA19-9 for pancreatic cancer, and CA125 for ovarian cancer (Sawyers, 2008). The

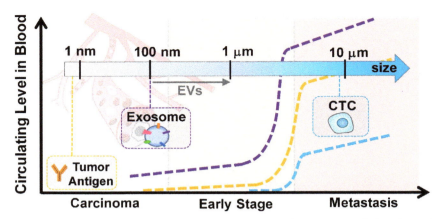

FIGURE 1 Illustration of circulating tumor antigens, exosomes, and circulating tumor cells (CTCs) associated with tumor status (not to scale). *Figure was adapted with permission He, M., & Zeng, Y. (2016). Microfluidic exosome analysis toward liquid biopsy for cancer.* Journal of Laboratory Automation, 21(4), 599–608.

lack of reliability and specificity of blood biomarkers prevents widespread utilization of liquid biopsy for cancers. The simple blood tests are increasingly needed along with the emergence of personalized medicine and point-of-care diagnosis.

Recently, it has been found that circulating exosomes derived from tumorous cells could lead to the simple blood tests for early detection of cancers (Jia et al., 2017; Munson & Shukla, 2015; Staals & Pruijn, 2010; Urbanelli et al., 2015). Exosomes are cellular endocytic pathway–derived extracellular nanovesicles (30–150 nm), which are systemically detectable in the blood of variable cancer patients and have been shown to correlate well with tumor progression, immune response suppression, angiogenesis, and metastasis (Hessvik & Llorente, 2017; Thery, Zitvogel, & Amigorena, 2002; Zhang et al., 2015). Exosomes are actively released from tumor cells with much higher concentration in the circulation ($\sim \times 10^{11}$ vesicles per mL in blood), compared with the circulating tumor cells (CTCs) (1–10 CTCs per mL of blood), as illustrated in Fig. 1 (Beach et al., 2014). Exosomes carry enriched genetic material and proteins from their cell of origin, thereby holding great promise for identifying early-stage tumors (Vlassov et al., 2012).

However, the effective isolation of exosome subpopulations is still challenging to date, due to the presence of other membrane-derived subcellular structures, such as apoptotic vesicles, exosome-like vesicles, membrane particles, and ectosomes (Akers et al., 2013; Choi et al., 2015; Coakley, Maizels, & Buck, 2015; Minciacchi, Freeman, & DI Vizio, 2015). Currently, the methods used to isolate the exosomes from the biological fluid often differ with a range of results, benefits, and disadvantages. In this chapter, we will discuss the techniques used for exosome isolation, including ultracentrifugation, filtration, microfluidics, and chromatography, and demonstrate the capability of tumor-derived exosomes for cancer diagnosis (Contreras-Naranjo, Wu, & Ugaz, 2017; He & Zeng, 2016; Li et al., 2017; Liga et al., 2015).

2. Methods and Tools for Exosome Isolation

In the recent 5 years, exosome research has been in exponential growth with technology development and discovery on manipulating exosomes and making them as easy to use as possible. There are four most common protocols that are currently in use for isolating exosomes from biological fluids, including several commercial-ready products such as ExoQuick (commercially available centrifuge-based kit) (Taylor, Zacharias, & Gercel-Taylor, 2011), ExoSearch (microfluidic-based isolation and diagnostic chip) (Zhao et al., 2016), and RNAPro-Sal (filtration-based isolation device) (Chiang et al., 2015).

2.1 Ultracentrifugation

The most common and widely used method for isolating exosomes from their host fluids is via the high-speed ultracentrifuge (Campoy et al., 2016; Greening et al., 2015; Helwa et al., 2017; Kenigsberg et al., 2017). Spinning at upwards of 10^5 gravity force, this separates out the fluid, debris particles, and leftover cells from the exosomes in a highly concentrated pellet. This process generally needs to be repeated multiple times, especially when high purity of the exosomes is desired. The isolation protocols are routine and easy to follow. However, high shear strength from spinning may cause exosome membrane fusion or damage (Greening et al., 2015). In addition, such isolation copurifies exosomes with other extracellular vesicles due to substantial size overlap (Bobrie et al., 2012).

2.2 ExoQuick

As an alternative to ultracentrifugation, there is a commercially available kit for exosome isolation known as ExoQuick from System Biosciences, which only uses cost-effective benchtop centrifuge (Helwa et al., 2017; Kenigsberg et al., 2017). Less time is needed to isolate the exosomes from a wide variety of biofluids, including saliva, urine, follicular fluid, plasma, serum, tissue culture media, breast milk, and malignant ascites. The exosome isolation is based on the reagent precipitation using volume-excluding polymers, which may precipitate free proteins presented in the sample and lead to overestimated protein contents. Meanwhile, each sample type requires a different precipitation kit.

2.3 Filtration

By taking advantage of exosomes in a specific size range, ultrafiltration can be used to isolate exosomes from biological fluids (Taylor et al., 2011). Generally, a semipermeable membrane is used within a thin layer of material capable of separating substances when a driving force is applied across the membrane. The porous membrane can trap exosomes as the sample fluid is forced through, where they are then eluted by an ethanol or NaOH buffer. Most common filtration membranes have pore sizes of 0.45 µm or 0.22 µm and can be used to collect exosomes between 400 nm and 200 nm. The ultrafiltration approach is scalable, however hard to process exosomes smaller than 100 nm.

2.4 Microfluidic

The microfluidic approach is one that can be very effective as a laboratory tool for exosome isolation (He & Zeng, 2016; Liga et al., 2015; Yang et al., 2017). For the past 5 years, research efforts have increased dramatically to develop microfluidic platforms for exosome isolation. Compared with benchtop methods, microfluidic technology offers fast isolation speed, high yield and efficiency, automation, and functional integration for streamlined exosome molecular analysis. The current reported microfluidic platforms demonstrated the efficient exosome isolation from either size selection or surface marker selection (immunoaffinity) to define a specific population of circulating exosomes. The immunoaffinity-based exosome isolation can be implemented into microfluidic devices by manipulating affinity particles/magnetic beads, or modifying microchannel surfaces with antibodies (Chen et al., 2010; He et al., 2014; Kanwar et al., 2014; Zhao et al., 2016). The size-based isolation has advantages of size uniformity without sample bias by integrating micropillars (Wang et al., 2013), nanoporous membranes (Lee et al., 2015), or using hydrodynamic fluidic filtration (Davies et al., 2012) in a microfluidic system. Only a very small volume of fluid is needed during the isolation, and the devices themselves are small and made from cheap materials (typically PDMS). The microfluidic channel design and the architecture allows for a wide variety of tests and biomolecular analysis to be conducted.

3. Tumor Exosomes for Cancer Diagnosis

Tumor cell–derived exosomes contain common exosomal proteins, as well as the tumor antigens that are reflective of tumor signatures (Atay & Godwin, 2014; Guo & Guo, 2015; Filipazzi et al., 2012; Kahlert & Kalluri, 2013; Kharaziha et al., 2012; Principe et al., 2013; Yang & Robbins, 2011). Recent studies have suggested that both tumor cells and normal cells secrete exosomes, although significantly higher amounts of exosomes have been observed from tumor cells. Therefore, specifically isolating, purifying, and characterizing tumor cell–derived exosomes is essential.

One example that highlights the microfluidic-based exosome isolation integrated with tumor exosome analysis is the ExoSearch chip we developed recently (Zhao et al., 2016). A simple and robust microfluidic continuous-flow platform (ExoSearch chip) was designed for rapid exosome isolation coupled with in situ, multiplexed detection of exospores, as shown in Fig. 2.

Several microfluidic approaches have been previously developed for exosome analysis, such as isolation, quantification, and molecular profiling (Im et al., 2014; Jeong et al., 2016; Rho et al., 2013; Shao et al., 2012, 2015; Vaidyanathan et al., 2014; Zhang, He, & Zeng, 2016). However, these platforms require either complicated fabrication or sophisticated sensing methods. Compared with other existing microfluidic methods, the ExoSearch chip possesses distinct features: first, continuous-flow operation affords dynamic scalability in processing sample volumes from microliter for on-chip analysis to milliliter-scale preparation of exosomes for benchtop measurements; second, it

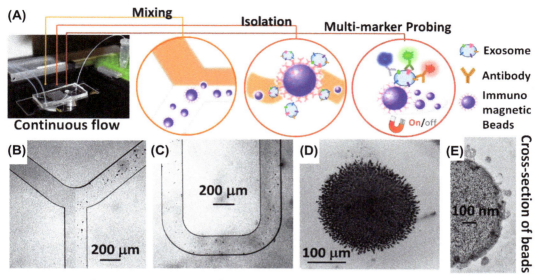

FIGURE 2 (A) Workflow of the ExoSearch chip for continuous mixing, isolation, and in situ, multiplexed detection of circulating exosomes. (B)–(C) Bright-field microscope images of immunomagnetic beads manipulated in microfluidic channel for mixing and isolation of exosomes. (D) Exosome-bound immunomagnetic beads aggregated in a microchamber with on/off switchable magnet for continuous collection and release of exosomes. (E) Transmission electron microscope image of exosome-bound immunomagnetic bead in a cross-sectional view. *Figure was adapted with permission Zhao, Z., Yang, Y., Zeng, Y., et al. (2016). A microfluidic ExoSearch chip for multiplexed exosome detection towards blood-based ovarian cancer diagnosis. Lab on a Chip, 16(3), 489–496.*

enables in situ, multiplexed quantification of marker combinations in one sample with a much improved speed (~40 min); last, because of the simplicity and robustness of the ExoSearch chip, it holds the potential to develop into a viable technology in cancer diagnosis in the clinical setting. Ovarian cancer diagnosis has been demonstrated using the microfluidic ExoSearch chip to quantify tumor-derived circulating exosomes in small-volume blood plasma samples (20 μL). The three exosomal biomarkers (EpCAM, CA-125, and CD24) have been quantified and showed significant diagnostic accuracy, which is comparable with a standard Bradford assay. These results would support the diagnostic potential of blood exosomes and develop minimally invasive diagnosis of cancer using ExoSearch chip (Zhao et al., 2016).

We characterized the specificity for microfluidic-based immunomagnetic isolation of exosomes from ovarian cancer patient blood plasma samples. The microfluidic isolation of variable exosome subpopulations was conducted by targeting both ovarian tumor–associated markers (EpCAM and CA-125) and common exosomal markers (CD9, CD81, and CD63). The EpCAM is a cargo protein in exosomes and is highly overexpressed in multiple types of carcinomas, including ovarian tumor. CA-125 antigen is the most commonly measured biomarker for epithelial ovarian tumors, which accounts for 85%–90% of ovarian cancer (Yuan et al., 2017). The exosome-bound beads were washed on the chip and then released and concentrated for morphology evaluation by counting intact exosomes

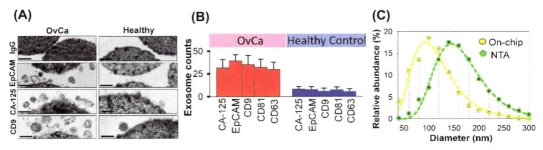

FIGURE 3 Microfluidic ExoSearch chip for specific isolation of ovarian cancer plasma–derived exosomes. (A) Transmission electron microscope (TEM) images of on-chip immunomagnetically isolated exosomes from ovarian cancer plasma, compared with healthy control. Scale bar is 100 nm. IgG-conjugated immunomagnetic beads were negative control beads. (B) Exosome counts analyzed from surfaces of variable capture beads (EpCAM+, CA-125+, CD9+, CD81+, CD63+) using TEM particle analysis (n = 25, CV = 2.8%–10%). Single bead diameter was 2.8 μm and sliced bead layer was 80-nm thick. (C) Size distribution of on-chip isolated exosomes (CD9+) using TEM particle analysis, compared with standard nanoparticle tracking analysis (NTA) analysis of ultracentrifugation-purified exosomes. *Dashed lines* were log-normal fit (R2 > 0.98). *Figure was adapted with permission Zhao, Z., Yang, Y., Zeng, Y., et al. (2016). A microfluidic ExoSearch chip for multiplexed exosome detection towards blood-based ovarian cancer diagnosis.* Lab on a Chip, 16(3), 489–496.

using a transmission electron microscope (TEM) (as shown in Fig. 3A). Significantly higher amounts of round membrane vesicles (smaller than 150 nm) were observed for EpCAM+, CA-125+, and CD9+ subpopulations from ovarian cancer plasma, compared with healthy controls. Negative control beads with IgG conjugation showed negative capture of vesicles, demonstrating a good specificity of immunomagnetic isolation. The relative expression levels of five surface markers were measured by counting the number of intact exosomes bound to beads (n = 25). The results showed a ~3–5-fold increase in expression level for selected five markers from ovarian cancer patients, compared with the healthy controls (Fig. 3B, $P = .001$). To verify the results of microfluidic isolation, we conducted nanoparticle tracking analysis (NTA) of ultracentrifugation-isolated exosomes to measure their size distribution and concentrations. In Fig. 3C, microfluidic isolated exosomes (CD9+) exhibited a notably narrower range with the log-normal fitted size distribution (R2 > 0.98). The smaller size than 150 nm is a commonly used criterion to differentiate exosomes from larger microvesicles. Compared with ultracentrifugation approaches, microfluidic immunoaffinity isolation yields a higher percentage of vesicles smaller than 150 nm (~79.7% vs. 60.7%), suggesting that developed ExoSearch chip offers high specificity in isolation of circulating exosomes.

The quantitative isolation and detection of exosomes were characterized in Fig. 4A, which shows the fluorescence images of exosomes isolated from serial dilutions of purified, fluorescently labeled plasma exosomes. The concentrations of purified plasma exosomes were determined by NTA measurements. Employing the same mixing and isolating conditions, increased fluorescence signals (ΔFL) were observed and proportional to exosome concentrations. Using fluorescently labeled anti-EpCAM as the detection antibody, exosome titration curves were obtained for a healthy plasma sample and an ovarian

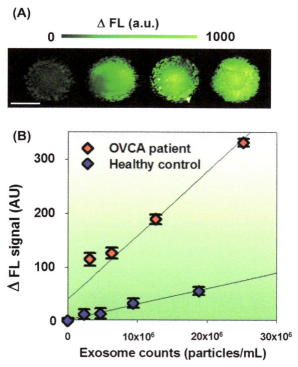

FIGURE 4 (A) The charge-coupled device (CCD) camera images of bead aggregates in ExoSearch chip captured with fluorescence-labeled plasma exosomes in serial dilutions (from left to right: 5×10^5, 1×10^6, 5×10^6, 1×10^7 particles/mL). Scale bar was 100 μm. (B) Calibration curves for quantitative detection of intact exosomes (R2 >0.98, CV = ~5%). Exosomes were purified from one healthy control plasma and one ovarian cancer patient plasma using ultracentrifugation. Concentrations were measured by nanoparticle tracking analysis. *Figure was adapted with permission Zhao, Z., Yang, Y., Zeng, Y., et al. (2016). A microfluidic ExoSearch chip for multiplexed exosome detection towards blood-based ovarian cancer diagnosis.* Lab on a Chip, 16*(3), 489–496.*

cancer plasma, which exhibited good linear response as seen in Fig. 4B (R2 >0.98, CV ~5%). Moreover, a much higher ΔFL signal (~30-fold increase) was observed for the ovarian cancer sample, compared with the healthy control under the same concentration.

These results demonstrate the ability of the ExoSeach chip in quantitative measurement of exosome surface markers for differentiating changes associated with disease. The quantitative detection of intact exosomes was achieved with a limit of detection of 7.5×10^5 particles/mL (LOD, S/N = 3), which is 1000-fold sensitive than Western blotting. Nonspecific adsorption and antibody cross-reactivity were characterized in Fig. 5, showing a low level of nonspecific interference.

In situ, multiplexed biomarker detection was then developed for rapid and quantitative microfluidic analysis of ovarian tumor–derived plasma exosomes. We chose common exosome marker CD9 as the capture antibody for selective isolation of exosomes because of the consistently high expression of CD9 observed from human plasma–derived exosomes. In addition to the established ovarian cancer biomarker CA-125, human epididymis protein

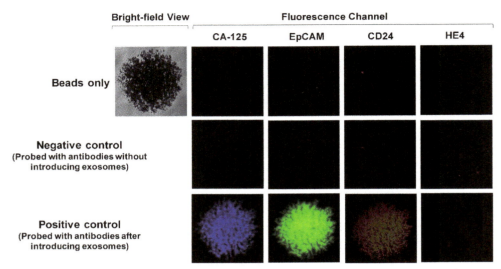

FIGURE 5 CCD-captured microscopic images of beads aggregates under negative and positive control experimental conditions. Image size is 200 μm × 200 μm. *Figure was adapted with permission Zhao, Z., Yang, Y., Zeng, Y., et al. (2016). A microfluidic ExoSearch chip for multiplexed exosome detection towards blood-based ovarian cancer diagnosis.* Lab on a Chip, 16*(3), 489–496.*

4 (HE4) has been recognized for improving diagnostic specificity of CA-125 in pathological tests. We did not observe substantial expression of HE4 from the exosome surface in plasma samples (Fig. 5), which could be due to the secretion pathway of HE4 that is not involved with surface membrane. Previous observations have indicated that CD24 could be a significant marker in ovarian tumor prognosis and diagnosis (Nakamura et al., 2017). Therefore, we developed a multiplexed sandwich immunofluorescence assay to quantify the exosomes by targeting three markers, CA-125, EpCAM, and CD24, at the same time, as exemplified in Fig. 6. Quantitative tests of human plasma collected from 20 subjects ($n_{OvCa} = 15$, $n_{healthy} = 5$) were conducted for a three-marker classification of ovarian tumor–derived exosomes, and a distinctive three-marker expression pattern was observed for ovarian cancer patients (Fig. 6). The average expression level of each exosomal marker from ovarian cancer patients was significantly higher as compared with healthy controls (CD24: 3-fold increase, $P = .003$; EpCAM: 6.5-fold increase, $P = .0009$; CA-125: 12.4-fold increase, $P < .0001$).

Currently, there is no single marker that can detect early-stage ovarian cancer with desired sensitivity and specificity (>98%). A large number of combinations of biomarkers have been investigated to improve diagnostic sensitivity and specificity. Circulating exosomes, enriched with a group of tumor antigens, provide a unique opportunity for cancer diagnosis using multi-marker combination. To this end, we assessed the ExoSearch chip for blood-based diagnosis of ovarian cancer by simultaneously detecting three tumor antigens in the same exosome subpopulations from ovarian cancer and healthy human plasma. The standard Bradford assay of total protein levels in ultracentrifugation-purified exosomes from matched human subjects was performed for parallel comparison. A total of

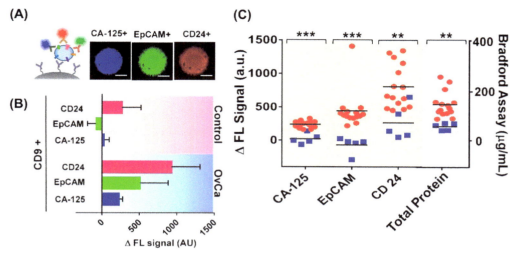

FIGURE 6 (A) CCD images of multiplexed three-color fluorescence detection of tumor markers (CA-125, EpCAM, CD24) from captured exosome subpopulation (CD9+). Scale bar was 50 μm, indicating bead aggregate size. (B) Average expression levels of three ovarian tumor markers measured by ExoSearch chip from 20 human subjects ($n_{OvCa} = 15$, $n_{healthy} = 5$). Error bars indicate standard deviations. (C) Scattering plots of expression levels of three tumor markers (CA-125, $P < 10\text{-}4$; EpCAM, $P = .0009$; CD24, $P = .003$) from blood plasma–derived exosomes ($n_{OvCa} = 15$, $n_{healthy} = 5$), compared with the standard Bradford assay of total proteins ($P = .0013$) in ultracentrifugation-purified exosomes from matched human subjects. *Black lines* indicate average expression levels of each group. Ovarian cancer patients were represented by *red dots* and healthy controls were represented by *blue dots*. *Figure was adapted with permission Zhao, Z., Yang, Y., Zeng, Y., et al. (2016). A microfluidic ExoSearch chip for multiplexed exosome detection towards blood-based ovarian cancer diagnosis.* Lab on a Chip, 16(3), 489–496.

Table 1   Diagnostic Accuracy Analysis Using the Receiver Operating Characteristic Curve

Test Variables	CA125	ExoSearch Chip EpCAM	CD24	Bradford Assay/Total Exosomal Protein	NTA/Particle Concentration
ROC curve area	1.000	1.000	0.9067	1.000	0.6750
Standard error	0.000	0.000	0.0903	0.000	0.1332
95% Confidence interval	1.000 to 1.000	1.000 to 1.000	0.729 to 1.084	1.000 to 1.000	0.413 to 0.936
P value	0.0010	0.0010	0.0078	0.0009	0.2477

20 human subjects ($n_{OvCa} = 15$, $n_{healthy} = 5$) were chosen for evaluating diagnostic accuracy, based on receiver operating characteristic (ROC) analysis of adequate sample size (Table 1). Both ExoSearch and Bradford assay showed a significantly increased level of exosome proteins from ovarian cancer patients, compared with healthy controls (Fig. 6, Bradford assay $P = .001$; ExoSearch chip $P < .001$). Particularly, the ExoSearch chip gave individual exosomal protein expression level and the levels of CA-125 and EpCAM showed extremely significant differences between ovarian cancer patients and healthy controls (EpCAM $P = .0009$; CA-125 $P < 10\text{-}4$). To determine the diagnostic accuracy of ExoSearch chip assay, we analyzed the true positives (sensitivity) and false positives (1-specificity) by ROC curves.

The areas under the curves (a.u.c.) obtained for CA-125, EpCAM, and CD24 were 1.0, 1.0, and 0.91, respectively, which are comparable with the standard Bradford assay (a.u.c. = 1.0, 95% CI) (Table 1). However, the diagnostic accuracy of using exosomal particle concentrations measured by NTA was relatively poor, with a.u.c. of only 0.67. It could be attributed to the variation of NTA measurement, which gives relatively large uncertainty in size and concentration. By ROC analysis (Table 1), the ExoSearch chip assay was highly accurate in discriminating plasma exosomes from ovarian cancer patients versus healthy individuals. The above results suggested that the ExoSearch chip enables sensitive multiplexed exosomal marker detection for blood-based diagnosis of ovarian cancer with significant predictive power. The combination of plasma exosomal markers CA-125, EpCAM, and CD24 provided desirable diagnostic accuracy for noninvasive, early detection of ovarian cancer.

4. Additional Diagnostic Potential of Exosomes

Although beyond the scope of this chapter, exosomes have greater potential beside tumor identification and cancer diagnosis. In addition to the identification of molecular contents (RNA, proteins, etc.), exosomes are also superior nanobiomaterials for drug delivery, immunotherapy, and biological regulation (Kooijmans et al., 2012). It has been shown that exosomes can be engineered for delivering therapeutic drugs and priming the immune system in cancer immunotherapy (Moore et al., 2017; Morishita et al., 2016; Yim et al., 2016). Although this is in the early stages of research, it is showing vast growth potential and will hopefully allow exosomes to become an exceptional medical tool.

5. Conclusion/Discussion

Tumor-derived circulating exosomes have attracted substantial interest for noninvasive cancer diagnosis and monitoring of treatment responses. Exosome secretion is a dynamic process, producing diverse populations with 5-fold differences in size and 10^4-fold differences in concentration; as a result, precise measurement and analysis of exosomes is challenging (Raposo & Stoorvogel, 2013). To date, standardized isolation protocols are still lacking, and the field of exosome research lags behind CTC research because the definition and characterization of exosome types are not yet firmly established. To increase understanding of exosomes and quantitatively decipher exosomal components, more novel technologies are needed for comprehensive characterization of surface and intravesicular compositions. Interconnections between exosomal RNA, surface protein topography, and posttranslational modification could aid identification of exosome associated with cancer phenotypes. To achieve clinical utilities in liquid biopsy for cancer, tremendous efforts are still needed to improve the adaptability of the microfluidic technologies to clinical settings and promote the commercialization of the systems. Cancer is a complicated and dynamic disease which requires reliable and novel liquid biopsy platform for advancing the patient care. We anticipate that microfluidic technology will play a game-changing role in exosome-based cancer diagnosis.

References

Akers, J. C., Gonda, D., Kim, R., et al. (2013). Biogenesis of extracellular vesicles (EV): Exosomes, microvesicles, retrovirus-like vesicles, and apoptotic bodies. *Journal of Neuro-Oncology, 113*(1), 1–11.

Atay, S., & Godwin, A. K. (2014). Tumor-derived exosomes: A message delivery system for tumor progression. *Communicative & Integrative Biology, 7*(1), e28231.

Beach, A., Zhang, H. G., Ratajczak, M. Z., et al. (2014). Exosomes: An overview of biogenesis, composition and role in ovarian cancer. *Journal of Ovarian Research, 7*, 14.

Bobrie, A., Colombo, M., Krumeich, S., et al. (2012). Diverse subpopulations of vesicles secreted by different intracellular mechanisms are present in exosome preparations obtained by differential ultracentrifugation. *Journal of Extracellular Vesicles, 1*.

Campoy, I., Lanau, L., Altadill, T., et al. (2016). Exosome-like vesicles in uterine aspirates: A comparison of ultracentrifugation-based isolation protocols. *Journal of Translational Medicine, 14*(1), 180.

Carter, J. V., Galbraith, N. J., Yang, D., et al. (2017). Blood-based microRNAs as biomarkers for the diagnosis of colorectal cancer: A systematic review and meta-analysis. *British Journal of Cancer, 116*(6), 762–774.

Chen, C., Skog, J., Hsu, C. H., et al. (2010). Microfluidic isolation and transcriptome analysis of serum microvesicles. *Lab on a Chip, 10*(4), 505–511.

Chiang, S. H., Thomas, G. A., Liao, W., et al. (2015). RNAPro*SAL: A device for rapid and standardized collection of saliva RNA and proteins. *Biotechniques, 58*(2), 69–76.

Choi, D. S., Kim, D. K., Kim, Y. K., et al. (2015). Proteomics of extracellular vesicles: Exosomes and ectosomes. *Mass Spectrometry Reviews, 34*(4), 474–490.

Coakley, G., Maizels, R. M., & Buck, A. H. (2015). Exosomes and other extracellular vesicles: The new communicators in parasite infections. *Trends in Parasitology, 31*(10), 477–489.

Contreras-Naranjo, J. C., Wu, H. J., & Ugaz, V. M. (2017). Microfluidics for exosome isolation and analysis: Enabling liquid biopsy for personalized medicine. *Lab on a Chip, 17*.

Davies, R. T., Kim, J., Jang, S. C., et al. (2012). Microfluidic filtration system to isolate extracellular vesicles from blood. *Lab on a Chip, 12*(24), 5202–5210.

Esposito, A., Criscitiello, C., Trapani, D., et al. (2017). The emerging role of "liquid biopsies," circulating tumor cells, and circulating cell-free tumor DNA in lung cancer diagnosis and identification of resistance mutations. *Current Oncology Reports, 19*(1), 1.

Filipazzi, P., Burdek, M., Villa, A., et al. (2012). Recent advances on the role of tumor exosomes in immunosuppression and disease progression. *Seminars in Cancer Biology, 22*(4), 342–349.

Greening, D. W., Xu, R., Ji, H., et al. (2015). A protocol for exosome isolation and characterization: Evaluation of ultracentrifugation, density-gradient separation, and immunoaffinity capture methods. *Methods in Molecular Biology, 1295*, 179–209.

Guo, L., & Guo, N. (2015). Exosomes: Potent regulators of tumor malignancy and potential bio-tools in clinical application. *Critical Reviews in Oncology, 95*(3), 346–358.

He, M., Crow, J., Roth, M., et al. (2014). Integrated immunoisolation and protein analysis of circulating exosomes using microfluidic technology. *Lab on a Chip, 14*(19), 3773–3780.

Helwa, I., Cai, J., Drewry, M. D., et al. (2017). A comparative study of serum exosome isolation using differential ultracentrifugation and three commercial reagents. *PLoS One, 12*(1), e0170628.

Hessvik, N. P., & Llorente, A. (2017). Current knowledge on exosome biogenesis and release. *Cellular and Molecular Life Sciences, 75*(2), 193–208.

He, M., & Zeng, Y. (2016). Microfluidic exosome analysis toward liquid biopsy for cancer. *Journal of Laboratory Automation, 21*(4), 599–608.

Im, H., Shao, H., Park, Y. I., et al. (2014). Label-free detection and molecular profiling of exosomes with a nano-plasmonic sensor. *Nature Biotechnology, 32*(5), 490–495.

Jeong, S., Park, J., Pathania, D., et al. (2016). Integrated magneto-electrochemical sensor for exosome analysis. *ACS Nano, 10*(2), 1802–1809.

Jia, Y., Chen, Y., Wang, Q., et al. (2017). Exosome: Emerging biomarker in breast cancer. *Oncotarget, 8*(25), 41717–41733.

Kahlert, C., & Kalluri, R. (2013). Exosomes in tumor microenvironment influence cancer progression and metastasis. *Journal of Molecular Medicine (Berlin), 91*(4), 431–437.

Kanwar, S. S., Dunlay, C. J., Simeone, D. M., et al. (2014). Microfluidic device (ExoChip) for on-chip isolation, quantification and characterization of circulating exosomes. *Lab on a Chip, 14*(11), 1891–1900.

Kenigsberg, S., Wyse, B. A., Librach, C. L., et al. (2017). Protocol for exosome isolation from small volume of ovarian follicular fluid: Evaluation of ultracentrifugation and commercial kits. *Methods in Molecular Biology, 1660*, 321–341.

Kharaziha, P., Ceder, S., Li, Q., et al. (2012). Tumor cell-derived exosomes: A message in a bottle. *Biochimica et Biophysica Acta, 1826*(1), 103–111.

Kooijmans, S. A., Vader, P., van Dommelen, S. M., et al. (2012). Exosome mimetics: A novel class of drug delivery systems. *International Journal of Nanomedicine, 7*, 1525–1541.

Lee, K., Shao, H., Weissleder, R., et al. (2015). Acoustic purification of extracellular microvesicles. *ACS Nano, 9*(3), 2321–2327.

Liga, A., Vliegenthart, A. D., Oosthuyzen, W., et al. (2015). Exosome isolation: A microfluidic road-map. *Lab on a Chip, 15*(11), 2388–2394.

Li, P., Kaslan, M., Lee, S. H., et al. (2017). Progress in exosome isolation techniques. *Theranostics, 7*(3), 789–804.

Minciacchi, V. R., Freeman, M. R., & Di Vizio, D. (2015). Extracellular vesicles in cancer: Exosomes, microvesicles and the emerging role of large oncosomes. *Seminars in Cell & Developmental Biology, 40*, 41–51.

Moore, C., Kosgodage, U., Lange, S., et al. (2017). The emerging role of exosome and microvesicle- (EMV-) based cancer therapeutics and immunotherapy. *International Journal of Cancer, 141*(3), 428–436.

Morishita, M., Takahashi, Y., Matsumoto, A., et al. (2016). Exosome-based tumor antigens-adjuvant co-delivery utilizing genetically engineered tumor cell-derived exosomes with immunostimulatory CpG DNA. *Biomaterials, 111*, 55–65.

Munson, P., & Shukla, A. (2015). Exosomes: Potential in cancer diagnosis and therapy. *Medicines (Basel), 2*(4), 310–327.

Nakamura, K., Terai, Y., Tanabe, A., et al. (2017). CD24 expression is a marker for predicting clinical outcome and regulates the epithelial-mesenchymal transition in ovarian cancer via both the Akt and ERK pathways. *Oncology Reports, 37*(6), 3189–3200.

Principe, S., Hui, A. B., Bruce, J., et al. (2013). Tumor-derived exosomes and microvesicles in head and neck cancer: Implications for tumor biology and biomarker discovery. *Proteomics, 13*(10–11), 1608–1623.

Raposo, G., & Stoorvogel, W. (2013). Extracellular vesicles: Exosomes, microvesicles, and friends. *The Journal of Cell Biology, 200*(4), 373–383.

Rho, J., Chung, J., Im, H., et al. (2013). Magnetic nanosensor for detection and profiling of erythrocyte-derived microvesicles. *ACS Nano, 7*(12), 11227–11233.

Sawyers, C. L. (2008). The cancer biomarker problem. *Nature, 452*(7187), 548–552.

Shao, H., Chung, J., Balaj, L., et al. (2012). Protein typing of circulating microvesicles allows real-time monitoring of glioblastoma therapy. *Nature Medicine, 18*(12), 1835–1840.

Shao, H., Chung, J., Lee, K., et al. (2015). Chip-based analysis of exosomal mRNA mediating drug resistance in glioblastoma. *Nature Communications, 6*, 6999.

Staals, R. H., & Pruijn, G. J. (2010). The human exosome and disease. *Advances in Experimental Medicine and Biology, 702*, 132–142.

Taylor, D. D., Zacharias, W., & Gercel-Taylor, C. (2011). Exosome isolation for proteomic analyses and RNA profiling. *Methods in Molecular Biology, 728*, 235–246.

Thery, C., Zitvogel, L., & Amigorena, S. (2002). Exosomes: Composition, biogenesis and function. *Nature Reviews Immunology, 2*(8), 569–579.

Urbanelli, L., Buratta, S., Sagini, K., et al. (2015). Exosome-based strategies for diagnosis and therapy. *Recent Patents on CNS Drug Discovery, 10*(1), 10–27.

Vaidyanathan, R., Naghibosadat, M., Rauf, S., et al. (2014). Detecting exosomes specifically: A multiplexed device based on alternating current electrohydrodynamic induced nanoshearing. *Analytical Chemistry, 86*(22), 11125–11132.

Vlassov, A. V., Magdaleno, S., Setterquist, R., et al. (2012). Exosomes: Current knowledge of their composition, biological functions, and diagnostic and therapeutic potentials. *Biochimica et Biophysica Acta, 1820*(7), 940–948.

Wang, Z., Wu, H. J., Fine, D., et al. (2013). Ciliated micropillars for the microfluidic-based isolation of nanoscale lipid vesicles. *Lab on a Chip, 13*(15), 2879–2882.

Yang, F., Liao, X., Tian, Y., et al. (2017). Exosome separation using microfluidic systems: Size-based, immunoaffinity-based and dynamic methodologies. *Biotechnology Journal, 12*(4).

Yang, C., & Robbins, P. D. (2011). The roles of tumor-derived exosomes in cancer pathogenesis. *Clinical & Developmental Immunology, 2011*, 842849.

Yim, N., Ryu, S. W., Choi, K., et al. (2016). Exosome engineering for efficient intracellular delivery of soluble proteins using optically reversible protein-protein interaction module. *Nature Communications, 7*, 12277.

Yuan, Q., Song, J., Yang, W., et al. (2017). The effect of CA125 on metastasis of ovarian cancer: Old marker new function. *Oncotarget, 8*(30), 50015–50022.

Zhang, P., He, M., & Zeng, Y. (2016). Ultrasensitive microfluidic analysis of circulating exosomes using a nanostructured graphene oxide/polydopamine coating. *Lab on a Chip, 16*(16), 3033–3042.

Zhang, J., Li, S., Li, L., et al. (2015). Exosome and exosomal microRNA: Trafficking, sorting, and function. *Genomics, Proteomics & Bioinformatics, 13*(1), 17–24.

Zhao, Z., Yang, Y., Zeng, Y., et al. (2016). A microfluidic ExoSearch chip for multiplexed exosome detection towards blood-based ovarian cancer diagnosis. *Lab on a Chip, 16*(3), 489–496.

Zhou, Y., Abel, G. A., Hamilton, W., et al. (2017). Diagnosis of cancer as an emergency: A critical review of current evidence. *Nature Reviews Clinical Oncology, 14*(1), 45–56.

10

Biodistribution of Cancer-Derived Exosomes

Luize G. Lima, Andreas Möller

QIMR BERGHOFER MEDICAL RESEARCH INSTITUTE, HERSTON, QLD, AUSTRALIA

CHAPTER OUTLINE

1. Introduction...**175**
   1.1 Exosome Definition 175
   1.2 Overview of Cancer-Derived Exosome Functions 176

2. Mechanisms Involved in Tumor-Derived Exosomes Homing to Specific Organs...................**177**

3. Impact of Diverse Factors on the Evaluation of Exosome Biodistribution In Vivo **179**
   3.1 Route of Injection...179
   3.2 Influence of the Cellular Source of Exosomes 180
   3.3 Kinetics of Exosome Distribution..180
   3.4 Labeling of Exosomes for Tracking Studies...181

4. Conclusions...**183**

Acknowledgments...**184**

References **184**

1. Introduction

1.1 Exosome Definition

Extracellular vesicles (EVs) are membrane-enclosed structures secreted by virtually all living cells. Based on their subcellular origin, EVs have been classified into different subtypes, including exosomes, microvesicles, and apoptotic bodies. Exosomes is the most common word used to define smaller vesicles (30–150 nm) pelleting at high centrifugal forces that are shed after fusion of multivesicular endosomal compartments with the plasma membrane. Nevertheless, we must stress that current exosome isolation methodologies are not selective for their endocytic origin and may not completely discriminate exosomes from larger microvesicles generated by the outward budding of cells' plasma membrane or other EV types.

In this chapter, we focus exclusively on cancer-derived exosomes, unless otherwise stated. The literature in the field is still divergent with the exosome definition. Several attempts to unify the nomenclature have been suggested, a reasonably accepted definition being a position editorial from the International Society for Extracellular Vesicles (Lötvall et al., 2014). However, recent findings in this fast developing area suggest that some of the marker proteins thought to be specific for exosomes, previously defined (Lötvall et al., 2014), are indeed shared among other EV subtypes of larger sizes (Kowal et al., 2016). Therefore, there is an ongoing confusion in the field, complicated by severe underreporting of basic characteristics of the EVs investigated (Consortium et al., 2017). This chapter aims at providing an overview of the biodistribution of cancer-derived exosomes and will qualify the vesicle characteristics if there is a reasonable doubt that these particles differ from the previously mentioned definition of exosomes (Lötvall et al., 2014).

1.2 Overview of Cancer-Derived Exosome Functions

The secretion of EVs, including exosomes, by different cell lineages has been described for over three decades (György et al., 2011). Recently, there is a notable surge in interest toward tumor-derived exosomes. This gain in traction could be partly due to an improved understanding of exosome roles in biological processes, as well as their potential use as diagnostic and prognostic markers. The wide-spectrum bioavailability of exosomes favors their utility as potential therapeutic targets for several diseases, including most cancer types (Azmi, Bao, & Sarkar, 2013). The presence of a repertoire of molecular contents (proteins, lipids, RNA [mRNA, miRNA, lncRNA], and DNA [dsDNA, ssDNA, mtDNA] (Thery, Zitvogel, & Amigorena, 2002)), encapsulated primarily from the cell or origin, suggests that exosomes could act as important mediators of intercellular communication. As such, cancer-derived exosomes have been described as an integral part of tumor–host interactions, where protumorigenic roles of tumor-derived exosomes have been comprehensively reported to contribute to initial tumor growth and subsequent metastasis (Azmi et al., 2013; Lobb, Lima, & Möller, 2017).

Exosomes have been isolated from diverse types of biological fluids obtained from cancer patients (An et al., 2015; Whiteside, 2015) and their molecular constituents been correlated to cancer patient outcomes (Hong, Muller, Whiteside, & Boyiadzis, 2014; Melo et al., 2015; Santiago-Dieppa et al., 2014; Taylor & Gercel-Taylor, 2008). These observations indicate that cancer cell–derived exosomes can stably reach the plasma circulation and distribute to other tissues distant from the primary cancer lesion. A recent report demonstrated that circulating cancer-derived exosomes play a role in the guidance of metastatic dissemination (Hoshino et al., 2015). Metastatic outgrowth of cancer cells at secondary organs is not a chance occurrence. Conversely, different cancer types are usually associated with specific sites of secondary tumor outgrowth, a phenomenon termed organotropism (Fidler, 2003). The metastatic process involves a highly coordinated sequence of biological events, including a wide variety of cancer cell–intrinsic and microenvironmental processes, some of which are dependent on receptor–ligand interactions between the tumor cells and the microenvironment at distant organs (Muller et al., 2001). Additionally, cancer-derived exosomes have been shown to play crucial roles in the generation of permissive sites for cancer establishment at secondary organs (termed "premetastatic niches") (Costa-Silva et al., 2015; Hector Peinado et al., 2012; Hood, San, & Wickline, 2011; Syn, Wang, Sethi, Thiery, & Goh, 2016; Wen et al., 2016), a process that occurs before cancer cell arrival (Hector Peinado et al., 2017; Sceneay, Smyth, & Möller, 2013). Despite our knowledge of the role exosomes play in the formation of premetastatic niches and subsequent metastatic outgrowth, how exosomes are specifically targeted to distinct secondary organs has yet to be fully elucidated.

Finally, there is recent evidence that exosomes can also serve as novel delivery vehicles for therapeutic molecules (Wang, Chen, Liu, & Tian, 2016). This evidence, together with similar observations, demonstrates the potential therapeutic clinical utility of exosomes. Notably, exosome studies to date have been mostly conducted using in vitro *cell-based* models or exosome injections into mice. Therefore, the generation of appropriate experimental cancer models is urgently required. These cancer models would facilitate a better understanding of exosome biodistribution and clearance dynamics in a physiologically precise manner.

In the following sections, the mechanisms that have been associated with the organotropic distribution of tumor-derived exosomes will be reviewed. An overview of the variables that influence the fate of exosomes after in vivo injections into animal models will be provided. Lastly, future directions on the use of novel systemic approaches to study the biology of cancer-derived exosomes in vivo and their implications in furthering our current understanding of the metastatic process will be discussed.

2. Mechanisms Involved in Tumor-Derived Exosomes Homing to Specific Organs

Metastasis is a highly coordinated event that involves the priming of secondary organs by tumor-derived factors before cancer cell arrival (Peinado, Lavotshkin, & Lyden, 2011; Sceneay, Parker, Smyth, & Möller, 2013). Among these factors, exosomes have been recently implicated as particular drivers of organotropic metastatic spread (Hoshino et al., 2015), suggesting that cancer-derived exosomes are also guided through the body in an organ-specific manner.

Melanoma-derived exosomes preferentially home to sentinel lymph nodes, whereas man-made control liposomes (similar sized particles with a comparable lipid content and structure) evenly distribute to regional and distant nodes (Hood et al., 2011). Accumulation of exosomes in these sites also induced microenvironment modifications such as extracellular matrix deposition and vascular proliferation, where changes were followed by a more prominent lymphatic dissemination of tumor cells to sentinel nodes (Hood et al., 2011). These results indicated that niche preparation and establishment by cancer-derived exosomes are preceded by specific guidance of these vesicles to premetastatic tissues.

An exosome-intrinsic organotropic mechanism was subsequently described (Hoshino et al., 2015) by assessing the differential in vivo distribution of exosomes derived from several cancer types displaying distinct organotropism. The organ specificity of tumor-derived exosomes recapitulated the metastatic dissemination pattern of their original cells (Hoshino et al., 2015). Notably, conditioning of naïve mice with organotropic exosomes redirected metastatic dissemination of cancer cells according to the tissue specificity of exosomes initially used for conditioning (Hoshino et al., 2015). Distinct tissue-resident cell types were involved in cancer exosome uptake in specific organs. Lung-tropic exosomes were preferentially taken up by fibroblasts and epithelial cells, whereas liver-tropic exosomes were predominantly found in Kupffer cells (Hoshino et al., 2015). Furthermore, the integrin repertoire of tumor-derived exosomes was shown to influence the overall exosome biodistribution (Hoshino et al., 2015). Mechanistically, it seems that exosomal surface integrins are involved in exosome adhesion to specific cell-associated extracellular matrix components, thereby promoting differential uptake of exosomes by distinct target organs and/or cell lineages (Hoshino et al., 2015). For instance, breast cancer exosomes positive for $\alpha6\beta4$ and $\alpha6\beta1$ integrins predominantly adhered to laminin-enriched lung microenvironments, whereas $\alpha v\beta5$ integrin-positive pancreatic exosomes preferentially interacted with fibronectin in the liver microenvironment (Hoshino et al., 2015). Remarkably, a correlation between the integrin composition of exosomes and organ-specific metastasis was established in various cancer patient samples (Hoshino et al., 2015). However, integrin profiles alone are insufficient to explain the organotropic metastasis of all types of cancer. Bone-tropic exosomes, for example, displayed a limited integrin repertoire and were unable to redirect lung-tropic cell metastasis (Hoshino et al., 2015). This limitation suggests that other organ-specific exosomal traits, besides integrin profile, still await discovery.

More recently, our group has shown that exosomes derived from murine breast cancer cells with differing metastatic capacity/organotropism exhibit a unique biodistribution pattern after intravenous (iv) injection into syngeneic mice (Wen et al., 2016). Irrespective of their cellular origin, cancer-derived exosomes predominantly accumulated in the lungs (Wen et al., 2016), a frequent site of breast cancer metastasis in patients and the syngeneic mouse models used (Wen et al., 2016). To a lesser degree, exosomes also localized within the liver and/or the spleen (Wen et al., 2016). Interestingly, breast cancer–derived exosomes were preferentially taken up by CD11b+ cells, including macrophage and dendritic cell subsets, in organs with high exosome accumulation. In contrast to the aforementioned report (Hoshino et al., 2015), murine breast cancer–derived exosomes did not exactly recapitulate the organotropic pattern of the source cells in syngeneic models (Wen et al., 2016). Despite the nonmetastatic phenotype of the 67NR cells, exosomes from this cell line significantly accumulated in both the lung and liver (Wen et al., 2016). Comparatively, exosomes from highly metastatic 4T1 cells, which rapidly form lung and liver metastases from orthotopic sites (Sceneay, Liu, et al., 2013), did not accumulate in the liver (Wen et al., 2016). Tissue-specific immune-suppressive effects after conditioning of naïve mice with exosomes showed that exosomes derived from the highly metastatic 4T1 cells induced a reduction in NK cell frequency and activity, coupled with an increase in the recruitment of myeloid-derived suppressor cells (Wen et al., 2016). The generation of an immune-suppressed premetastatic niche was accompanied by a promotion of metastatic outgrowth (Wen et al., 2016).

Collectively, these data support the hypothesis that cancer-derived exosomes play a crucial role in the formation of immune-suppressive premetastatic niches. Moreover, it indicates that by a partly described mechanism, circulating cancer-derived exosomes can influence cancer cell homing to specific target organs and could potentially be used as predictors of metastatic sites. It is likely that organotropism of tumor-derived exosomes is determined by an intricate combination of exosomal constituents rather than the sole expression of single surface molecules.

3. Impact of Diverse Factors on the Evaluation of Exosome Biodistribution In Vivo

Organ specificity of cancer exosomes was shown to be driven by the expression of different exosomal surface markers (Hoshino et al., 2015) that are associated to their cell of origin. However, it is likely that other cancer-extrinsic factors influence exosome biodistribution. As aforementioned, most discussions in this field are generated based on data collected after exogenous administration of exosomes into mice (Hoshino et al., 2015; Smyth et al., 2015; Suetsugu et al., 2013; Takahashi et al., 2013; Wen et al., 2016; Wiklander et al., 2015). As such, only a minority of studies have comprehensively assessed the implications of factors such as route of injection on systemic distribution of cancer-derived exosomes in vivo. Here, we discuss the multitude of variables that might impact the assessment of exosome accumulation into different organs.

3.1 Route of Injection

Cancer-derived exosome dissemination pattern to lung, liver, and/or bone was consistent regardless of injection via the tail vein (iv), the retro-orbital sinus, or intracardially (ic) (Hoshino et al., 2015). Exosome brain distribution, however, was only observed after ic injection (Hoshino et al., 2015). In contrast, injection of exosomes derived from nonmalignant HEK293T epithelial cells via different routes (iv, intraperitoneal [ip], and subcutaneous [sc] injections) led to different tissue distribution profiles (Wiklander et al., 2015). As compared with iv injection, ip and sc injections resulted in lower accumulation of exosomes in liver and spleen, whereas homing to both pancreas and gastrointestinal trait were increased (Wiklander et al., 2015).

Altogether, these results suggest that the distribution of cancer exosomes is not wholly dependent on the systemic delivery route (Hoshino et al., 2015; Wiklander et al., 2015). Nevertheless, the influence of the injection route on the fate of cancer-derived exosomes cannot be completely disregarded and should be further explored as a way to induce a more prominent exosome distribution to selected tissue targets. A recent work by our group has indicated that iv injection of tumor-derived exosomes is ideal to investigate the consequences of exosome accumulation in the lung (Wen et al., 2016), a frequent site of metastatic dissemination in several cancer types. On the other hand, injection of melanoma exosomes into the mouse footpad can result in their accumulation in the sentinel lymph nodes, representing a useful tool for the study of lymphatic dissemination of cancer-derived exosomes (Hood et al., 2011).

3.2 Influence of the Cellular Source of Exosomes

It has been shown that tumor-derived exosomes recapitulate, at least in part, the organotropism of the cell lines they were derived from (Hector Peinado et al., 2012; Hoshino et al., 2015; Wen et al., 2016). Interestingly, exosomes isolated from primary immature bone marrow–derived dendritic cells showed a higher accumulation in specific organs, particularly spleen, compared with other untransformed cells (i.e., HEK293T and C2C12 murine cell lines) (Wiklander et al., 2015). Altogether, these initial results suggest that acquisition of a repertoire of surface markers from parental cells can be partially associated with natural tropism of exosomes. Yet, this observation suggests an overarching general mechanism, not exclusive of cancer exosomes. Hence, further investigation is required to determine the underlying principles of exosomal biodistribution.

3.3 Kinetics of Exosome Distribution

Depending on the time point of analysis, different studies that monitor the dynamics and biodistribution of exogenously administered exosomes have gathered contrasting results (Smyth et al., 2015; Takahashi et al., 2013; Wiklander et al., 2015). Significant changes in the overall profile of exosome biodistribution can occur within 24-h postinjection (Wiklander et al., 2015). The observed differences in early phases of exosome distribution can be partially attributed to the use of lipophilic dyes, commonly used to label biological membranes and a mainstay in exosome distribution studies. Generally, these lipophilic dyes possess a longer half-life than exosomes and remain in the circulation for a long period of time. For example, iv injection of B16 murine melanoma–derived exosomes yielded an approximate 5% particle concentration in blood at 5 min, implying a rapid clearance of about 95% of particles occurred (Takahashi et al., 2013). In contrast, lipophilic dyes can have an in vivo half-life ranging from 100 h to more than 100 days (Gangadaran, Hong, & Ahn, 2017). Notably, an initial accumulation of exosomes after injection is usually observed in the liver and spleen, regardless of their cellular source (Smyth et al., 2015; Takahashi et al., 2013; Wiklander et al., 2015). Both organs are known to have specialized mononuclear phagocytic systems, so it should be considered that these vesicles could be predominantly trapped or cleared within these tissues before distribution to specific secondary sites.

3.4 Labeling of Exosomes for Tracking Studies

In vivo tracking of exosomes involves live imaging techniques or posthumous evaluations that require vesicular labeling with bioluminescent or fluorescent markers, such as luciferase reporters (Lai et al., 2014; Takahashi et al., 2013), fluorescent proteins (Suetsugu et al., 2013; Wiklander et al., 2015), and/or carbocyanine lipophilic dyes (Alvarez-Erviti et al., 2011; Hoshino et al., 2015; Ohno et al., 2013; Skog et al., 2008; Smyth et al., 2015; Wen et al., 2016; Wiklander et al., 2015). Among them, near-infrared (NIR) fluorescent dyes such as DiD and DiR, that readily fluoresces when incorporated into lipid membranes, have been broadly used (Hood et al., 2011; Hoshino et al., 2015; Lobb, van Amerongen, et al., 2017; Smyth et al., 2015; Wen et al., 2016; Wiklander et al., 2015). These cell tracers have been extensively applied in cell biology studies owing to important chemical features such as (1) no significant effect on the viability and other physiological properties of the target cell; (2) fast diffusion into and uniform labeling of the plasma membrane; and (3) markedly red-shifted fluorescence excitation and emission spectra, which results in a high tissue penetrance and low autofluorescence, thereby making them ideal for in vivo applications (Honig & Hume, 1989).

Electron microscopy–based and nanoparticle tracking analysis–based assessment of exosomes after labeling with DiR showed no significant alterations in particle morphology and size range (Wiklander et al., 2015). After fractionation of DiR-labeled exosome samples in sucrose gradients, DiR-related fluorescence was only observed in the expected density fraction for exosomes, which was also enriched in exosomal marker proteins (Wiklander et al., 2015). In contrast, fractionation of free DiR showed no significant levels of fluorescence emission in any of the analyzed density fractions, indicating that purified exosome samples did contain particle-bound DiR (Wiklander et al., 2015). More importantly, iv injection of free dye into mice resulted in negligible fluorescent signal in diverse organs as assessed by an in vivo imaging system and compared with DiR-exosome samples (Wiklander et al., 2015), thereby excluding the possibility that the binding of free dye to plasma molecules or cells in vivo could lead to a fluorescent emission regardless of DiR-exosome construct. As aforementioned, the usual long half-life of lipophilic dyes (including PKH and NIR dyes as DiR), however, can represent a potential disadvantage for their use as exosome tracers, especially when assessing exosome biodistribution at longer time points (Lai et al., 2014; Wiklander et al., 2015). A possible tissue accumulation of the free dye after degradation of exosomes would result in unspecific fluorescent signals, thereby causing false-positive signals for exosome accumulation or distribution.

In vivo visualization and tracking of GFP-positive exosomes isolated after cellular transfection of GFP-CD63 plasmids have also been demonstrated (Suetsugu et al., 2013; Wiklander et al., 2015). However, fluorescent signal from GFP-positive exosomes may be highly dependent on the fusion protein expression levels in the cells and on packaging efficiencies of these proteins into exosomes. Also, in mouse models, GFP in vivo imaging is difficult because of the usually high tissue background signals. In this way, imaging efficiency of GFP-containing exosomes can be quite low and thus below the detection sensitivities for current visualization methods. Compared with usage of DiR-labeled exosomes, only negligible levels of fluorescent signals could be detected in lung and kidney of mice after iv injection of GFP-positive exosomes (Wiklander et al., 2015). Interestingly, by multiplexing bioluminescent and fluorescent reporters in both normal epithelial and cancer-derived exosomes, it was shown that in vivo vesicle uptake and translation of exosome-delivered mRNAs occurs at rapid rates (Lai et al., 2015). This exosome labeling strategy enables imaging of time-dependant intravital vesicle distribution and uptake in a more specific and stable manner. Such specificity and stability in imaging labeled vesicles could be useful in future studies that aim to further interrogate the dynamics of exosome biogenesis, release, tissue distribution, and recipient cell uptake in respective in vivo models.

Additional in vivo tracking techniques, such as nuclear or magnetic resonance imaging (MRI), have also been applied to assess exosome distribution in vivo (Busato et al., 2016; Gangadaran et al., 2017; Hu, Wickline, & Hood, 2015; Hwang et al., 2015; Smyth et al., 2015). One particular advantage of radioisotope labeling and nuclear imaging is the possibility to quantitatively assess the accumulation of exosomes deeper into organs and tissues. ^{111}In-oxine-labeled exosomes isolated in vitro from breast and prostate cancer cell lines presented similar distribution patterns compared with fluorescently labeled ones (Smyth et al., 2015). Importantly, characteristics such as average size distribution and expression of exosome-enriched proteins showed no significant changes after radiolabeling of exosome-like vesicles with ^{99m}Tc-HMPAO (Hwang et al., 2015). In the same study, radiolabeled vesicles accumulated in distinct tissues were compared with the free labeling compound, indicating that the in vivo distribution of both samples can be easily distinguished (Hwang et al., 2015). Besides radioisotope labeling of exosomes, superparamagnetic iron oxide nanoparticle (SPION) distribution assessment by MRI has also been explored (Hu et al., 2015). SPION-labeled melanoma exosomes specifically homed to the draining lymph nodes, as previously demonstrated for fluorescently labeled exosomes (Hood et al., 2011). Using an indirect labeling approach, SPION-loaded exosomes were recovered from the supernatants of originally labeled adipose stem cells, thereby avoiding the manipulation of the exosome membrane by harsh methods, such as electroporation (Busato et al., 2016). Nevertheless, because of the low sensitivity of the MRI method, a large amount of vesicles and/or iron loading is required to achieve a significant signal intensity allowing for the generation of high-resolution images.

To our knowledge, very few papers have exploited the use of nuclear and MRI to assist in vivo exosomes tracking. We must stress that both modalities might provide new tools capable of overcoming some optical imaging–related restrictions, in particular, the disadvantage of signal retention after exosome uptake and putative subsequent recycling of exosomal content. Moreover, given the broad use of nuclear imaging technology in the clinic, it is reasonable to speculate that such technique would be more easily translated to clinical applications, thereby paving the way to the development of exosome-based diagnostic and therapeutic approaches.

4. Conclusions

In 1978, the presence of membranous particles of varied sizes in cell cultures established from spleen and lymph node samples of a Hodgkin's lymphoma patient was first reported (Friend et al., 1978). Since then, many studies have further supported the finding that both cancer and normal cells secrete exosomes, and these vesicles play a pivotal role in various steps during cancer progression (Azmi et al., 2013). Given that exosomes can be isolated from virtually all body fluids of cancer patients and healthy subjects, assessing their molecular composition in diverse conditions may have a significant impact on novel methods of cancer diagnosis, prognosis, and treatment monitoring. Exosomes have also been explored as specific delivery vehicles of therapeutic molecules to cancers. However, despite increasing interest in this area, a limited number of studies have monitored the systemic distribution of cancer-derived exosomes in vivo. Some of the techniques currently used for labeling, as well as following vesicle dynamics, are suboptimal and potentially biased with results likely to be partially flawed due to technical problems.

Incorporation of bioluminescent and/or fluorescent reporters into the exosomal membrane has allowed gathering data on exosome accumulation in different target organs and their uptake by diverse cell subtypes (Alvarez-Erviti et al., 2011; Hoshino et al., 2015; Lai et al., 2014; Ohno et al., 2013; Skog et al., 2008; Smyth et al., 2015; Suetsugu et al., 2013; Takahashi et al., 2013; Wen et al., 2016; Wiklander et al., 2015). Nevertheless, this method is restricted to exosomes purified in vitro. Whether tumor cell–derived exosomes secreted within the cancer lesions follow a similar tissue distribution pattern has yet to be demonstrated in a fully physiological context in vivo. Transfection of cancer cells with plasmid constructs containing exosomal transmembrane proteins, such as CD63 tagged to fluorescent proteins (i.e., GFP or other fluorescent proteins), enables incorporation of these fusion proteins into transfected cell–derived exosomes, allowing their tracing in in vivo models from a primary tumor (Suetsugu et al., 2013; Wiklander et al., 2015). However, a recent proteomic analysis showed that CD63 and other proteins generally considered as "exosome-specific" markers are actually present in different subsets of secreted EVs (Kowal et al., 2016). Syntenin-1 and TSG101, on the other hand, were shown to be specifically enriched in tetraspanin-positive endosome-derived small vesicles (Kowal et al., 2016) and should be regarded as new targets of fluorescent tagging on exosomes. Finally, we cannot disregard the eventual recycling of exosomal molecules by the recipient cells after vesicle uptake, which could lead to misinterpretation of fluorescence-based cellular and tissue analyses of exosomal biodistribution.

The development of a Cre-lox–based model of tracking exosomal RNA has further demonstrated the transfer of functional mRNA between tumor and host myeloid–derived suppressor cells, thereby promoting alterations in both their immunosuppressive phenotype and miRNA profile (Ridder et al., 2015). Transplanting tumor cells engineered to secrete exosomes containing Cre recombinase into mice with a variety of Cre-lox reporters will allow the monitoring of spontaneous uptake and transfer of specific tumor-derived exosome cargo within the cancer lesion and to distant tissues.

Together, these complimentary technologies will serve to support previous observations on the role of exosomes during tumor primary outgrowth and dissemination and to provide a better understanding of the systemic effects of these vesicles. Although challenging, the development of technical approaches permitting/facilitating accurate assessment of exosome trafficking in the body will be of high importance. Such development will assist in furthering our understanding on the various facets of cancer-derived exosomes in disease progression, thereby facilitating the translation of this knowledge into fundamental outcomes in the management of cancer patients.

Acknowledgments

The authors wish to thank the Tumor Microenvironment Laboratory members for critical input into this review. AM is supported by a National Health and Medical Research Council Australia grant (APP1068510).

References

Alvarez-Erviti, L., Seow, Y., Yin, H., Betts, C., Lakhal, S., & Wood, M. J. A. (2011). Delivery of siRNA to the mouse brain by systemic injection of targeted exosomes. *Nature Biotechnology*, *29*(4), 341–345.

An, T., Qin, S., Xu, Y., Tang, Y., Huang, Y., Situ, B., … Zheng, L. (2015). Exosomes serve as tumour markers for personalized diagnostics owing to their important role in cancer metastasis. *Journal of Extracellular Vesicles*, *4*. https://doi.org/10.3402/jev.v3404.27522.

Azmi, A. S., Bao, B., & Sarkar, F. H. (2013). Exosomes in cancer development, metastasis, and drug resistance: A comprehensive review. *Cancer and Metastasis Reviews*, *32*(3), 623–642.

Busato, A., Bonafede, R., Bontempi, P., Scambi, I., Schiaffino, L., Benati, D., … Mariotti, R. (2016). Magnetic resonance imaging of ultrasmall superparamagnetic iron oxide-labeled exosomes from stem cells: A new method to obtain labeled exosomes. *International Journal of Nanomedicine*, *2016*(11), 2481–2490.

Consortium, E.-T., Van Deun, J., Mestdagh, P., Agostinis, P., Akay, O., Anand, S., … Hendrix, A. (2017). EV-TRACK: Transparent reporting and centralizing knowledge in extracellular vesicle research. *Nature Methods*, *14*(3), 228–232.

Costa-Silva, B., Aiello, N. M., Ocean, A. J., Singh, S., Zhang, H., Thakur, B. K., … Lyden, D. (2015). Pancreatic cancer exosomes initiate pre-metastatic niche formation in the liver. *Nature Cell Biology*, *17*(6), 816–826.

Fidler, I. J. (2003). The pathogenesis of cancer metastasis: The 'seed and soil' hypothesis revisited. *Nature Reviews Cancer*, *3*(6), 453–458.

Friend, C., Marovitz, W., Henle, G., Henle, W., Tsuei, D., Hirschhorn, K., … Cuttner, J. (1978). Observations on cell lines derived from a patient with Hodgkin's disease. *Cancer Research*, *38*(8), 2581–2591.

Gangadaran, P., Hong, C. M., & Ahn, B.-C. (2017). Current perspectives on in vivo noninvasive tracking of extracellular vesicles with molecular imaging. *BioMed Research International*, *2017*, 11.

György, B., Szabó, T. G., Pásztói, M., Pál, Z., Misják, P., Aradi, B., … Buzás, E. I. (2011). Membrane vesicles, current state-of-the-art: Emerging role of extracellular vesicles. *Cellular and Molecular Life Sciences*, *68*(16), 2667–2688.

Hong, C.-S., Muller, L., Whiteside, T. L., & Boyiadzis, M. (2014). Plasma exosomes as markers of therapeutic response in patients with acute myeloid leukemia. *Frontiers in Immunology*, *5*, 160.

Honig, M. G., & Hume, R. I. (1989). DiI and DiO: Versatile fluorescent dyes for neuronal labelling and pathway tracing. *Trends in Neurosciences*, *12*(9), 333–341.

Hood, J. L., San, R. S., & Wickline, S. A. (2011). Exosomes released by melanoma cells prepare sentinel lymph nodes for tumor metastasis. *Cancer Research, 71*(11), 3792–3801.

Hoshino, A., Costa-Silva, B., Shen, T.-L., Rodrigues, G., Hashimoto, A., Tesic Mark, M., … Lyden, D. (2015). Tumour exosome integrins determine organotropic metastasis. *Nature, 527*(7578), 329–335.

Hu, L., Wickline, S. A., & Hood, J. L. (2015). Magnetic resonance imaging of melanoma exosomes in lymph nodes. *Magnetic Resonance in Medicine, 74*(1), 266–271.

Hwang, D. W., Choi, H., Jang, S. C., Yoo, M. Y., Park, J. Y., Choi, N. E., … Lee, D. S. (2015). Noninvasive imaging of radiolabeled exosome-mimetic nanovesicle using 99mTc-HMPAO. *Scientific Reports, 5,* 15636.

Kowal, J., Arras, G., Colombo, M., Jouve, M., Morath, J. P., Primdal-Bengtson, B., … Théry, C. (2016). Proteomic comparison defines novel markers to characterize heterogeneous populations of extracellular vesicle subtypes. *Proceedings of the National Academy of Sciences, 113*(8), E968–E977.

Lai, C. P., Kim, E. Y., Badr, C. E., Weissleder, R., Mempel, T. R., Tannous, B. A., & Breakefield, X. O. (2015). Visualization and tracking of tumour extracellular vesicle delivery and RNA translation using multiplexed reporters. *Nature Communications, 6,* 7029.

Lai, C. P., Mardini, O., Ericsson, M., Prabhakar, S., Maguire, C. A., Chen, J. W., … Breakefield, X. O. (2014). Dynamic biodistribution of extracellular vesicles in vivo using a multimodal imaging reporter. *ACS Nano, 8*(1), 483–494.

Lobb, R. J., Lima, L. G., & Möller, A. (2017). Exosomes: Key mediators of metastasis and pre-metastatic niche formation. *Seminars in Cell & Developmental Biology, 67,* 3–10.

Lobb, R. J., van Amerongen, R., Wiegmans, A., Ham, S., Larsen, J. E., & Möller, A. (2017). Exosomes derived from mesenchymal non-small cell lung cancer cells promote chemoresistance. *International Journal of Cancer, 141*(3), 614–620.

Lötvall, J., Hill, A. F., Hochberg, F., Buzás, E. I., Di Vizio, D., Gardiner, C., … Théry, C. (2014). Minimal experimental requirements for definition of extracellular vesicles and their functions: A position statement from the International Society for Extracellular Vesicles. *Journal of Extracellular Vesicles, 3.* https://doi.org/10.3402/jev.v3403.26913.

Melo, S. A., Luecke, L. B., Kahlert, C., Fernandez, A. F., Gammon, S. T., Kaye, J., … Kalluri, R. (2015). Glypican-1 identifies cancer exosomes and detects early pancreatic cancer. *Nature, 523*(7559), 177–182.

Muller, A., Homey, B., Soto, H., Ge, N., Catron, D., Buchanan, M. E., … Zlotnik, A. (2001). Involvement of chemokine receptors in breast cancer metastasis. *Nature, 410*(6824), 50–56.

Ohno, S-i., Takanashi, M., Sudo, K., Ueda, S., Ishikawa, A., Matsuyama, N., … Kuroda, M. (2013). Systemically injected exosomes targeted to EGFR deliver antitumor MicroRNA to breast cancer cells. *Molecular Therapy, 21*(1), 185–191.

Peinado, H., Aleckovic, M., Lavotshkin, S., Matei, I., Costa-Silva, B., Moreno-Bueno, G., … Lyden, D. (2012). Melanoma exosomes educate bone marrow progenitor cells toward a pro-metastatic phenotype through MET. *Nature Medicine, 18*(6), 883–891.

Peinado, H., Lavotshkin, S., & Lyden, D. (2011). The secreted factors responsible for pre-metastatic niche formation: Old sayings and new thoughts. *Seminars in Cancer Biology, 21*(2), 139–146.

Peinado, H., Zhang, H., Matei, I. R., Costa-Silva, B., Hoshino, A., Rodrigues, G., … Lyden, D. (2017). Pre-metastatic niches: Organ-specific homes for metastases. *Nature Reviews Cancer, 17*(5), 302–317.

Ridder, K., Sevko, A., Heide, J., Dams, M., Rupp, A.-K., Macas, J., … Momma, S. (2015). Extracellular vesicle-mediated transfer of functional RNA in the tumor microenvironment. *OncoImmunology, 4*(6), e1008371.

Santiago-Dieppa, D. R., Steinberg, J., Gonda, D., Cheung, V. J., Carter, B. S., & Chen, C. C. (2014). Extracellular vesicles as a platform for 'liquid biopsy' in glioblastoma patients. *Expert Review of Molecular Diagnostics, 14*(7), 819–825.

Sceneay, J., Liu, M. C. P., Chen, A., Wong, C. S. F., Bowtell, D. D. L., & Möller, A. (2013). The antioxidant N-Acetylcysteine prevents HIF-1 stabilization under hypoxia in vitro but does not affect tumorigenesis in multiple breast cancer models in vivo. *PLoS One, 8*(6), e66388.

Sceneay, J., Parker, B. S., Smyth, M. J., & Möller, A. (2013). Hypoxia-driven immunosuppression contributes to the pre-metastatic niche. *OncoImmunology, 2*(1), e22355.

Sceneay, J., Smyth, M. J., & Möller, A. (2013). The pre-metastatic niche: Finding common ground. *Cancer and Metastasis Reviews, 32*(3), 449–464.

Skog, J., Wurdinger, T., van Rijn, S., Meijer, D. H., Gainche, L., Curry, W. T., … Breakefield, X. O. (2008). Glioblastoma microvesicles transport RNA and proteins that promote tumour growth and provide diagnostic biomarkers. *Nature Cell Biology, 10*(12), 1470–1476.

Smyth, T., Kullberg, M., Malik, N., Smith-Jones, P., Graner, M. W., & Anchordoquy, T. J. (2015). Biodistribution and delivery efficiency of unmodified tumor-derived exosomes. *Journal of Controlled Release, 199*, 145–155.

Suetsugu, A., Honma, K., Saji, S., Moriwaki, H., Ochiya, T., & Hoffman, R. M. (2013). Imaging exosome transfer from breast cancer cells to stroma at metastatic sites in orthotopic nude-mouse models. *Advanced Drug Delivery Reviews, 65*(3), 383–390.

Syn, N., Wang, L., Sethi, G., Thiery, J.-P., & Goh, B.-C. (2016). Exosome-mediated metastasis: From epithelial–mesenchymal transition to escape from immunosurveillance. *Trends in Pharmacological Sciences, 37*(7), 606–617.

Takahashi, Y., Nishikawa, M., Shinotsuka, H., Matsui, Y., Ohara, S., Imai, T., & Takakura, Y. (2013). Visualization and in vivo tracking of the exosomes of murine melanoma B16-BL6 cells in mice after intravenous injection. *Journal of Biotechnology, 165*(2), 77–84.

Taylor, D. D., & Gercel-Taylor, C. (2008). MicroRNA signatures of tumor-derived exosomes as diagnostic biomarkers of ovarian cancer. *Gynecologic Oncology, 110*(1), 13–21.

Thery, C., Zitvogel, L., & Amigorena, S. (2002). Exosomes: Composition, biogenesis and function. *Nature Reviews Immunology, 2*(8), 569–579.

Wang, Z., Chen, J.-Q., Liu, J-l., & Tian, L. (2016). Exosomes in tumor microenvironment: Novel transporters and biomarkers. *Journal of Translational Medicine, 14*(1), 297.

Wen, S. W., Sceneay, J., Lima, L. G., Wong, C. S. F., Becker, M., Krumeich, S., … Möller, A. (2016). The biodistribution and immune suppressive effects of breast cancer–derived exosomes. *Cancer Research, 76*(23), 6816–6827.

Whiteside, T. L. (2015). The potential of tumor-derived exosomes for noninvasive cancer monitoring. *Expert Review of Molecular Diagnostics, 15*(10), 1293–1310.

Wiklander, O. P. B., Nordin, J. Z., Loughlin, A., Gustafsson, Y., Corso, G., Mäger, I., … El Andaloussi, S. (2015). Extracellular vesicle in vivo biodistribution is determined by cell source, route of administration and targeting. *Journal of Extracellular Vesicles, 4*. https://doi.org/10.3402/jev.v4.26316.

11

Exosome-Mediated Communication in the Tumor Microenvironment

Mei-Ju Su, Neha N. Parayath, Mansoor M. Amiji

NORTHEASTERN UNIVERSITY, BOSTON, MA, UNITED STATES

CHAPTER OUTLINE

1. Introduction... **187**
   1.1 Exosome Biogenesis in Tumor Cells and Uptake by Recipient Cells 187
   1.2 Exosome Uptake and Cellular Internalization Mechanisms... 191
   1.3 Direct Fusion 191
      1.3.1 Phagocytosis.. 194
      1.3.2 Macropinocytosis.. 195
      1.3.3 Receptor- and Raft-Mediated Endocytosis 196
      1.3.4 Soluble Signaling and Juxtacrine Signaling 198

2. Exosomes and the Tumor Microenvironment **198**
   2.1 Exosomes Transport Oncoproteins 199
   2.2 Exosomal Content Modulate Tumor Angiogenesis and Metastasis 201
   2.3 Exosomal Content Regulate Tumor Immunity 202

3. Exosome-Mediated Cellular Reprogramming **203**
   3.1 MicroRNA-Mediated Reprogramming of Tumor Cells 204
   3.2 The Role of MicroRNAs on Reprogramming Tumor-Associated Macrophages 205
      3.2.1 Modulation of Exosomal MicroRNA Content Affects Cellular Reprogramming 207

4. Future Prospects **210**

References **211**

1. Introduction

1.1 Exosome Biogenesis in Tumor Cells and Uptake by Recipient Cells

Cells can release diverse types of cytosolic components into extracellular environment via membrane vesicular carriers by a process called exocytosis. Constitutive (non-calcium-triggered) exocytosis occurs in either secreting extracellular matrix components or fusing newly synthesized proteins with transport vesicles and incorporating into the cellular plasma membrane (Beach, Zhang, Ratajczak, & Kakar, 2014). The nanovesicles that are

Diagnostic and Therapeutic Applications of Exosomes in Cancer. https://doi.org/10.1016/B978-0-12-812774-2.00011-0

generated within the endosomal compartments called multivesicular bodies (MVBs) in the secreting cells have been defined as exosomes; the ones that are generated from the plasma membrane have been called microvesicles (Raposo & Stoorvogel, 2013). Exosomes are nanoscale membranous vesicles present in the intracellular space and biological fluids. It is one of the mechanisms of exocytosis in which MVBs fuse with the plasma membrane and release their cargo subsequently (Momen-Heravi, Bala, Bukong, & Szabo, 2014). Exosomes may act locally or from a distance, through the secretion of soluble factors or cell–cell communication. Most cells, including those grown in culture and tumor cells, are constantly secreting exosomes during normal and pathological states. Their specific components contain a wide variety of proteins, lipids, microRNAs (miRNAs), and small RNAs reflecting their cellular origin (Beach et al., 2014). The characteristics of size and morphology are shown.

Initiation, endocytosis, MVBs formation, and exosome secretion are the four steps involved in the process of exosome biogenesis (Kharaziha, Ceder, Li, & Panaretakis, 2012). As seen in Figure 1, the exosome biogenesis process involves an endosomal system and the different stages and components of the endosomal system including endocytic vesicles, early endosomes, late endosomes, and lysosomes, which contain the distinguished shape and cellular location. Endocytic vesicles arise through nonclathrin-or clathrin-mediated endocytosis at the plasma membrane. They make monoubiquitinated proteins internalize and transport to early endosomes. Early endosomes are located at the outer margin of the cells with a tubular shape displayed. Late endosomes, on the other hand, display a spherical shape located close to the nucleus, which is developed from early endosomes by acidification, and changing in their protein content and tendency to fuse with other vesicles or plasma membranes (Stoorvogel, Strous, Geuze, Oorschot, & Schwartz, 1991). The formation of MVBs starts with the invagination of the late endosomal limiting membrane budding into the lumen, resulting in a continuous enrichment of internal luminal vesicles called reverse intracellular budding process (Novikoff, Essner, & Quintana, 1964). At this stage, MVBs contain surface proteins, proteins, and RNA derived from the cytoplasm. The proteins that are transported from the Golgi complex and endoplasmic reticulum or are internalized from the cell surface are ubiquitylated on their cytosolic domains may be sorted into the MVBs, which is guided by either sequential interaction of the endosomal-sorting complex required for transport (ESCRT) machinery or by a ceramide/tetraspanin-dependent pathway. The posttranslational modification of the cytosolic tail of the receptors can also cause sequential interaction of ESCRT machinery to target the receptors to the intraluminal vesicles (Kharaziha et al., 2012).

Although the formation of MVBs is dependent on ESCRT-I complex subunit TSG101, hepatocyte growth factor receptor substrate (Hrs), which is an ESCRT0 complex member, is needed for the accumulation of vesicles within these MVBs for intraluminal vesicles of MVBs formation (Bedoui et al., 2009). However, not all transmembrane proteins, such as MHC class II molecules, require ubiquitylation for targeting into vesicles but would require their sequestration in lipid domains enriched in the tetraspanins (Bobrie, Colombo, Raposo, & Thery, 2011). Once the ESCRT0 complex recognizes the ubiquitylated proteins on the cytosolic side of the MVBs membrane, it separates the proteins into microdomains and binds to the ESCRTI complex, which ends up recruiting ESCRTII subunits. ESCRTI and ESCRTII are the protein complexes to initiate the reverse budding of the intraluminal vesicles within MVBs. After this step, the cytosolic RNAs and proteins can enter into the interior of the forming vesicles. Next, the ESCRTII protein complex recruits ESCRTIII subunits inside the neck of the intraluminal vesicles, resulting in their cleavage into free vesicles. The free ESCRT subunits and ubiquitin molecules will then be released into the cytosol for recycling. Some certain proteins, such as proteolipid protein, are sorted into the intraluminal vesicles independently of the ESCRT machinery. This ESCRT-independent sorting signal is determined by enriched cholesterol-based microdomains in sphingolipids from which ceramides are formed by sphingomyelinases. Ceramide induces coalescence of the microdomains and can trigger exosome vesicles formation (Robbins & Morelli, 2014). With this mechanism, exosomes were also found to be enriched in raft-associated lipids such as cholesterol, ceramide, sphingolipids, and glycerophospholipids. Because budding of exosomes has been shown to be dependent on the conversion of sphingomyelin into ceramide by neutral sphingomyelinase (N-SMase) catalyzation, inhibiting N-SMase can lead to exosome release reduction. Although N-SMase seems to be required for exosome release, A-SMase (Acid-SMase) is dependent for microvesicle shedding. The two enzymes demonstrate the existence of separate regulation mechanisms for the release of the two distinct sets of vesicles (Bianco et al., 2009). These MVBs have several fates after vesicles accumulation: be released into extracellular space with plasma membrane, be recycled to the trans-Golgi network, be detected for degradation in lysosomes, or be fused with the plasma membrane causing the release of intraluminal vesicles, known as exosomes (Babst, 2005; Bedoui et al., 2009).

Typically, when the MVBs fuse with lysosomes, they cause their contents to be degraded by the hydrolysis process inside the lysosome. The lysosomal pathway is important for limiting the signaling of activated growth factor receptors (EGFR/ErbB1). Normal internalization without undergoing degradation happens when the cells are insufficient in decreasing signaling by the receptors. After internalizing into early endosomes, the C-terminal phosphorylated docking sites and kinase domain of the activated epidermal growth factor receptor retain access to the cytosol and for further signal transduction to occur. However, the exact mechanism of determining the sorting of the MVBs toward degradation or recycling is remaining unclear. A couple of sorting mechanisms have been proposed, such as sorting dependent of monoubiquitination of the membrane protein to be endocytosed, or depending on the ESCRT protein complexes (Babst, 2005).

Apart from fusing with the lysosome, the MVBs can also fuse with the plasma membrane and lead to the release of the internal vesicles into the extracellular environment in a proposed ion-dependent manner, and these vesicles are referred as exosomes. Exosomes are mainly secreted by two different mechanisms, including inducible release and constitutive release via the trans-Golgi network. A number of Rab guanosine triphosphatases (GTPases) family proteins, such as Rab27a and Rab27b, have been found to act as key regulators in the regulation of exosome secretion (Ostrowski et al., 2010), as Rab27a affected

MVBs size whereas Rab27b affected the location of MVBs (Ostrowski et al., 2010). Except for Rab 27a and 27b, other Rab family members, Rab 35 and Rab11 have also been found to regulate exosome secretion by interacting with GTPase-activating protein TBC1 domain family member 10A-C (TBC1D10A-C) (Hsu et al., 2010). Studies have shown that Rab35 regulates MVBs docking or tethering at the plasma membrane and inhibition of Rab35 impairs exosome secretion (Azmi, Bao, & Sarkar, 2013; Hsu et al., 2010). The Rab family proteins are often mutated or overexpressed in tumor cells. It has also been shown that when the tumor suppressor protein, p53, has been activated, it stimulates and increases the rate of exosome secretion by regulating transcription of several genes such as p53-regulated proteins, tumor suppressor–activated pathway 6 (TSAP6) and CHMP4C, which can activate and induce exosome production under stressed conditions (Yu, Harris, & Levine, 2006). When the cell experiences hypoxic or toxic stressed condition, it may result in DNA damage.

In addition, exosome secretion seems to be increased by K^+ channel–induced depolarization in neuronal cells, the cross-linking of CD3 with T cells, and can also be affected by perturbation in intracellular Ca^{2+} levels (Keller, Sanderson, Stoeck, & Altevogt, 2006; van Niel, Porto-Carreiro, Simoes, & Raposo, 2006). The accumulation of intracellular Ca^{2+} by monensin, a Na^+/H^+ exchanger, also results in increased exosome release (Savina, Furlan, Vidal, & Colombo, 2003). Secretion by exocytosis is a Ca^{2+}-dependent and highly regulated process. Expansion of MVBs caused by monensin indicated that also exosome secretion is regulated by intracellular Ca^{2+} levels. The importance of intracellular Ca^{2+} levels can be confirmed by treating with a Na^+/H^+ exchanger blocker, DMA, which results in inhibiting exosome secretion (Merendino et al., 2010). More studies suggested that autophagic compartments can fuse with MVBs in a Ca^{2+}-dependent manner, leading to an observed MVBs enlargement as seen with monensin treatment (Fader & Colombo, 2009). During autophagy induction, MVBs fuse with autophagosomes and reroute the MVBs toward lysosomal degradation. Therefore, secretion of exosomes is affected not only by Ca^{2+} levels but also by amino acid availability. Stress, rapamycin, nutrient starvation, and hormone treatment induce autophagy can result in blocking exosome secretion (Fader & Colombo, 2009).

Moreover, the change of other stimuli and intercellular pH has been shown to affect the secretion of exosomes from various cell types (Azmi et al., 2013; Parolini et al., 2009). Exosome secretion and uptake from recipient cells increase when the pH in the microenvironment is low (Parolini et al., 2009). Because the tumor core mass usually contains low pH due to limited vascularization, oxygen supplies, and nutrient, it may provide a possible explanation for the elevated exosome secretion observed in cancer patients. The clinical data have shown that pretreating melanoma cells with proton pump inhibitors (PPIs) can decrease exosome secretion and uptake as PPIs can induce acidification of the tumor cell cytosol.

Once exosomes are secreted by the parent cells, they carry out a variety of functions and may eventually be taken up by another type of cell. It has been reported that certain exosomes, such as the ones that have been secreted by tumor cells, carry phosphatidylserine

(PS) on their membranes. PS acts as a signal and has a large role in exosome uptake by certain cells (Hannafon & Ding, 2013). There are various specific activation mechanisms proposed at the four different stages of exosome biogenesis and exosome secretion, suggesting that exosome formation may be a fine control process. However, the mechanism generally about what specifically triggers exosome biogenesis and secretion still remain unclear.

1.2 Exosome Uptake and Cellular Internalization Mechanisms

Exosomes can carry a range of DNAs, RNAs, and proteins, which can have a significant impact on the phenotype of their recipients. For the phenotypic effect to occur, exosomes need to fuse with target cell membranes, either fusing with the plasma membrane directly or with the endosomal membrane after endocytic uptake and further elicit a cellular response through internal signaling pathways (Mulcahy, Pink, & Carter, 2014).

The reasons that exosomes hold a high therapeutic interest are because they are deregulated in diseases such as cancer and they could be used to deliver drugs to target cells. Thus, it is important to understand the molecular mechanisms by which exosomes are taken up into cells. However, it is still unclear that whether exosomes must be internalized by immune and nonimmune cells to elicit cellular responses. For instance, cellular responses evoke by exosomal RNAs rely on internalization. In comparison with cellular responses induced by membrane-bound or soluble apoptosis-inducing ligands FasL and TRAIL from exosomes, which do not require internalization but dependent on location and temporary adhesion for soluble signaling or juxtacrine (McKelvey, Powell, Ashton, Morris, & McCracken, 2015; Fig. 1).

How exosomes may interact with target cells is not yet fully understood, and several mechanisms have been hypothesized. Cells seem to take up exosomes by a range of endocytic uptake pathways, including direct fusion, phagocytosis, micropinocytosis, receptor- and raft-mediated internalization including clathrin-dependent endocytosis and clathrin-independent endocytosis, such as caveolin-mediated uptake. However, phagocytosis and micropinocytosis may represent mechanisms for the clearance of exosomes, rather than the elicitation of a cellular response. Moreover, it seems like that a heterogeneous population of exosomes may enter into a cell via more than one pathway. The uptake mechanism for a given exosome might depend on proteins and glycoproteins on the surface of both the exosome vesicle and the target cell. Further research is needed to understand the precise rules that underpin exosomes' entry into target cells (Mulcahy et al., 2014). A summary of the biogenesis and internalization mechanisms of exosomes and some of the proteins involved are provided in Fig. 2.

1.3 Direct Fusion

Vesicle–cell fusion is the process of a vesicle merges with the plasma membrane of a cell. Monocyte-derived microvesicles were demonstrated to bind and fuse with the plasma membrane of activated platelets by using a fluorescent lipid-mixing assay and membrane

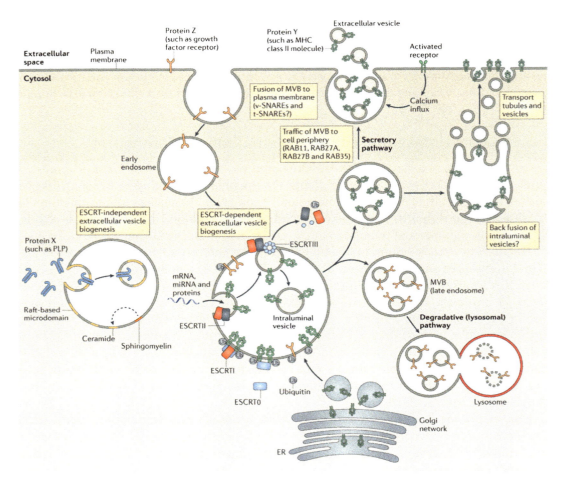

FIGURE 1 Biogenesis of extracellular vesicles. Exosomes are generated as internal luminal vesicles in multivesicular bodies (MVBs) by ESCRT-dependent or ESCRT-independent mechanisms. Proteins transported from the Golgi network (i.e., MHC class-II molecules) or internalized from the cellar surface (i.e., activated growth factor receptors) are ubiquitylated on their cytosolic domains. However, not all proteins required ubiquitinylation for targeting to vesicles. The MVBs then follow either the secretory or degradative pathway. In the former, MVBs traffic to the cell periphery and fuse with the cell membrane, releasing the extracellular vesicles constitutively, or after activation of surface receptors that trigger calcium influx. In the degradative route, MVBs released the internal luminal vesicles into lysosomes. The lysosomal pathway is critical for limiting signaling of activated growth factor receptors. *Reproduced with permission Robbins, P. D., & Morelli, A. E. (2014). Regulation of immune responses by extracellular vesicles.* Nature Reviews Immunology, 14(3), 195–208. https://doi.org/10.1038/nri3622.

fusion assay. The studies also found that these vesicles could transfer proteins, such as tissue factor and P-selectin glycoprotein ligand-1 (PSGL-1) to the recipient cell (Del Conde, Shrimpton, Thiagarajan, & Lopez, 2005). In a similar experiment, exosomes from metastatic melanoma cells fused with the plasma membrane of nontumor cells, which could

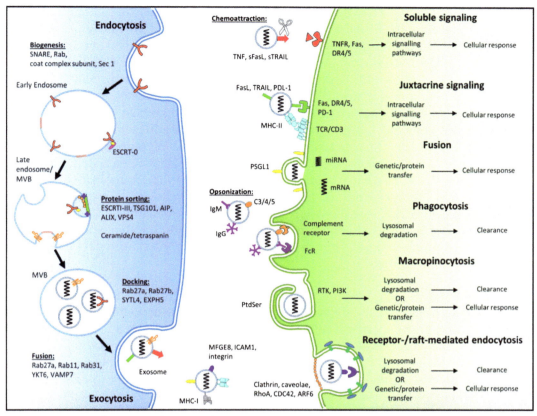

FIGURE 2 Schematic illustration of exosome biogenesis, internalization, and subsequent cellular response. The adhesion of exosomes to the recipient cell uses the interaction of various exosomal surface proteins and cellular receptors. Once bound, the exosome may (1) elicit transduction of the signal via intracellular signaling pathways and be released via juxtacrine signaling; (2) fuse with the cellular membrane transferring protein and release genetic contents into the cytoplasm of the recipient cell via direct fusion; or (3) be endocytosed via phagocytosis, receptor-mediated endocytosis, or macropinocytosis. *Reproduced with permission McKelvey, K. J., Powell, K. L., Ashton, A. W., Morris, J. M., & McCracken, S. A. (2015). Exosomes: Mechanisms of uptake.* Journal of Circulating Biomarkers, 4, 7.

have been inhibited by filipin. Another study of colocalization of exosomes with Rab53 or Lamp-1 suggested that exosomes are internalized and interact with cytoplasmic vesicles. Proteins were found to have a minor role during fusion (Parolini et al., 2009).

The different mechanisms of uptake and the downstream pathways for the endocytosis process have been investigated in detail. However, the mechanism of cell–cell fusion is not yet fully understood. This phenomenon may be regulated by tetraspanin complexes on target cells because tetraspanins were found to have a role in T cell activation and plasma membrane fusion between myoblasts (Tachibana & Hemler, 1999), sperm oocytes (Miyado et al., 2000; Rubinstein, Ziyyat, Wolf, Le Naour, & Boucheix, 2006), mononuclear phagocytes (Takeda et al., 2003), and mammalian viral–cell fusion (Fukudome et al., 1992).

Although tetraspanin CD81, which colocalizes with CD4, has been shown to be involved in exosome release from HIV-infected T cells (Booth et al., 2006; Grigorov et al., 2009), it has not yet been revealed if tetraspanins are involved in exosome–T cell fusion. In the condition of viral–cell fusion, tetraspanins inhibit cell fusion, in which CD9, CD63, CD81, CD82, CD151, and CD231 reduce HIV infection (Booth et al., 2006; Sato et al., 2008) by inhibiting cell fusion (Gordon-Alonso et al., 2006; Symeonides, Lambele, Roy, & Thali, 2014) and cell–cell transmission of viral particles (Krementsov, Weng, Lambele, Roy, & Thali, 2009). It also seems possible that integrins are involved in exosome adhesion/attachment to the target cell in a similar manner to leukocyte transendothelial migration, with tetraspanin-enriched microdomains facilitate exosome fusion (Hemler, 2003; Rana & Zoller, 2011).

Exosomes can also fuse with the recipient cell and lead to the nonselective transfer of RNAs and protein from exosome to the target cell. A study showed that exosome uptake from melanoma cells increases at a low pH level and the mechanism is dependent on the presence of sphingomyelin and ganglioside GM3. In addition, the fusion efficiency is higher with exosomes released from metastatic tumor cells than those derived from primary tumors or normal cells.

1.3.1 Phagocytosis

The process of phagocytosis involves the internalization of opsonized specific matter, including fragments of apoptotic cells and bacteria. Phagocytosis is a receptor-mediated mechanism that involves the continuous formation of invaginations to surround the material destined for internalization, with or without the involvement of enveloping membrane extensions, which is required for micropinocytosis (Doherty & McMahon, 2009; Swanson, 2008). Generally, phagocytosis mainly internalizes larger particles; however, it has been shown that particles as small as 85 nm in diameter have been internalized by this mechanism. Therefore, it is possible that exosomes could be internalized via this route (Rudt & Müller, 1993). Past study showed that the exosomes released by leukemia cells were shown to be internalized effectively by macrophages but were not taken up by other cell types (Feng et al., 2010). Phagocytosis is an actin-mediated process that requires the presence of specific opsonin receptors, such as Fc receptors and complement receptors, scavenger receptors or toll-like receptors (TLRs). This function is often performed by those "professional phagocytes" such as macrophages and dendritic cells (DCs). Although phagocytosis is typically performed by the specialized cells such as macrophages (Doherty & McMahon, 2009), it can also be performed by "nonprofessional" cells such as gamma delta T cells (γδ T cells), which have a distinctive T cell receptor on the surface (Wu et al., 2009). Phagocytosis has been proposed as a method of exosome internalization. Not surprisingly, studies found that monocytic and macrophagic cell lines could internalize exosomes derived from erythroleukemia and T cell leukemia cells more efficiently than "nonprofessional" phagocytic cell lines, including Jurkat T cells and 293T human embryonic kidney cells (Feng et al., 2010). Another exosome phagocytosis study used fluorescence labeled J774A.1 murine macrophage exosomes and labeled human pancreatic carcinoma (Panc-1) cells to reveal the intercellular tracking of exosomes in coculture system and demonstrated how the tumor-derived exosomes are secreted from cancer cells and be taken by the macrophages. In addition, luciferase-expressing Panc-1 cells cocultured with J774A.1 macrophages in bioluminescence imaging confirmed the presence of exosomes cumulated in the media and uptake by J77A.1 cells over time (Yamada et al., 2016). Although the experiment did not investigate the phagocytosis mechanisms deeply, it suggested that exosomes were engulfed effectively via the phagocytic process.

The phagocytosis of exosomes was shown to be dependent on the phosphatidylinositol 3-kinase (PI3K), actin cytoskeleton, and dynamin2 (Feng et al., 2010). In particular, PI3K, actin, and dynamin2 have all been implicated in both phagocytosis (Gold et al., 1999) and clathrin-mediated endocytosis (Grassart et al., 2014). PI3Ks play a major role in phagocytic processes, especially in enabling plasma membrane insertion to form phagosomes in the cytoplasm (Stephens, Ellson, & Hawkins, 2002). The PI3K inhibitors, LY294002, and wortmannin were used to assess the importance of functional PI3Ks in extracellular vesicles uptake, and both compounds were found to inhibit extracellular vesicles uptake in a dose-dependent manner (Feng et al., 2010). PS is required for initiating the removal of apoptotic bodies by phagocytosis (Fadok et al., 1992). It is also used by some viruses to enter cells by macropinocytosis mechanism (Shiratsuchi, Kaido, Takizawa, & Nakanishi, 2000). Incubation of macrophages with an antibody that blocks TIM4, a PS receptor involved in PS-dependent phagocytosis (Miyanishi et al., 2007), leads to exosome uptake reduction (Feng et al., 2010). Moreover, treatment of DCs with a competitive soluble PS analogue can reduce extracellular vesicles, including exosome uptake (Morelli et al., 2004). A study showed that treating extracellular vesicles with annexin-V, a protein that binds to PS, reduces the uptake of the vesicles into macrophages (Yuyama, Sun, Mitsutake, & Igarashi, 2012) and natural killer (NK) cells (Nolte-'t Hoen, Buschow, Anderton, Stoorvogel, & Wauben, 2009).

Internalized exosomes can colocalized with lysosomal-associated membrane protein 1 (LAMP-1), lysobisphosphatidic acid, and Rab7 (Feng et al., 2010) in late autophagosomes or endosomal and lysosomal vesicles (Jager et al., 2004). However, further research is required to determine whether phagocytosis represents a true mechanism of exosome internalization for intercellular communication or is merely a method of elimination.

1.3.2 Macropinocytosis

Macropinocytosis is a regulated form of endocytic mechanism that mediates the nonselective uptake of extracellular material such as solute molecules, nutrients, and antigens. It involves the formation of invaginated membrane ruffles that then pinch off into the intracellular compartment (Doherty & McMahon, 2009). When the vesicles carry extracellular components or fluid in close proximity to the region around the plasma membrane ruffles, ruffled extensions of the membrane protrude from the cell surface and surround an area of extracellular fluid, subsequently, this area is internalized entirely by fusing the membrane protrusions with themselves or back with the plasma membrane (Swanson, 2008). Although the uptake mechanism of macropinocytosis is similar to that of phagocytosis, direct contact with the internalized material is not required in macropinocytosis.

Plasma membrane protrusions are driven by actin filaments to form an invagination to endocytoses extracellular small particles and fluid nonspecifically. The mechanism is actin-, rac1-, and cholesterol-dependent and requires $Na^+–H^+$exchanger activity (Kerr & Teasdale, 2009). Cholesterol is involved in the recruitment of activated rac1 to sites of micropinocytosis (Grimmer, van Deurs, & Sandvig, 2002). It has been reported that PS on the surface of oligodendrocyte-derived exosomes played a role in activating macropinocytosis in a subset of microglia cells and macrophages without antigen-presenting capability (Fitzner et al., 2011). Phosphatidylinositol-3-kinase (PI3K), rac, and src activities have also been shown to stimulate micropinocytosis (Doherty & McMahon, 2009). Rac1 is a GTPase of Rho family, which plays an important role not only in micropinocytosis but also in the regulation of cell growth, protein kinase activation, cytoskeletal reorganization (Ridley, 2006), and epithelial–mesenchymal transition in cancer (Sanz-Moreno et al., 2008). A small molecule inhibitor of rac1, NSC23766, also inhibited exosome uptake by microglia (Fitzner et al., 2011). Macropinocytosis of exosomes is dependent on Na^+ and PI3K, as with the inhibition of $Na^+–H^+$ ion exchange and PI3K activity by the pharmacological inhibitors EIPA and LY294002, respectively, were found to reduce exosome uptake (Tian et al., 2014). Another study using alkalinizing drugs, such as monensin, chloroquine, bafilomycin A, showed the inhibition of microglial internalization of extracellular vesicles, consistent with a role for the acidification of vacuoles in macropinocytosis (Fitzner et al., 2011). Additionally, there are other studies using inhibitors which demonstrate the role for macropinocytosis in the uptake of extracellular vesicles (Christianson, Svensson, van Kuppevelt, Li, & Belting, 2013; Feng et al., 2010; Nanbo, Kawanishi, Yoshida, & Yoshiyama, 2013). These discoveries suggest that macropinocytosis is either a minor pathway used by cells to internalize extracellular vesicles or a mechanism used in particular cell types.

1.3.3 Receptor- and Raft-Mediated Endocytosis

As the names propose, receptor- and raft-mediated endocytosis require either a ligand on the exosomal surface to capture specific receptors on the cellular plasma membrane, or the presence of sphingolipid- and cholesterol-rich microdomains in the plasma membrane, respectively. Receptor-mediated endocytosis is also called clathrin-mediated endocytosis, which uses clathrin and adaptor protein 2 (AP2) adaptor complexes, which is an integral component of the clathrin coat to work on coating the membrane and inducing the invagination of the membrane into a vesicle. The mechanism involves cellular internalization of molecules through sequential and progressive assembly of clathrin-coated vesicles that contain a variety of transmembrane receptors and their ligands. The clathrin-coated vesicles strategically deform the membrane to have it collapse into a vesicular bud, mature and cut-off. The following intracellular vesicle undergoes clathrin uncoating and then fuses with the endosome to deposit its contents (Kirchhausen, 2000). Various studies implicate clathrin-mediated endocytosis involved in the uptake of extracellular vesicles including exosomes. Epidermal growth factor receptor pathway substrate clone 15 (EPS15) is a component of clathrin-coated pits that are commonly associated with AP2 adaptor complexes (Benmerah, Bayrou, Cerf-Bensussan, & Dautry-Varsat, 1999).

Expression of a dominant-negative mutant of EPS15 inhibits clathrin-mediated endocytosis and leads to a reduction in exosome uptake (Feng et al., 2010). Dynamin 2 is a GTPase responsible for endocytosis and is required for the clathrin-mediated endocytosis process (Vallee, Herskovits, Aghajanian, Burgess, & Shpetner, 1993). Dynamin 2 is recruited to incipient clathrin-coated pits where it forms a collar-like structure at the neck of invaginated clathrin-coated pits (Ehrlich et al., 2004; Merrifield, Feldman, Wan, & Almers, 2002; Taylor, Lampe, & Merrifield, 2012), followed by GTP hydrolysis–mediated changes in dynamin 2 conformation resulting in membrane fission and clathrin-coated vesicle release (Chappie, Acharya, Leonard, Schmid, & Dyda, 2010; Damke, Baba, Warnock, & Schmid, 1994; Marks et al., 2001). Dynamin 2 also facilitates membrane binding (Achiriloaie, Barylko, & Albanesi, 1999; Lee, Frank, Marks, & Lemmon, 1999; Vallis, Wigge, Marks, Evans, & McMahon, 1999) and curvature (Ramachandran et al., 2009) during clathrin-mediated endocytosis as inhibition of dynamin 2 (Vallee et al., 1993) prevents almost all extracellular vesicles internalization activity in phagocytic cells(Barres et al., 2010; Feng et al., 2010; Fitzner et al., 2011). Studies found that exosomes released from cultured rat adrenal gland medulla (PC12) tumor cells are partially internalized by clathrin-mediated endocytosis, as demonstrated by using pharmacological inhibitors such as chlorpromazine (CPZ) and siRNA knockdown of clathrin. (Tian et al., 2014; Wang, Rothberg, & Anderson, 1993). The clathrin-mediated endocytosis inhibition by CPZ also decreased uptake of extracellular vesicles by ovarian cancer recipient cells (Escrevente, Keller, Altevogt, & Costa, 2011) and some phagocytic recipient cells (Feng et al., 2010). In another report, the internalization of glioblastoma-derived exosomes was found to be involved nonclassical, lipid raft-mediated endocytosis, and required ERK1/2- HSP27 signaling pathway, as the negative regulation of ERK1/2 by caveolin-1 inhibited exosome internalization (Christianson et al., 2013). These results suggest that clathrin-mediated endocytosis plays at least some part in the uptake of extracellular vesicles, including exosomes.

Although clathrin-mediated endocytosis has been widely studied for several years, it becomes increasingly apparent that numerous of clathrin-independent endocytotic mechanisms exist in eukaryotic cells (Doherty & McMahon, 2009), such as raft-mediated endocytosis. Raft-mediated endocytosis includes caveolae-dependent endocytosis and the clathrin- and caveolae-independent endocytosis mechanisms that use distinct combinations of dynamin, flotillin, or Rab proteins in the process. It is so called RhoA-, CDC42-, and ARF6-regulated endocytosis (Mayor & Pagano, 2007). The particles from the external environment or plasma membrane under this endocytosis process may be sent to lysosomes for degradation or be recycled back to the plasma membrane. Caveolae are small cave-like invaginations on the plasma membrane that can be internalized into the cell, same as clathrin-coated pits. They are subdomains of glycolipid rafts, which are rich in caveolins, cholesterols, and sphingolipids on the plasma membrane. Thus, caveolae-dependent endocytosis is sensitive to cholesterol depletion agents such as filipin and methyl-b-cyclodextrin (MbCD) (Anderson, 1998; Kurzchalia & Parton, 1999; Nabi & Le, 2003). Caveolin-1 is a protein that is required for the formation of caveolae (Doherty & McMahon, 2009), and a sufficient amount of caveolin can be found clustered within membrane invaginations. Oligomerization of caveolins, which can be facilitated by caveolin oligomerization domains, mediates the formation of caveolin-rich rafts in the plasma membrane. The increased levels of cholesterol accompanied by caveolin scaffolding domains attach to the plasma membrane, and increased levels of dynamin2 activity, which is also required for clathrin-mediated endocytosis, enable assembly and expansion of caveolae-mediated endocytic vesicles (Doherty & McMahon, 2009; Nabi & Le, 2003; Parton & Simons, 2007).

Blocking dynamin2 by its specific inhibitor dynasore (Newton, Kirchhausen, & Murthy, 2006) leads to significantly reduced internalization of exosomes (Nanbo et al., 2013) or larger microvesicles (Menck et al., 2013), suggesting a role for caveolae-mediated endocytosis in vesicular uptake. However, as dynamin2 is also required for clathrin-mediated endocytosis, it is not possible to rule out a role for clathrin-coated vesicles in the experiments (Lee et al., 1999). Nevertheless, taken together these results imply some particular function for caveolae-mediated endocytosis in exosome uptake, though the precise role of this mechanism may vary between different extracellular vesicles and cell types. Clathrin- and caveolin-mediated endocytosis pathways shown to participate in exosome uptake by target cells were demonstrated in Fig. 3.

1.3.4 Soluble Signaling and Juxtacrine Signaling

Soluble signaling involves the proteolytic cleavage of ligands from the exosomal surface or alternative splicing, yet juxtacrine signaling requires the juxtaposition of ligands and receptors on the surfaces of the exosome and target cell. Exosomal membrane proteins can interact with the target cell in a juxtacrine manner, acting as a ligand for target cell surface receptors. Membrane-bound Fas ligand (FasL), tumor necrosis factor (TNF)-related apoptosis-inducing ligand (TRAIL), and TNF can be cleaved by metalloproteinases to form soluble cytokines. When they are in soluble form, FasL, TRAIL, and TNF have a reduced proapoptotic activity compared with those of the membrane-bound form (Schneider et al., 1998; Wajant et al., 2001). Exosomes from cultured placental explants or plasma from pregnant women have FasL and TRAIL on their membrane, and induce apoptosis in Jurkat T cells via NF-κB, CD3ζ, and JAK3 downregulation (Sabapatha, Gercel-Taylor, & Taylor, 2006; Stenqvist, Nagaeva, Baranov, & Mincheva-Nilsson, 2013; Taylor, Akyol, & Gercel-Taylor, 2006). The same mechanism has been demonstrated for exosomes derived from tumors (Taylor, Gercel-Taylor, Lyons, Stanson, & Whiteside, 2003).

2. Exosomes and the Tumor Microenvironment

The tumor microenvironment that consists of wide variety of cells other than tumor cells such as fibroblasts, macrophages, endothelial cells, and cells of the immune system is now known to have an essential part in cancer progression. The signals exchanged within the tumor microenvironment, which facilitates the tumor development through processes such as enhanced angiogenesis, metastasis, and suppression of the immune system.

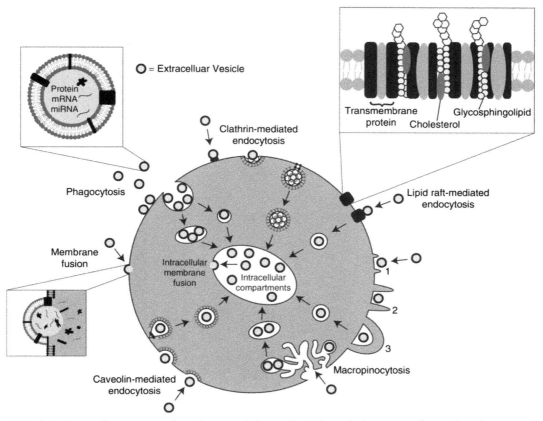

FIGURE 3 Pathways shown to participate in extracellular vesicle (EV) uptake by target cells. EVs have been shown to be internalized by cells through clathrin- and caveolin-mediated endocytosis. Lipid rafts are involved in both clathrin- and caveolin-mediated endocytosis. EVs including exosomes may also deliver their protein, mRNA, and miRNA cargo by phagocytosis or fusion with the plasma membrane. Alternatively, intraluminal EVs may fuse with the endosomal limiting membrane after endocytosis to enable their EV contents to elicit a phenotypic response. *Reproduced with permission Mulcahy, L. A., Pink, R. C., & Carter, D. R. (2014). Routes and mechanisms of extracellular vesicle uptake. Journal of Extracellular Vesicles, 3. https://doi.org/10.3402/jev.v3.24641.*

Exosomes form the crucial component of these signal exchanges within the tumor microenvironment. Thus exosomes in the tumor microenvironment combine to form a diverse group of signals, secreted by different cell types, which form an intricate system for tumor growth and development (Table 1).

2.1 Exosomes Transport Oncoproteins

Exosomes of the tumor microenvironment facilitate the exchange of oncogenic proteins secreted by the tumor cells into the neighboring cells. Gliomas, which are one of the most heterogeneous groups of tumors, often express a truncated form of epidermal

Table 1 Effect of Exosomal Content Transfer on the Various Cells That Are Present in the Tumor Microenvironment

Exosome Source	Recipient Cells	Effect on the Tumor Microenvironment	References
EGFR mutant glioma cells	Nonmutant glioma cells	Transfer of oncoproteins—transfer of mutant EGFRvIII	Bianco et al. (2009)
DKO-1 colorectal cancer cells (mutant KRAS	DKs-8 colorectal cancer cells (wild-type KRAS)	< three-dimensional growth in nonmutant cells similar to mutant cells	Beckler et al. (2013)
Lymphoblastoid B cells	–	Carry latent membrane protein 1 out of the cells thereby restraining downstream NF-κB activation	Verweij et al. (2011)
DU145 (docetaxel-resistant prostate cancer cells)	LNCap (nonresistant cell lines)	Transfer of MDR-1/p-gp resulted in docetaxel resistance	Corcoran et al. (2012)
Chronic myeloid leukemia cells K562	Human umbilical endothelial cell	Inhibit angiogenesis through src-mediated pathway	Mineo et al. (2012)
Breast cancer cells	Adipose tissue–derived mesenchymal stem cells	Differentiate to myofibroblast-like phenotype	Cho, Park, Lim, and Lee (2012)
Prostate cancer cells	Mesenchymal stem cells	Differentiate to myofibroblastic cells, resulting in proangiogenic functions	Chowdhury et al. (2015)
CD105-positive renal carcinoma cells	Endothelial cells	Angiogenic phenotype	Grange et al. (2011)
Ascites fluid of patients with metastatic malignancy	Mesothelial cells	Metastasis—conversion of mesothelial cells to carcinoma-associated fibroblasts by TGF-β1-induced mesothelial–mesenchymal transition	Wei et al. (2017)
Melanoma cells	Bone marrow–derived cells	Promote lung endothelial permeability and lung metastasis	Peinado et al. (2012)
Fibroblast	Breast cancer cells	Metastasis via Wnt-planar cell polarity signaling	Luga et al. (2012)
Epithelial ovarian cancer cells	Natural killer (NK) cell	Suppress the NK cell surface expression of NKG2D facilitating immune escape and tumor progression	Labani-Motlagh et al. (2016)
Lung cancer cells	Macrophages	miR-mediated silencing of transcripts associated with toll-like receptor, resulting in enhanced proinflammatory cytokines, supporting tumor dissemination	Fabbri et al. (2012)
Epithelial cancer cells	Macrophages	Polarization of macrophages to tumor-promoting phenotype	Ying et al. (2016)
Tumor cells expressing Fas ligand	CD8+ T lymphocytes	Induce apoptosis in CD8+ T lymphocytes	Morelli et al. (2004)
Nasopharyngeal cancer cells	CD4+CD25− T cells	Transformation of CD4+CD25− T cells into Treg cells and facilitated tumor recruitment of Treg cells mediated through CCL20	Mrizak et al. (2015)
Colorectal tumor cells	T cells	Transform T cells to Treg cell phenotype by activating TGF-β/SMAD signaling and inhibiting SAPK signaling	Yamada et al. (2016)
Lung carcinoma cells	T cells	Expansion of Treg cells by downregulating phosphatase and tensin homologue and resulting in tumor progression	Yin et al. (2014)

growth factor receptor known as EGFRvIII in some cancer cells whereas rest cancer cell population express the nonmutant form of EGFR protein (Bianco et al., 2009). Al-Nedawi et al. (2008) demonstrated exosome-mediated sharing of this mutated EGFRvIII between the mutant and nonmutant cells, thereby resulting in changes in expression of EGFRvIII-regulated genes in nonmutant cells. Other mutated proteins such as KRAS have been also been shown to be transferred through exosomes from mutant cells to nonmutant cells. Beckler et al. showed that DKs-8 colorectal cancer cells (wild-type KRAS) internalize exosomes secreted by DKO-1 colorectal cancer cells (mutant KRAS), resulting in greater three-dimensional growth in these nonmutant cells similar to mutant cells (Beckler et al., 2013).

Furthermore, Verweij et al. (2011) demonstrated that lymphoblastoid B cell line–derived exosomes carry latent membrane protein 1 out of the cells thereby restraining downstream NF-κB activation. Additionally, apart from the oncoproteins, exosomes are known to assist transfer of resistance-associated proteins from resistant tumor cells to the nonresistant tumor cells. Corcoran et al. (2012) demonstrated that exosome-mediated transfer of MDR-1/p-gp from DU145 (docetaxel-resistant prostate cancer cells) resulted in docetaxel resistance in DU145 and LNCap (nonresistant cell lines). Ciravolo et al. (2012) demonstrated that exosomes derived from HER2-overexpressing breast cancer cells, SKBR3, and BT474 express activated HER2 and thus HER2-targeted therapy would have differential effect on the cells in the presence of these exosomes. As these studies are based on in vitro tumor models, a further validation in preclinical murine tumor models becomes essential to ratify the role of exosomes in cancer progression.

2.2 Exosomal Content Modulate Tumor Angiogenesis and Metastasis

Exosomes secreted from cancer cells are further extensively studied for their involvement in the modulation of stromal cells of the tumor microenvironment. Myofibroblasts, which form a major component of the tumor stroma and mediate angiogenesis, can be modulated by the cancer cells through exosomal secretions (Vong & Kalluri, 2011). A study by Mineo et al. (2012) demonstrated that exosomes derived from chronic myeloid leukemia cells K562 can inhibit angiogenesis through src-mediated pathway through human umbilical endothelial cell–based angiogenesis model. In solid tumors, the stroma is rich in myofibroblasts supporting vascularization and tumor development. Webber, Steadman, Mason, Tabi, and Clayton (2010) demonstrated that when fibroblasts are exposed to exosomes derived from cancer cells, it results in fibroblast differentiation into myofibroblasts involving TGF-β-mediated SMAD-dependent signaling process. Another study by Cho et al. (2012) showed that adipose tissue–derived mesenchymal stem cells could be differentiated to myofibroblast-like phenotype with enhanced expression of α-SMA and tumor-promoting factors, such as TGF-β, VEGF, SDF-1, and CCL5, in the presence of exosomes derived from breast cancer cells. Chowdhury et al. (2015) demonstrated that prostate cancer cell–derived exosomes cause differentiation of mesenchymal stem cells to myofibroblastic cells with increased levels of VEGF-A resulting in proangiogenic functions. Furthermore, cancer cell (activated EGFR)–derived exosomes result in increased VEGF expression in endothelial cells (Al-Nedawi, Meehan, Kerbel, Allison, & Rak, 2009).

Along with angiogenesis, exosome-mediated signaling can also contribute to tumor metastasis. In a study in rat pancreatic adenocarcinoma model demonstrated the role of exosomes along with CD44 rich matrix in the formation of premetastatic niche (Jung et al., 2009). Another study, which reported that exosomes secreted from CD105-positive renal carcinoma cells resulted in angiogenic phenotype of endothelial cells (organize capillary-like structures on Matrigel), also showed that treating SCID mice with these exosomes resulted in enhanced lung metastasis of the renal cells (Grange et al., 2011). Thus, the role of tumor-derived exosomes has been studied extensively in promoting cancer cell metastasis. A recent study by Wei et al. (2017) demonstrated the role of exosomes derived from ascites fluid of patients with ovarian and gastric cancers in stimulating metastasis. They observed that treatment with exosomes derived from ascites fluid of patients with metastatic malignancies resulted in the increased proliferation of mesothelial cells and an increase in the expression of carcinoma-associated fibroblast-specific proteins such as fibroblast activation protein, alpha-smooth muscle actin, and fibronectin. They suggested that these exosomes from ascites may facilitate metastasis by promoting conversion of mesothelial cells to carcinoma-associated fibroblasts by TGF-β1-induced mesothelial–mesenchymal transition (Wei et al., 2017).

Furthermore, a study demonstrated that melanoma-derived exosomes could use bone marrow–derived cells to promote lung endothelial permeability and lung metastasis (Peinado et al., 2012). However, studies have also demonstrated the role of exosomes derived from cells of the tumor microenvironment on the cancer cells. In orthotopic mouse breast cancer model, fibroblast-derived exosomes facilitate metastasis via Wnt-planar cell polarity signaling in breast cancer cells (Luga et al., 2012).

2.3 Exosomal Content Regulate Tumor Immunity

The compromised immune system of the tumor microenvironment is a result of communication between the cancer cells and the immune cells. Cancer cells recruit immune cells such as neutrophils, DCs, NK cells, macrophages, T-lymphocytes, and Treg cells to generate a protumorigenic immune response. Exosomal release from tumor cells is regulated by small GTPases such as RAB27A and RAB27B. Angélique Bobrie et al. (2012) demonstrated that blocking of Rab27a in mammary carcinoma cells resulted in decreased neutrophil mobilization, which is required for reduced primary tumor growth and decreased lung metastasis. Studies have also demonstrated other mechanisms of immunosuppression by tumor-derived exosomes. Labani-Motlagh et al. (2016) showed that epithelial ovarian cancer cell–derived exosomes contain MHC class I-related chain A and B ligands for the NK cell–activating receptor NKG2D, which can suppress the NK cell surface expression of NKG2D facilitating immune escape and tumor progression. Furthermore, miRNAs bind to their corresponding messenger RNA (mRNA) and regulate gene expression of the recipient cells. Lung cancer–derived exosomes secrete miRs, which can silence the transcripts associated with TLR family in macrophages, resulting in enhanced proinflammatory cytokines, supporting tumor dissemination (Fabbri et al., 2012). Another study by Ying et al. (2016)

showed that epithelial cancer–derived exosomes could promote polarization of macrophages to tumor promoting phenotype. Even in this study, the identified miR-222-3p as a component of exosomes, responsible for macrophage polarization. Thus the involvement of MiRs in macrophage repolarization is extensively studied and is discussed in detail in Section 3 of this chapter.

Additionally, as mentioned above cancer cells show exosome-mediated communication with the immune system to inhibit the antitumor functions of the immune system. Myeloid-derived suppressor cells (MDSCs) of the tumor microenvironment can suppress T cell activation. A study demonstrated that the suppressive activity of MDSCs in the tumor microenvironment is determined by stat-3-mediated interaction between Hsp-72 of the exosomes and MDSCs (Chalmin et al., 2010). Additionally, exosomes secreted by tumor cells expressing Fas ligand, induce apoptosis in CD8[+] T lymphocytes of the tumor microenvironment, in turn suppressing the immune system (A. Morelli, 2006; Wieckowski et al., 2009) (Abusamra et al., 2005).

Apart from T cells, cancer cell–derived exosomes can regulate Treg cells. Szajnik, Czystowska, Szczepanski, Mandapathil, and Whiteside (2010) demonstrated that exosomes derived from ascites of ovarian cancer patients, blood of human head and neck squamous cell carcinoma patients could suppress proliferation of recipient cells. Another study by Mrizak et al. (2015) demonstrated that exosomes from nasopharyngeal cancer cells influenced the transformation of CD4[+]CD25[−] T cells into Treg cells and facilitated tumor recruitment of Treg cells mediated through CCL20. This ability of tumor cell–derived exosomes to convert T cells to Treg cell phenotype, resulting in tumor progression has been observed in multiple cancers. Yamada et al. (2016) showed exosomes derived from colorectal tumor cells can transform T cells to Treg cell phenotype by activating TGF-β/ SMAD signaling and inhibiting SAPK signaling. Another study demonstrated the role of miR-214 in exosomes derived from lung carcinoma cells in promoting the expansion of Treg cells by downregulating phosphatase and tensin homologue and resulting in tumor progression (Yin et al., 2014).

3. Exosome-Mediated Cellular Reprogramming

Increasing scientific evidence indicates that exosomes play an important role and act as mediators in cell–cell communication and "reprogram" the recipient cells by transferring proteins, mRNA, and miRNA into them. Several biological functions have been assigned to exosomes, such as intercellular communication, immune regulation, stromal remodeling, signaling pathway activation through growth factor/receptor transfer, oncoprotein and genetic intercellular exchange, induction of angiogenesis and modulation of response to therapy (Vlassov, Magdaleno, Setterquist, & Conrad, 2012). The exchange of molecular information between cells is facilitated by exosomes' unique composition, which is enriched with RNAs, enzymes, structural proteins, and lipid rafts (Zhang & Grizzle, 2011). Numerous evidence shows that tumor cells have the ability to secrete an excessive number of exosomes compared with nontumor cells. Thus, these vesicles can be used as diagnostic markers, and their active secretion has functional implications, although whether they are tumor suppressing or promoting is remaining unclear.

In particular, the interplay via the exchange of exosomes between tumor cells and the tumor stroma may promote the transfer of oncogenes, such as β-catenin, CEA, HER2, and Melan-A/Mart-1 onco-miRNAs, such as let-7, miR-1, miR-15, miR-16, and miR-375 from one cell to another, leading to the reprogramming of the recipient cells (Kharaziha et al., 2012). Previous studies have shown that by secreting exosomes, tumor cells can reprogram cells in the tumor microenvironment with the aim to promote tumor initiation, invasion, and metastasis. For example, tumor cell exosomes profoundly modulate fibroblast phenotype and function via exosomal TGF-β that also promotes the differentiation of fibroblast into myofibroblasts, leading to an altered stroma that is rich in myofibroblastic cells to support tumor growth, vascularization and metastasis. The tumor cell–derived exosomes could also promote angiogenesis and are capable of reprogramming endothelial cells to secrete VEGF in an autocrine mode. Moreover, diverse immunosuppressive effects of tumor-derived exosomes have been identified recently. The effect of tumor-derived exosomes on bone marrow cells is thought to be a coevolutionary strategy of the primary tumor and tumor microenvironment. Alteration of bone marrow cell behavior by tumor-derived exosomes can be mediated by proteins or by the transfer of genetic materials, such as mRNA or miRNA, between tumor cells and bone marrow cells to influence the function of future populations of bone marrow cells.

3.1 MicroRNA-Mediated Reprogramming of Tumor Cells

miRNAs are small noncoding RNAs with diverse functions, which can regulate gene expression at the posttranslational level by binding to the 3′ untranslated region of their target mRNAs to repress translation or direct cleavage. Studies indicate that these small noncoding RNAs undergo a complex, but finely tuned regulation in the early stages of cell reprogramming, for example, during the transition from a differentiated oocyte to a pluripotent blastomere. A single miRNA can influence multiple genes to alter the expression profile of a cell. Moreover, recent discoveries of secreted miRNAs in nearly all body fluids have shown that these miRNAs can influence remote cells and function at a long distance, a feature that is critical for tumor cells metastatic (Calin & Croce, 2006). Tumor-secreted miRNAs were first discovered in the serum of patients with diffuse large B cell lymphoma in which high levels of miR-21 correlated with improved relapse-free survival (Schwarzenbach, Nishida, Calin, & Pantel, 2014). The biogenesis of miRNAs is tightly regulated, and dysregulation of miRNAs is linked to cancer. In cancer, changes in miRNA expression are frequently associated with changes at fragile chromosome sites. miRNAs are thought to play two distinct roles in carcinogenesis, functioning as "oncomirs," such as miR-155 or members of the miR-17-92 clusters as well as tumor suppressors, such as miR-15a and miR-16-1(Strachan et al., 2013). Moreover, about 50% of miRNA genes are located in cancer-associated genomic regions, which further amplifies the evidence that miRNAs play a critical role in cancer. More importantly, roles of miRNAs are tissue- and tumor-specific, as an overexpressed "oncomir" in one certain cancer type may be downregulated in another cancer type, which possesses tumor-suppressive functions (Calin & Croce, 2006). For instance, the multifunctional miR-155 has involved in cancer development, inflammation, immune response, and hematopoietic lineage differentiation. miR-155 is overexpressed in some solid cancers of epithelial origin, in leukemia and in lymphoma but downregulated in some endocrine tumors, melanoma, ovarian, and gastric cancer (Aleckovic & Kang, 2015).

3.2 The Role of MicroRNAs on Reprogramming Tumor-Associated Macrophages

Macrophages are lymphocytes of the myeloid cell lineage, which are derived from myeloid progenitor cells. They can then either circulate as inflammatory monocytes and further differentiate into macrophages in inflamed tissue or migrate into tissues and differentiate into resident tissue macrophages (Rogers & Holen, 2011). Both inflammatory and resident macrophages perform a range of essential biological functions and are activated in response to environmental signals, including microbial products and cytokines. Activated macrophages possess phenotypic plasticity that can be divided into M1- and M2-polarized macrophages with different functions and characteristics within the body and immune system (Quatromoni & Eruslanov, 2012). M1 macrophages, also known as classically activated macrophages, play various roles in both arms of the immune system. In the innate immune system, they guard against infection by engulfing and digesting invasive microorganisms, also defending against tumor cells and eliciting tissue disruptive reactions by releasing tumoricidal agents such TNF-α and interleukin-12 (IL-12). In the adaptive immune system, they work as lymphocyte activators via presenting antigens to polarized Type I T lymphocytes and promote T-helper-1 (Th1) responses and secreting immune-modulatory and proinflammatory cytokines in the inflamed environment (Rogers & Holen, 2011). In contrast, M2 macrophages, also known as alternatively activated macrophages, are better adapted to scavenging debris and release growth factors that promote angiogenesis. They possess reduced immune activity such as poor antigen-presenting capabilities, and T cells' and NK cells' activity suppression. Despite they are still highly phagocytic, the major roles of M2 macrophages are helping repair sites of cellular injury by engulfing cell debris, regulating tissue remodeling, and repairing normal cell turnover (Rogers & Holen, 2011). Apart from playing a role in parasite clearance and wound healing, M2 macrophages also polarize T lymphocytes to Th2 helper cells and attenuate immune responses. Macrophages in tumor stroma usually differentiate into tumor-associated macrophages (TAMs) and predominantly express the M2 phenotype. These TAM's are involved in various stages of tumor progression (Mantovani & Sica, 2010). Therefore, as tumor-promoting TAMs rather resemble the M2-like phenotype in the case of solid cancers, reprogramming of TAMs toward the M1 phenotype may hold great promises in the treatment of cancers.

The crucial role of miRNAs in inflammation response is highlighted by studies in which deregulation of miRNAs was demonstrated to occur with diseases associated with uncontrolled inflammation. miR-155 is a multifunctional miRNA that possesses important functions in hematopoiesis, inflammation, immunity, and cancer (Strachan et al., 2013). miR-155 is processed from *BIC* gene, a noncoding transcript highly expressed in both activated B and T cells and in monocytes/macrophages. The levels of this miRNA change dynamically during both hematopoietic lineage differentiation and the course of the immune response. Different mouse miR-155 expression is controlled by a wide range of inflammatory factors, such as cytokines, TLR-, BCR-, and TCR-signaling as well as JNK or NF-κ B pathways. miR-155 is upregulated in response to lipopolysaccharides (LPS) or interferon (IFN) signaling in both macrophages and monocytes of mice or human origin, suggesting that this miRNA plays an important role in the innate immune response to infections. It has emerged as a key factor in cancer and cancer immunity development with both pro- and antitumoral effects of miR-155, depending on the cell type, disease type in which its expression is altered (Esquela-Kerscher & Slack, 2006). miR-155 upregulation has been associated with B cell cancer and breast carcinomas (Esquela-Kerscher & Slack, 2006; Zonari et al., 2013). On the other hand, miR-155 knockdown in a myeloid compartment of a breast cancer mouse model has shown to hasten breast tumor growth. Additionally, it has been proved to have a critical association with macrophage polarization, as the knockdown resulted in the transitioning of TAMs to an M2/Th2 response (Zonari et al., 2013).

miRNA-125b (miR-125b) is the human orthologue of lin-4, one of the very first miRNAs identified in *Caenorhabditis elegans*. It is enriched in hematopoietic stem cells and that increased miR-125b enhances hematopoietic engraftment. Further increasing the levels of miR-125b can cause an aggressive myeloproliferative disorder that leads to leukemia (Chaudhuri et al., 2011). miR-125b is a ubiquitously expressed miRNA, which is aberrantly expressed in a great variety of tumors. It is a "double-edged" miRNA that has multiple targets in which control proapoptotic and proproliferative signaling pathways and which needs to be tightly regulated under physiological conditions (Giray et al., 2014). If this level of regulation is lost during carcinogenesis, oncogenic or tumor suppressive pathways are activated or blocked, for instance, example, in some tumor types, such as colon cancer and hematopoietic tumors, miR-125b is upregulated and displays oncogenic potential, as it induces cell growth and proliferation and blocks the apoptotic machinery. On the contrary, in other tumor entities, such as mammary tumors and hepatocellular carcinoma, miR-125b is heavily downregulated, which is accompanied by derepression of cellular proliferation and antiapoptotic programs, contributing to malignant transformation. Moreover, it has been shown that when miR-125b is overexpressed in macrophages, it enhances surface activation markers basally and in response to IFN-γ (Banzhaf-Strathmann & Edbauer, 2014). miR-125b is enriched in the M1 phenotype macrophage and has been associated with improved antigen presentation, enhanced T cell activation and tumor destruction (Banzhaf-Strathmann & Edbauer, 2014).

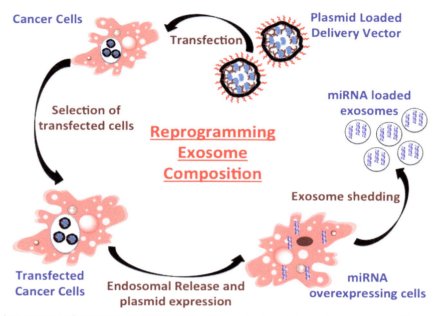

FIGURE 4 **The overview of exosome-mediated cellular communication between human pancreatic cancer (Panc-1) cells and macrophages (J771.A1) using a Transwell coculture system.** After characterization of exosome-mediated cellular communication and protumoral baseline M2 macrophage polarization, the Panc-1 cells were transfected with microRNA-155 (miR-155) and microRNA-125b-2 (miR-125b2) expressing plasmid DNA using hyaluronic acid-poly(ethylene imine)/hyaluronic acid-poly(ethylene glycol) (HA-PEI/HA-PEG) self-assembling nanoparticle-based nonviral vectors. On successful transfection of Panc-1 cells, the exosome content was altered leading to differential communication and reprogramming of the J774.A1 cells to an M1 phenotype. *Reproduced with permission Su, Aldawsari, and Amiji (2016).*

The action of miR-125b is via inhibition of interferon regulatory factor 4 (IRF-4), which is a negative regulator of proinflammatory macrophage activation (Chaudhuri et al., 2011; Graff, Dickson, Clay, McCaffrey, & Wilson, 2012). The overview of exosome-mediated cellular communication between Panc-1 cells and macrophages (J771.A1) was shown in Fig. 4.

3.2.1 Modulation of Exosomal MicroRNA Content Affects Cellular Reprogramming

It was confirmed by a study that the human pancreatic carcinoma, epithelial-like cell line (Panc-1)-derived exosomes were able to trigger murine macrophage (J774A.1) repolarization from M1 phenotype to M2 phenotype. Change of Panc-1-derived exosome cargo contained highly miR-155 and miR-125b-2 expression to achieve macrophage reprogramming effect from M2 back to M1 phenotype was investigated in the same study. The modified exosome-treated groups were carried out dosing 160 µg of miR-155 or miR-125b-2 modified exosomes coculture with M2 phenotype macrophages for 48 h after 6 h IL-4 treatment. J774A.1 macrophages treated with IL-4 for 6 h and kept for

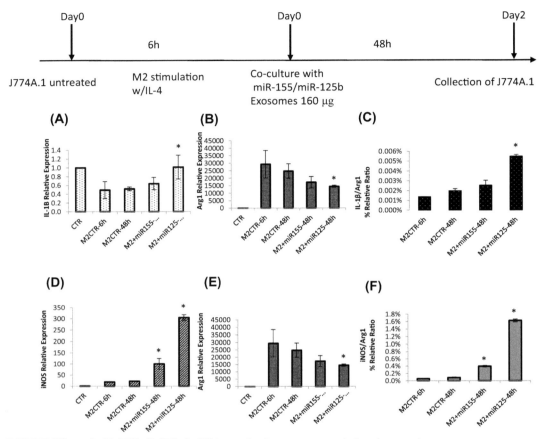

FIGURE 5 Effect of miR-155/miR-125b-2 pDNA transfection on exosome-induced macrophage reprogramming. Quantitative determination of (A) IL-1β and (B) Arg1 expression in M2 J774A.1 macrophages after treating 160 μg of miR-155/miR-125b-2-modified exosomes for 48 h. (C) IL-1β/Arg1 (M1/M2) ratio of M2 J774A.1 macrophages treated with miR-155/miR-125b-2-modified exosomes for 48 h. Quantitative determination of (D) iNOS and (E) Arg1 expression in M2 J774A.1 macrophages after treating 160 μg of miR-155/miR-125b-2-modified exosomes for 48 h. (C) iNOS/Arg1 (M1/M2) ratio of M2 J774A.1 macrophages treated with miR-155/miR-125b-2-modified exosomes for 48 h. n = 3, *P < .05 compared with M2 macrophage treated with IL-4 for 6 h and keep for 48 h. *Reproduced with permission Su et al. (2016).*

48 h were conducted as M2 phenotype controls. The miR-155 and miR-125b-2 expression level in J774A.1 macrophages were then being investigated by TaqMan qPCR assay with specific miR-155 and miR-125b-2 TaqMan primers. The result showed that the levels of miR-155 and miR-125b-2 in J774A.1 macrophages increased 2.22- and 11.5-fold, respectively 48 h posttreatment compared with the untreated groups. To investigate if high expression of miR-155 and miR-125b-2 in exosomes could repolarize the macrophages from M2 to M1 phenotypes, qPCR was used to measure IL-1β/Arg1 (Fig. 5A–C) and iNOS/Arg1 (Fig. 6D–F) ratio in the M2 phenotype treated with miR-155 or miR-125b-2 modified exosomes for 48 h. (IL-1β and iNOS were used as M1 macrophage

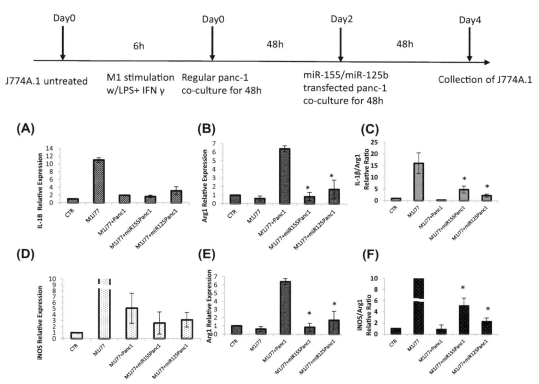

FIGURE 6 Effect of miR-155/miR-125b-2 pDNA transfection on exosome-induced macrophage reprogramming. Quantitative determination of (A) IL-1β and (B) Arg1 expression in M2 J774A.1 macrophages after coculturing with miR-155/miR-125b-2-modified transfected Panc-1 cells for 48 h. (C) IL-1β/Arg1 (M1/M2) ratio of M2 J774A.1 macrophages coculturing with miR-155/miR-125b-2-modified transfected Panc-1 cells for 48 h. Quantitative determination of (D) iNOS and (E) Arg1 expression in M2 J774A.1 macrophages coculturing with miR-155/miR-125b-2-modified transfected Panc-1 cells for 48 h. (C) iNOS/Arg1 (M1/M2) ratio of M2 J774A.1 macrophages coculturing with miR-155/miR-125b-2-modified transfected Panc-1 cells for 48 h. n = 3, *P < .05 compared with M1 macrophage coculture with regular Panc-1 cell group. *Reproduced with permission Su et al. (2016).*

endogenous markers, whereas Arg1 was used as M2 macrophage endogenous marker.) Owing to high expression of iNOS, IL-1β and low expression of Arg1 in M1 macrophages, the high iNOS/Arg1 and IL-1β/Arg1 ratio was used to represent the majority of macrophage in M1 phenotype. This ratio was slightly increased in miR-155-modified exosome–treated group and significantly increased in miR-125b-2-modified exosome–treated group. The results indicated that J774A.1 macrophages were effectively repolarized from M2 to M1 state in the presence of either 160 μg of miR-155 or miR-125b-2 modified exosomes for 48 h comparing with the IL-4 treated M2 macrophage group (Yamada et al., 2016).

After the macrophages' reprogramming effect of miRNA-modified exosomes on J774A.1 macrophages, a further study was carried out using miR-155 and miR-125b-2 plasmid DNA transfected Panc-1 cells in indirect Transwell coculture system to assess the ability of macrophages reprogramming effect from the exosomes secreted by these transfected Panc-1 cells. Untreated J774A.1 macrophages were stimulated with IFN-γ and LPS for 6 h to induce their polarization to M1 phenotype. The Transwell inserts containing nontransfected Panc-1 cells were placed into the culture dishes containing J774A.1 macrophages and incubated up to 48 h to enable the macrophages to polarize to M2 state. The miR-155 and miR-125b-2 plasmid DNA–transfected Panc-1 cells were seeded on new Transwell inserts after hyaluronic acid–based nanoparticle transfection for miRNA delivery. The Transwell inserts were then used to replace the inserts containing regular Panc-1 cells in coculture system for 48 h. The effect of miR-155 and miR-125b plasmid DNA–transfected Panc-1 cells on macrophages polarity was determined by qPCR for endogenous gene expression. The result revealed that the expression of Arg1 decreased significantly and the expression of IL-1β increased slightly after coculturing with miR-155 and miR-125b-2 plasmid DNA–transfected Panc-1 cells (Fig. 6A and B), as compared with regular Panc-1 cells coculture group. It could be clearly seen that IL-1β/Arg1 ratio increased in both miR-155 and miR-125b-2 plasmid DNA–transfected Panc-1 cells coculture groups. However, the expression level of another M1 marker iNOS in miR-155 and miR-125b-2 plasmid DNA–transfected Panc-1 cells coculture groups showed a reduction in comparison with control coculture group (Fig. 6D). The study again determined the macrophage polarity based on its iNOS/arg1 ratio and after normalizing with their own arg1 value, the iNOS/arg1 ratio of miR-155 and miR-125b-2 transfected Panc-1 cells coculture group both increased compared with control coculture group (Fig. 6F). The increase of IL-1β/Arg1 and iNOS/Arg1 ratio both confirmed that modified exosomes secreted by miR-155 and miR-125b-2 plasmid DNA–transfected Panc-1 cells contribute to the macrophage reprogramming from M2 phenotype back to M1 phenotype.

4. Future Prospects

The presence of exosomes to facilitate transport of messenger molecules between cancer cells and tumor microenvironment to induce cancer progression and metastasis has been proven in most cancers. Tumor-derived exosomes containing messenger molecules such as cell-specific makers and tumor antigens can serve as appropriate biomarkers for cancer prognosis and response to therapy. Additionally, the use of exosomes as drug delivery vehicle has numerous advantages such as biocompatibility, overcoming biological barriers and distribution in the tumor microenvironment. Thus, extensive efforts are invested in loading the exosomes with suitable therapeutic agents and targeting these exosomes to the cells of interest. Furthermore, a combination of exosomes and nanotechnology could result in generating targeted cancer therapies, which can be personalized to suit the disease state based on extensive tumor profiling. Thus, further understanding of the intricacies of exosomal biogenesis and transport will help in timely diagnosis and devising treatment strategies to modulate the effects of exosomes in tumor microenvironment.

References

Abusamra, A. J., Zhong, Z., Zheng, X., Li, M., Ichim, T. E., Chin, J. L., & Min, W.-P. (2005). Tumor exosomes expressing Fas ligand mediate CD8+ T-cell apoptosis. *Blood Cells, Molecules & Diseases*, *35*(2), 169–173.

Achiriloaie, M., Barylko, B., & Albanesi, J. P. (1999). Essential role of the dynamin pleckstrin homology domain in receptor-mediated endocytosis. *Molecular and Cellular Biology*, *19*(2), 1410–1415.

Al-Nedawi, K., Meehan, B., Kerbel, R. S., Allison, A. C., & Rak, J. (2009). Endothelial expression of autocrine VEGF upon the uptake of tumor-derived microvesicles containing oncogenic EGFR. *Proceedings of the National Academy of Sciences*, *106*(10), 3794–3799.

Al-Nedawi, K., Meehan, B., Micallef, J., Lhotak, V., May, L., Guha, A., & Rak, J. (2008). Intercellular transfer of the oncogenic receptor EGFRvIII by microvesicles derived from tumour cells. *Nature Cell Biology*, *10*(5), 619.

Aleckovic, M., & Kang, Y. (2015). Regulation of cancer metastasis by cell-free miRNAs. *Biochimica et Biophysica Acta*, *1855*(1), 24–42. https://doi.org/10.1016/j.bbcan.2014.10.005.

Anderson, R. G. (1998). The caveolae membrane system. *Annual Review of Biochemistry*, *67*, 199–225. https://doi.org/10.1146/annurev.biochem.67.1.199.

Azmi, A. S., Bao, B., & Sarkar, F. H. (2013). Exosomes in cancer development, metastasis, and drug resistance: A comprehensive review. *Cancer and Metastasis Reviews*, *32*(3–4), 623–642. https://doi.org/10.1007/s10555-013-9441-9.

Babst, M. (2005). A protein's final ESCRT. *Traffic*, *6*(1), 2–9. https://doi.org/10.1111/j.1600-0854.2004.00246.x.

Banzhaf-Strathmann, J., & Edbauer, D. (2014). Good guy or bad guy: The opposing roles of microRNA 125b in cancer. *Cell Communication and Signaling: CCS*, *12*, 30. https://doi.org/10.1186/1478-811X-12-30.

Barres, C., Blanc, L., Bette-Bobillo, P., Andre, S., Mamoun, R., Gabius, H. J., & Vidal, M. (2010). Galectin-5 is bound onto the surface of rat reticulocyte exosomes and modulates vesicle uptake by macrophages. *Blood*, *115*(3), 696–705. https://doi.org/10.1182/blood-2009-07-231449.

Beach, A., Zhang, H. G., Ratajczak, M. Z., & Kakar, S. S. (2014). Exosomes: An overview of biogenesis, composition and role in ovarian cancer. *Journal of Ovarian Research*, *7*, 14. https://doi.org/10.1186/1757-2215-7-14.

Beckler, M. D., Higginbotham, J. N., Franklin, J. L., Ham, A.-J., Halvey, P. J., Imasuen, I. E., … Coffey, R. J. (2013). Proteomic analysis of exosomes from mutant KRAS colon cancer cells identifies intercellular transfer of mutant KRAS. *Molecular & Cellular Proteomics*, *12*(2), 343–355.

Bedoui, S., Prato, S., Mintern, J., Gebhardt, T., Zhan, Y., Lew, A. M., … Segura, E. (2009). Characterization of an immediate splenic precursor of CD8+ dendritic cells capable of inducing antiviral T cell responses. *The Journal of Immunology*, *182*(7), 4200–4207. https://doi.org/10.4049/jimmunol.0802286.

Benmerah, A., Bayrou, M., Cerf-Bensussan, N., & Dautry-Varsat, A. (1999). Inhibition of clathrin-coated pit assembly by an Eps15 mutant. *Journal of Cell Science*, *112*(Pt 9), 1303–1311.

Bianco, F., Perrotta, C., Novellino, L., Francolini, M., Riganti, L., Menna, E., … Verderio, C. (2009). Acid sphingomyelinase activity triggers microparticle release from glial cells. *The EMBO Journal*, *28*(8), 1043–1054. https://doi.org/10.1038/emboj.2009.45.

Bobrie, A., Colombo, M., Raposo, G., & Thery, C. (2011). Exosome secretion: Molecular mechanisms and roles in immune responses. *Traffic*, *12*(12), 1659–1668. https://doi.org/10.1111/j.1600-0854.2011.01225.x.

Bobrie, A., Krumeich, S., Reyal, F., Recchi, C., Moita, L. F., Seabra, M. C., … Théry, C. (2012). Rab27a supports exosome-dependent and -independent mechanisms that modify the tumor microenvironment and can promote tumor progression. *Cancer Research*, *72*(19), 4920–4930.

Booth, A. M., Fang, Y., Fallon, J. K., Yang, J. M., Hildreth, J. E., & Gould, S. J. (2006). Exosomes and HIV Gag bud from endosome-like domains of the T cell plasma membrane. *The Journal of Cell Biology*, *172*(6), 923–935. https://doi.org/10.1083/jcb.200508014.

Calin, G. A., & Croce, C. M. (2006). MicroRNA-cancer connection: The beginning of a new tale. *Cancer Research*, *66*(15), 7390–7394. https://doi.org/10.1158/0008-5472.CAN-06-0800.

Chalmin, F., Ladoire, S., Mignot, G., Vincent, J., Bruchard, M., Remy-Martin, J.-P., … Lanneau, D. (2010). Membrane-associated Hsp72 from tumor-derived exosomes mediates STAT3-dependent immuno-suppressive function of mouse and human myeloid-derived suppressor cells. *The Journal of Clinical Investigation*, *120*(2), 457.

Chappie, J. S., Acharya, S., Leonard, M., Schmid, S. L., & Dyda, F. (2010). G domain dimerization controls dynamin's assembly-stimulated GTPase activity. *Nature*, *465*(7297), 435–440. https://doi.org/10.1038/nature09032.

Chaudhuri, A. A., So, A. Y., Sinha, N., Gibson, W. S., Taganov, K. D., O'Connell, R. M., & Baltimore, D. (2011). MicroRNA-125b potentiates macrophage activation. *The Journal of Immunology*, *187*(10), 5062–5068. https://doi.org/10.4049/jimmunol.1102001.

Cho, J. A., Park, H., Lim, E. H., & Lee, K. W. (2012). Exosomes from breast cancer cells can convert adipose tissue-derived mesenchymal stem cells into myofibroblast-like cells. *International Journal of Oncology*, *40*(1), 130–138.

Chowdhury, R., Webber, J. P., Gurney, M., Mason, M. D., Tabi, Z., & Clayton, A. (2015). Cancer exosomes trigger mesenchymal stem cell differentiation into pro-angiogenic and pro-invasive myofibroblasts. *Oncotarget*, *6*(2), 715.

Christianson, H. C., Svensson, K. J., van Kuppevelt, T. H., Li, J. P., & Belting, M. (2013). Cancer cell exosomes depend on cell-surface heparan sulfate proteoglycans for their internalization and functional activity. *Proceedings of the National Academy of Sciences of the United States of America*, *110*(43), 17380–17385. https://doi.org/10.1073/pnas.1304266110.

Ciravolo, V., Huber, V., Ghedini, G. C., Venturelli, E., Bianchi, F., Campiglio, M., … Menard, S. (2012). Potential role of HER2-overexpressing exosomes in countering trastuzumab-based therapy. *Journal of Cellular Physiology*, *227*(2), 658–667.

Corcoran, C., Rani, S., O'Brien, K., O'Neill, A., Prencipe, M., Sheikh, R., … Crown, J. (2012). Docetaxel-resistance in prostate cancer: Evaluating associated phenotypic changes and potential for resistance transfer via exosomes. *PLoS One*, *7*(12), e50999.

Damke, H., Baba, T., Warnock, D. E., & Schmid, S. L. (1994). Induction of mutant dynamin specifically blocks endocytic coated vesicle formation. *The Journal of Cell Biology*, *127*(4), 915–934.

Del Conde, I., Shrimpton, C. N., Thiagarajan, P., & Lopez, J. A. (2005). Tissue-factor-bearing microvesicles arise from lipid rafts and fuse with activated platelets to initiate coagulation. *Blood*, *106*(5), 1604–1611. https://doi.org/10.1182/blood-2004-03-1095.

Doherty, G. J., & McMahon, H. T. (2009). Mechanisms of endocytosis. *Annual Review of Biochemistry*, *78*, 857–902. https://doi.org/10.1146/annurev.biochem.78.081307.110540.

Ehrlich, M., Boll, W., Van Oijen, A., Hariharan, R., Chandran, K., Nibert, M. L., & Kirchhausen, T. (2004). Endocytosis by random initiation and stabilization of clathrin-coated pits. *Cell*, *118*(5), 591–605. https://doi.org/10.1016/j.cell.2004.08.017.

Escrevente, C., Keller, S., Altevogt, P., & Costa, J. (2011). Interaction and uptake of exosomes by ovarian cancer cells. *BMC Cancer*, *11*, 108. https://doi.org/10.1186/1471-2407-11-108.

Esquela-Kerscher, A., & Slack, F. J. (2006). Oncomirs - microRNAs with a role in cancer. *Nature Reviews Cancer*, *6*(4), 259–269. https://doi.org/10.1038/nrc1840.

Fabbri, M., Paone, A., Calore, F., Galli, R., Gaudio, E., Santhanam, R., … Nuovo, G. J. (2012). MicroRNAs bind to Toll-like receptors to induce prometastatic inflammatory response. *Proceedings of the National Academy of Sciences*, *109*(31), E2110–E2116.

Fader, C. M., & Colombo, M. I. (2009). Autophagy and multivesicular bodies: Two closely related partners. *Cell Death and Differentiation, 16*(1), 70–78. https://doi.org/10.1038/cdd.2008.168.

Fadok, V. A., Voelker, D. R., Campbell, P. A., Cohen, J. J., Bratton, D. L., & Henson, P. M. (1992). Exposure of phosphatidylserine on the surface of apoptotic lymphocytes triggers specific recognition and removal by macrophages. *The Journal of Immunology, 148*(7), 2207–2216.

Feng, D., Zhao, W. L., Ye, Y. Y., Bai, X. C., Liu, R. Q., Chang, L. F., … Sui, S. F. (2010). Cellular internalization of exosomes occurs through phagocytosis. *Traffic, 11*(5), 675–687. https://doi.org/10.1111/j.1600-0854.2010.01041.x.

Fitzner, D., Schnaars, M., van Rossum, D., Krishnamoorthy, G., Dibaj, P., Bakhti, M., … Simons, M. (2011). Selective transfer of exosomes from oligodendrocytes to microglia by macropinocytosis. *Journal of Cell Science, 124*(Pt 3), 447–458. https://doi.org/10.1242/jcs.074088.

Fukudome, K., Furuse, M., Imai, T., Nishimura, M., Takagi, S., Hinuma, Y., & Yoshie, O. (1992). Identification of membrane antigen C33 recognized by monoclonal antibodies inhibitory to human t-cell leukemia virus type 1 (HTLV-1)-induced syncytium formation: Altered glycosylation of C33 antigen in HTLV-1-positive T cells. *Journal of Virology, 66*(3), 1394–1401.

Giray, B. G., Emekdas, G., Tezcan, S., Ulger, M., Serin, M. S., Sezgin, O., … Tiftik, E. N. (2014). Profiles of serum microRNAs; miR-125b-5p and miR223-3p serve as novel biomarkers for HBV-positive hepatocellular carcinoma. *Molecular Biology Reports, 41*(7), 4513–4519. https://doi.org/10.1007/s11033-014-3322-3.

Gold, E. S., Underhill, D. M., Morrissette, N. S., Guo, J., McNiven, M. A., & Aderem, A. (1999). Dynamin 2 is required for phagocytosis in macrophages. *The Journal of Experimental Medicine, 190*(12), 1849–1856.

Gordon-Alonso, M., Yanez-Mo, M., Barreiro, O., Alvarez, S., Munoz-Fernandez, M. A., Valenzuela-Fernandez, A., & Sanchez-Madrid, F. (2006). Tetraspanins CD9 and CD81 modulate HIV-1-induced membrane fusion. *The Journal of Immunology, 177*(8), 5129–5137.

Graff, J. W., Dickson, A. M., Clay, G., McCaffrey, A. P., & Wilson, M. E. (2012). Identifying functional microRNAs in macrophages with polarized phenotypes. *Journal of Biological Chemistry, 287*(26), 21816–21825. https://doi.org/10.1074/jbc.M111.327031.

Grange, C., Tapparo, M., Collino, F., Vitillo, L., Damasco, C., Deregibus, M. C., … Camussi, G. (2011). Microvesicles released from human renal cancer stem cells stimulate angiogenesis and formation of lung premetastatic niche. *Cancer Research, 71*(15), 5346–5356.

Grassart, A., Cheng, A. T., Hong, S. H., Zhang, F., Zenzer, N., Feng, Y., … Drubin, D. G. (2014). Actin and dynamin2 dynamics and interplay during clathrin-mediated endocytosis. *The Journal of Cell Biology, 205*(5), 721–735. https://doi.org/10.1083/jcb.201403041.

Grigorov, B., Attuil-Audenis, V., Perugi, F., Nedelec, M., Watson, S., Pique, C., … Muriaux, D. (2009). A role for CD81 on the late steps of HIV-1 replication in a chronically infected T cell line. *Retrovirology, 6*, 28. https://doi.org/10.1186/1742-4690-6-28.

Grimmer, S., van Deurs, B., & Sandvig, K. (2002). Membrane ruffling and macropinocytosis in A431 cells require cholesterol. *Journal of Cell Science, 115*(Pt 14), 2953–2962.

Hannafon, B. N., & Ding, W. Q. (2013). Intercellular communication by exosome-derived microRNAs in cancer. *International Journal of Molecular Sciences, 14*(7), 14240–14269. https://doi.org/10.3390/ijms140714240.

Hemler, M. E. (2003). Tetraspanin proteins mediate cellular penetration, invasion, and fusion events and define a novel type of membrane microdomain. *Annual Review of Cell and Developmental Biology, 19*, 397–422. https://doi.org/10.1146/annurev.cellbio.19.111301.153609.

Hsu, C., Morohashi, Y., Yoshimura, S., Manrique-Hoyos, N., Jung, S., Lauterbach, M. A., … Simons, M. (2010). Regulation of exosome secretion by Rab35 and its GTPase-activating proteins TBC1D10A-C. *The Journal of Cell Biology, 189*(2), 223–232. https://doi.org/10.1083/jcb.200911018.

Jager, S., Bucci, C., Tanida, I., Ueno, T., Kominami, E., Saftig, P., & Eskelinen, E. L. (2004). Role for Rab7 in maturation of late autophagic vacuoles. *Journal of Cell Science*, *117*(Pt 20), 4837–4848. https://doi.org/10.1242/jcs.01370.

Jung, T., Castellana, D., Klingbeil, P., Hernández, I. C., Vitacolonna, M., Orlicky, D. J., ... Zöller, M. (2009). CD44v6 dependence of premetastatic niche preparation by exosomes. *Neoplasia*, *11*(10), 1093IN1013–1105IN1017.

Keller, S., Sanderson, M. P., Stoeck, A., & Altevogt, P. (2006). Exosomes: From biogenesis and secretion to biological function. *Immunology Letters*, *107*(2), 102–108. https://doi.org/10.1016/j.imlet.2006.09.005.

Kerr, M. C., & Teasdale, R. D. (2009). Defining macropinocytosis. *Traffic*, *10*(4), 364–371. https://doi.org/10.1111/j.1600-0854.2009.00878.x.

Kharaziha, P., Ceder, S., Li, Q., & Panaretakis, T. (2012). Tumor cell-derived exosomes: A message in a bottle. *Biochimica et Biophysica Acta*, *1826*(1), 103–111. https://doi.org/10.1016/j.bbcan.2012.03.006.

Kirchhausen, T. (2000). Clathrin. *Annual Review of Biochemistry*, *69*, 699–727. https://doi.org/10.1146/annurev.biochem.69.1.699.

Krementsov, D. N., Weng, J., Lambele, M., Roy, N. H., & Thali, M. (2009). Tetraspanins regulate cell-to-cell transmission of HIV-1. *Retrovirology*, *6*, 64. https://doi.org/10.1186/1742-4690-6-64.

Kurzchalia, T. V., & Parton, R. G. (1999). Membrane microdomains and caveolae. *Current Opinion in Cell Biology*, *11*(4), 424–431.

Labani-Motlagh, A., Israelsson, P., Ottander, U., Lundin, E., Nagaev, I., Nagaeva, O., ... Mincheva-Nilsson, L. (2016). Differential expression of ligands for NKG2D and DNAM-1 receptors by epithelial ovarian cancer-derived exosomes and its influence on NK cell cytotoxicity. *Tumour Biology*, *37*(4), 5455–5466.

Lee, A., Frank, D. W., Marks, M. S., & Lemmon, M. A. (1999). Dominant-negative inhibition of receptor-mediated endocytosis by a dynamin-1 mutant with a defective pleckstrin homology domain. *Current Biology*, *9*(5), 261–264.

Luga, V., Zhang, L., Viloria-Petit, A. M., Ogunjimi, A. A., Inanlou, M. R., Chiu, E., ... Wrana, J. L. (2012). Exosomes mediate stromal mobilization of autocrine Wnt-PCP signaling in breast cancer cell migration. *Cell*, *151*(7), 1542–1556.

Mantovani, A., & Sica, A. (2010). Macrophages, innate immunity and cancer: Balance, tolerance, and diversity. *Current Opinion in Immunology*, *22*(2), 231–237. https://doi.org/10.1016/j.coi.2010.01.009.

Marks, B., Stowell, M. H., Vallis, Y., Mills, I. G., Gibson, A., Hopkins, C. R., & McMahon, H. T. (2001). GTPase activity of dynamin and resulting conformation change are essential for endocytosis. *Nature*, *410*(6825), 231–235. https://doi.org/10.1038/35065645.

Mayor, S., & Pagano, R. E. (2007). Pathways of clathrin-independent endocytosis. *Nature Reviews Molecular Cell Biology*, *8*(8), 603–612. https://doi.org/10.1038/nrm2216.

McKelvey, K. J., Powell, K. L., Ashton, A. W., Morris, J. M., & McCracken, S. A. (2015). Exosomes: Mechanisms of uptake. *Journal of Circulating Biomarkers*, *4*, 7.

Menck, K., Klemm, F., Gross, J. C., Pukrop, T., Wenzel, D., & Binder, C. (2013). Induction and transport of Wnt 5a during macrophage-induced malignant invasion is mediated by two types of extracellular vesicles. *Oncotarget*, *4*(11), 2057–2066. https://doi.org/10.18632/oncotarget.1336.

Merendino, A. M., Bucchieri, F., Campanella, C., Marciano, V., Ribbene, A., David, S., ... Cappello, F. (2010). Hsp60 is actively secreted by human tumor cells. *PLoS One*, *5*(2), e9247. https://doi.org/10.1371/journal.pone.0009247.

Merrifield, C. J., Feldman, M. E., Wan, L., & Almers, W. (2002). Imaging actin and dynamin recruitment during invagination of single clathrin-coated pits. *Nature Cell Biology*, *4*(9), 691–698. https://doi.org/10.1038/ncb837.

Mineo, M., Garfield, S., Taverna, S., Flugy, A., De Leo, G., Alessandro, R., & Kohn, E. (2012). Exosomes released by K562 chronic myeloid leukemia cells promote angiogenesis in a Src-dependent fashion. *Angiogenesis*, *15*(1), 33–45.

Miyado, K., Yamada, G., Yamada, S., Hasuwa, H., Nakamura, Y., Ryu, F., … Mekada, E. (2000). Requirement of CD9 on the egg plasma membrane for fertilization. *Science, 287*(5451), 321–324.

Miyanishi, M., Tada, K., Koike, M., Uchiyama, Y., Kitamura, T., & Nagata, S. (2007). Identification of Tim4 as a phosphatidylserine receptor. *Nature, 450*(7168), 435–439. https://doi.org/10.1038/nature06307.

Momen-Heravi, F., Bala, S., Bukong, T., & Szabo, G. (2014). Exosome-mediated delivery of functionally active miRNA-155 inhibitor to macrophages. *Nanomedicine, 10*(7), 1517–1527. https://doi.org/10.1016/j.nano.2014.03.014.

Morelli, A. (2006). The immune regulatory effect of apoptotic cells and exosomes on dendritic cells: Its impact on transplantation. *American Journal of Transplantation, 6*(2), 254–261.

Morelli, A. E., Larregina, A. T., Shufesky, W. J., Sullivan, M. L., Stolz, D. B., Papworth, G. D., … Thomson, A. W. (2004). Endocytosis, intracellular sorting, and processing of exosomes by dendritic cells. *Blood, 104*(10), 3257–3266. https://doi.org/10.1182/blood-2004-03-0824.

Mrizak, D., Martin, N., Barjon, C., Jimenez-Pailhes, A.-S., Mustapha, R., Niki, T., … Busson, P. (2015). Effect of nasopharyngeal carcinoma-derived exosomes on human regulatory T cells. *Journal of the National Cancer Institute, 107*(1).

Mulcahy, L. A., Pink, R. C., & Carter, D. R. (2014). Routes and mechanisms of extracellular vesicle uptake. *Journal of Extracellular Vesicles, 3*. https://doi.org/10.3402/jev.v3.24641.

Nabi, I. R., & Le, P. U. (2003). Caveolae/raft-dependent endocytosis. *The Journal of Cell Biology, 161*(4), 673–677. https://doi.org/10.1083/jcb.200302028.

Nanbo, A., Kawanishi, E., Yoshida, R., & Yoshiyama, H. (2013). Exosomes derived from Epstein-Barr virus-infected cells are internalized via caveola-dependent endocytosis and promote phenotypic modulation in target cells. *Journal of Virology, 87*(18), 10334–10347. https://doi.org/10.1128/JVI.01310-13.

Newton, A. J., Kirchhausen, T., & Murthy, V. N. (2006). Inhibition of dynamin completely blocks compensatory synaptic vesicle endocytosis. *Proceedings of the National Academy of Sciences of the United States of America, 103*(47), 17955–17960. https://doi.org/10.1073/pnas.0606212103.

Nolte-'t Hoen, E. N., Buschow, S. I., Anderton, S. M., Stoorvogel, W., & Wauben, M. H. (2009). Activated T cells recruit exosomes secreted by dendritic cells via LFA-1. *Blood, 113*(9), 1977–1981. https://doi.org/10.1182/blood-2008-08-174094.

Novikoff, A. B., Essner, E., & Quintana, N. (1964). Golgi apparatus and lysosomes. *Federation Proceedings, 23*, 1010–1022.

Ostrowski, M., Carmo, N. B., Krumeich, S., Fanget, I., Raposo, G., Savina, A., … Thery, C. (2010). Rab27a and Rab27b control different steps of the exosome secretion pathway. *Nature Cell Biology, 12*(1), 19–30. https://doi.org/10.1038/ncb2000. sup pp. 11–13.

Parolini, I., Federici, C., Raggi, C., Lugini, L., Palleschi, S., De Milito, A., … Fais, S. (2009). Microenvironmental pH is a key factor for exosome traffic in tumor cells. *Journal of Biological Chemistry, 284*(49), 34211–34222. https://doi.org/10.1074/jbc.M109.041152.

Parton, R. G., & Simons, K. (2007). The multiple faces of caveolae. *Nature Reviews Molecular Cell Biology, 8*(3), 185–194. https://doi.org/10.1038/nrm2122.

Peinado, H., Alečković, M., Lavotshkin, S., Matei, I., Costa-Silva, B., Moreno-Bueno, G., … Ghajar, C. M. (2012). Melanoma exosomes educate bone marrow progenitor cells toward a pro-metastatic phenotype through MET. *Nature Medicine, 18*(6), 883–891.

Quatromoni, J. G., & Eruslanov, E. (2012). Tumor-associated macrophages: Function, phenotype, and link to prognosis in human lung cancer. *American Journal of Translational Research, 4*(4), 376–389.

Ramachandran, R., Pucadyil, T. J., Liu, Y. W., Acharya, S., Leonard, M., Lukiyanchuk, V., & Schmid, S. L. (2009). Membrane insertion of the pleckstrin homology domain variable loop 1 is critical for dynamin-catalyzed vesicle scission. *Molecular Biology of the Cell, 20*(22), 4630–4639. https://doi.org/10.1091/mbc.E09-08-0683.

Rana, S., & Zoller, M. (2011). Exosome target cell selection and the importance of exosomal tetraspanins: A hypothesis. *Biochemical Society Transactions*, *39*(2), 559–562. https://doi.org/10.1042/BST0390559.

Raposo, G., & Stoorvogel, W. (2013). Extracellular vesicles: Exosomes, microvesicles, and friends. *The Journal of Cell Biology*, *200*(4), 373–383. https://doi.org/10.1083/jcb.201211138.

Ridley, A. J. (2006). Rho GTPases and actin dynamics in membrane protrusions and vesicle trafficking. *Trends in Cell Biology*, *16*(10), 522–529. https://doi.org/10.1016/j.tcb.2006.08.006.

Robbins, P. D., & Morelli, A. E. (2014). Regulation of immune responses by extracellular vesicles. *Nature Reviews Immunology*, *14*(3), 195–208. https://doi.org/10.1038/nri3622.

Rogers, T. L., & Holen, I. (2011). Tumour macrophages as potential targets of bisphosphonates. *Journal of Translational Medicine*, *9*, 177. https://doi.org/10.1186/1479-5876-9-177.

Rubinstein, E., Ziyyat, A., Wolf, J. P., Le Naour, F., & Boucheix, C. (2006). The molecular players of sperm-egg fusion in mammals. *Seminars in Cell & Developmental Biology*, *17*(2), 254–263. https://doi.org/10.1016/j.semcdb.2006.02.012.

Rudt, S., & Müller, R. (1993). In vitro phagocytosis assay of nano-and microparticles by chemiluminescence. III. Uptake of differently sized surface-modified particles, and its correlation to particle properties and in vivo distribution. *European Journal of Pharmaceutical Sciences*, *1*(1), 31–39.

Sabapatha, A., Gercel-Taylor, C., & Taylor, D. D. (2006). Specific isolation of placenta-derived exosomes from the circulation of pregnant women and their immunoregulatory consequences. *American Journal of Reproductive Immunology*, *56*(5–6), 345–355. https://doi.org/10.1111/j.1600-0897.2006.00435.x.

Sanz-Moreno, V., Gadea, G., Ahn, J., Paterson, H., Marra, P., Pinner, S., … Marshall, C. J. (2008). Rac activation and inactivation control plasticity of tumor cell movement. *Cell*, *135*(3), 510–523. https://doi.org/10.1016/j.cell.2008.09.043.

Sato, K., Aoki, J., Misawa, N., Daikoku, E., Sano, K., Tanaka, Y., & Koyanagi, Y. (2008). Modulation of human immunodeficiency virus type 1 infectivity through incorporation of tetraspanin proteins. *Journal of Virology*, *82*(2), 1021–1033. https://doi.org/10.1128/JVI.01044-07.

Savina, A., Furlan, M., Vidal, M., & Colombo, M. I. (2003). Exosome release is regulated by a calcium-dependent mechanism in K562 cells. *Journal of Biological Chemistry*, *278*(22), 20083–20090. https://doi.org/10.1074/jbc.M301642200.

Schneider, P., Holler, N., Bodmer, J. L., Hahne, M., Frei, K., Fontana, A., & Tschopp, J. (1998). Conversion of membrane-bound Fas(CD95) ligand to its soluble form is associated with downregulation of its proapoptotic activity and loss of liver toxicity. *The Journal of Experimental Medicine*, *187*(8), 1205–1213.

Schwarzenbach, H., Nishida, N., Calin, G. A., & Pantel, K. (2014). Clinical relevance of circulating cell-free microRNAs in cancer. *Nature Reviews Clinical Oncology*, *11*(3), 145–156. https://doi.org/10.1038/nrclinonc.2014.5.

Shiratsuchi, A., Kaido, M., Takizawa, T., & Nakanishi, Y. (2000). Phosphatidylserine-mediated phagocytosis of influenza A virus-infected cells by mouse peritoneal macrophages. *Journal of Virology*, *74*(19), 9240–9244.

Stenqvist, A. C., Nagaeva, O., Baranov, V., & Mincheva-Nilsson, L. (2013). Exosomes secreted by human placenta carry functional Fas ligand and TRAIL molecules and convey apoptosis in activated immune cells, suggesting exosome-mediated immune privilege of the fetus. *The Journal of Immunology*, *191*(11), 5515–5523. https://doi.org/10.4049/jimmunol.1301885.

Stephens, L., Ellson, C., & Hawkins, P. (2002). Roles of PI3Ks in leukocyte chemotaxis and phagocytosis. *Current Opinion in Cell Biology*, *14*(2), 203–213.

Stoorvogel, W., Strous, G. J., Geuze, H. J., Oorschot, V., & Schwartz, A. L. (1991). Late endosomes derive from early endosomes by maturation. *Cell*, *65*(3), 417–427.

Strachan, D. C., Ruffell, B., Oei, Y., Bissell, M. J., Coussens, L. M., Pryer, N., & Daniel, D. (2013). CSF1R inhibition delays cervical and mammary tumor growth in murine models by attenuating the turnover of tumor-associated macrophages and enhancing infiltration by CD8+ T cells. *OncoImmunology, 2*(12), e26968. https://doi.org/10.4161/onci.26968.

Su, M. J., Aldawsari, H., Amiji, M. (2016). Pancreatic cancer cell exosome-mediated macrophage reprogramming and the role of microRNAs 155 and 125b2 transfection using nanoparticle delivery systems. *Scientific Reports, 6*, 30110.

Swanson, J. A. (2008). Shaping cups into phagosomes and macropinosomes. *Nature Reviews Molecular Cell Biology, 9*(8), 639–649. https://doi.org/10.1038/nrm2447.

Symeonides, M., Lambele, M., Roy, N. H., & Thali, M. (2014). Evidence showing that tetraspanins inhibit HIV-1-induced cell-cell fusion at a post-hemifusion stage. *Viruses, 6*(3), 1078–1090. https://doi.org/10.3390/v6031078.

Szajnik, M., Czystowska, M., Szczepanski, M. J., Mandapathil, M., & Whiteside, T. L. (2010). Tumor-derived microvesicles induce, expand and up-regulate biological activities of human regulatory T cells (Treg). *PLoS One, 5*(7), e11469.

Tachibana, I., & Hemler, M. E. (1999). Role of transmembrane 4 superfamily (TM4SF) proteins CD9 and CD81 in muscle cell fusion and myotube maintenance. *The Journal of Cell Biology, 146*(4), 893–904.

Takeda, Y., Tachibana, I., Miyado, K., Kobayashi, M., Miyazaki, T., Funakoshi, T., … Mekada, E. (2003). Tetraspanins CD9 and CD81 function to prevent the fusion of mononuclear phagocytes. *The Journal of Cell Biology, 161*(5), 945–956. https://doi.org/10.1083/jcb.200212031.

Taylor, D. D., Akyol, S., & Gercel-Taylor, C. (2006). Pregnancy-associated exosomes and their modulation of T cell signaling. *The Journal of Immunology, 176*(3), 1534–1542.

Taylor, D. D., Gercel-Taylor, C., Lyons, K. S., Stanson, J., & Whiteside, T. L. (2003). T-cell apoptosis and suppression of T-cell receptor/CD3-zeta by Fas ligand-containing membrane vesicles shed from ovarian tumors. *Clinical Cancer Research, 9*(14), 5113–5119.

Taylor, M. J., Lampe, M., & Merrifield, C. J. (2012). A feedback loop between dynamin and actin recruitment during clathrin-mediated endocytosis. *PLoS Biology, 10*(4), e1001302. https://doi.org/10.1371/journal.pbio.1001302.

Tian, T., Zhu, Y. L., Zhou, Y. Y., Liang, G. F., Wang, Y. Y., Hu, F. H., & Xiao, Z. D. (2014). Exosome uptake through clathrin-mediated endocytosis and macropinocytosis and mediating miR-21 delivery. *Journal of Biological Chemistry, 289*(32), 22258–22267. https://doi.org/10.1074/jbc.M114.588046.

Vallee, R. B., Herskovits, J. S., Aghajanian, J. G., Burgess, C. C., & Shpetner, H. S. (1993). Dynamin, a GTPase involved in the initial stages of endocytosis. *Ciba Foundation Symposia, 176*, 185–193 discussion 193–187.

Vallis, Y., Wigge, P., Marks, B., Evans, P. R., & McMahon, H. T. (1999). Importance of the pleckstrin homology domain of dynamin in clathrin-mediated endocytosis. *Current Biology, 9*(5), 257–260.

van Niel, G., Porto-Carreiro, I., Simoes, S., & Raposo, G. (2006). Exosomes: A common pathway for a specialized function. *Journal of Biochemistry, 140*(1), 13–21. https://doi.org/10.1093/jb/mvj128.

Verweij, F. J., Van Eijndhoven, M. A., Hopmans, E. S., Vendrig, T., Wurdinger, T., Cahir-McFarland, E., … Neefjes, J. (2011). LMP1 association with CD63 in endosomes and secretion via exosomes limits constitutive NF-κB activation. *The EMBO Journal, 30*(11), 2115–2129.

Vlassov, A. V., Magdaleno, S., Setterquist, R., & Conrad, R. (2012). Exosomes: Current knowledge of their composition, biological functions, and diagnostic and therapeutic potentials. *Biochimica et Biophysica Acta, 1820*(7), 940–948. https://doi.org/10.1016/j.bbagen.2012.03.017.

Vong, S., & Kalluri, R. (2011). The role of stromal myofibroblast and extracellular matrix in tumor angiogenesis. *Genes & Cancer, 2*(12), 1139–1145.

Wajant, H., Moosmayer, D., Wuest, T., Bartke, T., Gerlach, E., Schonherr, U., … Pfizenmaier, K. (2001). Differential activation of TRAIL-R1 and -2 by soluble and membrane TRAIL allows selective surface antigen-directed activation of TRAIL-R2 by a soluble TRAIL derivative. *Oncogene, 20*(30), 4101–4106. https://doi.org/10.1038/sj.onc.1204558.

Wang, L. H., Rothberg, K. G., & Anderson, R. G. (1993). Mis-assembly of clathrin lattices on endosomes reveals a regulatory switch for coated pit formation. *J Cell Biol, 123*(5), 1107–1117.

Webber, J., Steadman, R., Mason, M. D., Tabi, Z., & Clayton, A. (2010). Cancer exosomes trigger fibroblast to myofibroblast differentiation. *Cancer Research, 70*(23), 9621–9630.

Wei, M., Yang, T., Chen, X., Wu, Y., Deng, X., He, W., … Wang, Z. (2017). Malignant ascites-derived exosomes promote proliferation and induce carcinoma-associated fibroblasts transition in peritoneal mesothelial cells. *Oncotarget, 8*(26), 42262.

Wieckowski, E. U., Visus, C., Szajnik, M., Szczepanski, M. J., Storkus, W. J., & Whiteside, T. L. (2009). Tumor-derived microvesicles promote regulatory T cell expansion and induce apoptosis in tumor-reactive activated CD8+ T lymphocytes. *The Journal of Immunology, 183*(6), 3720–3730.

Wu, Y., Wu, W., Wong, W. M., Ward, E., Thrasher, A. J., Goldblatt, D., … Gustafsson, K. (2009). Human gamma delta T cells: A lymphoid lineage cell capable of professional phagocytosis. *The Journal of Immunology, 183*(9), 5622–5629. https://doi.org/10.4049/jimmunol.0901772.

Yamada, N., Kuranaga, Y., Kumazaki, M., Shinohara, H., Taniguchi, K., & Akao, Y. (2016). Colorectal cancer cell-derived extracellular vesicles induce phenotypic alteration of T cells into tumor-growth supporting cells with transforming growth factor-β1-mediated suppression. *Oncotarget, 7*(19), 27033.

Yin, Y., Cai, X., Chen, X., Liang, H., Zhang, Y., Li, J., … Yokoyama, S. (2014). Tumor-secreted miR-214 induces regulatory T cells: A major link between immune evasion and tumor growth. *Cell Research, 24*(10), 1164.

Ying, X., Wu, Q., Wu, X., Zhu, Q., Wang, X., Jiang, L., … Wang, X. (2016). Epithelial ovarian cancer-secreted exosomal miR-222-3p induces polarization of tumor-associated macrophages. *Oncotarget, 7*(28), 43076.

Yu, X., Harris, S. L., & Levine, A. J. (2006). The regulation of exosome secretion: A novel function of the p53 protein. *Cancer Research, 66*(9), 4795–4801. https://doi.org/10.1158/0008-5472.CAN-05-4579.

Yuyama, K., Sun, H., Mitsutake, S., & Igarashi, Y. (2012). Sphingolipid-modulated exosome secretion promotes clearance of amyloid-beta by microglia. *Journal of Biological Chemistry, 287*(14), 10977–10989. https://doi.org/10.1074/jbc.M111.324616.

Zhang, H. G., & Grizzle, W. E. (2011). Exosomes and cancer: A newly described pathway of immune suppression. *Clinical Cancer Research, 17*(5), 959–964. https://doi.org/10.1158/1078-0432.CCR-10-1489.

Zonari, E., Pucci, F., Saini, M., Mazzieri, R., Politi, L. S., Gentner, B., & Naldini, L. (2013). A role for miR-155 in enabling tumor-infiltrating innate immune cells to mount effective antitumor responses in mice. *Blood, 122*(2), 243–252. https://doi.org/10.1182/blood-2012-08-449306.

12 ⊞

Exosomes and Tunneling Nanotube Conduits: Synergistic Interaction That Facilitates Intercellular Communication Between Malignant and Stromal Cells in the Tumor Microenvironment

Emil Lou, William Sperduto, Subbaya Subramanian

UNIVERSITY OF MINNESOTA, MINNEAPOLIS, MN, UNITED STATES

CHAPTER OUTLINE

1. Introduction 219

2. Interplay Between Tunneling Nanotubes and Exosomes 220

3. Other Tunneling Nanotube–Like Nanostructures and Interaction With Exosomes 223

4. Identifying Metabolic Conditions and Physiologic Factors That Mutually Promote TNT Formation and Exosome Activity 225

5. Pharmacologic Suppression of Tunneling Nanotubes 226

6. The Search for Specific Molecular and Structural Biomarkers of TNT Formation in Cancer. 227

7. Conclusion and Perspectives 229

References 230

1. Introduction

Extracellular vesicles (EVs)—a broad category of membranous vesicles that encompasses exosomes, microvesicles (MVs), microparticles, and apoptotic bodies—facilitate long-range cross talk between cells in a way that was not as greatly appreciated until the past decade. During this time, there has been a great deal of research focusing on the ability of EVs (exosomes in particular) to harbor cellular cargo that can be transmitted to other cells and induce molecular and physiologic changes in recipient hosts. Exosome delivery is predicated on diffusion of these vesicles across the intercellular space. For this reason, the payload of exosomes must be quantitatively high enough to account for the potential randomness of this diffusion and the inevitability that not all exosomes will reach an ultimate destination group of individual cells or organs. Thus, in some ways, exosomes represent an indirect form of intercellular communication. Gap junctions composed of connexin proteins have been well established as a form of direct physical communication, but their role has been limited to cells immediately adjacent to each other spatially. In contrast, tunneling nanotubes (TNTs) represent a more direct form of long-range (e.g., on average spanning 30–200 µm or longer) intercellular communication. TNTs are defined as filamentous actin–based cytoplasmic extensions that are open-ended, nonadherent when cultured in vitro and have been shown to function by mediating direct transfer of various proteins, organelles, and other vital cargo from cell to cell (Lou et al., 2012). Since Rustom, Saffrich, Markovic, Walther, & Gerdes (2004) first identified and reported the existence of TNTs in vitro in 2004, studies have focused on clarifying the role of these unique cellular protrusions in noncancer and cancer cells in intercellular communication and transport. As no biologic or physiologic system exists in isolation, it is logical to expect that exosomes and TNTs may play parallel and at times complementary roles in cellular communication in health and disease, rather than be completely exclusive in nature. This chapter will discuss findings to date that support mutually beneficial interactions between exosomes and TNTs in several cell systems and the possibility that exosome–TNT synergistic interactions in long-range communication may depend on time and the ever-changing, dynamic state of the tumor microenvironment.

2. Interplay Between Tunneling Nanotubes and Exosomes

Given previous examination of exosomes as possible cancer treatment targets and communicatory mediators, and increasing knowledge of the role of TNTs in cancer biology, investigating the potential interactions of exosomes with TNTs provides a unique perspective to understanding intercellular communication (Hegmans et al., 2004; Koumangoye, Sakwe, Goodwin, Patel, & Ochieng, 2011). Hood, Pan, Lanza, & Wickline (2009) were the first to use dynamic light scattering combined with fluorescent exosome labeling to efficiently isolate and track melanoma-derived exosomes. This work revealed a new relationship between exosomes and cellular nanotubes. Exosomal messengers appeared to assist long-range tumor communication by inducing paracrine signaling. Melanoma exosomes or the effector molecules inside the exosomes led to increased tubule branching and were shown to stimulate angiogenesis in a dose-dependent manner.

Our group had previously reported the discovery and function of TNTs in cell lines and also in primary cells from patients with malignant pleural mesothelioma, an especially invasive and difficult-to-treat form of cancer that carries a poor prognosis (Lou et al., 2012). Following our initial work, and in the context of prior work that confirmed

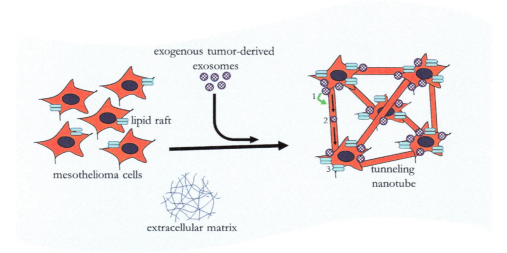

FIGURE 1 Schematic representation illustrating (1) uptake of exosomes that stimulate tunneling nanotube (TNT) formation; (2) TNTs acting as intercellular conduits to facilitate cell-to-cell transport of exosomes; (3) TNT-forming cells demonstrating increased lipid raft expression.

the existence of mesothelioma-derived exosomes (Hegmans et al., 2004), we hypothesized that TNTs and exosomes would likely not be mutually exclusive in mesothelioma and other cancers. To test this hypothesis, we again used mesothelioma as a model system. To visually determine potential close physical interactions between these entities, we performed electron microscopy (EM) and discovered exosomes localizing around and near TNTs, specifically at their base point of entry into the cellular membrane (Thayanithy, Babatunde, et al., 2014) (Fig. 1). We further performed time-lapse fluorescence microscopic imaging to label exosomes that were then added to mesothelioma cells in culture to track their patterns of movement. A series of time-lapse videos demonstrated cellular uptake of exosomes and subsequent entry of these exosomes into TNTs, which enabled their direct intercellular transfer between TNT-connected cells (Fig. 2). These results provide two important postulates: TNTs transport exosomes and exosomes may in turn stimulate TNT formation. The implications provided a foundation for supporting further consideration that exosomes and TNTs are not mutually exclusive factors in long-range cellular communication, but in fact they play a previously undiscovered complementary role in optimizing each other's function.

Further experiments clarified the biophysical composition of the areas of the cell that attracted exosomes at the base of TNTs by detecting an enriched population of lipid rafts in cellular regions associated with greater TNT activity (Fig. 3). Since lipid rafts have long been associated with sorting lipids and proteins into exosomes and exosomal formation and transport (de Gassart, Geminard, Fevrier, Raposo, & Vidal, 2003; Valapala & Vishwanatha, 2011), this three-way association of lipid rafts–exosomes–TNTs may help clarify how exosomes can successfully facilitate long-range communication, regardless of exact payload

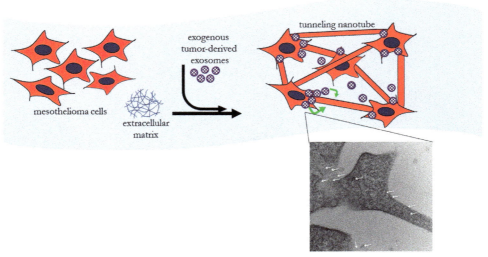

FIGURE 2 Schematic representation of exogenous, tumor-derived exosomes stimulating tunneling nanotube (TNT) formation. *Microscopy image in inset was originally published in Thayanithy, V., Babatunde, V., Dickson, E. L., Wong, P., Oh, S., Ke, X., … Lou, E. (2014). Tumor exosomes induce tunneling nanotubes in lipid raft-enriched regions of human mesothelioma cells.* Experimental Cell Research, 323(1), 178–188. https://doi.org/10.1016/j.yexcr.2014.01.014.

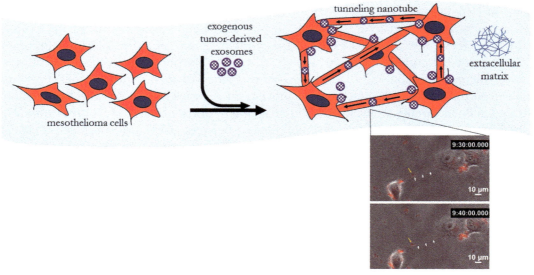

FIGURE 3 Schematic representation of TNTs acting as direct intercellular conduits that transport exosomes from host to recipient cells. *Microscopy image in inset was originally published in Thayanithy, V., Babatunde, V., Dickson, E. L., Wong, P., Oh, S., Ke, X., … Lou, E. (2014). Tumor exosomes induce tunneling nanotubes in lipid raft–enriched regions of human mesothelioma cells.* Experimental Cell Research, 323(1), 178–188. https://doi.org/10.1016/j.yexcr.2014.01.014.

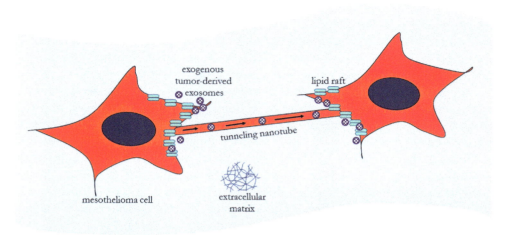

FIGURE 4 Schematic representation of TNT-forming cells enriched for increased lipid raft expression.

quantities (Thayanithy, Babatunde, et al., 2014) (Fig. 4). Future investigation of lipid rafts as potential TNT biomarkers is needed, but these preliminary findings have provided insights into TNT function and growth by providing a direct link to simultaneous cell processes vital to cell communication in the immediate tumor microenvironment and beyond.

The interaction of cancer-derived exosomes and TNTs is not limited exclusively to solid tumor malignancies. There are ample biologic situations in hematologic malignancies that could also benefit from this interaction. Using a chronic myeloid leukemia (CML) cell line, Mineo et al. (2012) isolated exosomes and investigated their effects on TNT formation in human umbilical vein endothelial cells (HUVECs). Live-cell imaging and three-dimensional (3D) confocal image reconstruction revealed nanotube formation and exosome localization within the nanotube structures. Time-lapse video microscopy and fluorescent markers were used to visualize exosome transfer between HUVECs. Treating CML cells with two tyrosine kinase inhibitors, imatinib and dasatinib, significantly reduced exosomal release. However, given an exosome-mediated Src activation and subsequent phosphorylation of FAK, Erk, AKT protein, only dasatinib inhibited HUVEC tubular differentiation and exosome vascularization. The Src signaling is implicated in leukemic cells and the surrounding angiogenic microenvironment, which may explain the role of dasatinib in CML therapy and warrant further research into exosomes as therapeutic targets.

3. Other Tunneling Nanotube–Like Nanostructures and Interaction With Exosomes

Researchers in the field of TNT biology continue to actively investigate vital structural and functional components of TNTs. It is clear that there is some heterogeneity in physical characteristics as well as the role of these protrusions, with most of this heterogeneity likely to vary between cell types of origin and between disease and healthy states. For that reason, there are some forms of actin-based cell extensions that appropriately are not categorized as filopodia, invadopodia, or lamellipodia, or another microtubule-related phenomenon known as "microtentacles" (Balzer et al., 2010; Matrone, Whipple, Balzer, & Martin, 2010; Matrone, Whipple, Thompson, et al., 2010; Whipple et al., 2008, 2010; Yoon et al., 2011), a cellular protrusion that is associated with increased risk of distant metastasis in breast cancer. These forms of cellular protrusions also do not fit the current definition of TNTs or tumor microtubes (TMs), an entity recently described in an animal model of glioblastoma that essentially is a wider, longer variant of TNTs (Osswald et al., 2015). Physical characteristics of TNTs, as visualized in cell culture, are nonadherent to the substratum, have a diameter in nanometers, and extend from the plasma membrane at roughly a 90-degree angle. TMs, as identified in a series of recent publications, are on average wider than 1 µm in thickness and longer than most TNTs that have been described in vitro but essentially perform the same functions (Osswald et al., 2015).

There have been several further studies examining interactions of exosomes with TNT-like nanostructures, as discussed here. One such study examined samples of cerebrospinal fluid (CSF), which serves a vital role in cushioning the brain and spine. Harrington et al. (2009) hypothesized that CSF may facilitate humoral signaling in a manner separate from synaptic transmission. Researchers performed light liquid chromatography mass spectrometry, transmission EM, electrophoresis, Western blotting, and immunostaining to identify thousands of proteins and nanostructures. An observed abundance of membrane proteins suggested either they are cleaved by proteases and transported into the CSF or the CSF contains membranes (Zougman et al., 2008). These investigators detected TNT-like nanostructures that showed variation in CSF nanostructures that with further research may explain their roles in modulating brain function and dysfunction.

P66Shc regulates mast cell activation. Vav and paxillin contribute to GTPase activation, F-actin rearrangement, and granule release. By analyzing Vav and paxillin immunoblots, Masi et al. (2014) discovered that p66Shc impairs Vav and paxillin phosphorylation and F-actin disassembly. Given SEM results, P66Shc also acted to prevent microvesicle budding from the plasma membrane, which limited the release of granules. These findings demonstrate the importance of p66Shc in actin regulation and mechanisms of vesicle-mediated secretion and may contribute to our knowledge of TNTs since they are composed of F-actin.

Cytonemes are another well-studied form of intercellular extension implicated in cellular communication, but they are distinct from TNTs. One characteristic difference is their growth and extension guided by morphogen gradients; they have often been well characterized in Drosophila optic wing disc models (Zhou et al., 2012). These projections can exchange signaling proteins critical to cell growth in this context. Fluorescent labeling and live images indicate Hedgehog (Hh) especially may be localized by and transported with other Hh pathway components in EVs. Gradilla et al. (2014) used EV analysis for *Drosophila* cells to determine a fraction of Hh secretion completed through exosome-like vesicles. RNA interference (RNAi)—essentially knocking out EV production genes—identified proteins involved in EV secretion and formation and exosome release, which are pertinent to Hh signaling. In vivo images seemed to suggest Hh-containing EVs may be transported by cytonemes. These results imply that morphogen transport may achieve a proper gradient via cytonemes. To advance the current knowledge of developmental processes as well as gene regulation in cancer requires a more comprehensive look into the distribution of signaling molecules like Hh.

4. Identifying Metabolic Conditions and Physiologic Factors That Mutually Promote TNT Formation and Exosome Activity

To date, identification of the underlying stimuli that provoke TNT formation has been limited. TNT cellular networks are upregulated between cells (including cancer cells and between tumor and stromal cells) under conditions of physiologic and metabolic stress, including exposure to hydrogen peroxide, serum deprivation, and hyperglycemia (Lou et al., 2012; Wang, Cui, Sun, & Zhang, 2011; Zhu et al., 2005); this finding implicates the potential role of TNTs in facilitating tumor–stromal interactions in the complex, dynamic, and ever-changing heterogeneous tumor microenvironment. One example of a condition that affects both TNTs and exosomes is an acidic culture microenvironment, which has been shown in vitro to stimulate exosomal release and increased TNT formation (Hood et al., 2009; Lou et al., 2012; Parolini et al., 2009; Rustom et al., 2004). There is a growing body of knowledge of additional factors that lead to this result in cancer specifically, including hypoxic microenvironments in which the mTOR pathway is upregulated (Desir et al., 2016). Suppression of TNT formation by targeting mTOR signaling—either directly with everolimus or indirectly with metformin—demonstrated that TNT suppression negatively regulates intercellular transfer in mesothelioma and ovarian cancers (Lou et al., 2012; Thayanithy, Dickson, Steer, Subramanian, & Lou, 2014), implying that this pathway may be important to TNTs in those cell types and perhaps others.

At this time, it has become well established that exosomes can harbor and mediate intercellular transfer of microRNAs (miRNAs) between distant cells and organs (Lee, El Andaloussi, & Wood, 2012; Mittelbrunn et al., 2011). Our group investigated the ability of TNTs to do the same and found that intercellular trafficking of oncogenic miRNAs associated with platinum chemoresistance could utilize TNT conduits for horizontal transfer between cancer cells and also between cancer and stromal cells (Thayanithy, Dickson, et al., 2014). More recently, we have even shown evidence that oncolytic viruses can harness TNTs as a novel mode of intercellular spread of infection and subsequent bystander effects following administration of viral thymidine kinase–activated drugs. This discovery further implies that a better understanding of the routes and patterns of these TNT "intercellular highways" may allow for more efficacious delivery of oncolytic viruses (Ady et al., 2016). In this context, identification of patients whose tumors have more prolific TNTs may permit more selective identification of patients who will benefit most from this therapeutic modality. In both of these examples, whether cells are transferring miRNAs to induce malignant transformation or drug resistance or therapeutic viruses are hijacking TNTs to spread from cell to cell, exosomes may prove to be a potentially useful, sensitive, and clinically testable biomarker for identifying such patients in whom TNTs are most prolific. Further research will be needed to characterize the interactions of exosomes and TNTs in the tumor microenvironment to provide strong support for this rationale. Altogether, these findings open new possibilities into targeting TNTs using a novel therapeutic strategy: disrupting horizontal cell-to-cell transfer of cellular cargo that stimulates increased cellular invasive capabilities and chemoresistance.

5. Pharmacologic Suppression of Tunneling Nanotubes

The most common drug compounds used to disrupt or prevent TNT formation have limited utility, as they are adequate for proof-of-principle in vitro evaluation, but have little to no relevance for potential clinical use. Specifically, actin-depolymerizing agents have been used most commonly to block TNT formation in vitro (Bukoreshtliev et al., 2009; Jung, Park, Joo, Kim, & Ha, 2011; Rudnicka et al., 2009); other compounds such as azide, colchicines, and tubulin inhibitors such as nocodazole have also been used (Onfelt et al., 2006; Rudnicka et al., 2009). One point of caution is that to correctly evaluate the effect of potential TNT inhibitors, it is critical to ensure to the extent possible that the structures being evaluated are truly TNTs.

The agents described are helpful for basic in vitro studies. However, the overarching goal of TNT biology should be to identify its function in cancer and other diseases; to prevent this potentially critical disease-promoting function, it would not be viable or feasible to use these drugs in the clinical setting. With advancement in knowledge of TNTs and their potential pathophysiologic roles, a strong need has emerged to clinically evaluate drugs and compounds that could theoretically be used in human patients if TNT function is proven to be vital and targetable.

Microtubules (MTs) are worth mentioning briefly in this context. MTs are already well-studied, characterized, and known cancer drug targets, as MT inhibitors are part of standard-of-care treatment in a number of invasive malignancies. MTs as well as actin have been implicated for their role in exosome secretion, along with important molecular factors such as specific members of the Rab GTPase family (e.g., Rab27) (Raposo & Stoorvogel, 2013). Drugs that target MTs, including chemotherapeutic agents that prevent MT depolymerization (e.g., paclitaxel), are commonly used for treating cancer patients and have been associated with decreased exosome release in vitro (Iero et al., 2008).

In the context of TNTs, which largely form around filamentous actin bridges but which at variable times may also contain MTs, it will be important to determine the role of the latter for drug therapy. Examination of MTs and actin-based structures using two-dimensional in vitro cultured conditions may not provide an accurate assessment of these structures; for this reason, 3D assessment providing a confined space alters MT polymerization and depolymerization (Balzer et al., 2012). Consideration of using 3D models, such as organoid models, will be important to more accurately examine cell interactions in the 3D tumor microenvironment. The use of paclitaxel at high enough concentrations can cause cells to detach and fail to readhere in routine in vitro culture. We discovered that this detachment deforms TNTs and prevents formation of new ones. For example, an assay of diluted paclitaxel on MSTO-211H cells treated with paclitaxel at a wide range of doses (83 µg/mL, 41.5 µg/mL, 20 µg/mL, 10 µg/mL, and 5 µg/mL) led to detachment of most cells treated with the highest two doses by 24 h of culture; cells treated with 20, 10, or 5 µg/mL, however, mostly remained attached by this time point and remained viable for further evaluation (unpublished data).

Although microtubule-targeting drugs are used routinely in other cancers but not in gliomas, targeting MTs using tumor-treating fields (TTF) provides an alternate tactic. In recent years, this approach has gained FDA approval for treatment of recurrent glioblastoma (Rehman, Elmore, & Mattei, 2015; Stupp et al., 2012, 2015), and in recent months, objective demonstration of significant clinical benefit, and overall survival improvement for patients receiving first-line TTF. Could this significant benefit be due to the ability of TTF to effectively prevent TNT/tumor microtube–mediated cellular networks from forming—mitigating invasive capability of burgeoning tumors? If so, what effect would exosomes have in this context, considering they too have been shown to transport RNA and proteins in this disease, promoting tumor growth and proliferation (Skog et al., 2008)? This is but one potential area of exploration as the field of neuro-oncology advances and embraces cellular biomarkers as well as known and undiscovered, relevant molecular markers.

6. The Search for Specific Molecular and Structural Biomarkers of TNT Formation in Cancer

To date, few studies have successfully identified biomarkers specifically used to identify TNTs in culture. As mentioned earlier, the strongest potential limitation of such studies is the concept that there may be a significant yet undetermined amount of TNT heterogeneity between types of cells, tissues of origin, and different disease processes. Nonetheless, within the context of a particular form of disease, methods of differential gene expression of putative biomarkers can be examined under favorable and unfavorable conditions for TNT formation. Ideally, these methods will include effective knockdown of TNTs to prove that their absence does not lead to intercellular transport and the usual downstream effects when TNTs are present.

As a first examination of this topic, our group examined differential gene expression in malignant mesothelioma under conditions promoting TNT formation and compared these results to standard culture conditions in which there were fewer TNTs. Interestingly, our results indicated that mesothelioma cells grown in TNT medium show significant upregulation of TNT-enriched genes, such as M-Sec and LST1, two markers previously reported to be associated with TNTs in other cell types (Hase et al., 2009; Schiller et al., 2013). M-Sec (also known in the scientific literature as B94 or as TNFaip2, or tumor necrosis factor-alpha-induced protein 2) has been reported to regulate de novo membrane nanotube formation by interacting with Ral and the exocyst complex (Hase et al., 2009). M-Sec is upregulated in macrophage cells exposed to inflammatory stimuli such as lipopolysaccharide (Hase et al., 2009), which has been shown to stimulate TNT formation (Chinnery, Pearlman, & McMenamin, 2008); our results demonstrating upregulation of M-Sec expression in MSTO cells cultured in similar conditions of metabolic stress (i.e., low-serum, high glucose) that also stimulate TNT formation are consistent with and support these data as well. It was previously thought that M-Sec expression was restricted to cells of myeloid lineage. However, we found significant M-Sec expression in malignant nonmyeloid cells, which supports the possibility that this protein can play a role in TNTs across a spectrum of cell types with variable expression depending on environmental and culture conditions. Thus, M-Sec is not necessarily restricted to myeloid cells. Interestingly, while addition of TNF-α to macrophages is known to induce M-Sec/TNFaip2 expression, the above study led to negligible or no relative increase in M-Sec (Hase et al., 2009). Notably, a separate study has effectively demonstrated stimulation of TNT formation by TNF-α in human peritoneal mesothelial cells (Ranzinger et al., 2011). LST1 is known to induce the formation of TNTs by recruiting RalA and exocyst complex proteins (Schiller et al., 2013).

We also found that TNT-inducing medium significantly lowers the expression of key genes involved in the cell cycle, including E2F1, CCNA2, CDC20, and CDKN3. E2F1 is expressed at maximal levels at the G1 stage of the cell cycle where it promotes the G1/S transition. E2F1 is a transcription factor that activates numerous genes involved in the cell cycle, including CCNA2 and CDC20 (Bracken, Ciro, Cocito, & Helin, 2004). Consistent with the downregulation of E2F1 expression, both CCNA2 and CDC20 were downregulated in mesothelioma cells grown in TNT medium. This observation supports the idea that E2F1 is the causative factor for the observed lower expression of CCNA2 and CDC20. We also tested the expression of CDKN3, a tumor suppressor that regulates mitosis (Nalepa et al., 2013). CDKN3 was expressed at significantly lower levels on TNT-conditioned cells. These findings support the notion that cells in TNT medium undergo lower rates of cell division.

In contrast to cell division, TNT-inducing medium significantly increased the expression of genes involved in cellular migration, invasion, and metastasis. We found that TNT medium increased expression of tenascin-C, CD44, osteopontin, fascin, and mesothelin. Tenascin-C, an extracellular matrix glycoprotein that shows prominent stromal expression in many solid tumors, promotes tumor cell invasion and growth in breast carcinoma cell lines (Guttery et al., 2010). It is especially overexpressed in human malignant pleural mesothelioma, in which it localizes within fibrotic stroma and especially at the leading edge of tumors and is associated with worse prognosis (Kaarteenaho, Sormunen, & Paakko, 2010). As mentioned, we have also used in vitro scratch assay models to demonstrate TNT formation at the leading edge of mesothelioma invasion (Lou et al., 2012). Further overexpression in in vitro studies of TNT-primed mesothelioma cells suggests that tenascin-C may play a role in supporting TNTs, especially at the leading edge of tumor invasion. CD44, a glycoprotein transmembrane receptor, is an adhesion molecule expressed in normal and neoplastic tissues that contributes to aggressive tumor behavior

(Pietras et al., 2014). CD44 has been shown to be highly expressed in a high percentage of human mesotheliomas (>90%), with epithelioid variants having the most relative expression and sarcomatoid variants having the least expression (Penno et al., 1995). For this reason, we performed our testing on MSTO-211H malignant mesothelioma cells, which display biphasic histology (i.e., elements of both epithelioid and sarcomatoid histologies) for relative comparison of CD44 levels following addition of TNT medium.

Osteopontin, a secreted integrin-binding glycophosphoprotein, is highly expressed in mesothelioma and mediates cell adhesion via interactions with CD44 and integrins (Ohashi et al., 2009). Fascin is an actin-bundling protein localized in filopodia; it functions in cell adhesion and motility and is a key molecule in tumor metastasis. Fascin expression is also induced in a variety of metastatic cancers including intestinal, breast, and gliomas (Al-Alwan et al., 2011; Hwang, Smith, Salhia, & Rutka, 2008; Jawhari et al., 2003). Our prior study provided visual demonstration of localization of fascin near the base of TNTs at the membrane of mesothelioma cells using IF staining (Lou et al., 2012). The ability of migrastatin, a fascin-targeting agent, to suppress TNT formation provides even further evidence of the role of fascin in TNT activity. Mesothelin is a cancer-associated glycoprotein antigen overexpressed in a variety of solid tumors such as mesothelioma, ovarian, and pancreatic cancers (Scholler et al., 1999). Its overexpression promotes mesothelioma cell invasion (Servais et al., 2012). Overexpression of mesothelin is additionally associated with worse prognosis and overall survival in early-stage lung adenocarcinoma (Kachala et al., 2014). Induction of these genes in cells grown in TNT media not only reinforces the connection between TNTs and tumor malignancy but also hints that TNT-related genes could be induced by the tumor microenvironment and thereby lead to a higher propensity to invade, migrate, and metastasize.

7. Conclusion and Perspectives

Intercellular communication is a key mechanism for cell signaling and biological functions. EVs and TNTs are effectively used by almost all cell types as long-range communication channels. While communication through EVs is extensively studied in various cellular and functional biology contexts, it is clear that their mode of communication is primarily indirect; i.e., vesicles secreted by cells have to be effectively internalized by the recipient cells. The internalization process may be passive and/or active depending on the proximity of cells and context of the cellular communication. Since EVs are released into the circulation, they are known to impart their effects on distant cells, tissues, and organs. Integrin protein family members are involved in specific EV communication (Hoshino et al., 2015). Further, the required number of EVs and cellular cargo should be delivered to induce an effective response on the recipient cells. In contrast, TNTs are mainly the source of direct and long-distance cell-to-cell communication, where the cellular cargo including the endosomes of various sizes is effectively delivered to a specific cell type for a determined period of time. The distance and duration of TNT communication are determined by the dynamics of TNT formation and connection. We have begun to appreciate the role of TNTs in the tissue microenvironment especially in the context of tumors. While direct communication between cancer cells, cancer–stromal cells, and stromal–immune cells is observed, the direct communication channels created by TNTs between various cell types in the tumor microenvironment are limitless. Currently, TNT function and biological context are predominantly studied by using in vitro models. Development of effective TNT imaging techniques coupled with TNT quantification in a given tissue space will increase our understanding of cellular communication channels. Novel techniques, such as single molecule tracking and Fluorescence Recovery After Photobleaching (FRAP), will provide insights into how TNTs deliver cellular cargo. The key unanswered questions in the field of cellular communications are as follows:

1. How does the cell choose one mode of communication over another?
2. Does the cell use all the modes of communication at a given time, if so what are the levels of usage of these various means of communication?
3. Is it possible to study only one mode of cellular communication by effectively blocking the other mechanisms?
4. Does the stoichiometry play a role? Is the effective delivery of cargo achieved by combining different modes of cellular communication over a period of time? Although speculative, it is possible that direct TNT-mediated transfer of cellular cargo can efficiently induce and affect several cellular functions including (1) cellular reprogramming, (2) malignant transformation, and (3) drug resistance.

Thus, both TNTs and exosomes are under active investigation as novel targets for cancer therapy. Published findings implicating tumor-derived exosomes in the process of angiogenesis (in a melanoma model) serve as just one example (Hood et al., 2009). To build off this work, future studies should focus on trying to disrupt the nanoscale cancer communication system to yield new insights into antiangiogenic therapy and other mechanisms of metastasis. While the biological functions of TNTs are implicated in tumor development and progression, their effect in the developmental process and tissue regeneration are also appreciated.

References

Ady, J., Thayanithy, V., Mojica, K., Wong, P., Carson, J., Rao, P., … Lou, E. (2016). Tunneling nanotubes: An alternate route for propagation of the bystander effect following oncolytic viral infection. *Molecular Therapy—Oncolytics, 3*, 16029. https://doi.org/10.1038/mto.2016.29. http://www.nature.com/articles/mto201629#supplementary-information.

Al-Alwan, M., Olabi, S., Ghebeh, H., Barhoush, E., Tulbah, A., Al-Tweigeri, T., … Adra, C. (2011). Fascin is a key regulator of breast cancer invasion that acts via the modification of metastasis-associated molecules. *PLoS One, 6*(11), e27339. https://doi.org/10.1371/journal.pone.0027339.

Balzer, E. M., Tong, Z., Paul, C. D., Hung, W. C., Stroka, K. M., Boggs, A. E., … Konstantopoulos, K. (2012). Physical confinement alters tumor cell adhesion and migration phenotypes. *The FASEB Journal, 26*(10), 4045–4056. https://doi.org/10.1096/fj.12-211441.

Balzer, E. M., Whipple, R. A., Thompson, K., Boggs, A. E., Slovic, J., Cho, E. H., … Martin, S. S. (2010). c-Src differentially regulates the functions of microtentacles and invadopodia. *Oncogene, 29*(48), 6402–6408. https://doi.org/10.1038/onc.2010.360.

Bracken, A. P., Ciro, M., Cocito, A., & Helin, K. (2004). E2F target genes: Unraveling the biology. *Trends in Biochemical Sciences, 29*(8), 409–417. https://doi.org/10.1016/j.tibs.2004.06.006.

Bukoreshtliev, N. V., Wang, X., Hodneland, E., Gurke, S., Barroso, J. F., & Gerdes, H. H. (2009). Selective block of tunneling nanotube (TNT) formation inhibits intercellular organelle transfer between PC12 cells. *FEBS Letters, 583*(9), 1481–1488. https://doi.org/10.1016/j.febslet.2009.03.065. pii:S0014-5793(09)00262-2.

Chinnery, H. R., Pearlman, E., & McMenamin, P. G. (2008). Cutting edge: Membrane nanotubes in vivo: A feature of MHC class II+ cells in the mouse cornea. *The Journal of Immunology, 180*(9), 5779–5783 180/9/5779 [pii].

Desir, S., Dickson, E. L., Vogel, R. I., Thayanithy, V., Wong, P., Teoh, D., … Lou, E. (2016). Tunneling nanotube formation is stimulated by hypoxia in ovarian cancer cells. *Oncotarget, 7*(28), 43150–43161. https://doi.org/10.18632/oncotarget.9504.

de Gassart, A., Geminard, C., Fevrier, B., Raposo, G., & Vidal, M. (2003). Lipid raft-associated protein sorting in exosomes. *Blood, 102*(13), 4336–4344. https://doi.org/10.1182/blood-2003-03-0871.

Gradilla, A. C., Gonzalez, E., Seijo, I., Andres, G., Bischoff, M., Gonzalez-Mendez, L., … Guerrero, I. (2014). Exosomes as hedgehog carriers in cytoneme-mediated transport and secretion. *Nature Communications, 5*, 5649. https://doi.org/10.1038/ncomms6649.

Guttery, D. S., Hancox, R. A., Mulligan, K. T., Hughes, S., Lambe, S. M., Pringle, J. H., … Shaw, J. A. (2010). Association of invasion-promoting tenascin-C additional domains with breast cancers in young women. *Breast Cancer Research, 12*(4), R57. https://doi.org/10.1186/bcr2618.

Harrington, M. G., Fonteh, A. N., Oborina, E., Liao, P., Cowan, R. P., McComb, G., … Huhmer, A. F. (2009). The morphology and biochemistry of nanostructures provide evidence for synthesis and signaling functions in human cerebrospinal fluid. *Cerebrospinal Fluid Research, 6*, 10. https://doi.org/10.1186/1743-8454-6-10.

Hase, K., Kimura, S., Takatsu, H., Ohmae, M., Kawano, S., Kitamura, H., … Ohno, H. (2009). M-Sec promotes membrane nanotube formation by interacting with Ral and the exocyst complex. *Nature Cell Biology, 11*(12), 1427–1432. https://doi.org/10.1038/ncb1990.

Hegmans, J. P., Bard, M. P., Hemmes, A., Luider, T. M., Kleijmeer, M. J., Prins, J. B., … Lambrecht, B. N. (2004). Proteomic analysis of exosomes secreted by human mesothelioma cells. *The American Journal of Pathology, 164*(5), 1807–1815. https://doi.org/10.1016/S0002-9440(10)63739-X.

Hood, J. L., Pan, H., Lanza, G. M., & Wickline, S. A. (2009). Paracrine induction of endothelium by tumor exosomes. *Laboratory Investigation, 89*(11), 1317–1328. https://doi.org/10.1038/labinvest.2009.94.

Hoshino, A., Costa-Silva, B., Shen, T. L., Rodrigues, G., Hashimoto, A., Tesic Mark, M., … Lyden, D. (2015). Tumour exosome integrins determine organotropic metastasis. *Nature, 527*(7578), 329–335. https://doi.org/10.1038/nature15756.

Hwang, J. H., Smith, C. A., Salhia, B., & Rutka, J. T. (2008). The role of fascin in the migration and invasiveness of malignant glioma cells. *Neoplasia, 10*(2), 149–159.

Iero, M., Valenti, R., Huber, V., Filipazzi, P., Parmiani, G., Fais, S., & Rivoltini, L. (2008). Tumour-released exosomes and their implications in cancer immunity. *Cell Death and Differentiation, 15*(1), 80–88. https://doi.org/10.1038/sj.cdd.4402237.

Jawhari, A. U., Buda, A., Jenkins, M., Shehzad, K., Sarraf, C., Noda, M., … Adams, J. C. (2003). Fascin, an actin-bundling protein, modulates colonic epithelial cell invasiveness and differentiation in vitro. *The American Journal of Pathology, 162*(1), 69–80. https://doi.org/10.1016/S0002-9440(10)63799-6.

Jung, S. H., Park, J. Y., Joo, J. H., Kim, Y. M., & Ha, K. S. (2011). Extracellular ultrathin fibers sensitive to intracellular reactive oxygen species: formation of intercellular membrane bridges. *Experimental Cell Research, 317*(12), 1763–1773. https://doi.org/10.1016/j.yexcr.2011.02.010.

Kaarteenaho, R., Sormunen, R., & Paakko, P. (2010). Variable expression of tenascin-C, osteopontin and fibronectin in inflammatory myofibroblastic tumour of the lung. *Acta Pathologica, Microbiologica et Immunologica Scandinavica, 118*(2), 91–100. https://doi.org/10.1111/j.1600-0463.2009.02566.x.

Kachala, S. S., Bograd, A. J., Villena-Vargas, J., Suzuki, K., Servais, E. L., Kadota, K., … Adusumilli, P. S. (2014). Mesothelin overexpression is a marker of tumor aggressiveness and is associated with reduced recurrence-free and overall survival in early-stage lung adenocarcinoma. *Clinical Cancer Research, 20*(4), 1020–1028. https://doi.org/10.1158/1078-0432.CCR-13-1862.

Koumangoye, R. B., Sakwe, A. M., Goodwin, J. S., Patel, T., & Ochieng, J. (2011). Detachment of breast tumor cells induces rapid secretion of exosomes which subsequently mediate cellular adhesion and spreading. *PLoS One, 6*(9), e24234. https://doi.org/10.1371/journal.pone.0024234.

Lee, Y., El Andaloussi, S., & Wood, M. J. (2012). Exosomes and microvesicles: Extracellular vesicles for genetic information transfer and gene therapy. *Human Molecular Genetics, 21*(R1), R125–R134. https://doi.org/10.1093/hmg/dds317.

Lou, E., Fujisawa, S., Morozov, A., Barlas, A., Romin, Y., Dogan, Y., … Moore, M. A. (2012). Tunneling nanotubes provide a unique conduit for intercellular transfer of cellular contents in human malignant pleural mesothelioma. *PLoS One, 7*(3), e33093. https://doi.org/10.1371/journal.pone.0033093.

Masi, G., Mercati, D., Vannuccini, E., Paccagnini, E., Riparbelli, M. G., Lupetti, P., … Ulivieri, C. (2014). p66Shc regulates vesicle-mediated secretion in mast cells by affecting F-actin dynamics. *Journal of Leukocyte Biology, 95*(2), 285–292. https://doi.org/10.1189/jlb.0313178.

Matrone, M. A., Whipple, R. A., Balzer, E. M., & Martin, S. S. (2010). Microtentacles tip the balance of cytoskeletal forces in circulating tumor cells. *Cancer Research, 70*(20), 7737–7741. https://doi.org/10.1158/0008-5472.CAN-10-1569.

Matrone, M. A., Whipple, R. A., Thompson, K., Cho, E. H., Vitolo, M. I., Balzer, E. M., … Martin, S. S. (2010). Metastatic breast tumors express increased tau, which promotes microtentacle formation and the reattachment of detached breast tumor cells. *Oncogene, 29*(22), 3217–3227. https://doi.org/10.1038/onc.2010.68.

Mineo, M., Garfield, S. H., Taverna, S., Flugy, A., De Leo, G., Alessandro, R., & Kohn, E. C. (2012). Exosomes released by K562 chronic myeloid leukemia cells promote angiogenesis in a Src-dependent fashion. *Angiogenesis, 15*(1), 33–45. https://doi.org/10.1007/s10456-011-9241-1.

Mittelbrunn, M., Gutierrez-Vazquez, C., Villarroya-Beltri, C., Gonzalez, S., Sanchez-Cabo, F., Gonzalez, M. A., … Sanchez-Madrid, F. (2011). Unidirectional transfer of microRNA-loaded exosomes from T cells to antigen-presenting cells. *Nature Communications, 2,* 282. https://doi.org/10.1038/ncomms1285.

Nalepa, G., Barnholtz-Sloan, J., Enzor, R., Dey, D., He, Y., Gehlhausen, J. R., … Clapp, W. (2013). The tumor suppressor CDKN3 controls mitosis. *The Journal of Cell Biology, 201*(7), 997–1012. https://doi.org/10.1083/jcb.201205125.

Ohashi, R., Tajima, K., Takahashi, F., Cui, R., Gu, T., Shimizu, K., … Takahashi, K. (2009). Osteopontin modulates malignant pleural mesothelioma cell functions in vitro. *Anticancer Research, 29*(6), 2205–2214.

Onfelt, B., Nedvetzki, S., Benninger, R. K., Purbhoo, M. A., Sowinski, S., Hume, A. N., … Davis, D. M. (2006). Structurally distinct membrane nanotubes between human macrophages support long-distance vesicular traffic or surfing of bacteria. *The Journal of Immunology, 177*(12), 8476–8483.

Osswald, M., Jung, E., Sahm, F., Solecki, G., Venkataramani, V., Blaes, J., … Winkler, F. (2015). Brain tumour cells interconnect to a functional and resistant network. *Nature, 528*(7580), 93–98. https://doi.org/10.1038/nature16071.

Parolini, I., Federici, C., Raggi, C., Lugini, L., Palleschi, S., De Milito, A., … Fais, S. (2009). Microenvironmental pH is a key factor for exosome traffic in tumor cells. *Journal of Biological Chemistry, 284*(49), 34211–34222. https://doi.org/10.1074/jbc.M109.041152.

Penno, M. B., Askin, F. B., Ma, H., Carbone, M., Vargas, M. P., & Pass, H. I. (1995). High CD44 expression on human mesotheliomas mediates association with hyaluronan. *Cancer Journal from Scientific American, 1*(3), 196–203.

Pietras, A., Katz, A. M., Ekstrom, E. J., Wee, B., Halliday, J. J., Pitter, K. L., … Holland, E. C. (2014). Osteopontin-CD44 signaling in the glioma perivascular niche enhances cancer stem cell phenotypes and promotes aggressive tumor growth. *Cell Stem Cell, 14*(3), 357–369. https://doi.org/10.1016/j.stem.2014.01.005.

Ranzinger, J., Rustom, A., Abel, M., Leyh, J., Kihm, L., Witkowski, M., … Schwenger, V. (2011). Nanotube action between human mesothelial cells reveals novel aspects of inflammatory responses. *PLoS One*, *6*(12), e29537. https://doi.org/10.1371/journal.pone.0029537.

Raposo, G., & Stoorvogel, W. (2013). Extracellular vesicles: Exosomes, microvesicles, and friends. *The Journal of Cell Biology*, *200*(4), 373–383. https://doi.org/10.1083/jcb.201211138.

Rehman, A. A., Elmore, K. B., & Mattei, T. A. (2015). The effects of alternating electric fields in glioblastoma: Current evidence on therapeutic mechanisms and clinical outcomes. *Neurosurgicalical Focus*, *38*(3), E14. https://doi.org/10.3171/2015.1.FOCUS14742.

Rudnicka, D., Feldmann, J., Porrot, F., Wietgrefe, S., Guadagnini, S., Prevost, M. C., … Schwartz, O. (2009). Simultaneous cell-to-cell transmission of human immunodeficiency virus to multiple targets through polysynapses. *Journal of Virology*, *83*(12), 6234–6246. https://doi.org/10.1128/JVI.00282-09.

Rustom, A., Saffrich, R., Markovic, I., Walther, P., & Gerdes, H. H. (2004). Nanotubular highways for intercellular organelle transport. *Science*, *303*(5660), 1007–1010. https://doi.org/10.1126/science.1093133. pii:303/5660/1007.

Schiller, C., Diakopoulos, K. N., Rohwedder, I., Kremmer, E., von Toerne, C., Ueffing, M., … Weiss, E. H. (2013). LST1 promotes the assembly of a molecular machinery responsible for tunneling nanotube formation. *Journal of Cell Science*, *126*(Pt 3), 767–777. https://doi.org/10.1242/jcs.114033.

Scholler, N., Fu, N., Yang, Y., Ye, Z., Goodman, G. E., Hellstrom, K. E., & Hellstrom, I. (1999). Soluble member(s) of the mesothelin/megakaryocyte potentiating factor family are detectable in sera from patients with ovarian carcinoma. *Proceedings of the National Academy of Sciences of the United States of America*, *96*(20), 11531–11536.

Servais, E. L., Colovos, C., Rodriguez, L., Bograd, A. J., Nitadori, J., Sima, C., … Adusumilli, P. S. (2012). Mesothelin overexpression promotes mesothelioma cell invasion and MMP-9 secretion in an orthotopic mouse model and in epithelioid pleural mesothelioma patients. *Clinical Cancer Research*, *18*(9), 2478–2489. https://doi.org/10.1158/1078-0432.CCR-11-2614.

Skog, J., Wurdinger, T., van Rijn, S., Meijer, D. H., Gainche, L., Sena-Esteves, M., … Breakefield, X. O. (2008). Glioblastoma microvesicles transport RNA and proteins that promote tumour growth and provide diagnostic biomarkers. *Nature Cell Biology*, *10*(12), 1470–1476. https://doi.org/10.1038/ncb1800.

Stupp, R., Taillibert, S., Kanner, A. A., Kesari, S., Steinberg, D. M., Toms, S. A., … Ram, Z. (2015). Maintenance therapy with tumor-treating fields plus temozolomide vs temozolomide alone for glioblastoma: A randomized clinical trial. *JAMA*, *314*(23), 2535–2543. https://doi.org/10.1001/jama.2015.16669.

Stupp, R., Wong, E. T., Kanner, A. A., Steinberg, D., Engelhard, H., Heidecke, V., … Gutin, P. H. (2012). NovoTTF-100A versus physician's choice chemotherapy in recurrent glioblastoma: A randomised phase III trial of a novel treatment modality. *European Journal of Cancer*, *48*(14), 2192–2202. https://doi.org/10.1016/j.ejca.2012.04.011.

Thayanithy, V., Babatunde, V., Dickson, E. L., Wong, P., Oh, S., Ke, X., … Lou, E. (2014). Tumor exosomes induce tunneling nanotubes in lipid raft-enriched regions of human mesothelioma cells. *Experimental Cell Research*, *323*(1), 178–188. https://doi.org/10.1016/j.yexcr.2014.01.014.

Thayanithy, V., Dickson, E. L., Steer, C., Subramanian, S., & Lou, E. (2014). Tumor-stromal cross talk: Direct cell-to-cell transfer of oncogenic microRNAs via tunneling nanotubes. *Translational Research*, *164*(5), 359–365. https://doi.org/10.1016/j.trsl.2014.05.011.

Valapala, M., & Vishwanatha, J. K. (2011). Lipid raft endocytosis and exosomal transport facilitate extracellular trafficking of annexin A2. *Journal of Biological Chemistry*, *286*(35), 30911–30925. https://doi.org/10.1074/jbc.M111.271155.

Wang, Y., Cui, J., Sun, X., & Zhang, Y. (2011). Tunneling-nanotube development in astrocytes depends on p53 activation. *Cell Death and Differentiation*, *18*(4), 732–742. https://doi.org/10.1038/cdd.2010.147.

Whipple, R. A., Balzer, E. M., Cho, E. H., Matrone, M. A., Yoon, J. R., & Martin, S. S. (2008). Vimentin filaments support extension of tubulin-based microtentacles in detached breast tumor cells. *Cancer Research*, *68*(14), 5678–5688. https://doi.org/10.1158/0008-5472.CAN-07-6589.

Whipple, R. A., Matrone, M. A., Cho, E. H., Balzer, E. M., Vitolo, M. I., Yoon, J. R., ... Martin, S. S. (2010). Epithelial-to-mesenchymal transition promotes tubulin detyrosination and microtentacles that enhance endothelial engagement. *Cancer Research, 70*(20), 8127–8137. https://doi.org/10.1158/0008-5472.CAN-09-4613.

Yoon, J. R., Whipple, R. A., Balzer, E. M., Cho, E. H., Matrone, M. A., Peckham, M., & Martin, S. S. (2011). Local anesthetics inhibit kinesin motility and microtentacle protrusions in human epithelial and breast tumor cells. *Breast Cancer Research and Treatment, 129*(3), 691–701. https://doi.org/10.1007/s10549-010-1239-7.

Zhou, S., Lo, W. C., Suhalim, J. L., Digman, M. A., Gratton, E., Nie, Q., & Lander, A. D. (2012). Free extracellular diffusion creates the Dpp morphogen gradient of the Drosophila wing disc. *Current Biology, 22*(8), 668–675. https://doi.org/10.1016/j.cub.2012.02.065.

Zhu, D., Tan, K. S., Zhang, X., Sun, A. Y., Sun, G. Y., & Lee, J. C. (2005). Hydrogen peroxide alters membrane and cytoskeleton properties and increases intercellular connections in astrocytes. *Journal of Cell Science, 118*(Pt 16), 3695–3703. https://doi.org/10.1242/jcs.02507.

Zougman, A., Pilch, B., Podtelejnikov, A., Kiehntopf, M., Schnabel, C., Kumar, C., & Mann, M. (2008). Integrated analysis of the cerebrospinal fluid peptidome and proteome. *Journal of Proteome Research, 7*(1), 386–399. https://doi.org/10.1021/pr070501k.

13

Exosomes in Tumor Angiogenesis— Multifunctional Messengers With Mixed Intentions

Liang Zhang

CITY UNIVERSITY OF HONG KONG, KOWLOON TONG, HONG KONG SAR, CHINA

CHAPTER OUTLINE

1. Introduction **235**
    1.1 Angiogenesis in Cancer 235
    1.2 Biology of Exosomes.. 236

2. Exosomes in Tumor Angiogenesis **237**
    2.1 Cancer-Derived Exosomes in Tumor Angiogenesis... 237
    2.2 Microenvironment-Derived Exosomes in Tumor Angiogenesis 239
    2.3 Regulation of Exosomes During Tumor Angiogenesis 240

3. Future Perspectives **241**

References **243**

1. Introduction

1.1 Angiogenesis in Cancer

All metabolically active tissues need to be in close vicinity (less than $100\,\mu m$) with blood supplies to ensure exchanges of oxygen, nutrients, metabolites, waste, etc. Generally, proliferating cells fulfill this obligation by inducing angiogenesis, which describes the formation of new blood capillaries from the preexisting vasculature. At the cellular level, angiogenesis depends on the differentiation, growth, and migration of endothelial cells that line the lumen of blood vessels, accompanied by the coordinated growth of parenchyma. All these processes are intricately regulated by a balanced network of pro- and antiangiogenic signals mediated by various growth factors and morphogens, as well as cell adhesion molecules (Adams & Alitalo, 2007; Bergers & Benjamin, 2003; Chung & Ferrara, 2011; Potente, Gerhardt, & Carmeliet, 2011). Coordinated angiogenesis is essential in a wide range of physiological events, such as embryonic development, wound healing, and the menstruation cycle. On the other hand, aberrant regulation of angiogenesis also plays an important role in many pathological processes, including the malignant development of cancer (Chung & Ferrara, 2011; Hanahan & Weinberg, 2000; Kerbel, 2008).

Sustained angiogenesis is a hallmark of cancer (Hanahan & Weinberg, 2000, 2011). Cancer cells need blood vessels not only for nutrient supply and waste removal but also for spreading to nearby and distant organs (Folkman, 1971; Steeg, 2016). The latter process is termed metastasis, which accounts for over 90% of cancer-related deaths. Given the pivotal significance of angiogenesis in cancer progression, extensive studies have been conducted to elucidate the molecular basis of malignant angiogenesis and to identify novel targets for cancer treatments (Ellis & Hicklin, 2008; Potente et al., 2011). However, antiangiogenic therapy achieved little success in cancer patients, primarily due to the intricate mechanisms underlying tumor angiogenesis (Bergers & Hanahan, 2008; Giuliano & Pages, 2013). Aggregations of evidence indicate that the incipient tumors already possess proangiogenic ability to recruit blood supply before fast growth and metastasis (Hanahan & Weinberg, 2000). This intrinsic ability encompasses changing the balance between pro- and antiangiogenic signals in favor of forming new blood vessels at the tumor site.

At the molecular level, the angiogenic balance is regulated by a plethora of inducers and inhibitors of different nature. For example, dozens of proteins have been characterized as angiogenic inducers, including vascular endothelial growth factor (VEGF), basic fibroblast growth factors (bFGFs), tumor necrosis factor α (TNFα), transforming growth factor β (TGFβ), matrix metalloproteinases (MMPs), etc. At the same time, the proangiogenic function of these inducers is counteracted by a variety of inhibitors, such as thrombospondin, angiostatin, endostatin, tissue inhibitors of metalloproteinase, etc. (Adams & Alitalo, 2007; Bergers & Benjamin, 2003; Chung & Ferrara, 2011). In addition to protein molecules, noncoding RNAs, including miRNAs and lncRNAs, also have essential roles in regulating the angiogenic balance (Landskroner-Eiger, Moneke, & Sessa, 2013). It is important to note that cancer cells are not the sole origin of these regulators; cells in the tumor microenvironment, such as fibroblasts and endothelial cells themselves, are also important sources of angiogenic factors. Cancer angiogenesis is the result of a complex network of cancer–microenvironment communication that is mediated by inducers and inhibitors of angiogenesis (De Palma, Biziato, & Petrova, 2017; Moserle & Casanovas, 2013; Watnick, 2012). Although in principle these angiogenic regulators can freely diffuse from the source cells to the target cells to mediate the dialogue, there is a growing recognition of extracellular vesicles as a platform that transfer pro- and antiangiogenic signaling molecules during angiogenesis (Ribeiro, Zhu, Millard, & Fan, 2013; Todorova, Simoncini, Lacroix, Sabatier, & Dignat-George, 2017). Here we focus on exosomes, a specific type of extracellular vesicle, and discuss their emerging role in cancer angiogenesis.

1.2 Biology of Exosomes

Exosomes are a specific population of extracellular vesicles of endocytic origin that are released by many, and perhaps all, eukaryotic cells. The current consensus is that exosomes originate within the organelle named multivesicular body (MVB) along the endocytic pathway (Harding, Heuser, & Stahl, 2013; Thery, 2011). Inward invagination of the endosomal membrane leads to the formation of intraluminal vesicles within MVBs. Upon fusion of MVBs with the plasma membrane, certain intraluminal vesicles are released as exosomes into the extracellular milieu, carrying a variety of signaling contents that mediate cell–cell communication. Typically, exosomes are described as 30–120 nm cup-shaped vesicles under a transmission electron microscope (TEM). However, recent TEM characterizations revealed diverse morphologies of exosomes, suggesting that procedures of sample preparation may play an important role in determining the appearance of these vesicles (Wu, Deng, & Klinke, 2015; Zabeo et al., 2017). In addition to TEM, other innovative technologies have been developed to analyze exosomes, such as nanoparticical tracking analysis, dynamic light scattering, and resistive pulse sensing (van der Pol, Coumans, Varga, Krumrey, & Nieuwland, 2013). Overall, these technologies illustrate the physical features of exosomes from complementary angles.

Early studies proposed that cells utilize exosomes as a vehicle to discard "unwanted" biomolecules. For example, exosomes were first described as the membrane vesicles that facilitate the shedding process of transferrin receptor during reticulocyte maturation (Harding et al., 2013). Recently, increasing studies have indicated that exosomes are signaling vehicles that transmit a broad spectrum of bioactive molecules, including proteins, lipids, lncRNA, miRNA, mRNA, DNA, etc. Exosomes are now well accepted as essential transporters for cell–cell communications, and the signaling activities of exosomes in numerous physiological conditions have been well documented (Thery, 2011; Tkach & Thery, 2016). For example, exosomes released from the mesenchymal stem cells (MSCs) have the capacity to modulate tissue repair and regeneration (Pashoutan Sarvar, Shamsasenjan, & Akbarzadehlaleh, 2016). The list of beneficial roles of exosomes is growing at a fast pace. However, exosomes can also be hijacked as "malignant messengers" in many pathophysiological processes, including cancer angiogenesis (Azmi, Bao, & Sarkar, 2013; De Toro, Herschlik, Waldner, & Mongini, 2015). Depending on the cellular sources and biological contexts during tumor progression, exosomes may possess various pro- or antiangiogenic molecular contents, which constitute the focus of this chapter. Understanding these pathological implications of exosomes will help develop novel therapeutic targets/tools for cancer treatment.

2. Exosomes in Tumor Angiogenesis

2.1 Cancer-Derived Exosomes in Tumor Angiogenesis

A fundamental process in angiogenesis involves activation of endothelial cells, which undergo differentiation, proliferation, and migration to lead the formation of new capillaries (Chung & Ferrara, 2011). Recent studies demonstrated that cancer exosomes can effectively stimulate endothelial cells in vitro and in vivo. For example, exosomes from melanoma cells have been shown to enhance the formation of endothelial spheroids and capillary sprouts in a dose-dependent manner (Hood et al., 2009). Mechanistically, melanoma exosomes stimulate endothelial production of proangiogenic cytokines, including

IL-1α, bFGF, TNF-α, etc. Interestingly, exosomes directly from endothelial cells do not possess such activity, suggesting that exosomes specifically mediate the communication between tumor cells and endothelial cells in a paracrine manner (Hood et al., 2009). Similarly, exosomes from chronic myelogenous leukemia cells were shown to promote in vitro vascular remodeling through inducing the expression of IL-8 and a series of cell adhesion molecules in endothelial cells (Taverna et al., 2012). However, the molecular identity of the functional agent(s) on cancer exosomes was not determined. In another study by Grange et al. exosomes from renal carcinoma were shown to promote the proliferation and vessel formation of normal endothelial cells implanted into the immunodeficient mice, supporting the in vivo activity of cancer exosomes (Grange et al., 2011). Interestingly, the authors reported that the cellular source of these exosomes is CD105-positive cancer stem cells, rather than other populations, suggesting clonal heterogeneity in the function of cancer-derived exosomes. To understand how exosomes promote cancer angiogenesis, it is necessary to explicitly characterize exosomes from different population of cells.

Studies have identified a variety of mechanisms underlying stimulation of endothelial cells by cancer exosomes. One mechanism involves direct activation of proangiogenic pathways in recipient cells. For example, Al-Nedawi et al. have reported that exosomes from different cancer cell lines could transport epidermal growth factor receptor to endothelial cells, leading to activation of downstream MAPK and Akt signaling and production of autocrine VEGF (Al-Nedawi, Meehan, Kerbel, Allison, & Rak, 2009). These findings suggest that cancer cells could use exosome to transfer oncogenic growth factors or receptors to directly rewire the signaling pathways in endothelial cells, leading to amplification of angiogenic responses.

Notch signaling is an evolutionarily conserved pathway that has a critical role in angiogenesis (Phng & Gerhardt, 2009). The interaction of a notch receptor with its corresponding ligand is generally mediated by direct cell–cell contact. However, it has been reported that Delta-like 4 (Dll4), a notch ligand, can be incorporated onto exosomes after overexpression in glioblastoma cells (Sharghi-Namini, Tan, Ong, Ge, & Asada, 2014; Sheldon et al., 2010). Cancer exosomes can transfer Dll4 to endothelial cells and integrate it into the cell membrane, resulting in inhibition of Notch signaling and increase of vessel branching. Interestingly, Shelton et al. demonstrated that endothelial exosomes can also transfer Dll4 and have the same effect on angiogenesis (Sheldon et al., 2010). It is therefore possible that cancer cells hijack a paracrine mechanism of Notch signaling in endothelial cells to promote angiogenesis.

In addition to direct transfer of active proteins, it is increasingly recognized that cancer exosomes carry a large variety of noncoding RNAs (ncRNAs) that modulate angiogenic signaling in recipient cells. For example, studies have identified many proangiogenic miRNAs in cancer exosomes (Falcone, Felsani, & D'Agnano, 2015). Being the most extensively studied type of exosomal ncRNA, miRNAs are typically 17~24 nucleotides in length and mediate gene silencing by binding to target mRNAs in the 3′-untranslated region or open reading frame region. Zhuang et al. reported that cancer cells deliver several microRNAs in vesicles to endothelial cells and promote angiogenic response. As a mechanism, the authors demonstrated that cancer-derived miR-9 downregulates the suppressor of cytokine signaling 5 and activates the JAK-STAT signaling pathway in endothelial cells (Zhuang et al., 2012). In addition, Umezu et al. reported that exosomes of leukemia cells contain miR-92a, a microRNA of the miR17-92 cluster that targets antiangiogenic factors (Umezu, Ohyashiki, Kuroda, & Ohyashiki, 2013). The authors showed that exosomal miR-92a of cancer cells reduces the expression of integrin α5 in endothelial cells, resulting in enhanced cell migration and tube formation. Moreover, metastatic breast cancer cells have been shown to release miR-210-enriched exosomes, which stimulate angiogenic response in endothelial cells (Kosaka et al., 2013). This horizontal transfer of exosomal miR-210 was shown to promote in vivo cancer metastasis.

Overall, a series of studies have demonstrated that cancer exosomes are heterogeneous in molecular contents and functional effects. Although we still do not understand how specific molecules are selectively packaged into exosomes, multiple investigations have revealed that cancer cells utilize exosomes as an effective way to transfer protein and miRNA factors that regulate the angiogenic balance in endothelial cells. Future investigations should explore the potential utility of such cancer exosomes as clinical biomarkers.

2.2 Microenvironment-Derived Exosomes in Tumor Angiogenesis

While research has conventionally focused on the proangiogenic factors derived from cancer cells, it is increasingly recognized that the complex microenvironment also plays an essential role in regulating tumor angiogenesis (De Palma et al., 2017; Hanahan & Weinberg, 2011; Roma-Rodrigues, Fernandes, & Baptista, 2014; Watnick, 2012). The tumor microenvironment hosts various types of stromal cells, such as fibroblasts, endothelial cells, immune cells, etc. Paracrine signaling interactions between stromal and cancer cells make significant contributions to tumor initiation, growth, and metastasis. Studies have revealed that stromal exosomes are a key component in the regulation of tumor angiogenesis and metastasis. Depending on the sources and functional contexts, stromal exosomes may have a positive or negative impact on the angiogenic balance.

During tumor angiogenesis, endothelial cells are the primary target of cancer-derived signals that regulate the angiogenic balance. In the other direction, studies have shown that endothelial cells also secrete exosomes containing factors that affect cancer cells. For example, Bovy et al. reported that endothelial exosomes transfer miR-503 to breast cancer cells and inhibit their proliferative and invasive capacities through targeting CCND2 and CCND3 (Bovy et al., 2015). The authors propose that this could underlie the antitumor effect of neoadjuvant chemotherapy, which lead to increased levels of plasmatic miR-503 in breast cancer patients (Bovy et al., 2015). Interestingly, incubation of endothelial cells with a cancer cell–conditioned medium suppressed the secretion of exosomal miR-503, suggesting that cancer cells release factor(s) to regulate endothelial exosomes (Bovy et al., 2015). Although the identity of the factor(s) remains elusive, exosomes are increasingly recognized as a robust vehicle that cancer cells and endothelial cells utilize to communicate with each other.

In addition to their significant roles in coagulation, platelets are also recognized as a circulating storage of both positive and negative regulators of angiogenesis (Folkman, Browder, & Palmblad, 2001). Recent studies show that exosomes can serve as an effective way that platelets release or sequester these regulators during cancer progression. For example, Janowska-Wieczorek et al. have reported that platelet-derived exosomes transfer integrin CD41 to lung cancer cells to promote metastasis and angiogenesis (Janowska-Wieczorek et al., 2005). The mechanisms involve activation of MAPK p42/44 and of induced expression of various proangiogenic factors, including MT1-MMP, MMP9, VEGF, IL8, and HGF, etc. Consequently, cancer cells displayed high activity in inducing angiogenesis. Interestingly, platelet-derived exosomes may have antiangiogenic effects when directly acting on endothelial cells. For example, platelets activated by NO and LPS release exosomes that induce apoptosis of endothelial cells and lead to vascular dysfunction (Gambim et al., 2007). Therefore understanding the biology of platelet-derived exosomes with respect to molecular contents and functions has important implications for targeting tumor angiogenesis.

MSCs are bone marrow–derived cells that have the capacity to differentiate into a series of mesenchymal cells including fibroblasts, adipocytes, osteoblasts, pericytes, muscles cells, etc. (Pittenger et al., 1999). MSCs can be recruited to the tumor sites by a variety of cancer-derived growth factors, and a positive correlation between the presence of MSCs and tumor angiogenesis has been reported in different types of cancers (Watnick, 2012). At the molecular level, the proangiogenic activity of MSCs has been at least partially attributed to the secretion of VEGF (Watnick, 2012). Interestingly, exosomes derived from MSCs have been generally reported to inhibit tumor angiogenesis instead. For example, Lee et al. reported that MSCs secrete exosomes containing abundant miR-16 that downregulates the expression of VEGF in breast cancer cells and inhibits tumor angiogenesis both in vitro and in vivo (Chen et al., 2015). In addition, MSC-derived exosomes are demonstrated to induce dormancy of breast cancer cells that disseminated to the bone marrow in a mouse model (Ono et al., 2014). Mechanistically, the authors illustrated that miR-23b enriched in MSC exosomes could inhibit the expression of myristoylated alanine-rich C-kinase substrate, which has been associated with the angiogenic and metastatic potential of breast cancer cells. Therefore although MSCs in tumor may directly activate endothelial cells and angiogenesis with secretion of VEGF, exosomes derived from these cells appear to function in the opposite way via imposing antiangiogenic activities on cancer cells.

In addition, the tumor microenvironment also contains other types of stromal cell, such as fibroblasts and immune cells, which secrete exosomes that have significant contributions in tumor initiation, growth, metastasis, and drug resistance. However, the functional involvement of fibroblast- or immune cell–derived exosomes in tumor angiogenesis is understudied and is certain to be an active area for future investigations.

2.3 Regulation of Exosomes During Tumor Angiogenesis

Understanding the regulatory mechanisms of tumor angiogenesis holds the promise for successfully limiting cancer progression. While the signaling pathways regulating the expression of various pro- and antiangiogenic proteins have been extensively characterized in many human cancers, we just begin to uncover the mechanisms underlying the generation of exosomes with various activities in tumor angiogenesis. In general, many key players that have significant impacts on angiogenesis have demonstrated activities in regulating angiogenic exosomes during cancer progression.

The hypoxic (low availability of oxygen) state in the tumor microenvironment is a common feature of many solid tumors and has been proposed as a key factor that drives tumor angiogenesis. A well-characterized pathway stimulated by hypoxia is the activation of hypoxia-induced factor 1 (HIF-1), which is a transcription factor that promotes the expression of a variety of proangiogenic factors, such as VEGF, bFGF, and TNFα (Masoud & Li, 2015). This hypoxia/HIF-1-signaling axis has been implicated in the generation and activity of exosomes that regulate tumor angiogenesis. For example, Hsu et al. reported that lung cancer cells cultured under hypoxic conditions release more exosomes than cells of normoxic cultures (Hsu et al., 2017). The authors demonstrated that hypoxic cancer exosomes contain increased level of miR-23a, which stimulates endothelial cells via activating HIF-1, as well as increases vascular permeability by downregulating tight junction protein ZO-1. Moreover, hypoxic conditions have been shown to enhance the HIF-1-dependent release of exosomes from breast cancer cells. These exosomes are enriched in miR-210 that promotes the proangiogenic activation of endothelial cells as mentioned above (King, Michael, & Gleadle, 2012). In another study, Kosaka et al. demonstrated that secretion of exosomal miR-210 is dependent on the activity of the neutral sphingomyelinase 2 (nSMase2) (Kosaka et al., 2013). However, it is not clear that whether the activity of nSMase2 is under the regulation of the hypoxia/HIF-1 signaling.

Compared with the hypoxia-stimulated release of proangiogenic exosomes from cancer cells, much less is known about the angiogenic-modulating exosomes from stromal cells. In addition, some hypoxia-independent mechanisms may regulate the generation and functions of exosomes during tumor angiogenesis. Exploring these directions will further advance our understanding of the regulation of exosome biogenesis and activity during tumor angiogenesis.

3. Future Perspectives

Angiogenesis is essential for tumor progression and is controlled by the complex interactions between cancer cells and the tumor microenvironment. Exosomes are increasingly recognized as an important signaling platform that mediates cancer–stroma interaction and regulates the angiogenic balance (Fig. 1). Studies have revealed that exosomes can deliver a myriad of factors with pro- or antiangiogenic activities, depending on their cellular origin and contextual signals during cancer progression (Azmi et al., 2013; Ribeiro et al., 2013; Roma-Rodrigues et al., 2014; Todorova et al., 2017). It is therefore reasonable to propose exosomes as a specific biomarker to indicate the angiogenic capacity of a tumor or as a promising tool to deliver antiangiogenic signaling molecules for cancer therapy. However, many important issues need to be addressed before the transition from bench to bedside can be achieved.

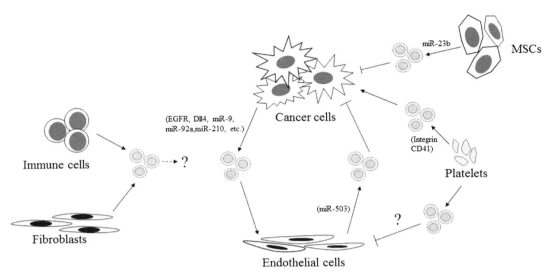

FIGURE 1 Exosomes mediate cancer–stroma interactions in tumor angiogenesis. Cancer cells secrete exosomes with various signaling molecules that regulate the angiogenic balance of endothelial cells. Interestingly, endothelial cells have been shown to utilize exosomes to deliver miRNAs that inhibit angiogenic signaling in cancer cells. In addition, stromal cells such as mesenchymal stem cells and platelets release exosomes that carry protein or miRNAs with pro- or antiangiogenic activity, depending on the identity of target cells and functional contexts. Finally, other stromal cells including fibroblasts and immune cells are also important sources of exosomes, but their functions in tumor angiogenesis remain obscure.

Exosomes are recognized as heterogeneous cellular entities that display large complexity in their molecular contents. The advances in genomic and proteomic technologies have enabled characterizing exosomes at the "-omics" level. It has facilitated the identification of key exosomal molecules, especially miRNAs and proteins that regulate various signaling processes, including tumor angiogenesis (Thery, 2011; Tkach & Thery, 2016; Todorova et al., 2017). However, most of our knowledge is obtained from analyses that only capture limited "snapshots" of the tumor–microenvironment communication of discrete cancer types. Currently the primary challenge is to understand how exosome-mediated signaling is dynamically integrated with the tumor angiogenesis process. To overcome this, we need systematic profiling of cancer- and stroma-derived exosomes at progressive stages of tumor angiogenesis. This will lead to identifications of stage-specific biomarkers of cancer progression. Importantly, such biomarkers need to be validated in multiple cancer model systems to investigate their general applicability.

In addition, many basic aspects of exosome biology remain elusive. Before we can utilize exosomes as a vehicle to deliver antiangiogenic therapeutics, we need to elucidate the mechanisms that underlie the biogenesis, trafficking, and processing of exosomes. For example, to package antiangiogenic factors (e.g., miR16 and miR-23b) into exosomes, current research is focusing on how distinct factors (miRNAs and proteins) are specifically and effectively incorporated onto exosomes and released to the extracellular milieu (Villarroya-Beltri, Baixauli, Gutierrez-Vazquez, Sanchez-Madrid, & Mittelbrunn, 2014). Moreover, to help reduce off-target effects, it is necessary to investigate how exosomes interact specifically with recipient cells and to develop technologies that enable in vivo monitoring of exosome delivery. In summary, gaining insights into the basic biology of exosomes may ultimately lead to innovative strategies to specifically inhibit tumor angiogenesis.

References

Adams, R. H., & Alitalo, K. (2007). Molecular regulation of angiogenesis and lymphangiogenesis. *Nature Reviews. Molecular Cell Biology, 8*(6), 464–478. https://doi.org/10.1038/nrm2183.

Al-Nedawi, K., Meehan, B., Kerbel, R. S., Allison, A. C., & Rak, J. (2009). Endothelial expression of autocrine VEGF upon the uptake of tumor-derived microvesicles containing oncogenic EGFR. *Proceedings of the National Academy of Sciences of the United States of America, 106*(10), 3794–3799. https://doi.org/10.1073/pnas.0804543106.

Azmi, A. S., Bao, B., & Sarkar, F. H. (2013). Exosomes in cancer development, metastasis, and drug resistance: A comprehensive review. *Cancer and Metastasis Reviews, 32*(3–4), 623–642. https://doi.org/10.1007/s10555-013-9441-9.

Bergers, G., & Benjamin, L. E. (2003). Tumorigenesis and the angiogenic switch. *Nature Reviews. Cancer, 3*(6), 401–410. https://doi.org/10.1038/nrc1093.

Bergers, G., & Hanahan, D. (2008). Modes of resistance to anti-angiogenic therapy. *Nature Reviews. Cancer, 8*(8), 592–603. https://doi.org/10.1038/nrc2442.

Bovy, N., Blomme, B., Freres, P., Dederen, S., Nivelles, O., Lion, M., … Struman, I. (2015). Endothelial exosomes contribute to the antitumor response during breast cancer neoadjuvant chemotherapy via microRNA transfer. *Oncotarget, 6*(12), 10253–10266. https://doi.org/10.18632/oncotarget.3520.

Chen, C. H., Cheng, C. T., Yuan, Y., Zhai, J., Arif, M., Fong, L. W., … Ann, D. K. (2015). Elevated MARCKS phosphorylation contributes to unresponsiveness of breast cancer to paclitaxel treatment. *Oncotarget, 6*(17), 15194–15208. https://doi.org/10.18632/oncotarget.3827.

Chung, A. S., & Ferrara, N. (2011). Developmental and pathological angiogenesis. *Annual Review of Cell and Developmental Biology, 27*, 563–584. https://doi.org/10.1146/annurev-cellbio-092910-154002.

De Palma, M., Biziato, D., & Petrova, T. V. (2017). Microenvironmental regulation of tumour angiogenesis. *Nature Reviews. Cancer, 17*(8), 457–474. https://doi.org/10.1038/nrc.2017.51.

De Toro, J., Herschlik, L., Waldner, C., & Mongini, C. (2015). Emerging roles of exosomes in normal and pathological conditions: New insights for diagnosis and therapeutic applications. *Frontiers in Immunology, 6*, 203. https://doi.org/10.3389/fimmu.2015.00203.

Ellis, L. M., & Hicklin, D. J. (2008). VEGF-targeted therapy: Mechanisms of anti-tumour activity. *Nature Reviews. Cancer, 8*(8), 579–591. https://doi.org/10.1038/nrc2403.

Falcone, G., Felsani, A., & D'Agnano, I. (2015). Signaling by exosomal microRNAs in cancer. *Journal of Experimental & Clinical Cancer Research: CR, 34*, 32. https://doi.org/10.1186/s13046-015-0148-3.

Folkman, J. (1971). Tumor angiogenesis: Therapeutic implications. *The New England Journal of Medicine, 285*(21), 1182–1186. https://doi.org/10.1056/NEJM197111182852108.

Folkman, J., Browder, T., & Palmblad, J. (2001). Angiogenesis research: Guidelines for translation to clinical application. *Thrombosis and Haemostasis, 86*(1), 23–33.

Gambim, M. H., do Carmo Ade, O., Marti, L., Verissimo-Filho, S., Lopes, L. R., & Janiszewski, M. (2007). Platelet-derived exosomes induce endothelial cell apoptosis through peroxynitrite generation: Experimental evidence for a novel mechanism of septic vascular dysfunction. *Critical Care, 11*(5), R107. https://doi.org/10.1186/cc6133.

Giuliano, S., & Pages, G. (2013). Mechanisms of resistance to anti-angiogenesis therapies. *Biochimie, 95*(6), 1110–1119. https://doi.org/10.1016/j.biochi.2013.03.002.

Grange, C., Tapparo, M., Collino, F., Vitillo, L., Damasco, C., Deregibus, M. C., … Camussi, G. (2011). Microvesicles released from human renal cancer stem cells stimulate angiogenesis and formation of lung premetastatic niche. *Cancer Research, 71*(15), 5346–5356. https://doi.org/10.1158/0008-5472.CAN-11-0241.

Hanahan, D., & Weinberg, R. A. (2000). The hallmarks of cancer. *Cell, 100*(1), 57–70.

Hanahan, D., & Weinberg, R. A. (2011). Hallmarks of cancer: The next generation. *Cell, 144*(5), 646–674. https://doi.org/10.1016/j.cell.2011.02.013.

Harding, C. V., Heuser, J. E., & Stahl, P. D. (2013). Exosomes: Looking back three decades and into the future. *Journal of Cell Biology, 200*(4), 367–371. https://doi.org/10.1083/jcb.201212113.

Hood, J. L., Pan, H., Lanza, G. M., Wickline, S. A., & Consortium for Translational Research in Advanced, I., & Nanomedicine (2009). Paracrine induction of endothelium by tumor exosomes. *Laboratory Investigation; a Journal of Technical Methods and Pathology, 89*(11), 1317–1328. https://doi.org/10.1038/labinvest.2009.94.

Hsu, Y. L., Hung, J. Y., Chang, W. A., Lin, Y. S., Pan, Y. C., Tsai, P. H., … Kuo, P. L. (2017). Hypoxic lung cancer-secreted exosomal miR-23a increased angiogenesis and vascular permeability by targeting prolyl hydroxylase and tight junction protein ZO-1. *Oncogene, 36*(34), 4929–4942. https://doi.org/10.1038/onc.2017.105.

Janowska-Wieczorek, A., Wysoczynski, M., Kijowski, J., Marquez-Curtis, L., Machalinski, B., Ratajczak, J., & Ratajczak, M. Z. (2005). Microvesicles derived from activated platelets induce metastasis and angiogenesis in lung cancer. *International Journal of Cancer, 113*(5), 752–760. https://doi.org/10.1002/ijc.20657.

Kerbel, R. S. (2008). Tumor angiogenesis. *The New England Journal of Medicine, 358*(19), 2039–2049. https://doi.org/10.1056/NEJMra0706596.

King, H. W., Michael, M. Z., & Gleadle, J. M. (2012). Hypoxic enhancement of exosome release by breast cancer cells. *BMC Cancer, 12*, 421. https://doi.org/10.1186/1471-2407-12-421.

Kosaka, N., Iguchi, H., Hagiwara, K., Yoshioka, Y., Takeshita, F., & Ochiya, T. (2013). Neutral sphingomyelinase 2 (nSMase2)-dependent exosomal transfer of angiogenic microRNAs regulate cancer cell metastasis. *The Journal of Biological Chemistry, 288*(15), 10849–10859. https://doi.org/10.1074/jbc.M112.446831.

Landskroner-Eiger, S., Moneke, I., & Sessa, W. C. (2013). miRNAs as modulators of angiogenesis. *Cold Spring Harbor Perspectives in Medicine, 3*(2), a006643. https://doi.org/10.1101/cshperspect.a006643.

Masoud, G. N., & Li, W. (2015). HIF-1alpha pathway: Role, regulation and intervention for cancer therapy. *Acta Pharmaceutica Sinica B, 5*(5), 378–389. https://doi.org/10.1016/j.apsb.2015.05.007.

Moserle, L., & Casanovas, O. (2013). Anti-angiogenesis and metastasis: A tumour and stromal cell alliance. *Journal of Internal Medicine, 273*(2), 128–137. https://doi.org/10.1111/joim.12018.

Ono, M., Kosaka, N., Tominaga, N., Yoshioka, Y., Takeshita, F., Takahashi, R. U., … Ochiya, T. (2014). Exosomes from bone marrow mesenchymal stem cells contain a microRNA that promotes dormancy in metastatic breast cancer cells. *Science Signaling [electronic Resource], 7*(332), ra63. https://doi.org/10.1126/scisignal.2005231.

Pashoutan Sarvar, D., Shamsasenjan, K., & Akbarzadehlaleh, P. (2016). Mesenchymal stem cell-derived exosomes: New opportunity in Cell-Free Therapy. *Advanced Pharmaceutical Bulletin, 6*(3), 293–299. https://doi.org/10.15171/apb.2016.041.

Phng, L. K., & Gerhardt, H. (2009). Angiogenesis: A team effort coordinated by notch. *Developmental Cell, 16*(2), 196–208. https://doi.org/10.1016/j.devcel.2009.01.015.

Pittenger, M. F., Mackay, A. M., Beck, S. C., Jaiswal, R. K., Douglas, R., Mosca, J. D., … Marshak, D. R. (1999). Multilineage potential of adult human mesenchymal stem cells. *Science, 284*(5411), 143–147.

Potente, M., Gerhardt, H., & Carmeliet, P. (2011). Basic and therapeutic aspects of angiogenesis. *Cell, 146*(6), 873–887. https://doi.org/10.1016/j.cell.2011.08.039.

Ribeiro, M. F., Zhu, H., Millard, R. W., & Fan, G. C. (2013). Exosomes function in pro- and anti-angiogenesis. *Current Angiogenes*, *2*(1), 54–59. https://doi.org/10.2174/22115528113020020001.

Roma-Rodrigues, C., Fernandes, A. R., & Baptista, P. V. (2014). Exosome in tumour microenvironment: Overview of the crosstalk between normal and cancer cells. *BioMed Research International*, *2014*, 179486. https://doi.org/10.1155/2014/179486.

Sharghi-Namini, S., Tan, E., Ong, L. L., Ge, R., & Asada, H. H. (2014). Dll4-containing exosomes induce capillary sprout retraction in a 3D microenvironment. *Scientific Reports*, *4*, 4031. https://doi.org/10.1038/srep04031.

Sheldon, H., Heikamp, E., Turley, H., Dragovic, R., Thomas, P., Oon, C. E., … Harris, A. L. (2010). New mechanism for Notch signaling to endothelium at a distance by Delta-like 4 incorporation into exosomes. *Blood*, *116*(13), 2385–2394. https://doi.org/10.1182/blood-2009-08-239228.

Steeg, P. S. (2016). Targeting metastasis. *Nature Reviews. Cancer*, *16*(4), 201–218. https://doi.org/10.1038/nrc.2016.25.

Taverna, S., Flugy, A., Saieva, L., Kohn, E. C., Santoro, A., Meraviglia, S., … Alessandro, R. (2012). Role of exosomes released by chronic myelogenous leukemia cells in angiogenesis. *International Journal of Cancer*, *130*(9), 2033–2043. https://doi.org/10.1002/ijc.26217.

Thery, C. (2011). Exosomes: Secreted vesicles and intercellular communications. *F1000 Biology Reports*, *3*, 15. https://doi.org/10.3410/B3-15.

Tkach, M., & Thery, C. (2016). Communication by extracellular vesicles: Where We Are and Where We Need to Go. *Cell*, *164*(6), 1226–1232. https://doi.org/10.1016/j.cell.2016.01.043.

Todorova, D., Simoncini, S., Lacroix, R., Sabatier, F., & Dignat-George, F. (2017). Extracellular vesicles in angiogenesis. *Circulation Research*, *120*(10), 1658–1673. https://doi.org/10.1161/CIRCRESAHA.117.309681.

Umezu, T., Ohyashiki, K., Kuroda, M., & Ohyashiki, J. H. (2013). Leukemia cell to endothelial cell communication via exosomal miRNAs. *Oncogene*, *32*(22), 2747–2755. https://doi.org/10.1038/onc.2012.295.

van der Pol, E., Coumans, F., Varga, Z., Krumrey, M., & Nieuwland, R. (2013). Innovation in detection of microparticles and exosomes. *Journal of Thrombosis and Haemostasis: JTH*, *11*(Suppl 1), 36–45. https://doi.org/10.1111/jth.12254.

Villarroya-Beltri, C., Baixauli, F., Gutierrez-Vazquez, C., Sanchez-Madrid, F., & Mittelbrunn, M. (2014). Sorting it out: Regulation of exosome loading. *Seminars in Cancer Biology*, *28*, 3–13. https://doi.org/10.1016/j.semcancer.2014.04.009.

Watnick, R. S. (2012). The role of the tumor microenvironment in regulating angiogenesis. *Cold Spring Harbor Perspectives in Medicine*, *2*(12), a006676. https://doi.org/10.1101/cshperspect.a006676.

Wu, Y., Deng, W., & Klinke, D. J., 2nd. (2015). Exosomes: Improved methods to characterize their morphology, RNA content, and surface protein biomarkers. *Analyst*, *140*(19), 6631–6642. https://doi.org/10.1039/c5an00688k.

Zabeo, D., Cvjetkovic, A., Lasser, C., Schorb, M., Lotvall, J., & Hoog, J. L. (2017). Exosomes purified from a single cell type have diverse morphology. *Journal of Extracellular Vesicles*, *6*(1), 1329476. https://doi.org/10.1080/20013078.2017.1329476.

Zhuang, G., Wu, X., Jiang, Z., Kasman, I., Yao, J., Guan, Y., … Ferrara, N. (2012). Tumour-secreted miR-9 promotes endothelial cell migration and angiogenesis by activating the JAK-STAT pathway. *The EMBO Journal*, *31*(17), 3513–3523. https://doi.org/10.1038/emboj.2012.183.

14

Role of Exosomes in Development of Premetastatic Niche

Sagar Bhayana, Marshleen Yadav, Naduparambil K. Jacob

THE OHIO STATE UNIVERSITY COMPREHENSIVE CANCER CENTER, COLUMBUS, OH, UNITED STATES

CHAPTER OUTLINE

1. Introduction 247

2. Role of Cancer Cell–Derived Exosomes in Initiation and Progression of Metastasis 248

3. Role of Exosomal Noncoding RNAs in Initiation and Priming for Metastasis 250

4. Role of Exosomal Proteins in Priming Metastasis 251

5. Role of Toll-Like Receptors in Exosome-Mediated Metastasis 252

6. Epithelial–Mesenchymal Transition, a Hallmark of Premetastatic Niche Formation 253

7. Pharmacologic and Genetic Strategies Targeting Exosome-Mediated Metastasis 255

References 256

1. Introduction

The leading cause of cancer-related deaths is metastasis. Research during the last decade has provided significant insights into the metastatic processes, defining the sequential events involved in metastasis that include invasion and outgrowth of the primary tumor, dissemination of tumor cells to the circulatory system, extravasation and formation of premetastatic niche at the distant organ, followed by metastatic outgrowth at the secondary site (Mehlen & Puisieux, 2006; Valastyan & Weinberg, 2011). It is basically the compatibility and the interaction of disseminated tumor cells (referred to as seed on the basis of Paget's "seed and soil hypothesis") with the environment in the distant organ (referred to as soil) that brings about the metastatic spreading of cancer (Fidler, 2003; Steeg, 2016). An essential cellular event that occurs during the metastasis is the formation of premetastatic niche, which primes and increases the receptiveness of distant organs for the incoming tumor cells. Several elegant studies have identified key molecules and pathways that initiate and accelerate the process of premetastatic niche formation. Originally, formation of premetastatic environment in the distant organ was shown to be mediated by bone marrow–derived cells (BMDCs) that expressed vascular endothelial growth factor receptor 1 (VEGFR1). These VEGFR1+ cells home tumor-specific metastatic sites and then colonize by interaction of VLA-4 (integrin α4β1) with its ligand fibronectin (Fig. 1), providing a conducive premetastatic environment before tumor cells arrive (Kaplan et al., 2005). Recent studies underscore the role and significance of exosomes, small extracellular vesicles, that are secreted from primary tumor and tumor-associated cells and released to the circulation. Upon contact with target cells, exosomes play critical roles in initiation and progression of programmed steps in metastasis.

2. Role of Cancer Cell–Derived Exosomes in Initiation and Progression of Metastasis

Although discovered almost 30 years ago (Johnstone, Adam, Hammond, Orr, & Turbide, 1987), the role of exosomes in spreading of tumor to different parts of the body was explored only recently. The studies on the structure, composition, and biogenesis of exosomes have broadened the scope of their involvement in metastasis. Exosomes facilitate transport of mRNAs, miRNAs, long noncoding RNAs (lncRNAs), and proteins, including transducers and effectors from tumor cells for priming the early events in metastasis. Their high abundance in blood and ability to carry the signature of the originating cells and ability to engage with surface markers of specific recipient cells make them the obvious choice for signaling metastasis. Components of exosomal cargo are capable of carrying the message for invasion to the surrounding tissues, promoting vascular leakiness, increasing the levels of protumorigenic adhesion molecules, and modulating inflammation by recruiting BMDCs, thus culminating in the formation of the premetastatic niche. Arguably, because the heterogeneity of exosomes originate from diverse cell types at the primary tumor site, including tumor cells and cells comprising the tumor microenvironment, they may exert pro- as well as antitumorigenic functions. Metastatic potential and aggressiveness is primarily dictated by signals emanating from the malignant cells at the primary site. Exosomes derived from mesenchymal stromal cells or fibroblast in the tumor microenvironment also facilitate tumor progression and immunosuppression as reported in the case of multiple myeloma, colorectal cancer, and gastric cancer (GC) (Hu et al., 2015; Ji et al., 2015; Roccaro et al., 2013; J. Wang et al., 2014). Exosomes derived from malignant cells facilitate the recruitment of BMDCs by increasing proinflammatory molecules at premetastatic sites and promoting cell proliferation and angiogenesis (Fan, 2014). Preconditioning the bone marrow with exosomes from a highly metastatic mouse cancer cell line (B16–F10) has been shown to increase the metastatic tumor burden (Peinado et al., 2012). Moreover, exosome-mediated receptor tyrosine kinase MET (hepatic growth factor receptor) signaling was found to be a principal mediator of bone marrow priming when protein content of exosomes that originated from highly metastatic and poorly metastatic melanoma cells were compared. MET oncogene has a well-established role in mediating cellular transformation and tumor cell proliferation, survival, motility, invasion,

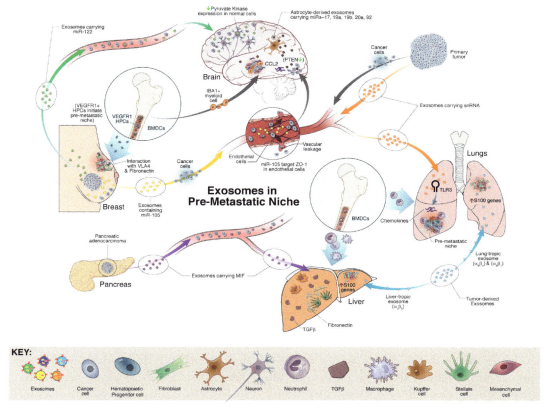

FIGURE 1 Role of Exosomes in the development of Premetastatic Niche. Exosomes are one of the crucial tools in specific tissue homing and subsequently in the formation of premetastatic niche formation in distant organ. The premetastatic niche study was originally initiated from an observation wherein the BMDCs that express vascular endothelial growth factor receptor 1 (VEGFR1) home to tumor-specific premetastatic sites and form cellular clusters before the arrival of tumor cells. Several follow-up studies encompassed the role of exosomes in premetastatic events. One of the important oncogenic signature molecules that is transported by exosomes are miRNAs. For instances, the process shown via *yellow arrows* depicts the release of exosomes carrying miR-105 from breast tumor cells, targeting endothelial cells in blood vessels, leading to vascular leakage. The process shown via *green arrows* depicts the release of exosomes carrying miR-122 from breast tumor cells, targeting pyruvate kinase expression in normal brain cells, thus reducing the intake of glucose in normal brain cells. The process shown via *black arrows* depicts the release of cancer cells from primary tumor to the brain wherein the astrocyte releases exosomes containing tumor-suppressor miRNAs targeting PTEN expression in cancer cells, thus releasing chemokine CCL2 that acts as an attractant for IBA1+ myeloid cells from bone marrow, leading to formation of premetastatic niche in brain. The process shown via *purple arrows* depicts the induction of liver premetastatic niche via pancreatic ductal adenocarcinoma exosomes carrying macrophage MIF. Uptake of MIF carrying exosomes by Kupffer cells caused transforming growth factor β secretion and upregulation of fibronectin production by hepatic stellate cells. This fibrotic microenvironment enhanced recruitment of bone marrow–derived macrophages and neutrophils. The process shown via *blue arrows* depicts the organotropic premetastasis via organ-specific exosomes. Lung-specific exosomes expressed unique integrins $\alpha_6\beta_4$ and $\alpha_6\beta_1$, whereas liver-specific exosomes expressed $\alpha_v\beta_5$; fusion of these organ-specific integrin exosomes with resident cells led to induction of proinflammatory S100 expression, leading to premetastatic niche development. The process shown via *orange arrows* depicted toll-like receptor, TLR3, with a role in premetastasis events. Depending on the origin of primary tumor, exosomes carrying small nuclear RNA (snRNA) entered lungs where snRNA activated TLR3, leading to release of chemokines and neutrophil infiltration.

and metastasis (Birchmeier, Birchmeier, Gherardi, & Vande Woude, 2003; Boccaccio & Comoglio, 2006; Christensen, Burrows, & Salgia, 2005; Peruzzi & Bottaro, 2006; Scott et al., 2011; Tesio et al., 2011). The upregulation of MET in bone marrow progenitor cells by tumor-derived exosomes was proposed as a mechanism that promotes education, mobilization, and prometastatic behavior of BMDCs. MET was found highly expressed in both bone marrow progenitor cells and those that promote vasculogenesis isolated from the blood of high-grade melanoma patients. Furthermore, expression of specific Ras-related (Rab) proteins (*RAB5B*, *RAB7*, and *RAB27A*), which are functionally involved in exosome secretion, was reported to be higher in metastatic tumors (Peinado et al., 2012). Thus exosomes of metastatic cancer patients carry protein signature mechanistically connected to various processes involved in metastasis, including initiation and development of premetastatic niche.

3. Role of Exosomal Noncoding RNAs in Initiation and Priming for Metastasis

The pattern of expression of over 2000 miRNAs reported from human genome varies with cell types and disease states. Genetic studies in animal models have shown that exosomal miRNAs play a critical role in the development of premetastatic niche. Distinct expression pattern of miRNAs and their ability to barcode the exosomes during biogenesis and release to the circulation provide unique opportunity for communication of cancer cells to distant sites. It was demonstrated that miR-105 from exosomes originated from breast cancer cells targets and destroys the tight junction zonula occludens 1 (ZO-1), leading to leakage in the vascular endothelial barrier (Zhou et al., 2014), which further facilitates the movement of cancer cells to the circulatory system, causing extravasation to distant secondary sites (Fig. 1, shown via *yellow arrow*). Exosomes have been shown to be capable of transporting miRs that could target tumor suppressor genes, contributing to priming of premetastatic niche. Studies have shown that exosomes derived from astrocytes can transfer clusters of miR-17, miR-19a, miR-19b, miR-20a, and miR-92 to inhibit the tumor suppressor PTEN gene in disseminated tumor cells localized in the brain (L. Zhang et al., 2015). This priming of the premetastatic niche by PTEN loss leads to an increased secretion of chemokine CCL2, which recruits IBA1-expressing myeloid cells that enhances the outgrowth of brain metastatic tumor cells (L. Zhang et al., 2015) (Fig. 1, shown via *black arrow*). Exosome-enriched miRNAs such as miR-29a, miR-650, and miR-151 are found to be associated with tumor invasion and metastasis in various tumor models (Gebeshuber, Zatloukal, & Martinez, 2009; Luedde, 2010; X. Zhang et al., 2010). miR-155 transported by acute myeloid leukemic cell–derived exosomes has been shown to invade into normal hematopoietic stem and progenitor cells and target c-Myb expression to impair normal hematopoiesis (Hornick et al., 2016). Exosomal transport of miR-23b from bone marrow MSCs in breast cancer models has been shown to induce breast cancer cell dormancy in the premetastatic niche by suppressing target gene MARCKS (myristoylated alanine-rich C-kinase substrate) (Ono et al., 2014). Breast cancer cell–derived exosomes are able to reprogram the premetastatic niche stromal cells in the brain to suppress glucose uptake by downregulating the glycolytic enzyme pyruvate kinase by delivering inhibitory miR-122 (Fong et al., 2015) (Fig. 1, shown via *green arrow*). The reduced levels of exosomal miR-34a and miR-192 have been proposed as potential indicators of cancer progression and metastasis stage in prostate cancer and lung adenocarcinoma, respectively (Corcoran, Rani, & O'Driscoll, 2014; Valencia et al., 2014). In human renal cell carcinoma, tumor-initiating cells expressing CD105 was shown to release exosomes that triggered angiogenesis and promoted the formation of premetastatic niche (Grange et al., 2011). These tumor cell–derived exosomal miRNAs such as miR-200c, miR-92, and miR-141 was reported to be associated with unfavorable prognosis in ovarian (Iorio et al., 2007; Motoyama et al., 2009; Taylor & Gercel-Taylor, 2008) and prostate cancer (Brase et al., 2011).

Exosomal cargo also contains small nuclear RNAs (snRNAs) and lncRNAs that are implicated in the processes of metastasis. For instance, expression of lncRNA MALAT1 (metastasis-associated lung adenocarcinoma transcript 1) showed strong correlation with tumor stage and lymphatic metastasis in nonsmall cell lung carcinoma (NSCLC) patients (R. Zhang et al., 2017). Serum exosome–derived MALAT was proposed as a strong circulating prognostic marker for NSCLC (R. Zhang et al., 2017). Exosomal lncRNA ZFAS1 was reported to be involved in gastric cancer progression by enhancing proliferation and migration (Pan et al., 2017). In the mechanistic studies, the recipient cells, MKN-28 cells, showed increased migration and proliferation, suggesting a role of exosomal ZFAS1 in metastasis (Pan et al., 2017). Exosome derived from tumor cells contain high content of snRNA, which is functionally associated with proinflammatory signaling facilitating the development of premetastatic niche (Liu et al., 2016).

4. Role of Exosomal Proteins in Priming Metastasis

The proteins present either on the surface or inside the exosomes impact the activation of S100 genes or promote the activation of proinflammatory signaling pathways, two events leading to metastasis. The lipid bilayer of exosomes is able to mediate fusion with the receiving cells, while molecules on the surface of exosomes can also directly engage with receptors present on receiving cells and induce intracellular signaling (Colombo, Raposo, & Thery, 2014). Such interactions can be specific and directional as evident from the report on the tropism of integrin repertoire present on exosomes guiding to metastatic sites (Hoshino et al., 2015). Notably, exosomes expressing $ITG\alpha_v\beta_5$ specifically bind to Kupffer cells, mediating liver tropism, whereas exosomal $ITG\alpha_6\beta_4$ and $ITG\alpha_6\beta_1$ bind lung-resident fibroblasts and epithelial cells, governing lung tropism (Hoshino et al., 2015). Exosome uptake by either of these organ-specific cells promotes promigratory and proinflammatory S100 gene upregulation (Fig. 1, shown via *blue arrow*) which in turn promote metastatic outgrowth (Hoshino et al., 2015). Another study analyzing the premetastatic environment in pancreatic ductal adenocarcinomas showed the role of pancreatic cancer exosomes in establishing a premetastatic niche in the liver (Costa-Silva et al., 2015). Selective uptake of exosomes consisting of migration inhibitory factor (MIF) by Kupffer cells in the liver releases transforming growth factor β (TGFβ) which in turn increases the fibronectin production by hepatic stellate cells (Costa-Silva et al., 2015). TGFβ and fibronectin promotes an influx of bone marrow–derived macrophages and neutrophils, initiating an inflammation cascade (Fig. 1, shown via *purple arrow*). This circuit from uptake of exosomes to activation of resident macrophages in liver leads to the formation of premetastatic niche (Costa-Silva et al., 2015). Exosomes derived from chronic lymphocytic leukemia (CLL) have been shown to transfer exosomal proteins to mesenchymal stem cells, inducing an inflammatory phenotype similar to cancer-associated fibroblasts (CAFs) (Paggetti et al., 2015). Of all the cytokines secreted by these CAF-like cells, CCL2 (MCP1) and CXCL16 primarily attracted macrophages and T-cells, modulating the immune cells to establish the niche for metastatic outgrowth (Paggetti et al., 2015). Another example of exosome-mediated transfer of protumorigenic protein was shown by the comprehensive analysis of circulating exosomal cargo in progressive CLL patients. It was found out that S100-A9 protein present primarily in exosomes of progressive patients than in indolent patients (Prieto et al., 2017) and this protein is capable of activating the NF-κB pathway in primary CLL cells (Prieto et al., 2017). This indicated that the exosomal cargo with S100-A9 protein could be a potential mechanism of cross talk between leukemic cells and the other organs. The diversity in molecular cargo carried by tumor cell originated exosomes is evident from a recent report demonstrating levels of caspase 8 control the secretion of an inflammatory enzyme, lysyl-tRNA synthetase (S. B. Kim et al., 2017). Exosomes carrying lysyl-tRNA synthetase have been shown to induce macrophage migration, and their TNF-α secretion seems to modulate the development of premetastatic niche.

5. Role of Toll-Like Receptors in Exosome-Mediated Metastasis

Toll-like receptors (TLRs) have been reported to mediate tumor growth and progression by diverse mechanisms, including inhibition of apoptosis and activation of proliferative signals (Cherfils-Vicini et al., 2010; He et al., 2007; Pradere, Dapito, & Schwabe, 2014; L. F. Wang, Chien, Kuo, Tai, & Juo, 2008). It is well established that host TLRs are vitally involved in the tumorigenesis process by promoting chronic inflammation (Dapito et al., 2012; Pradere et al., 2014; Scheeren et al., 2014). Of the 10 TLRs identified in human beings, TLR1, TLR2, TLR5, and TLR6 have been reported to be expressed on the cell surface, while TLR3, TLR7, TLR8, and TLR9 have been shown to express on the endosome membrane. The biogenesis of exosomes requires fusion with endosomal membrane, so it is conceivable that exosomes carry the endosome signature. One of the important characteristics of TLRs is pattern recognition receptors (PRRs), which allows them to sense molecular components from invading microorganisms and damaged cells (Akira & Takeda, 2004; Cao, 2016). The role of PRR in cancer was shown biochemically by screening culture medium of Lewis lung carcinoma (LLC) that led to an identification of extracellular matrix proteoglycan versican. Versican is able to activate macrophages through activation of TLR2 (specific to proteoglycans) and the coreceptor TLR6, confirming the role of extracellular factors in metastasis via TLR activation (S. Kim et al., 2009). Small RNAs, including miRNAs, could act as an agonist for activation of TLR7 and TLR8 receptors (specific to single-stranded RNAs), thus playing a critically important role in the intercellular communication in tumor microenvironment (Heil et al., 2004). miR-21 and miR-29a containing exosomes, released from lung tumor cells, have been identified as activators of immune response via murine TLR7 and human TLR8 receptors in immune cells, resulting in activation of NF-κB and secretion of inflammatory cytokines, which ultimately supports the tumor growth and premetastasis niche (Fabbri et al., 2012). Sequence specificity of RNAs in the TLR7-dependent immune response has been reported in PBMCs, where GU-rich sequences have been shown as strong inducers of immune responses (Forsbach et al., 2008). TLR3, which is present in the endosome membrane, is reported as a key PRR player in sensing the incoming exosomes to initiate downstream pathways in breast cancer cell lines (Boelens et al., 2014). In an elegant genetic study, wherein LLC and B16/F10 melanoma cells were subcutaneously injected in littermates of $Tlr3^{-/-}$, $Tlr4^{-/-}$, $Tlr9^{-/-}$, and lung metastasis analysis was then performed (Liu et al., 2016). Interestingly, only $Tlr3^{-/-}$ littermate showed reduced lung metastasis in comparison with wild-type littermate (Liu et al., 2016). In this model, premetastasis niche regulating genes such as Bv8, S100a8, S100a9, and MMP9 were found to be downregulated in the $Tlr3^{-/-}$ littermate lungs (Wu et al., 2015; Liu et al., 2016). Furthermore, exosomal RNA sequencing revealed that exosomes derived from tumor contain high content of snRNA (Liu et al., 2016). These stem loops structure in snRNAs potentially serve as a platform for TLR3 recognition and activation via NF-κB and MAPK pathways in AT-II (alveolar lung epithelial) cells leading to chemokine production and neutrophil recruitment from bone marrow to form premetastasis niche (Liu et al., 2016) (Fig. 1, shown via *orange arrow*).

6. Epithelial–Mesenchymal Transition, a Hallmark of Premetastatic Niche Formation

Epithelial–mesenchymal transition (EMT) is a crucial process in embryonic development wherein intrinsic and external factors stimulate genetic and epigenetic reprogramming in cells to gain motility and invasiveness, facilitating the shaping of different tissues and organs. Tumor cells acquire such phenotypic transition to intravasate and disseminate into the circulatory system to invade and colonize at distant organ. Molecular characterization of tumor cells–derived exosomes revealed presence

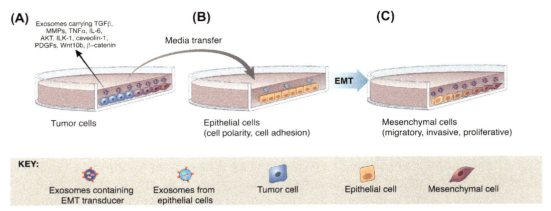

FIGURE 2 Role of exosomal epithelial–mesenchymal transition inducers in premetastasis. Exosomes activate epithelial–mesenchymal transition (EMT) signaling mechanism by carrying EMT inducers and activators from primary tumor to distant organ. For instance, several in vitro media transfer experiments have shown the release of EMT molecules from tumor cells (A) such as transforming growth factor β (TGFβ), matrix metalloproteinases (MMPs), tumor necrosis factor α (TNFα), interleukin-6 (IL-6), protein kinase B (AKT), integrin-linked kinase (ILK-1), caveolin-1, platelet-derived growth factors (PDGFs), Wnt10b, and β-catenin have remodeled epithelial cells (B) to migratory, invasive, and proliferative mesenchymal cells, (C) promoting tumor-supportive processes.

of EMT inducers and transcriptional regulators that influences diverse pro-EMT signaling pathways. Cancer cells seem to use intrinsic and extrinsic cues for inducing EMT for orchestrating the premetastatic niche formation (Banyard & Bielenberg, 2015; Heerboth et al., 2015). For instance, appreciable amounts of TGFβ are found in exosomes secreted by carcinoma cells that induced differentiation of tumor promoting stromal myofibroblasts in the recipient cells (Webber et al., 2015). Also, matrix-metalloproteinase 13 (MMP13)-containing exosomes has been shown to promote nasopharyngeal carcinoma (NPC) metastasis (You et al., 2015; Radisky & Radisky, 2010). MMP13 containing exosomes when added to NPC-cultured media, an enhanced extracellular matrix degradation was observed suggesting a role for MMP13 containing exosomes in metastasis. Exosomes released from hypoxic tumors are enriched in potent EMT-transducing signaling molecules such as TGFβ, MMPs, tumor necrosis factor alpha (TNFα), interleukin-6 (IL-6), protein kinase B (AKT), integrin-linked kinase 1 (ILK-1), caveolin-1, platelet-derived growth factors (PDGFs), Wnt10b, and β-catenin (Fig. 2A). When those EMT exosomes are taken up, it alters the cell transcriptome and proteome of recipient cells (Fig. 2B), further enhancing the EMT process and tumorigenicity (Fig. 2C) (Chen et al., 2017; King, Michael, & Gleadle, 2012; Kucharzewska et al., 2013; Ramteke et al., 2015; Svensson et al., 2011; Tadokoro, Umezu, Ohyashiki, Hirano, & Ohyashiki, 2013). Recent studies also show that tumor viruses such as Epstein–Barr virus (EBV) and Kaposi sarcoma–associated herpesvirus (KSHV) can manipulate the tumor microenvironment by secreting specific viral and cellular components via exosomes that activates the EMT transition in recipient cells. KSHV has been shown to induce the expression of EMT factor SNAI1, and EBV exosomes, when taken up by recipient cells, regulate PTEN-dependent tumor suppressor pathways, contributing to migration and invasion of NPC cells in vitro (Cai et al., 2015; Jha et al., 2016). Arguably, chemotherapeutic agents and radiation commonly used for local control might be contributing to the activation of EMT program and release of exosome cargo that promote development of premetastatic niche at distant sites.

7. Pharmacologic and Genetic Strategies Targeting Exosome-Mediated Metastasis

As detailed in preceding paragraphs, exosomes released from primary tumor and tumor microenvironment play crucial roles in development of premetastatic niche at distant locations. Thus targeting exosomes or using exosomes as a vehicle for delivery of drug or biomolecules represent a new platform for therapy targeting metastatic cancer. One such approach was demonstrated recently using iExosomes targeting against Kras mutant tumors in mouse model for pancreatic cancer (Kamerkar et al., 2017). Exosomes derived from normal fibroblast-like mesenchymal cells were engineered to carry short interfering RNA specific to oncogenic Kras^{G12D}, a common driver mutation found in pancreatic cancer. The targeting specificity after systemic administration lies in the presence of CD47 on engineered exosomes (iExosomes) that protects these exosomes from getting phagocytosed, providing longer stability compared with that achieved with liposomes (Kamerkar et al., 2017). Modulating exosome function, targeting biogenesis and secretion by inhibiting the genes involved in exosome biogenesis, also have been tested in preclinical models. For instance, specific RAB genes (RAB27A) overexpressed in highly metastatic melanoma, when targeted, led to decreased exosome production, hindering BMDC mobilization, tumor growth, and metastasis (Peinado et al., 2012). TSG101, protein integral to exosome production and release, was found to be high in triple-negative breast cancer cell lines, contributing to enhanced exosome-mediated communication. Targeting TSG101 was shown to inhibit exosome production (Sharma et al., 2014), whereas targeting the integrins $\alpha_6\beta_4$ and $\alpha_v\beta_5$ decreased exosome uptake, with concomitant decrease, respectively, in lung and liver metastasis (Hoshino et al., 2015). Studies have shown that antileukemic activity induced by curcumin could be attributed to the selective packaging of microRNAs in exosomes, providing a route for cellular disposal of oncogenic miR-21. This resulted in decreased Akt phosphorylation and vascular endothelial growth factor (VEGF) expression (Taverna et al., 2015). Moreover, proinflammatory signals from exosomal miRNAs and proteins could be exploited to enhance the efficiency of cancer cells to immune therapy. Exosome-mediated delivery of nucleic acids targeting oncomirs such as miR-21 and miR-105, which promote barriers for cancer cell dissemination to blood cells, has been tested in experimental models (Zhou et al., 2014). Ongoing studies on understanding the biology, the unique signature for tropism, and the effort to engineer the exosomes capable of delivering the drugs will provide immense opportunities to control distant cancer progression.

References

Akira, S., & Takeda, K. (2004). Toll-like receptor signalling. *Nature Reviews. Immunology, 4*(7), 499–511. https://doi.org/10.1038/nri1391.

Banyard, J., & Bielenberg, D. R. (2015). The role of EMT and MET in cancer dissemination. *Connective Tissue Research, 56*(5), 403–413. https://doi.org/10.3109/03008207.2015.1060970.

Birchmeier, C., Birchmeier, W., Gherardi, E., & Vande Woude, G. F. (2003). Met, metastasis, motility and more. *Nature Reviews. Molecular Cell Biology, 4*(12), 915–925. https://doi.org/10.1038/nrm1261.

Boccaccio, C., & Comoglio, P. M. (2006). Invasive growth: A MET-driven genetic programme for cancer and stem cells. *Nature Reviews. Cancer, 6*(8), 637–645. https://doi.org/10.1038/nrc1912.

Boelens, M. C., Wu, T. J., Nabet, B. Y., Xu, B., Qiu, Y., Yoon, T., … Minn, A. J. (2014). Exosome transfer from stromal to breast cancer cells regulates therapy resistance pathways. *Cell, 159*(3), 499–513. https://doi.org/10.1016/j.cell.2014.09.051.

Brase, J. C., Johannes, M., Schlomm, T., Falth, M., Haese, A., Steuber, T., … Sultmann, H. (2011). Circulating miRNAs are correlated with tumor progression in prostate cancer. *International Journal of Cancer, 128*(3), 608–616. https://doi.org/10.1002/ijc.25376.

Cai, L., Ye, Y., Jiang, Q., Chen, Y., Lyu, X., Li, J., … Li, X. (2015). Epstein–Barr virus-encoded microRNA BART1 induces tumour metastasis by regulating PTEN-dependent pathways in nasopharyngeal carcinoma. *Nature Communications, 6*, 7353. https://doi.org/10.1038/ncomms8353.

Cao, X. (2016). Self-regulation and cross-regulation of pattern-recognition receptor signalling in health and disease. *Nature Reviews. Immunology, 16*(1), 35–50. https://doi.org/10.1038/nri.2015.8.

Chen, Y., Zeng, C., Zhan, Y., Wang, H., Jiang, X., & Li, W. (2017). Aberrant low expression of p85alpha in stromal fibroblasts promotes breast cancer cell metastasis through exosome-mediated paracrine Wnt10b. *Oncogene, 36*, 4705–4792. https://doi.org/10.1038/onc.2017.100.

Cherfils-Vicini, J., Platonova, S., Gillard, M., Laurans, L., Validire, P., Caliandro, R., … Cremer, I. (2010). Triggering of TLR7 and TLR8 expressed by human lung cancer cells induces cell survival and chemoresistance. *Journal of Clinical Investigation, 120*(4), 1285–1297. https://doi.org/10.1172/JCI36551.

Christensen, J. G., Burrows, J., & Salgia, R. (2005). c-Met as a target for human cancer and characterization of inhibitors for therapeutic intervention. *Cancer Letters, 225*(1), 1–26. https://doi.org/10.1016/j.canlet.2004.09.044.

Colombo, M., Raposo, G., & Thery, C. (2014). Biogenesis, secretion, and intercellular interactions of exosomes and other extracellular vesicles. *Annual Review of Cell and Developmental Biology, 30*, 255–289. https://doi.org/10.1146/annurev-cellbio-101512-122326.

Corcoran, C., Rani, S., & O'Driscoll, L. (2014). miR-34a is an intracellular and exosomal predictive biomarker for response to docetaxel with clinical relevance to prostate cancer progression. *The Prostate, 74*(13), 1320–1334. https://doi.org/10.1002/pros.22848.

Costa-Silva, B., Aiello, N. M., Ocean, A. J., Singh, S., Zhang, H., Thakur, B. K., … Lyden, D. (2015). Pancreatic cancer exosomes initiate pre-metastatic niche formation in the liver. *Nature Cell Biology, 17*(6), 816–826. https://doi.org/10.1038/ncb3169. http://www.nature.com/ncb/journal/v17/n6/abs/ncb3169.html#supplementary-information.

Dapito, D. H., Mencin, A., Gwak, G. Y., Pradere, J. P., Jang, M. K., Mederacke, I., … Schwabe, R. F. (2012). Promotion of hepatocellular carcinoma by the intestinal microbiota and TLR4. *Cancer Cell, 21*(4), 504–516. https://doi.org/10.1016/j.ccr.2012.02.007.

Fabbri, M., Paone, A., Calore, F., Galli, R., Gaudio, E., Santhanam, R., … Croce, C. M. (2012). MicroRNAs bind to Toll-like receptors to induce prometastatic inflammatory response. *Proceedings of the National Academy of Sciences of the United States of America, 109*(31), E2110–E2116. https://doi.org/10.1073/pnas.1209414109.

Fan, G. C. (2014). Hypoxic exosomes promote angiogenesis. *Blood, 124*(25), 3669–3670. https://doi.org/10.1182/blood-2014-10-607846.

Fidler, I. J. (2003). The pathogenesis of cancer metastasis: The 'seed and soil' hypothesis revisited. *Nature Reviews. Cancer, 3*(6), 453–458.

Fong, M. Y., Zhou, W., Liu, L., Alontaga, A. Y., Chandra, M., Ashby, J., … Wang, S. E. (2015). Breast-cancer-secreted miR-122 reprograms glucose metabolism in premetastatic niche to promote metastasis. *Nature Cell Biology, 17*(2), 183–194. https://doi.org/10.1038/ncb3094.

Forsbach, A., Nemorin, J. G., Montino, C., Muller, C., Samulowitz, U., Vicari, A. P., … Vollmer, J. (2008). Identification of RNA sequence motifs stimulating sequence-specific TLR8-dependent immune responses. *The Journal of Immunology: Official Journal of the American Association of Immunologists, 180*(6), 3729–3738.

Gebeshuber, C. A., Zatloukal, K., & Martinez, J. (2009). miR-29a suppresses tristetraprolin, which is a regulator of epithelial polarity and metastasis. *EMBO Reports, 10*(4), 400–405. https://doi.org/10.1038/embor.2009.9.

Grange, C., Tapparo, M., Collino, F., Vitillo, L., Damasco, C., Deregibus, M. C., … Camussi, G. (2011). Microvesicles released from human renal cancer stem cells stimulate angiogenesis and formation of lung premetastatic niche. *Cancer Research, 71*(15), 5346–5356. https://doi.org/10.1158/0008-5472.CAN-11-0241.

Heerboth, S., Housman, G., Leary, M., Longacre, M., Byler, S., Lapinska, K., … Sarkar, S. (2015). EMT and tumor metastasis. *Clinical and Translational Medicine, 4*, 6. https://doi.org/10.1186/s40169-015-0048-3.

Heil, F., Hemmi, H., Hochrein, H., Ampenberger, F., Kirschning, C., Akira, S., … Bauer, S. (2004). Species-specific recognition of single-stranded RNA via toll-like receptor 7 and 8. *Science, 303*(5663), 1526–1529. https://doi.org/10.1126/science.1093620.

He, W., Liu, Q., Wang, L., Chen, W., Li, N., & Cao, X. (2007). TLR4 signaling promotes immune escape of human lung cancer cells by inducing immunosuppressive cytokines and apoptosis resistance. *Molecular Immunology, 44*(11), 2850–2859. https://doi.org/10.1016/j.molimm.2007.01.022.

Hornick, N. I., Doron, B., Abdelhamed, S., Huan, J., Harrington, C. A., Shen, R., … Kurre, P. (2016). AML suppresses hematopoiesis by releasing exosomes that contain microRNAs targeting c-MYB. *Science Signaling, 9*(444), ra88. https://doi.org/10.1126/scisignal.aaf2797.

Hoshino, A., Costa-Silva, B., Shen, T.-L., Rodrigues, G., Hashimoto, A., Tesic Mark, M., … Lyden, D. (2015). Tumour exosome integrins determine organotropic metastasis. *Nature, 527*(7578), 329–335. https://doi.org/10.1038/nature15756.   http://www.nature.com/nature/journal/v527/n7578/abs/nature15756.html#supplementary-information.

Hu, G. W., Li, Q., Niu, X., Hu, B., Liu, J., Zhou, S. M., … Deng, Z. F. (2015). Exosomes secreted by human-induced pluripotent stem cell-derived mesenchymal stem cells attenuate limb ischemia by promoting angiogenesis in mice. *Stem Cell Research & Therapy, 6*, 10. https://doi.org/10.1186/scrt546.

Iorio, M. V., Visone, R., Di Leva, G., Donati, V., Petrocca, F., Casalini, P., … Croce, C. M. (2007). MicroRNA signatures in human ovarian cancer. *Cancer Research, 67*(18), 8699–8707. https://doi.org/10.1158/0008-5472.CAN-07-1936.

Jha, H. C., Sun, Z., Upadhyay, S. K., El-Naccache, D. W., Singh, R. K., Sahu, S. K., & Robertson, E. S. (2016). KSHV-Mediated Regulation of Par3 and SNAIL contributes to B-cell proliferation. *Plos Pathogensens, 12*(7), e1005801. https://doi.org/10.1371/journal.ppat.1005801.

Ji, R., Zhang, B., Zhang, X., Xue, J., Yuan, X., Yan, Y., … Xu, W. (2015). Exosomes derived from human mesenchymal stem cells confer drug resistance in gastric cancer. *Cell Cycle, 14*(15), 2473–2483. https://doi.org/10.1080/15384101.2015.1005530.

Johnstone, R. M., Adam, M., Hammond, J. R., Orr, L., & Turbide, C. (1987). Vesicle formation during reticulocyte maturation. Association of plasma membrane activities with released vesicles (exosomes). *Journal of Biological Chemistry, 262*(19), 9412–9420.

Kamerkar, S., LeBleu, V. S., Sugimoto, H., Yang, S., Ruivo, C. F., Melo, S. A., … Kalluri, R. (2017). Exosomes facilitate therapeutic targeting of oncogenic KRAS in pancreatic cancer. *Nature, 546*(7659), 498–503. https://doi.org/10.1038/nature22341. advance online publication http://www.nature.com/nature/journal/vaop/ncurrent/abs/nature22341.html#supplementary-information.

Kaplan, R. N., Riba, R. D., Zacharoulis, S., Bramley, A. H., Vincent, L., Costa, C., … Lyden, D. (2005). VEGFR1-positive haematopoietic bone marrow progenitors initiate the pre-metastatic niche. *Nature, 438*(7069), 820–827. https://doi.org/10.1038/nature04186.

Kim, S. B., Kim, H. R., Park, M. C., Cho, S., Goughnour, P. C., Han, D., … Kim, S. (2017). Caspase-8 controls the secretion of inflammatory lysyl-tRNA synthetase in exosomes from cancer cells. *The Journal of Cell Biology, 216*(7), 2201–2216. https://doi.org/10.1083/jcb.201605118.

Kim, S., Takahashi, H., Lin, W. W., Descargues, P., Grivennikov, S., Kim, Y., … Karin, M. (2009). Carcinoma-produced factors activate myeloid cells through TLR2 to stimulate metastasis. *Nature, 457*(7225), 102–106. https://doi.org/10.1038/nature07623.

King, H. W., Michael, M. Z., & Gleadle, J. M. (2012). Hypoxic enhancement of exosome release by breast cancer cells. *BMC Cancer, 12*, 421. https://doi.org/10.1186/1471-2407-12-421.

Kucharzewska, P., Christianson, H. C., Welch, J. E., Svensson, K. J., Fredlund, E., Ringnér, M., … Belting, M. (2013). Exosomes reflect the hypoxic status of glioma cells and mediate hypoxia-dependent activation of vascular cells during tumor development. *Proceedings of the National Academy of Sciences of the United States of America, 110*(18), 7312–7317. https://doi.org/10.1073/pnas.1220998110.

Liu, Y., Gu, Y., Han, Y., Zhang, Q., Jiang, Z., Zhang, X., … Cao, X. (2016). Tumor exosomal RNAs promote lung pre-metastatic niche formation by activating alveolar epithelial TLR3 to recruit neutrophils. *Cancer Cell, 30*(2), 243–256. https://doi.org/10.1016/j.ccell.2016.06.021.

Luedde, T. (2010). MicroRNA-151 and its hosting gene FAK (focal adhesion kinase) regulate tumor cell migration and spreading of hepatocellular carcinoma. *Hepatology, 52*(3), 1164–1166. https://doi.org/10.1002/hep.23854.

Mehlen, P., & Puisieux, A. (2006). Metastasis: A question of life or death. *Nature Reviews. Cancer, 6*(6), 449–458.

Motoyama, K., Inoue, H., Takatsuno, Y., Tanaka, F., Mimori, K., Uetake, H., … Mori, M. (2009). Over- and under-expressed microRNAs in human colorectal cancer. *International Journal of Oncology, 34*(4), 1069–1075.

Ono, M., Kosaka, N., Tominaga, N., Yoshioka, Y., Takeshita, F., Takahashi, R. U., … Ochiya, T. (2014). Exosomes from bone marrow mesenchymal stem cells contain a microRNA that promotes dormancy in metastatic breast cancer cells. *Science Signaling, 7*(332), ra63. https://doi.org/10.1126/scisignal.2005231.

Paggetti, J., Haderk, F., Seiffert, M., Janji, B., Distler, U., Ammerlaan, W., … Moussay, E. (2015). Exosomes released by chronic lymphocytic leukemia cells induce the transition of stromal cells into cancer-associated fibroblasts. *Blood, 126*(9), 1106–1117. https://doi.org/10.1182/blood-2014-12-618025.

Pan, L., Liang, W., Fu, M., Huang, Z. H., Li, X., Zhang, W., … Zhang, X. (2017). Exosomes-mediated transfer of long noncoding RNA ZFAS1 promotes gastric cancer progression. *Journal of Cancer Research and Clinical Oncology, 143*(6), 991–1004. https://doi.org/10.1007/s00432-017-2361-2.

Peinado, H., Aleckovic, M., Lavotshkin, S., Matei, I., Costa-Silva, B., Moreno-Bueno, G., … Lyden, D. (2012). Melanoma exosomes educate bone marrow progenitor cells toward a pro-metastatic phenotype through MET. *Nature Medicine, 18*(6), 883–891. https://doi.org/10.1038/nm.2753.

Peruzzi, B., & Bottaro, D. P. (2006). Targeting the c-Met signaling pathway in cancer. *Clinical Cancer Research, 12*(12), 3657–3660. https://doi.org/10.1158/1078-0432.CCR-06-0818.

Pradere, J. P., Dapito, D. H., & Schwabe, R. F. (2014). The Yin and Yang of toll-like receptors in cancer. *Oncogene, 33*(27), 3485–3495. https://doi.org/10.1038/onc.2013.302.

Prieto, D., Sotelo, N., Seija, N., Sernbo, S., Abreu, C., Duran, R., … Oppezzo, P. (2017). S100–A9 protein in exosomes from chronic lymphocytic leukemia cells promotes NF-kappaB activity during disease progression. *Blood, 130*, 777–788. https://doi.org/10.1182/blood-2017-02-769851.

Radisky, E. S., & Radisky, D. C. (2010). Matrix metalloproteinase-induced epithelial-mesenchymal transition in breast cancer. *Journal of Mammary Gland Biology and Neoplasia*, *15*(2), 201–212. https://doi.org/10.1007/s10911-010-9177-x.

Ramteke, A., Ting, H., Agarwal, C., Mateen, S., Somasagara, R., Hussain, A., … Deep, G. (2015). Exosomes secreted under hypoxia enhance invasiveness and stemness of prostate cancer cells by targeting adherens junction molecules. *Molecular Carcinogenesis*, *54*(7), 554–565. https://doi.org/10.1002/mc.22124.

Roccaro, A. M., Sacco, A., Maiso, P., Azab, A. K., Tai, Y. T., Reagan, M., … Ghobrial, I. M. (2013). BM mesenchymal stromal cell-derived exosomes facilitate multiple myeloma progression. *Journal of Clinical Investigation*, *123*(4), 1542–1555. https://doi.org/10.1172/JCI66517.

Scheeren, F. A., Kuo, A. H., van Weele, L. J., Cai, S., Glykofridis, I., Sikandar, S. S., … Clarke, M. F. (2014). A cell-intrinsic role for TLR2-MYD88 in intestinal and breast epithelia and oncogenesis. *Nature Cell Biology*, *16*(12), 1238–1248. https://doi.org/10.1038/ncb3058.

Scott, K. L., Nogueira, C., Heffernan, T. P., van Doorn, R., Dhakal, S., Hanna, J. A., … Chin, L. (2011). Proinvasion metastasis drivers in early-stage melanoma are oncogenes. *Cancer Cell*, *20*(1), 92–103. https://doi.org/10.1016/j.ccr.2011.05.025.

Sharma, S. D. J., Gubbins, L., Weiner-Gorzel, K., Simpson, J., McCann, A., & Kell, M. R. (2014). The impact of TSG101 in triple-negative breast cancers. *Journal of Clinical Oncology*, *32*(15_suppl), 1114. https://doi.org/10.1200/jco.2014.32.15_suppl.1114.

Steeg, P. S. (2016). Targeting metastasis. *Nature Reviews. Cancer*, *16*(4), 201–218. https://doi.org/10.1038/nrc.2016.25.

Svensson, K. J., Kucharzewska, P., Christianson, H. C., Sköld, S., Löfstedt, T., Johansson, M. C., … Belting, M. (2011). Hypoxia triggers a proangiogenic pathway involving cancer cell microvesicles and PAR-2–mediated heparin-binding EGF signaling in endothelial cells. *Proceedings of the National Academy of Sciences of the United States of America*, *108*(32), 13147–13152. https://doi.org/10.1073/pnas.1104261108.

Tadokoro, H., Umezu, T., Ohyashiki, K., Hirano, T., & Ohyashiki, J. H. (2013). Exosomes derived from hypoxic leukemia cells enhance tube formation in endothelial cells. *The Journal of Biological Chemistry*, *288*(48), 34343–34351. https://doi.org/10.1074/jbc.M113.480822.

Taverna, S., Giallombardo, M., Pucci, M., Flugy, A., Manno, M., Raccosta, S., … Alessandro, R. (2015). Curcumin inhibits in vitro and in vivo chronic myelogenous leukemia cells growth: A possible role for exosomal disposal of miR-21. *Oncotarget*, *6*(26), 21918–21933. https://doi.org/10.18632/oncotarget.4204.

Taylor, D. D., & Gercel-Taylor, C. (2008). MicroRNA signatures of tumor-derived exosomes as diagnostic biomarkers of ovarian cancer. *Gynecologic Oncology*, *110*(1), 13–21. https://doi.org/10.1016/j.ygyno.2008.04.033.

Tesio, M., Golan, K., Corso, S., Giordano, S., Schajnovitz, A., Vagima, Y., … Lapidot, T. (2011). Enhanced c-Met activity promotes G-CSF-induced mobilization of hematopoietic progenitor cells via ROS signaling. *Blood*, *117*(2), 419–428. https://doi.org/10.1182/blood-2009-06-230359.

Valastyan, S., & Weinberg, R. A. (2011). Tumor metastasis: Molecular insights and evolving paradigms. *Cell*, *147*(2), 275–292. https://doi.org/10.1016/j.cell.2011.09.024.

Valencia, K., Luis-Ravelo, D., Bovy, N., Anton, I., Martinez-Canarias, S., Zandueta, C., … Lecanda, F. (2014). miRNA cargo within exosome-like vesicle transfer influences metastatic bone colonization. *Molecular Oncology*, *8*(3), 689–703. https://doi.org/10.1016/j.molonc.2014.01.012.

Wang, L. F., Chien, C. Y., Kuo, W. R., Tai, C. F., & Juo, S. H. (2008). Matrix metalloproteinase-2 gene polymorphisms in nasal polyps. *Archives of Otolaryngology - Head and Neck Surgery*, *134*(8), 852–856. https://doi.org/10.1001/archotol.134.8.852.

Wang, J., Hendrix, A., Hernot, S., Lemaire, M., De Bruyne, E., Van Valckenborgh, E., … Menu, E. (2014). Bone marrow stromal cell-derived exosomes as communicators in drug resistance in multiple myeloma cells. *Blood*, *124*(4), 555–566. https://doi.org/10.1182/blood-2014-03-562439.

Webber, J. P., Spary, L. K., Sanders, A. J., Chowdhury, R., Jiang, W. G., Steadman, R., … Clayton, A. (2015). Differentiation of tumour-promoting stromal myofibroblasts by cancer exosomes. *Oncogene, 34*(3), 290–302. https://doi.org/10.1038/onc.2013.560.

Wu, C. F., Andzinski, L., Kasnitz, N., Kroger, A., Klawonn, F., Lienenklaus, S., … Jablonska, J. (2015). The lack of type I interferon induces neutrophil-mediated pre-metastatic niche formation in the mouse lung. *International Journal of Cancer, 137*(4), 837–847. https://doi.org/10.1002/ijc.29444.

You, Y., Shan, Y., Chen, J., Yue, H., You, B., Shi, S., … Cao, X. (2015). Matrix metalloproteinase 13-containing exosomes promote nasopharyngeal carcinoma metastasis. *Cancer Science, 106*(12), 1669–1677. https://doi.org/10.1111/cas.12818.

Zhang, R., Xia, Y., Wang, Z., Zheng, J., Chen, Y., Li, X., … Ming, H. (2017). Serum long non coding RNA MALAT-1 protected by exosomes is up-regulated and promotes cell proliferation and migration in non-small cell lung cancer. *Biochemical and Biophysical Research Communications, 490*(2), 406–414. https://doi.org/10.1016/j.bbrc.2017.06.055.

Zhang, L., Zhang, S., Yao, J., Lowery, F. J., Zhang, Q., Huang, W. C., … Yu, D. (2015). Microenvironment-induced PTEN loss by exosomal microRNA primes brain metastasis outgrowth. *Nature, 527*(7576), 100–104. https://doi.org/10.1038/nature15376.

Zhang, X., Zhu, W., Zhang, J., Huo, S., Zhou, L., Gu, Z., & Zhang, M. (2010). MicroRNA-650 targets ING4 to promote gastric cancer tumorigenicity. *Biochemical and Biophysical Research Communications, 395*(2), 275–280. https://doi.org/10.1016/j.bbrc.2010.04.005.

Zhou, W., Fong, M. Y., Min, Y., Somlo, G., Liu, L., Palomares, M. R., … Wang, S. E. (2014). Cancer-secreted miR-105 destroys vascular endothelial barriers to promote metastasis. *Cancer Cell, 25*(4), 501–515. https://doi.org/10.1016/j.ccr.2014.03.007.

15

Exosomes: Key Supporters of Tumor Metastasis

Girijesh K. Patel, Haseeb Zubair, Mohammad A. Khan, Sanjeev K. Srivastava, Aamir Ahmad, Mary C. Patton, Seema Singh, Moh'd Khushman, Ajay P. Singh

UNIVERSITY OF SOUTH ALABAMA-MITCHELL CANCER INSTITUTE, MOBILE, AL, UNITED STATES

CHAPTER OUTLINE

1. Introduction 262

2. **Exosomes in Metastasis Progression: Events at Primary Tumor Site** 262
   2.1 Exosomes in Epithelial-to-Mesenchymal Transition 263
   2.2 Role of Exosomes in Angiogenesis 265
   2.3 Exosomes in Invasion and Migration... 267
   2.4 Exosomes in Transformation of Normal Fibroblast Cells Into Cancer-Associated
       Fibroblast 267
   2.5 Tumor Microenvironment Cell–Derived Exosomes in Cancer Metastasis......... 268

3. **Exosome-Mediated Events at Metastatic Sites**... 269
   3.1 Exosomes in Metastatic Niche Formation 269
   3.2 Exosomes in Metabolic Shift 270
   3.3 Immunosuppression 271
   3.4 Mesenchymal-to-Epithelial Transition, Survival, and Maintenance of
       Metastasized Cells 272

4. **Strategies Against Exosome-Mediated Metastasis** 272
   4.1 Inhibition of Exosome Biogenesis.. 273
   4.2 Inhibition of Packaging the Exosomal Cargo 273
   4.3 Inhibition of Exosome Internalization 274
   4.4 Removal of Tumor Exosomes 275
   4.5 Exosome as a Therapeutic Vehicle.. 275

5. **Conclusions and Perspective** 276

Acknowledgments... 276

Conflicts of Interest 276

References 277

Diagnostic and Therapeutic Applications of Exosomes in Cancer. https://doi.org/10.1016/B978-0-12-812774-2.00015-8

1. Introduction

Metastasis is a multistep process in which tumor cells detach from the primary tumor site, invade the extracellular matrix (ECM), travel through the blood or lymph system, migrate over a long distance through circulation, adhere and colonize at different organ sites to develop secondary tumors. These events require tumor cells to achieve certain phenotypic characteristics, which they do by acquiring progressive genetic and epigenetic alterations (Khan et al., 2017; Mardin, Haier, & Mees, 2013; Zubair & Ahmad, 2017). It has now become evident that the tumor microenvironment (TME) plays an important role in these processes as well, and tumor cell-TME cross talk has a major impact on tumor cell behavior by altering their endogenous gene expression. In recent years, it has also been recognized that TME can influence tumor cell properties through intercellular vesicular transfer, mainly through exosomal mRNAs, miRNAs, long noncoding RNA, proteins, and lipids (Allenson et al., 2017; Braicu et al., 2015; Patel, Patton, Singh, Khushman, & Singh, 2016; Raposo & Stoorvogel, 2013). Some reports also confirm the presence of DNA fragment in the exosomes (Allenson et al., 2017; Montermini et al., 2015). Exosomes are extracellular nanovesicles (30–150 nm) of endocytic origin that are shed by almost all cell types under normal and pathophysiological conditions (Johnstone, Bianchini, & Teng, 1989). Exosomes can travel to short and far distances to mediate the horizontal transfer of molecular information to the recipient cells (Salomon et al., 2014; Umezu, Ohyashiki, Kuroda, & Ohyashiki, 2013; Zhao et al., 2016). It has been shown that exosomes play important roles in tumor growth and metastasis by influencing cell proliferation, survival, invasiveness, stemness, angiogenesis, and immune suppression (Meckes et al., 2010; Patel et al., 2017; Qu et al., 2016; Richards et al., 2017; Taylor, Gercel-Taylor, Lyons, Stanson, & Whiteside, 2003; Thery, Ostrowski, & Segura, 2009; Yoshizaki et al., 2013). Recent reports have also suggested a role of exosomes in premetastatic niche formation and organotropic metastasis (Costa-Silva et al., 2015; Hoshino et al., 2015), thus implicating their multifaceted roles in tumor metastasis.

In this chapter, we provide an in-depth knowledge, and the associated mechanistic details, of the significance of exosomes in cancer metastasis. We also highlight the strategies being examined to inhibit exosome-mediated cancer progression and metastasis.

2. Exosomes in Metastasis Progression: Events at Primary Tumor Site

As previously stated, exosomes serve as carriers for a variety of biomolecules, such as proteins, DNA, mRNA, miRNA, etc., in their lumen and/or on the surface. Different molecules transferred through exosomes that play a role in cancer metastasis are listed in Tables 1 and 2. After shedding, exosomes communicate with recipient cells either by interaction and/or transfer the exosomal content to facilitate cell signaling and promote different events related to tumor progression and metastasis (Webber, Steadman, Mason, Tabi, & Clayton, 2010), as depicted in Fig. 1. In the subsections that follow, we provide a detailed discussion on the significance of exosomes in metastasis progression.

Table 1 List of Different Proteins Transferred via Exosomes and Involved in Metastasis

Protein	Function	Cancer Type	References
EGFR	Liver metastasis	Gastric cancer	Zhang et al. (2017)
EGFRvIII	Oncogenic transformation	Glioma	Al-Nedawi et al. (2008)
Amphiregulin	Invasion	Breast and colorectal	Higginbotham et al. (2011)
MIF	Metastatic niche formation	Pancreatic	Costa-Silva et al. (2015)
TGF-β	Myofibroblast differentiation	Prostate	Webber et al. (2010)
Bcl-2 and cyclin D1	Activate Akt and ERK pathway	Bladder	Yang et al. (2013)
Met	Premetastatic niche formation	Melanoma	Peinado et al. (2012)
CD63, CD9, CD81	Invasion	Different malignancies	Lafleur et al. (2009) and Thery et al. (2009)
KRAS	Oncogenic potential	Colon cancer	Demory Beckler et al. (2013)
Dll-4	Angiogenesis	In 3D culture	Sharghi-Namini et al. (2014)
Integrins	Organotropic metastasis	Pancreas	Hoshino et al. (2015)
αvβ6	Promotes cell migration	Prostate	Fedele et al. (2015)
HIF 1-α	Invasion and migration	Nasopharyngeal	Aga et al. (2014)
LIN28A	EMT and invasion and migration	Ovarian	Enriquez et al. (2015)
Wnt4	Angiogenesis and migration	Colorectal	Huang and Feng (2017)
EMMPRIN	Invasion and migration	Lung	Sidhu et al. (2004)

Table 2 List of microRNA, lncRNA, and DNA Transferred via Exosomes and Involved in Metastasis

MicroRNA	Function	Cancer Type	References
miR-9	Angiogenesis, migration. CAF transformation	Pancreatic, lung GBM, colorectal	Zhuang et al. (2012) Baroni et al. (2016)
miR-10b	Cell migration	Breast cancer	Singh et al. (2014)
miR-19a	Induce metastasis	Brain	Zhang et al. (2015)
miR-21, miR-27, miR-29a	Proinflammatory	Lung	Fabbri et al. (2012)
miR-1246	Angiogenesis	Colorectal	Yamada et al. (2014)
miR-181c	Metastasis	Breast	Tominaga et al. (2015)
miR-135b	Angiogenesis	Multiple myeloma	Umezu et al. (2014)
miR-100-5p, miR-21-5p, and miR-139-5p	Migration and metastatic niche formation	Prostate	Sanchez et al. (2016)
lncRNA H19	Angiogenesis	Liver	Conigliaro et al. (2015)
lncRNA ZFAS1	Lymphatic metastasis	Gastric	Pan et al. (2017)
Mutated *KRAS* DNA	Correlates with cancer progression and metastasis	Pancreatic	Allenson et al. (2017)

2.1 Exosomes in Epithelial-to-Mesenchymal Transition

Epithelial-to-mesenchymal transition (EMT) is a process by which epithelial cells lose their cell polarity and acquire a migratory and invasive phenotype. Accumulating evidence suggest a critical role of exosomes in facilitating the process of EMT. Exosomes serve as a vehicle to deliver various EMT-inducing proteins, such as Notch-1, IL6, matrix

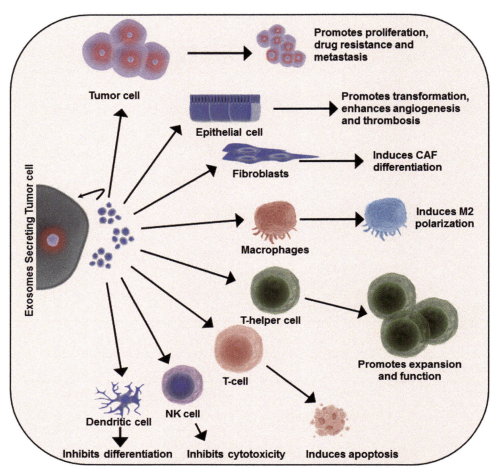

FIGURE 1 Diverse role of exosomes at primary tumor sites. Tumor cell–derived exosomes act in an autocrine and paracrine manner. The interaction of tumor cell and other cells within the tumor microenvironment through exosomes may result in tumor growth, proliferation, metastasis, and drug resistance. Tumor cell–derived exosomes are involved in immune suppression, macrophage polarization, transformation of fibroblasts to cancer-associated fibroblasts, etc.

metalloproteinases (MMPs), β-catenin, AKT, miR-100, LMP1, caveolin 1, HIF-1α, annexin A2, casein kinase II (CK2), integrin 3, etc. (Aga et al., 2014; Ang, Fang, Ashman, & Frauman, 2010; Franzen et al., 2015; Maji et al., 2017). Al-Nedawi and coworkers demonstrated that EGFRvIII, an oncogenic form of EGFR (epidermal growth factor receptor), can be transferred through extracellular vesicles to tumor cells. Transfer of EGFRvIII activates MAPK and AKT pathways, leading to metastatic potential and anchorage-independent phenotypes (Al-Nedawi et al., 2008). Additionally, colon cancer cells, possessing mutant *KRAS*, secrete exosomes with a mutant KRAS protein, along with other oncogenic molecules, including EGFR, SRC-family kinases, and integrins, which further underscores the role of exosomes in tumor progression (Demory Beckler et al., 2013). Franzen et al. (2015) reported that exosomes derived from bladder cancer cells induce EMT in urothelial cells. Exposure of urothelial cells to exosomes upregulated the expression of mesenchymal markers, such as snail, whereas reduction in epithelial markers E-cadherin and β-catenin was seen (Franzen et al., 2015). Exosomes secreted by prostate cancer cells under hypoxic conditions could enhance invasiveness and stem cell–like phenotypes. Moreover, these exosomes also affected the prostate stromal cells by transforming them into cancer-associated fibroblast (CAF) phenotypes (Ramteke et al., 2015). The RNA-binding protein LIN28A was found to be overexpressed in ovarian cancer cells and in the secreted exosomes which, after internalization into nontumorogenic HEK293 cells, induced the expression of EMT-related genes and promoted the invasion and migration of HEK293 cells (Enriquez et al., 2015). Rahman and coworkers reported that exosomes isolated from the blood of metastatic lung cancer patients, as well as from aggressive lung cancer cells, overexpress vimentin and promote EMT in human bronchial epithelial cells (Rahman et al., 2016).

Tetraspanins are expressed on the cell surface of all tissues as well as on exosomes (Canel, Serrels, Frame, & Brunton, 2013; Huttenlocher & Horwitz, 2011). They are known to play a crucial role in tissue differentiation, tumor progression, and cell migration by interacting with integrins, E-cadherins, and MMPs (Ang et al., 2010; Lafleur, Xu, & Hemler, 2009; Nakamura et al., 2017; Powner, Kopp, Monkley, Critchley, & Berditchevski, 2011; Tejera et al., 2013). Considering these significant roles of tetraspanins in cancer progression, and their abundance in exosomes, it can be speculated that, after being transferred to recipient cells, these tetraspanins may facilitate tumor progression. This hypothesis is supported by several studies. For example, knockdown of CD81 enhanced focal adhesion formation and suppressed cell migration (Tejera et al., 2013). In additional studies, upregulation of CD9, CD63, CD81, and CD151 was reported to be associated with cancer cell motility in different malignancies (De Bruyne et al., 2006; Powner et al., 2011; Tejera et al., 2013; Yanez-Mo et al., 1998). The overexpression of CD9 is known to promote bone metastasis in breast cancer and promotes chemoresistance in non–small-cell lung cancer (Kischel et al., 2012). The expression of CD63 is associated with melanoma progression (Donoso et al., 1986). Similarly, higher expression of CD63 and CD9 was observed in pancreatic cancer tissues, compared with normal ones, and inversely correlated with therapeutic outcome and overall survival (Khushman et al., 2017). It has been shown that CD44 present in ovarian cancer–derived exosomes induces the expression of MMP9, resulting in cytoskeletal degradation and thereby promoting invasion (Nakamura et al., 2017).

2.2 Role of Exosomes in Angiogenesis

Development of new blood vessels from preexisting ones, called "angiogenesis," is an essential step in cancer metastasis. The process of angiogenesis is controlled by various angiogenic factors such as interleukins, cytokines, vascular endothelial growth factor (VEGF), etc. (Inoue, Hager, Ferrara, Gerber, & Hanahan, 2002). With increasing interest in exosomes, their role in angiogenesis has also gained significant interest. Studies suggest that the chronic myeloid leukemia (CML)–derived exosomes promote angiogenesis via IL8-mediated VCAM-1 activation (Khan et al., 2015; Mineo et al., 2012). Furthermore, Hood and coworkers demonstrated that melanoma-derived exosomes also possess the ability to induce angiogenesis in the lymph nodes (Hood, San, & Wickline, 2011). Similarly, Peinado et al. (2012) demonstrated that melanoma-derived exosomes, after delivery into naïve mice, educate bone marrow cells and induce neo-angiogenesis. These exosomes contain different proteins (HSP70, TYRP2, MET, VLA4, RAB27a, etc.) that play different roles in tumor progression. They found that RAB family protein (RAB27a) was enriched in highly metastatic cells (B16-F10), which is a known regulator of exosome biosynthesis. Suppression of RAB27a expression significantly decreased the metastatic potential and neo-angiogenesis (Peinado et al., 2012). Annexin II is overexpressed in breast cancer cells as well as present in their secreted exosomes. The annexin II–expressing exosomes result in macrophage-mediated induction of the p38MAPK, NF-κB, and STAT3 signaling pathways, and ultimately induce the secretion of TNFα and IL-6, thus promoting angiogenesis and lung metastasis (Maji et al., 2017). In another study, activation of Notch signaling was reported to be induced by delta-like ligand 4 (Dll4) of the exosomes. The Dll4-containing exosomes cause tip cells to lose their filopodia, inducing capillary sprout retraction in a collagen matrix, implicated in angiogenesis and vein–artery differentiation (Sharghi-Namini, Tan, Ong, Ge, & Asada, 2014).

The growth of solid tumors results in hypoxia due to the lack of nutrient and oxygen supply. This limiting microenvironment induces vasculature formation to maintain tissue homeostasis. Kucharzewska and coworkers demonstrated that the hypoxia-induced exosomes from brain tumor glioblastoma multiforme (GBM) are enriched with hypoxia-regulated mRNAs and proteins, including MMPs, IL-8, platelet-derived growth factors, caveolin 1, and lysyl oxidase. After uptake of these exosomes, endothelial cells secrete different cytokines and growth factors which induce angiogenesis (Kucharzewska et al., 2013). Hypoxic myeloma cell–derived exosomes also induce angiogenesis. These exosomes contain miR-135b that targets the inhibitors of HIF1α, thus activating HIF1-α and upregulating the expression of angiogenesis-promoting genes (Umezu et al., 2014). Cancer cells also induce angiogenesis indirectly through microenvironment modulation by exosomes (Ribeiro, Zhu, Millard, & Fan, 2013). In a separate study, exosomes derived from colorectal cancer cells contained TGF-β and miR-1246, and those exosomes were shown to be contributing factors for angiogenesis. The miR-1246 directly targets promyelocytic leukemia gene, inhibits Smad2/3 signaling in endothelial cells, and promotes angiogenesis (Yamada et al., 2014). It is demonstrated that the miR-9 containing exosomes secreted from cells of different cancers such as NSCLC (H1299), melanoma (MDA435), pancreatic cancer (Panc1), glioblastoma (SF-539), and colorectal cancer (HM7) is transferred to endothelial cells. The transferred miR-9 targets *SOCS5* gene and induces the phosphorylation and activation of JAK-STAT pathway, which promotes migration of endothelial cells as well as tumor angiogenesis (Zhuang et al., 2012). The exosomes derived from CD90+ liver cancer cells show enhanced secretion of long noncoding RNA (lncRNA) H19, which modulates endothelial cells and promotes angiogenesis (Conigliaro et al., 2015).

2.3 Exosomes in Invasion and Migration

Tumor cell invasion is a complex process facilitated by various factors such as alteration of cellular dynamics, ECM, junction proteins, expression of protease, cytokines, followed by cell migration to neighboring tissues (Friedl & Alexander, 2011). Higginbotham and coworkers identified the significance of amphiregulin (AREG)-containing exosomes in cancer cell invasion. A mechanistic study revealed that accumulated AREG in recipient cell (through exosomal transfer) induces EGFR signaling by aggregating and oligomerizing EGFR, thus stimulating various signaling pathways necessary for tumor cell invasion and metastasis (Higginbotham et al., 2011).

Exosomes from the hypoxic colorectal cancer cells induce proliferation and migration of endothelial cells. Activation of HIF1-α in hypoxic conditions enhances the Wnt4 enrichment in exosomes, helps nuclear localization of β-catenin, and promotes cellular migration (Huang & Feng, 2017). CD44 is highly enriched in ovarian cancer cell–derived exosomes and is transferred to human peritoneal mesothelial cells which, in turn, are induced to secrete MMP9 to clean the mesothelial barrier and enhance cancer cell invasion (Nakamura et al., 2017). It is reported that Rab27 helps the release of exosomal HSP90, which activates MMP2, leading to proteolysis of ECM components and induction of cancer cell invasion (Hendrix et al., 2010). Sidhu and coworkers reported that exosomes shed by cancer cells possess extracellular matrix metalloproteinase inducer (EMMPRIN), which was delivered to fibroblasts and induced secretion of MMPs. These MMPs alter ECM via proteolysis, thus promoting tumor cell invasion and metastasis (Sidhu, Mengistab, Tauscher, LaVail, & Basbaum, 2004). The cancer cell exosomes are an abundant source of different miRNAs. Different miRNA such as miR-100-5p, miR-21-5p, and miR-139-5p are enriched in prostate cancer exosomes which, after internalization into prostate fibroblast cells, enhance the expression of MMP- 2, 9, 13 and RANKL and help in fibroblast migration for metastatic niche formation (Sanchez et al., 2016). The lncRNA ZFAS1 was overexpressed in gastric cancer cells and was also present in secreted exosomes. A high level of ZFAS1 was shown to be significantly correlated with lymphatic metastasis (Pan et al., 2017). Ramteke and coworkers observed that prostate cancer–derived exosomes could efficiently enhance migration and invasion. Using PC3 cell line, authors demonstrated that cells grown in the presence of exosomes derived under hypoxic conditions significantly enhanced the migration and wound healing potential of tumor cells; in fact, almost a complete closure of the wound was seen (Ramteke et al., 2015). Moreover, Harris and coworkers found that aggressive breast cancer cell lines release exosomes that have potential in promoting cell invasion and migration (Harris et al., 2015).

2.4 Exosomes in Transformation of Normal Fibroblast Cells Into Cancer-Associated Fibroblast

The tumor cell–derived exosomes modulate surrounding cells by alteration of different gene expression to support tumor cell growth. Studies are emerging to suggest the role of exosomes in cellular transformation of fibroblasts into CAFs. CAFs are reported to have a cancer cell–supportive role, including alterations in metabolism, immunosuppression, growth and proliferation, angiogenesis, and help disease progression and metastasis (Gascard & Tlsty, 2016; Kalluri, 2016). Gu et al., (2012) reported that cancer cell–derived exosomes transform human umbilical cord–derived mesenchymal stem cells (MSCs) into CAFs (Gu et al., 2012). Exosome derived from chronic lymphocytic leukemia cells were shown to transfer different proteins and miRNAs, which induce inflammatory phenotype in the endothelial cells and MSCs. After internalization of exosomes, these cells acquire the features of CAFs (Paggetti et al., 2015). In another study, Webber et al. (2010) demonstrated that mesothelioma cell–derived exosomes possess TGF-β on exosome surface in association with transmembrane betaglycan and induce SMAD signaling, promoting fibroblast to myofibroblast differentiation (Webber et al., 2010). Baroni and coworkers demonstrated that triple negative breast cancer cell–derived exosomes are enriched with miR-9. They further identified the significance of exosomal miR-9 in affecting the properties of human breast fibroblasts, leading to the switching to CAF phenotype, while inhibition of miR-9 impaired this transformation to CAFs (Baroni et al., 2016).

2.5 Tumor Microenvironment Cell–Derived Exosomes in Cancer Metastasis

Exosomes mediate bidirectional cross talk between tumor- and stromal cells and thus facilitate cancer metastasis. Exosomes derived from cells of TME have also been identified to support cancer metastasis in different ways. These exosomes possess intact metabolites, including amino acids, lipids, and TCA cycle intermediates, which are received by cancer cells for central carbon metabolism, promoting tumor growth under nutrient-deprived conditions (Zhao et al., 2016).

The exosomes from the CAF cell promote metastasis by the Wnt signaling pathway in breast cancer cell (Luga et al., 2012). Likewise, macrophage-derived exosomes and MSC-derived exosomes show an enhanced rate of invasion and migration in breast cancer cells by activation of Wnt signaling (Lin, Wang, & Zhao, 2013; Menck et al., 2013). Atay and coworkers reported that gastrointestinal stromal cells secrete exosomes with oncogenic tyrosine kinase. These exosomes, after internalization into progenitor smooth muscle cells, transform these cells into tumor-promoting cells, acquiring CAF-like phenotypes, which secrete MMP1 and promote tumor cell invasion (Atay et al., 2014). In another study, the pancreatic stellate cell–derived exosomes were analyzed and found to contain various proteins and miRNAs. These exosomes induce pancreatic cancer growth by stimulating the CXCL1 and CXCL2 chemokines, which promote proliferation and migration of pancreatic cancer cells (Takikawa et al., 2017). Moreover, astrocyte-derived exosomes are reported to possess miR-19, which targets PTEN in metastasized tumor cells after internalization. These metastatic tumor cells secrete CCL2 and help to infiltrate the myeloid cells for the growth and proliferation of metastasized cells at the secondary site (Zhang et al., 2015). Additionally, exosomes from adipocytes contain different enzymes and proteins involved in fatty acid oxidation in melanoma cells and also in promoting invasion and migration (Lazar et al., 2016).

3. Exosome-Mediated Events at Metastatic Sites

Cancer metastasis is a complex process and occurs through a series of sequential steps. Exosome-mediated different processes are described below in detail.

3.1 Exosomes in Metastatic Niche Formation

Development of a suitable environment at secondary sites before the circulating cancer cells reach, adhere, adapt, and set outgrowth at secondary sites is termed as "metastatic niche formation" (Kaplan et al., 2005). Emerging research evidence underscores the importance of tumor exosomes in metastatic niche formation in different malignancies. It is shown that exosomes, in combination with the soluble matrix assembled by CD44v6, can influence the metastatic organ sites and help in metastatic niche formation in lymph nodes as well as in lung tissues of pancreatic adenocarcinoma in the rat model (Jung et al., 2009). Melanoma exosomes were found to educate the bone marrow cells by transferring different oncogenic proteins such as TYRP2, VLA-4, HSP70, HSP-90, and MET, and thus promoting premetastatic niche formation to support the growth and proliferation of metastatic cells (Peinado et al., 2011, 2012). Another study highlighted the role of pancreatic cancer exosomes in premetastatic niche formation in the liver for future metastasis (Costa-Silva et al., 2015). These exosomes contained elevated levels of macrophage migration inhibitory factor (MIF), and after selective uptake of exosomes by liver Kupffer cells, MIF induces the release of transforming growth factor β (TGF-β) from Kupffer cells, which, in turn, promotes fibronectin production by hepatic stellate cells and gets deposited in the liver. Fibronectin deposits, subsequently, promote the infiltration and arrest of bone marrow–derived macrophages and neutrophils in the liver and establish the premetastatic niche formation with proinflammatory environment, which ultimately supports liver metastasis (Costa-Silva et al., 2015). A detailed study of exosome-mediated organ-specific metastasis has been described recently; the study demonstrated the role of different integrins in metastatic distribution (Hoshino et al., 2015). The quantitative mass spectroscopic- and Western blot analysis of exosomal integrin patterns showed the correlation for brain-, liver-, and lung metastasis. It was reported that the exosomes expressing $\alpha_6\beta_4$ and $\alpha_6\beta_1$ colocalized with S100A4-positive fibroblast cells in laminin-rich lung microenvironments and mediate the lung metastasis, while $\alpha_v\beta_5$-type integrin–enriched exosomes colocalized with F4/80[+] macrophages and mediate liver metastasis in the presence of fibronectin-enriched microenvironments. Moreover, these exosomes promoted vascular leakiness by overexpression of a subset of S100 proteins and activating Src kinase signaling (Hoshino et al., 2015). Similarly, the integrin $\alpha_v\beta_5$ is enriched in prostate cancer cell–derived exosomes, transferred to the normal pancreatic cells, establishing a role in fibronectin-dependent migration (Fedele, Singh, Zerlanko, Iozzo, & Languino, 2015). Another study demonstrated that breast cancer exosomal annexin II helps in premetastatic niche formation and subsequent metastasis by activating the STAT3, p38MAPK, and NF-κB pathways. The tumor exosomes also activate macrophages which induce the secretion of IL6, IL10, and TNF-α and direct breast cancer cell organotropisms to lungs

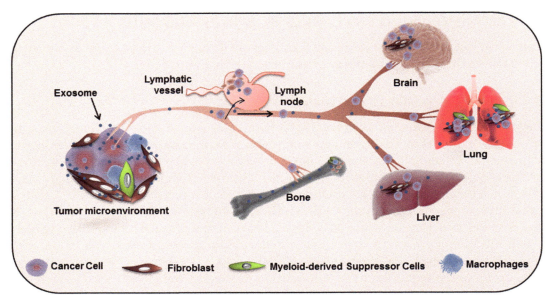

FIGURE 2 Role of exosomes at metastatic sites. Tumor- and/or cancer-associated cell-derived exosomes mediate the communication between neighboring cells. They also circulate through the bloodstream and interact with cells of various organs, such as lymph node, lung, bone, liver, and brain, that lead to recruitment of various supporting cell and premetastatic niche formation for the subsequent colonization and survival of metastasized cells.

(Maji et al., 2017). Furthermore, Zhang and coworkers reported that gastric cancer cell–derived exosomes contain EGFR, which is transferred and integrated into the plasma membrane of liver stromal cells where it activates hepatocyte growth factor (HGF) by suppressing miR-26a/b expression. The upregulated HGF binds to c-MET receptor on the migrated cancer cells, thus facilitating the landing and proliferation of metastatic cancer cells for liver metastasis (Zhang et al., 2017). Altogether, the above information suggests the crucial role of exosomes in metastatic niche formation and organotropism in different cancers, which is illustrated in Fig. 2.

3.2 Exosomes in Metabolic Shift

For continuous growth and proliferation, cancer cells need continuous supply of the different building blocks and energy. Therefore, cancer cells adapt to the aerobic glycolysis–based energy generation for a quick supply along with different intermediates serving as basic building blocks of cells (Warburg, 1956). These changes occur through reprogrammed energy metabolism and aberrant expression of oncogenes that leads to cancer growth and progression (Zubair et al., 2016). It is also reported that HIF1α is transferred by hypoxic cancer cells to other cells via exosomes (Aga et al., 2014), which can induce the glucose uptake by the expression of Glut3 and other glycolytic genes, including ALDA, PKM, PGK1 (Semenza, Roth, Fang, & Wang, 1994). Recently, Fong and coworkers reported that breast cancer cell–derived extracellular vesicles (mainly exosomes) are enriched with miR-122 and transferred to the normal cells of metastatic niche. The miR-122 suppresses glucose uptake in normal cells of the metastatic niche by targeting the glycolytic enzyme pyruvate kinase and increases glucose availability for the incoming metastatic cells, thus favoring metastasis (Fong et al., 2015). In another study, the CAF-derived exosomes modulate glucose metabolism in prostate cancer cells. These exosomes inhibited mitochondrial oxidative phosphorylation and increased glycolysis and glutamine-dependent reductive carboxylation in prostate cancer cells. Moreover, these CAF exosomes also delivered different amino acids (glutamine, lysine, phenylalanine, leucine, and glutamic acid) to cancer cells (Zhao et al., 2016). The mass spectroscopic analysis of bladder cancer exosomes showed the presence of higher level of MUC1 proteins (Welton et al., 2010), which may be involved in metabolic alteration. Recent studies have also highlighted the significance of MUC1 in the regulation of glucose metabolism (Gunda et al., 2017; Shukla et al., 2017). This raises the possibility that MUC1 may be transferred to the recipient cells via exosomes and regulates glucose metabolism. In pancreatic cancer, the role of mutated *KRAS*[G12D] in the reprogramming of glucose metabolism by activating the MAPK and Myc pathways has been documented (Ying et al., 2012). Similarly, other studies also report the transfer of mutated KRAS DNA (Allenson et al., 2017), RNA, and protein (Demory Beckler et al., 2013) via exosomes, and thereby altering glucose metabolism in the recipient cells and supporting the tumor cell survival.

3.3 Immunosuppression

The growing body of evidence suggests that the tumor-derived exosomes play a crucial role in the promotion of the inflammatory response. These exosomes are involved in the immune cell dysfunction by suppressing the antitumor activity. It is shown that Fas ligand–positive exosomes from ovarian and oral cancer induce apoptosis in T lymphocytes (Kim et al., 2005; Taylor et al., 2003), while in another study, exosomes, isolated from melanoma and colorectal carcinoma, possess Fas and TRAIL protein on the membrane surface with inhibitory effects on the T-cell–mediated apoptosis of cancer cells (Iero et al., 2008). Lung cancer cell (A-549)–derived exosomes were found to have various miRNAs, among which miR-21, -27b, and -29a were found to be significantly elevated, and these miRNAs activate TLR8 receptor in immune cells. TLR activation promotes the NF-κB-mediated secretion of proinflammatory cytokines and helps in metastasis (Fabbri et al., 2012). In a mouse model, exosomes derived from murine mammary tumor cells have been implicated in the suppression of natural killer (NK) cell cytotoxic activity in vitro and ex vivo by activation of the Jak3-mediated pathways (Liu et al., 2006). Similarly, the exosomes from breast cancer cell suppress the IL2-mediated NK cell cytotoxicity (Wen et al., 2016; H. G. Zhang et al., 2007). Various exosomal miRNAs are also reported to suppress the T-cell response and induce immune evasion (Clayton & Mason, 2009; Liu et al., 2006; Okoye et al., 2014).

Immune cells, especially macrophages, play a crucial role in tumor pathogenesis. Intercellular milieu induces macrophage polarization either in M1 or in M2; M2 macrophages secret immunosuppressive cytokines (TGF-β and IL-10) which helps tumor cells circumvent immune surveillance (Mantovani, Marchesi, Malesci, Laghi, & Allavena, 2017). Beside cytokines, now exosomes also emerge as novel inducers of macrophage polarization. Studies suggested that tumor cell–derived exosomal microRNAs act as a key regulator of macrophage polarization, M2 type, which help in creating a protumorigenic environment and tumor progression (Chen et al., 2017; Ying et al., 2016).

3.4 Mesenchymal-to-Epithelial Transition, Survival, and Maintenance of Metastasized Cells

The mesenchymal-to-epithelial transition (MET) is a common event during embryogenesis and development. It is a reverse process of EMT, where motile, multipolar, mesenchymal cells transform into a planar array of polarized cells for the deployment to different tissues and organs. In cancer, some growth factors, integrins, cell adhesion molecules, and different proto-oncogenes such as c-ret, c-ros, and c-met, and Rho family GTPases, namely Cdc42 and Rac1, are known to be involved in the MET process (Nakaya, Kuroda, Katagiri, Kaibuchi, & Takahashi, 2004; Yao, Dai, & Peng, 2011). Exosomes are reported to modulate the expression of these molecules that may have a role in the MET process and still need to be elucidated.

The metastasis of the cancerous cell to distant sites/organs is very inefficient as these cells need to adapt and survive in different microenvironments. In this process, cells at secondary sites show differential gene expression patterns compared with the cells of primary tumor site. During this process, cells at secondary sites evade apoptosis. Recently, Zhang et al. (2015) reported that PTEN was suppressed at both mRNA and protein levels in brain metastasized cells compared with that of primary breast cancer cells. Authors reported that astrocyte-derived exosomes enable the intercellular transfer of the PTEN targeting microRNA-19a. The loss of the PTEN in metastatic brain tumor cells induces the enhanced secretion of CCL2 chemokine, which promotes growth and proliferation and increases cell survival by inhibition of apoptosis (Zhang et al., 2015). Similarly, Yang, Wu, Wang, Luo, and Chen (2013) reported that the bladder cancer cell–derived exosomes induce the overexpression of Bcl-2 and cyclin D1 proteins and suppress the expression of Bax and caspase-3 proteins, enhancing the phosphorylation of Akt and extracellular signal–regulated protein kinase (ERK) and inhibiting the bladder tumor cell apoptosis (Yang et al., 2013).

4. Strategies Against Exosome-Mediated Metastasis

The growing body of research evidence has substantiated the role of exosomes in cancer pathogenesis and metastasis. So, it is apparent to develop effective therapeutic strategies against exosome-mediated cancer progression. For this, selective blocking of exosome secretion, inhibition of packaging of pathogenic/oncogenic miRNAs or proteins, or inhibition of exosomal uptake by recipient cells by blocking or disruption of interaction, or selective removal of the cancer exosomes may help to restrict the disease spread and drug resistance, thus enhancing thus may enhance the survival of cancer patients.

4.1 Inhibition of Exosome Biogenesis

The exosome formation and secretion is a complex process; several components have been identified, which play a pivotal role in exosome biogenesis and secretion in extracellular spaces. The ESCART and small GTPases, RAB27a and RAB27b, play a crucial role in sorting and movement of multivesicular bodies toward the cell membrane to fuse and release the exosomes. The targeting of Rab family of proteins using inhibitors or delivery of the miRNAs may hamper exosomal secretion and reduce exosome-mediated disease progression and metastasis. A Rab27a blockade in mammary carcinoma cells decrease secretion of exosomes along with MMP9 expression, thereby suppressing tumor growth and metastasis (Bobrie et al., 2012).

GW4869 is a pharmacological molecule known to effectively inhibit neutral sphingomyelinase in a noncompetitive manner and suppress the secretion of exosomes (Montermini et al., 2015; Singh, Pochampally, Watabe, Lu, & Mo, 2014). The nontargeted nature of GW4869 may induce undesirable effects by hampering exosome-mediated signaling in normal cells. The use of imatinib and dasatinib are also reported to have an inhibitory effect on exosome secretion in CML cells and suppression of antigenic activity in HUVEC cells (Mineo et al., 2012).

All together, these evidence suggests that the selective targeting of exosome secretion pathway may have an added advantage to suppress exosome-mediated disease progression and metastasis.

4.2 Inhibition of Packaging the Exosomal Cargo

The cancer cell exosomes are reported to package different oncogenic molecules, including proteins, intact and mutated mRNA, and miRNAs. Taverna et al. (2015) reported the enhanced PTEN expression and an increased packaging of miR-21 into exosomes for disposal (Taverna et al., 2015). Different phytochemicals are known to inhibit carcinogenesis, including curcumin (Zubair et al., 2017). It is known that curcumin treatment suppresses the AKT phosphorylation and VEGF expression, leading to decreased migration potential (Taverna et al., 2015). Various reports have shown that exosomal miR-21 is responsible for cancer progression and aggressiveness; therefore, curcumin-induced exosomal secretion needs additional study to identify the signaling pathways causing such alterations. It is reported that breast cancer cell–derived exosomes decrease the IL-2-dependent NK cell activation. The pretreatment of breast cancer cells with curcumin (dietary phenol) enhances the antioncogenic proteins in exosomes and promotes the NK cell cytotoxic activity via Jak3-mediated activation of Stat5 (H. G. Zhang et al., 2007). Similarly, histone deacetylase inhibitor (MS-275), an epigenetic drug, has been found to increase the packaging of antitumor molecules into exosomes and activate NK cells (Xiao et al., 2013).

4.3 Inhibition of Exosome Internalization

The internalization of exosomes into recipient cells has been reported to occur through different modes, including fusion, pinocytosis, phagocytosis, and receptor-mediated endocytosis, etc. So, critical evaluation and identification of the specific mode of exosome internalization can be exploited as a therapeutic target to inhibit the exosome internalization, which may help in enhanced therapeutic outcome. The specific molecules bearing exosomes have the ability to internalize into specific types of cells. In a mouse model, intravenously administered B16BL6–derived exosomes were efficiently internalized into macrophages of hepatic and splenic tissue, while in the lung, exosomes are preferentially internalized into the endothelial cells (Imai et al., 2015). In another study, it was reported that the exosomes derived from MDA-MB-231 cells are preferentially taken up by macrophages in the lung and brain (Chow et al., 2014). Moreover, dendritic cell–derived exosomes are reported to preferentially internalize into macrophages (Kim, Bianco, Shufesky, Morelli, & Robbins, 2007). Hoshino et al. (2015) reported that the integrin patterns on the exosome surface determine the target cell for internalization (Hoshino et al., 2015). These exosomes are taken up either by lung fibroblasts, epithelial cells, liver Kupffer cells, or brain endothelial cells. So, the targeting of these molecules may inhibit the exosome-mediated metastasis, leading to effective therapeutic response.

The internalization of melanoma cell exosomes is reported to occur by fusion and their secretion, as well as uptake is enhanced at acidic pH. The use of proton pump inhibitor (omeprazole) decreased the exosome internalization and reversed the exosome-mediated effect. Moreover, the treatment with omeprazole sensitizes melanoma cells to cisplatin (Luciani et al., 2004; Parolini et al., 2009). Feng et al. (2010) reported that exosomal uptake is dependent on the actin cytoskeleton and phosphatidylinositol 3-kinase (PI3K) and is inhibited by PI3K inhibitors (Wortmannin and LY294002) in a dose-dependent manner (Feng et al., 2010). The inhibition of exosome uptake by ovarian cancer cells, using different inhibitors, including, chlorpromazine, 5-ethyl-N-isopropyl amiloride, cytochalasin D, and methyl-beta-cyclodextrin, showed multiple modes of endocytic pathways (Escrevente, Keller, Altevogt, & Costa, 2011). Another molecule, heparin sulfate proteoglycan, functions as a receptor for exosome internalization. The selective enzymatic degradation or targeting of endogenous proteoglycan biosynthesis pathway, using xyloside or heparin, inhibits exosomal internalization into GBM cells (Christianson, Svensson, van Kuppevelt, Li, & Belting, 2013). Moreover, lipid raft–mediated uptake of GBM cell–derived exosomes was efficiently attenuated by filipin (Svensson et al., 2013). Various antibodies (ICAM-1, TIM-4, Tspan8, CD-91, CD106, CD-151, and others) and pharmacological molecules, reported to inhibit exosome uptake, have been reviewed by Mulcahy, Pink, and Carter (2014). Such molecules can be exploited as a therapy adjuvant after critical evaluation (Mulcahy et al., 2014). Some of the effective inhibitors which inhibit exosome-mediated metastasis are listed in Table 3.

Table 3 List of Different Molecules Used to Inhibit Exosome Secretion or Uptake

Name of Inhibitor	Mode of Action	Target Molecule	References
GW4869	Inhibit exosome release	Sphingomyelinase	Montermini et al. (2015) and Singh et al. (2014)
Proton pump inhibitor	Proton pump inhibition	Omeprazole	Parolini et al. (2009) and Luciani et al. (2004)
Heparin and xyloside	Inhibit endocytosis	Heparin sulfate proteoglycan	Christianson et al. (2013)
Filipin	Lipid raft–mediated endocytosis	Cholesterol	Svensson et al. (2013)
Chlorpromazine and others	Different modes of endocytosis	Multiple	Escrevente et al. (2011)
Phagocytosis	Wortmannin and LY294002	Phosphoinositide 3-kinase	Feng et al. (2010)

4.4 Removal of Tumor Exosomes

Current updates on exosome research indicate their involvement in different aspects of cancer pathogenesis and therapy resistance. It is also known that the cancer cells aberrantly secrete more number of exosomes compared with normal cells and also contains more oncogenic cargo (Keerthikumar et al., 2015). So, the removal of the cancer exosomes will also be an added advantage in cancer therapy to reduce malignant behavior. Marleau, Chen, Joyce, and Tullis (2012) proposed that affinity-based lectin and antibody can be used in extracorporeal hemofiltration for selective removal of <200 nm-sized particles (exosomes) from patient blood (Marleau et al., 2012). Although this approach has potential positive implications, the continuous secretion of exosomes by cells makes the removal process a very tedious and expensive job. Researchers at Aethlon Medical have developed an affinity-based device using lectins, antibodies, and aptamers named adaptive dialysis-like affinity platform technology (ADAPT) for hemofiltration, which can be fit into continuous renal replacement therapy machines where particles <200 nm can be selectively separated. This device is developed following the same strategy used when clearing hepatitis C virus load by using Aethlon Hemopurifier (Tullis et al., 2009). Altogether, inhibition of biogenesis, secretion, and uptake of exosomes may therefore be a novel strategy in future treatment of different malignancies.

4.5 Exosome as a Therapeutic Vehicle

Exosomes are an attractive entity to load desired cargo (drug, microRNA, proteins) and can be used to deliver these at desired sites after surface modification. The exosomal delivery not only increases the stability and targeted delivery of therapeutic molecules but also reduces the immunogenic response and also allows for therapy delivery across the blood–brain barrier (Tominaga et al., 2015). Exosome-based delivery was found to have significant antiproliferative activity against pancreatic cancer cells (CFPAC-1) where MDR gene (P-gp) was overexpressed (Pascucci et al., 2014). Recently, it has been reported that exosomes containing CD47 on their surface escape the phagocytosis by circulating monocytes, thus increasing stability (Kamerkar et al., 2017). So, these findings represent a promising role for exosomes as an effective, stable, and targeted drug delivery vehicle with no/reduced side effects.

5. Conclusions and Perspective

Alterations in the TME, invasion, migration, premetastatic niche formation, immune modulation, and adaptation at secondary sites are important events in cancer metastasis. An understanding of mechanisms underlying cancer metastasis is important for designing effective strategies to reduce cancer-related deaths. Existing and emerging data continue to suggest important multifaceted roles of tumor- and stromal cell–derived exosomes in cancer metastasis. Notably, the data suggesting a role of exosomes in premetastatic niche formation is very intriguing from a clinical and translational perspective (early cancer detection and metastasis prevention). A role of exosomes in organ-specific metastasis has also instigated research for the exploitation of exosomes for tumor-directed drug delivery. Therefore, exosome research is expected to grow at an increasing pace in the coming years. However, there are many challenges in exosome research that need to be adequately addressed. The standard exosome isolation procedure needs to be refined further to address the concerns related to the size overlap in different vesicles and to improve exosome yield and quality. Exosomes are prone to clumping/aggregation and membrane rupturing in freeze–thaw cycles; therefore we need better approaches for their storage and handling. We also need more reliable and specific markers to differentiate exosomes derived from one cell type to another, which is critical for clearly assigning the biological significance of circulating exosomes and for their clinical exploitation. Moreover, differential detection of exosomes is also critical to selectively target pathologically relevant exosomes, while sparing the ones required for intercellular communications to maintain normal physiological functions. Despite these hurdles, there is no doubt that exosomes holds significant potential to impact current clinical management of cancers, and in coming years, we are going to witness tremendous growth in the field of exosome research.

Acknowledgments

The authors would like to thankfully acknowledge the Editors of this book who provided us the opportunity to write this chapter. This work is financially supported from NIH/NCI [R01CA175772, and U01CA185490 (to APS)] and USA-MCI.

Conflicts of Interest

The authors declare no conflict of interest.

References

Aga, M., Bentz, G. L., Raffa, S., Torrisi, M. R., Kondo, S., Wakisaka, N., … Shackelford, J. (2014). Exosomal HIF1alpha supports invasive potential of nasopharyngeal carcinoma-associated LMP1-positive exosomes. *Oncogene*, *33*(37), 4613–4622.

Al-Nedawi, K., Meehan, B., Micallef, J., Lhotak, V., May, L., Guha, A., & Rak, J. (2008). Intercellular transfer of the oncogenic receptor EGFRvIII by microvesicles derived from tumour cells. *Nature Cell Biology*, *10*(5), 619–624.

Allenson, K., Castillo, J., San Lucas, F. A., Scelo, G., Kim, D. U., Bernard, V., … Alvarez, H. (2017). High prevalence of mutant KRAS in circulating exosome-derived DNA from early-stage pancreatic cancer patients. *Annals of Oncology*, *28*(4), 741–747.

Ang, J., Fang, B. L., Ashman, L. K., & Frauman, A. G. (2010). The migration and invasion of human prostate cancer cell lines involves CD151 expression. *Oncology Reports*, *24*(6), 1593–1597.

Atay, S., Banskota, S., Crow, J., Sethi, G., Rink, L., & Godwin, A. K. (2014). Oncogenic KIT-containing exosomes increase gastrointestinal stromal tumor cell invasion. *Proceedings of the National Academy of Sciences of the United States of America*, *111*(2), 711–716.

Baroni, S., Romero-Cordoba, S., Plantamura, I., Dugo, M., D'Ippolito, E., Cataldo, A., … Iorio, M. V. (2016). Exosome-mediated delivery of miR-9 induces cancer-associated fibroblast-like properties in human breast fibroblasts. *Cell Death & Disease*, *7*(7), e2312.

Bobrie, A., Krumeich, S., Reyal, F., Recchi, C., Moita, L. F., Seabra, M. C., … Thery, C. (2012). Rab27a supports exosome-dependent and -independent mechanisms that modify the tumor microenvironment and can promote tumor progression. *Cancer Research*, *72*(19), 4920–4930.

Braicu, C., Tomuleasa, C., Monroig, P., Cucuianu, A., Berindan-Neagoe, I., & Calin, G. A. (2015). Exosomes as divine messengers: Are they the Hermes of modern molecular oncology? *Cell Death and Differentiation*, *22*(1), 34–45.

Canel, M., Serrels, A., Frame, M. C., & Brunton, V. G. (2013). E-cadherin-integrin crosstalk in cancer invasion and metastasis. *Journal of Cell Science*, *126*(Pt 2), 393–401.

Chen, X., Ying, X., Wang, X., Wu, X., Zhu, Q., & Wang, X. (2017). Exosomes derived from hypoxic epithelial ovarian cancer deliver microRNA-940 to induce macrophage M2 polarization. *Oncology Reports*, *38*(1), 522–528.

Chow, A., Zhou, W., Liu, L., Fong, M. Y., Champer, J., Van Haute, D., … Wang, S. E. (2014). Macrophage immunomodulation by breast cancer-derived exosomes requires Toll-like receptor 2-mediated activation of NF-kappaB. *Scientific Reports*, *4*, 5750.

Christianson, H. C., Svensson, K. J., van Kuppevelt, T. H., Li, J. P., & Belting, M. (2013). Cancer cell exosomes depend on cell-surface heparan sulfate proteoglycans for their internalization and functional activity. *Proceedings of the National Academy of Sciences of the United States of America*, *110*(43), 17380–17385.

Clayton, A., & Mason, M. D. (2009). Exosomes in tumour immunity. *Current Oncology*, *16*(3), 46–49.

Conigliaro, A., Costa, V., Dico, A. L., Saieva, L., Buccheri, S., Dieli, F., … De Leo, G. (2015). CD90+ liver cancer cells modulate endothelial cell phenotype through the release of exosomes containing H19 lncRNA. *Molecular Cancer*, *14*(1), 155.

Costa-Silva, B., Aiello, N. M., Ocean, A. J., Singh, S., Zhang, H., Thakur, B. K., … Lyden, D. (2015). Pancreatic cancer exosomes initiate pre-metastatic niche formation in the liver. *Nature Cell Biology*, *17*(6), 816–826.

De Bruyne, E., Andersen, T. L., De Raeve, H., Van Valckenborgh, E., Caers, J., Van Camp, B., … Vanderkerken, K. (2006). Endothelial cell-driven regulation of CD9 or motility-related protein-1 expression in multiple myeloma cells within the murine 5T33MM model and myeloma patients. *Leukemia*, *20*(10), 1870–1879.

Demory Beckler, M., Higginbotham, J. N., Franklin, J. L., Ham, A. J., Halvey, P. J., Imasuen, I. E., … Coffey, R. J. (2013). Proteomic analysis of exosomes from mutant KRAS colon cancer cells identifies intercellular transfer of mutant KRAS. *Molecular & Cellular Proteomics*, *12*(2), 343–355.

Donoso, L. A., Felberg, N. T., Edelberg, K., Borlinghaus, P., & Herlyn, M. (1986). Metastatic uveal melanoma: an ocular melanoma associated antigen in the serum of patients with metastatic disease. *J Immunoassay*, *7*(4), 273–283.

Enriquez, V. A., Cleys, E. R., Da Silveira, J. C., Spillman, M. A., Winger, Q. A., & Bouma, G. J. (2015). High LIN28A expressing ovarian cancer cells secrete exosomes that induce invasion and migration in HEK293 cells. *BioMed Research International*, *2015*, 701390.

Escrevente, C., Keller, S., Altevogt, P., & Costa, J. (2011). Interaction and uptake of exosomes by ovarian cancer cells. *BMC Cancer*, *11*, 108.

Fabbri, M., Paone, A., Calore, F., Galli, R., Gaudio, E., Santhanam, R., … Croce, C. M. (2012). MicroRNAs bind to Toll-like receptors to induce prometastatic inflammatory response. *Proceedings of the National Academy of Sciences of the United States of America*, *109*(31), E2110–E2116.

Fedele, C., Singh, A., Zerlanko, B. J., Iozzo, R. V., & Languino, L. R. (2015). The alphavbeta6 integrin is transferred intercellularly via exosomes. *Journal of Biological Chemistry*, *290*(8), 4545–4551.

Feng, D., Zhao, W. L., Ye, Y. Y., Bai, X. C., Liu, R. Q., Chang, L. F., … Sui, S. F. (2010). Cellular internalization of exosomes occurs through phagocytosis. *Traffic*, *11*(5), 675–687.

Fong, M. Y., Zhou, W., Liu, L., Alontaga, A. Y., Chandra, M., Ashby, J., … Wang, S. E. (2015). Breast-cancer-secreted miR-122 reprograms glucose metabolism in premetastatic niche to promote metastasis. *Nature Cell Biology*, *17*(2), 183–194.

Franzen, C. A., Blackwell, R. H., Todorovic, V., Greco, K. A., Foreman, K. E., Flanigan, R. C., … Gupta, G. N. (2015). Urothelial cells undergo epithelial-to-mesenchymal transition after exposure to muscle invasive bladder cancer exosomes. *Oncogenesis*, *4*, e163.

Friedl, P., & Alexander, S. (2011). Cancer invasion and the microenvironment: Plasticity and reciprocity. *Cell*, *147*(5), 992–1009.

Gascard, P., & Tlsty, T. D. (2016). Carcinoma-associated fibroblasts: Orchestrating the composition of malignancy. *Genes & Development*, *30*(9), 1002–1019.

Gunda, V., Souchek, J. J., Abrego, J., Shukla, S. K., Goode, G. D., Vernucci, E., … Singh, P. K. (2017). MUC1-mediated metabolic alterations regulate response to radiotherapy in pancreatic cancer. *Clinical Cancer Research*, *23*(19), 5881–5891.

Gu, J., Qian, H., Shen, L., Zhang, X., Zhu, W., Huang, L., … Xu, W. (2012). Gastric cancer exosomes trigger differentiation of umbilical cord derived mesenchymal stem cells to carcinoma-associated fibroblasts through TGF-beta/Smad pathway. *PLoS One*, *7*(12), e52465.

Harris, D. A., Patel, S. H., Gucek, M., Hendrix, A., Westbroek, W., & Taraska, J. W. (2015). Exosomes released from breast cancer carcinomas stimulate cell movement. *PLoS One*, *10*(3), e0117495.

Hendrix, A., Maynard, D., Pauwels, P., Braems, G., Denys, H., Van den Broecke, R., … Westbroek, W. (2010). Effect of the secretory small GTPase Rab27B on breast cancer growth, invasion, and metastasis. *Journal of the National Cancer Institute*, *102*(12), 866–880.

Higginbotham, J. N., Demory Beckler, M., Gephart, J. D., Franklin, J. L., Bogatcheva, G., Kremers, G. J., … Coffey, R. J. (2011). Amphiregulin exosomes increase cancer cell invasion. *Current Biology*, *21*(9), 779–786.

Hood, J. L., San, R. S., & Wickline, S. A. (2011). Exosomes released by melanoma cells prepare sentinel lymph nodes for tumor metastasis. *Cancer Research*, *71*(11), 3792–3801.

Hoshino, A., Costa-Silva, B., Shen, T. L., Rodrigues, G., Hashimoto, A., Tesic Mark, M., … Lyden, D. (2015). Tumour exosome integrins determine organotropic metastasis. *Nature*, *527*(7578), 329–335.

Huang, Z., & Feng, Y. (2017). Exosomes derived from hypoxic colorectal cancer cells promote angiogenesis through wnt4-induced beta-catenin signaling in endothelial cells. *Oncology Research, 25*(5), 651–661.

Huttenlocher, A., & Horwitz, A. R. (2011). Integrins in cell migration. *Cold Spring Harbor Perspectives in Biology, 3*(9), a005074.

Iero, M., Valenti, R., Huber, V., Filipazzi, P., Parmiani, G., Fais, S., & Rivoltini, L. (2008). Tumour-released exosomes and their implications in cancer immunity. *Cell Death and Differentiation, 15*(1), 80–88.

Imai, T., Takahashi, Y., Nishikawa, M., Kato, K., Morishita, M., Yamashita, T., … Takakura, Y. (2015). Macrophage-dependent clearance of systemically administered B16BL6-derived exosomes from the blood circulation in mice. *Journal of Extracellular Vesicles, 4*, 26238.

Inoue, M., Hager, J. H., Ferrara, N., Gerber, H. P., & Hanahan, D. (2002). VEGF-A has a critical, nonredundant role in angiogenic switching and pancreatic beta cell carcinogenesis. *Cancer Cell, 1*(2), 193–202.

Johnstone, R. M., Bianchini, A., & Teng, K. (1989). Reticulocyte maturation and exosome release: Transferrin receptor containing exosomes shows multiple plasma membrane functions. *Blood, 74*(5), 1844–1851.

Jung, T., Castellana, D., Klingbeil, P., Cuesta Hernandez, I., Vitacolonna, M., Orlicky, D. J., … Zoller, M. (2009). CD44v6 dependence of premetastatic niche preparation by exosomes. *Neoplasia, 11*(10), 1093–1105.

Kalluri, R. (2016). The biology and function of fibroblasts in cancer. *Nature Reviews. Cancer, 16*(9), 582–598.

Kamerkar, S., LeBleu, V. S., Sugimoto, H., Yang, S., Ruivo, C. F., Melo, S. A., … Kalluri, R. (2017). Exosomes facilitate therapeutic targeting of oncogenic KRAS in pancreatic cancer. *Nature, 546*(7659), 498–503.

Kaplan, R. N., Riba, R. D., Zacharoulis, S., Bramley, A. H., Vincent, L., Costa, C., … Lyden, D. (2005). VEGFR1-positive haematopoietic bone marrow progenitors initiate the pre-metastatic niche. *Nature, 438*(7069), 820–827.

Keerthikumar, S., Gangoda, L., Liem, M., Fonseka, P., Atukorala, I., Ozcitti, C., … Mathivanan, S. (2015). Proteogenomic analysis reveals exosomes are more oncogenic than ectosomes. *Oncotarget, 6*(17), 15375–15396.

Khan, M. A., Azim, S., Zubair, H., Bhardwaj, A., Patel, G. K., Khushman, M., … Singh, A. P. (2017). Molecular drivers of pancreatic cancer Pathogenesis: Looking inward to move forward. *International Journal of Molecular Sciences, 18*(4).

Khan, M. A., Srivastava, S. K., Bhardwaj, A., Singh, S., Arora, S., Zubair, H., … Singh, A. P. (2015). Gemcitabine triggers angiogenesis-promoting molecular signals in pancreatic cancer cells: Therapeutic implications. *Oncotarget, 6*(36), 39140–39150.

Khushman, M., Bhardwaj, A., Patel, G. K., Laurini, J. A., Roveda, K., Tan, M. C., … Singh, A. P. (2017). Exosomal markers (CD63 and CD9) expression pattern using immunohistochemistry in resected malignant and nonmalignant pancreatic specimens. *Pancreas, 46*, 782–788.

Kim, S. H., Bianco, N. R., Shufesky, W. J., Morelli, A. E., & Robbins, P. D. (2007). Effective treatment of inflammatory disease models with exosomes derived from dendritic cells genetically modified to express IL-4. *The Journal of Immunology, 179*(4), 2242–2249.

Kim, J. W., Wieckowski, E., Taylor, D. D., Reichert, T. E., Watkins, S., & Whiteside, T. L. (2005). Fas ligand-positive membranous vesicles isolated from sera of patients with oral cancer induce apoptosis of activated T lymphocytes. *Clinical Cancer Research, 11*(3), 1010–1020.

Kischel, P., Bellahcene, A., Deux, B., Lamour, V., Dobson, R., DE Pauw, E., … Castronovo, V. (2012). Overexpression of CD9 in human breast cancer cells promotes the development of bone metastases. *Anticancer Research, 32*(12), 5211–5220.

Kucharzewska, P., Christianson, H. C., Welch, J. E., Svensson, K. J., Fredlund, E., Ringner, M., … Belting, M. (2013). Exosomes reflect the hypoxic status of glioma cells and mediate hypoxia-dependent activation of vascular cells during tumor development. *Proceedings of the National Academy of Sciences of the United States of America, 110*(18), 7312–7317.

Lafleur, M. A., Xu, D., & Hemler, M. E. (2009). Tetraspanin proteins regulate membrane type-1 matrix metalloproteinase-dependent pericellular proteolysis. *Molecular Biology of the Cell*, *20*(7), 2030–2040.

Lazar, I., Clement, E., Dauvillier, S., Milhas, D., Ducoux-Petit, M., LeGonidec, S., … Nieto, L. (2016). Adipocyte exosomes promote melanoma aggressiveness through fatty acid oxidation: A novel mechanism linking obesity and cancer. *Cancer Research*, *76*(14), 4051–4057.

Lin, R., Wang, S., & Zhao, R. C. (2013). Exosomes from human adipose-derived mesenchymal stem cells promote migration through Wnt signaling pathway in a breast cancer cell model. *Molecular and Cellular Biochemistry*, *383*(1–2), 13–20.

Liu, C., Yu, S., Zinn, K., Wang, J., Zhang, L., Jia, Y., … Zhang, H. G. (2006). Murine mammary carcinoma exosomes promote tumor growth by suppression of NK cell function. *The Journal of Immunology*, *176*(3), 1375–1385.

Luciani, F., Spada, M., De Milito, A., Molinari, A., Rivoltini, L., Montinaro, A., … Fais, S. (2004). Effect of proton pump inhibitor pretreatment on resistance of solid tumors to cytotoxic drugs. *Journal of the National Cancer Institute*, *96*(22), 1702–1713.

Luga, V., Zhang, L., Viloria-Petit, A. M., Ogunjimi, A. A., Inanlou, M. R., Chiu, E., … Wrana, J. L. (2012). Exosomes mediate stromal mobilization of autocrine Wnt-PCP signaling in breast cancer cell migration. *Cell*, *151*(7), 1542–1556.

Maji, S., Chaudhary, P., Akopova, I., Nguyen, P. M., Hare, R. J., Gryczynski, I., & Vishwanatha, J. K. (2017). Exosomal annexin II promotes angiogenesis and breast cancer metastasis. *Molecular Cancer Research*, *15*(1), 93–105.

Mantovani, A., Marchesi, F., Malesci, A., Laghi, L., & Allavena, P. (2017). Tumour-associated macrophages as treatment targets in oncology. *Nature Reviews. Clinical Oncology*, *14*(7), 399–416.

Mardin, W. A., Haier, J., & Mees, S. T. (2013). Epigenetic regulation and role of metastasis suppressor genes in pancreatic ductal adenocarcinoma. *BMC Cancer*, *13*, 264.

Marleau, A. M., Chen, C. S., Joyce, J. A., & Tullis, R. H. (2012). Exosome removal as a therapeutic adjuvant in cancer. *Journal of Translational Medicine*, *10*, 134.

Meckes, D. G., Jr., Shair, K. H., Marquitz, A. R., Kung, C. P., Edwards, R. H., & Raab-Traub, N. (2010). Human tumor virus utilizes exosomes for intercellular communication. *Proceedings of the National Academy of Sciences of the United States of America*, *107*(47), 20370–20375.

Menck, K., Klemm, F., Gross, J. C., Pukrop, T., Wenzel, D., & Binder, C. (2013). Induction and transport of Wnt 5a during macrophage-induced malignant invasion is mediated by two types of extracellular vesicles. *Oncotarget*, *4*(11), 2057–2066.

Mineo, M., Garfield, S. H., Taverna, S., Flugy, A., De Leo, G., Alessandro, R., & Kohn, E. C. (2012). Exosomes released by K562 chronic myeloid leukemia cells promote angiogenesis in a Src-dependent fashion. *Angiogenesis*, *15*(1), 33–45.

Montermini, L., Meehan, B., Garnier, D., Lee, W. J., Lee, T. H., Guha, A., … Rak, J. (2015). Inhibition of oncogenic epidermal growth factor receptor kinase triggers release of exosome-like extracellular vesicles and impacts their phosphoprotein and DNA content. *Journal of Biological Chemistry*, *290*(40), 24534–24546.

Mulcahy, L. A., Pink, R. C., & Carter, D. R. (2014). Routes and mechanisms of extracellular vesicle uptake. *Journal of Extracellular Vesicles*, *3*.

Nakamura, K., Sawada, K., Kinose, Y., Yoshimura, A., Toda, A., Nakatsuka, E., … Kimura, T. (2017). Exosomes promote ovarian cancer cell invasion through transfer of CD44 to peritoneal mesothelial cells. *Molecular Cancer Research*, *15*(1), 78–92.

Nakaya, Y., Kuroda, S., Katagiri, Y. T., Kaibuchi, K., & Takahashi, Y. (2004). Mesenchymal-epithelial transition during somitic segmentation is regulated by differential roles of Cdc42 and Rac1. *Developmental Cell*, *7*(3), 425–438.

Okoye, I. S., Coomes, S. M., Pelly, V. S., Czieso, S., Papayannopoulos, V., Tolmachova, T., … Wilson, M. S. (2014). MicroRNA-containing T-regulatory-cell-derived exosomes suppress pathogenic T helper 1 cells. *Immunity*, *41*(1), 89–103.

Paggetti, J., Haderk, F., Seiffert, M., Janji, B., Distler, U., Ammerlaan, W., … Moussay, E. (2015). Exosomes released by chronic lymphocytic leukemia cells induce the transition of stromal cells into cancer-associated fibroblasts. *Blood*, *126*(9), 1106–1117.

Pan, L., Liang, W., Fu, M., Huang, Z. H., Li, X., Zhang, W., … Zhang, X. (2017). Exosomes-mediated transfer of long noncoding RNA ZFAS1 promotes gastric cancer progression. *Journal of Cancer Research and Clinical Oncology*, *143*(6), 991–1004.

Parolini, I., Federici, C., Raggi, C., Lugini, L., Palleschi, S., De Milito, A., … Fais, S. (2009). Microenvironmental pH is a key factor for exosome traffic in tumor cells. *Journal of Biological Chemistry*, *284*(49), 34211–34222.

Pascucci, L., Cocce, V., Bonomi, A., Ami, D., Ceccarelli, P., Ciusani, E., … Pessina, A. (2014). Paclitaxel is incorporated by mesenchymal stromal cells and released in exosomes that inhibit in vitro tumor growth: A new approach for drug delivery. *Journal of Controlled Release*, *192*, 262–270.

Patel, G. K., Khan, M. A., Bhardwaj, A., Srivastava, S. K., Zubair, H., Patton, M. C., … Singh, A. P. (2017). Exosomes confer chemoresistance to pancreatic cancer cells by promoting ROS detoxification and miR-155-mediated suppression of key gemcitabine-metabolising enzyme, DCK. *British Journal of Cancer*, *116*(5), 609–619.

Patel, G. K., Patton, M. C., Singh, S., Khushman, M., & Singh, A. P. (2016). Pancreatic cancer exosomes: Shedding off for a meaningful journey. *Pancreatic Disorders & Therapy*, *6*(2), e148.

Peinado, H., Aleckovic, M., Lavotshkin, S., Matei, I., Costa-Silva, B., Moreno-Bueno, G., … Lyden, D. (2012). Melanoma exosomes educate bone marrow progenitor cells toward a pro-metastatic phenotype through MET. *Nature Medicine*, *18*(6), 883–891.

Peinado, H., Lavotshkin, S., & Lyden, D. (2011). The secreted factors responsible for pre-metastatic niche formation: Old sayings and new thoughts. *Seminars in Cancer Biology*, *21*(2), 139–146.

Powner, D., Kopp, P. M., Monkley, S. J., Critchley, D. R., & Berditchevski, F. (2011). Tetraspanin CD9 in cell migration. *Biochemical Society Transactions*, *39*(2), 563–567.

Qu, L., Ding, J., Chen, C., Wu, Z. J., Liu, B., Gao, Y., … Wang, L. H. (2016). Exosome-transmitted lncARSR promotes sunitinib resistance in renal cancer by acting as a competing endogenous RNA. *Cancer Cell*, *29*(5), 653–668.

Rahman, M. A., Barger, J. F., Lovat, F., Gao, M., Otterson, G. A., & Nana-Sinkam, P. (2016). Lung cancer exosomes as drivers of epithelial mesenchymal transition. *Oncotarget*, *7*(34), 54852–54866.

Ramteke, A., Ting, H., Agarwal, C., Mateen, S., Somasagara, R., Hussain, A., … Deep, G. (2015). Exosomes secreted under hypoxia enhance invasiveness and stemness of prostate cancer cells by targeting adherens junction molecules. *Molecular Carcinogenesis*, *54*(7), 554–565.

Raposo, G., & Stoorvogel, W. (2013). Extracellular vesicles: Exosomes, microvesicles, and friends. *Journal of Cell Biology*, *200*(4), 373–383.

Ribeiro, M. F., Zhu, H., Millard, R. W., & Fan, G. C. (2013). Exosomes function in pro- and anti-angiogenesis. *Current Angiogenes*, *2*(1), 54–59.

Richards, K. E., Zeleniak, A. E., Fishel, M. L., Wu, J., Littlepage, L. E., & Hill, R. (2017). Cancer-associated fibroblast exosomes regulate survival and proliferation of pancreatic cancer cells. *Oncogene*, *36*(13), 1770–1778.

Salomon, C., Torres, M. J., Kobayashi, M., Scholz-Romero, K., Sobrevia, L., Dobierzewska, A., … Rice, G. E. (2014). A gestational profile of placental exosomes in maternal plasma and their effects on endothelial cell migration. *PLoS One*, *9*(6), e98667.

Sanchez, C. A., Andahur, E. I., Valenzuela, R., Castellon, E. A., Fulla, J. A., Ramos, C. G., & Trivino, J. C. (2016). Exosomes from bulk and stem cells from human prostate cancer have a differential microRNA content that contributes cooperatively over local and pre-metastatic niche. *Oncotarget*, *7*(4), 3993–4008.

Semenza, G. L., Roth, P. H., Fang, H. M., & Wang, G. L. (1994). Transcriptional regulation of genes encoding glycolytic enzymes by hypoxia-inducible factor 1. *Journal of Biological Chemistry*, *269*(38), 23757–23763.

Sharghi-Namini, S., Tan, E., Ong, L. L., Ge, R., & Asada, H. H. (2014). Dll4-containing exosomes induce capillary sprout retraction in a 3D microenvironment. *Scientific Reports*, 4, 4031.

Shukla, S. K., Purohit, V., Mehla, K., Gunda, V., Chaika, N. V., Vernucci, E., … Singh, P. K. (2017). MUC1 and HIF-1alpha signaling crosstalk induces anabolic glucose metabolism to impart gemcitabine resistance to pancreatic cancer. *Cancer Cell*, 32(1), 71–87 e77.

Sidhu, S. S., Mengistab, A. T., Tauscher, A. N., LaVail, J., & Basbaum, C. (2004). The microvesicle as a vehicle for EMMPRIN in tumor-stromal interactions. *Oncogene*, 23(4), 956–963.

Singh, R., Pochampally, R., Watabe, K., Lu, Z., & Mo, Y. Y. (2014). Exosome-mediated transfer of miR-10b promotes cell invasion in breast cancer. *Molcular Cancer*, 13, 256.

Svensson, K. J., Christianson, H. C., Wittrup, A., Bourseau-Guilmain, E., Lindqvist, E., Svensson, L. M., … Belting, M. (2013). Exosome uptake depends on ERK1/2-heat shock protein 27 signaling and lipid Raft-mediated endocytosis negatively regulated by caveolin-1. *Journal of Biological Chemistry*, 288(24), 17713–17724.

Takikawa, T., Masamune, A., Yoshida, N., Hamada, S., Kogure, T., & Shimosegawa, T. (2017). Exosomes derived from pancreatic stellate cells: MicroRNA signature and effects on pancreatic cancer cells. *Pancreas*, 46(1), 19–27.

Taverna, S., Giallombardo, M., Pucci, M., Flugy, A., Manno, M., Raccosta, S., … Alessandro, R. (2015). Curcumin inhibits in vitro and in vivo chronic myelogenous leukemia cells growth: A possible role for exosomal disposal of miR-21. *Oncotarget*, 6(26), 21918–21933.

Taylor, D. D., Gercel-Taylor, C., Lyons, K. S., Stanson, J., & Whiteside, T. L. (2003). T-cell apoptosis and suppression of T-cell receptor/CD3-zeta by Fas ligand-containing membrane vesicles shed from ovarian tumors. *Clinical Cancer Research*, 9(14), 5113–5119.

Tejera, E., Rocha-Perugini, V., Lopez-Martin, S., Perez-Hernandez, D., Bachir, A. I., Horwitz, A. R., … Yanez-Mo, M. (2013). CD81 regulates cell migration through its association with Rac GTPase. *Molecular Biology of the Cell*, 24(3), 261–273.

Thery, C., Ostrowski, M., & Segura, E. (2009). Membrane vesicles as conveyors of immune responses. *Nature Reviews. Immunology*, 9(8), 581–593.

Tominaga, N., Kosaka, N., Ono, M., Katsuda, T., Yoshioka, Y., Tamura, K., … Ochiya, T. (2015). Brain metastatic cancer cells release microRNA-181c-containing extracellular vesicles capable of destructing blood-brain barrier. *Nature Communications*, 6, 6716.

Tullis, R. H., Duffin, R. P., Handley, H. H., Sodhi, P., Menon, J., Joyce, J. A., & Kher, V. (2009). Reduction of hepatitis C virus using lectin affinity plasmapheresis in dialysis patients. *Blood Purification*, 27(1), 64–69.

Umezu, T., Ohyashiki, K., Kuroda, M., & Ohyashiki, J. H. (2013). Leukemia cell to endothelial cell communication via exosomal miRNAs. *Oncogene*, 32(22), 2747–2755.

Umezu, T., Tadokoro, H., Azuma, K., Yoshizawa, S., Ohyashiki, K., & Ohyashiki, J. H. (2014). Exosomal miR-135b shed from hypoxic multiple myeloma cells enhances angiogenesis by targeting factor-inhibiting HIF-1. *Blood*, 124(25), 3748–3757.

Warburg, O. (1956). On the origin of cancer cells. *Science*, 123(3191), 309–314.

Webber, J., Steadman, R., Mason, M. D., Tabi, Z., & Clayton, A. (2010). Cancer exosomes trigger fibroblast to myofibroblast differentiation. *Cancer Research*, 70, 9621–9630.

Welton, J. L., Khanna, S., Giles, P. J., Brennan, P., Brewis, I. A., Staffurth, J., … Clayton, A. (2010). Proteomics analysis of bladder cancer exosomes. *Molecular & Cellular Proteomics*, 9(6), 1324–1338.

Wen, S. W., Sceneay, J., Lima, L. G., Wong, C. S., Becker, M., Krumeich, S., … Moller, A. (2016). The biodistribution and immune suppressive effects of breast cancer-derived exosomes. *Cancer Research*, 76(23), 6816–6827.

Xiao, W., Dong, W., Zhang, C., Saren, G., Geng, P., Zhao, H., … Ye, M. (2013). Effects of the epigenetic drug MS-275 on the release and function of exosome-related immune molecules in hepatocellular carcinoma cells. *European Journal of Medical Research, 18*, 61.

Yamada, N., Tsujimura, N., Kumazaki, M., Shinohara, H., Taniguchi, K., Nakagawa, Y., … Akao, Y. (2014). Colorectal cancer cell-derived microvesicles containing microRNA-1246 promote angiogenesis by activating Smad 1/5/8 signaling elicited by PML down-regulation in endothelial cells. *Biochimica et Biophysica Acta, 1839*(11), 1256–1272.

Yanez-Mo, M., Alfranca, A., Cabanas, C., Marazuela, M., Tejedor, R., Ursa, M. A., … Sanchez-Madrid, F. (1998). Regulation of endothelial cell motility by complexes of tetraspan molecules CD81/TAPA-1 and CD151/PETA-3 with alpha3 beta1 integrin localized at endothelial lateral junctions. *Journal of Cell Biology, 141*(3), 791–804.

Yang, L., Wu, X. H., Wang, D., Luo, C. L., & Chen, L. X. (2013). Bladder cancer cell-derived exosomes inhibit tumor cell apoptosis and induce cell proliferation in vitro. *Molecular Medicine Reports, 8*(4), 1272–1278.

Yao, D., Dai, C., & Peng, S. (2011). Mechanism of the mesenchymal-epithelial transition and its relationship with metastatic tumor formation. *Molecular Cancer Research, 9*(12), 1608–1620.

Ying, H., Kimmelman, A. C., Lyssiotis, C. A., Hua, S., Chu, G. C., Fletcher-Sananikone, E., … DePinho, R. A. (2012). Oncogenic Kras maintains pancreatic tumors through regulation of anabolic glucose metabolism. *Cell, 149*(3), 656–670.

Ying, X., Wu, Q., Wu, X., Zhu, Q., Wang, X., Jiang, L., … Wang, X. (2016). Epithelial ovarian cancer-secreted exosomal miR-222-3p induces polarization of tumor-associated macrophages. *Oncotarget, 7*(28), 43076–43087.

Yoshizaki, T., Kondo, S., Wakisaka, N., Murono, S., Endo, K., Sugimoto, H., … Ito, M. (2013). Pathogenic role of Epstein-Barr virus latent membrane protein-1 in the development of nasopharyngeal carcinoma. *Cancer Letters, 337*(1), 1–7.

Zhang, H., Deng, T., Liu, R., Bai, M., Zhou, L., Wang, X., … Ba, Y. (2017). Exosome-delivered EGFR regulates liver microenvironment to promote gastric cancer liver metastasis. *Nature Communications, 8*, 15016.

Zhang, H. G., Kim, H., Liu, C., Yu, S., Wang, J., Grizzle, W. E., … Barnes, S. (2007). Curcumin reverses breast tumor exosomes mediated immune suppression of NK cell tumor cytotoxicity. *Biochimica et Biophysica Acta, 1773*(7), 1116–1123.

Zhang, L., Zhang, S., Yao, J., Lowery, F. J., Zhang, Q., Huang, W. C., … Yu, D. (2015). Microenvironment-induced PTEN loss by exosomal microRNA primes brain metastasis outgrowth. *Nature, 527*(7576), 100–104.

Zhao, H., Yang, L., Baddour, J., Achreja, A., Bernard, V., Moss, T., … Nagrath, D. (2016). Tumor microenvironment derived exosomes pleiotropically modulate cancer cell metabolism. *Elife, 5*, e10250.

Zhuang, G., Wu, X., Jiang, Z., Kasman, I., Yao, J., Guan, Y., … Ferrara, N. (2012). Tumour-secreted miR-9 promotes endothelial cell migration and angiogenesis by activating the JAK-STAT pathway. *The EMBO Journal, 31*(17), 3513–3523.

Zubair, H., & Ahmad, A. (2017). Cancer metastasis: An introduction. In A. Ahmad (Ed.), *Introduction to cancer metastasis* (pp. 3–12). New York: Elsevier, E-Publishing Inc.

Zubair, H., Azim, S., Ahmad, A., Khan, M. A., Patel, G. K., Singh, S., & Singh, A. P. (2017). Cancer chemoprevention by phytochemicals: Nature's healing touch. *Molecules, 22*(3).

Zubair, H., Azim, S., Srivastava, S. K., Ahmad, A., Bhardwaj, A., Khan, M. A., … Singh, A. P. (2016). Glucose metabolism reprogrammed by overexpression of IKKepsilon promotes pancreatic tumor growth. *Cancer Research, 76*(24), 7254–7264.

16

The Emerging Roles and Clinical Potential of Exosomes in Cancer: Drug Resistance

Li Min, Cassandra Garbutt, Francis Hornicek, Zhenfeng Duan

MASSACHUSETTS GENERAL HOSPITAL AND HARVARD MEDICAL SCHOOL, BOSTON, MA, UNITED STATES

CHAPTER OUTLINE

1. Introduction 286
2. Mechanisms of Resistance to Current Drug Therapies 287
3. Mechanisms of Exosome in Drug Resistance 291
   3.1 Exosome-Mediated Direct Drug Efflux... 291
   3.2 Transportations of Drug Resistance via Exosomal Contents 292
   3.3 Cross-Talk Between Tumor Cells and the Stroma via Exosomes 293
4. Exosome-Associated Drug Resistance in Cancers 293
   4.1 Breast Cancer 293
   4.2 Hematologic Malignancies... 296
       4.2.1 Multiple Myeloma 296
       4.2.2 Lymphoma 297
       4.2.3 Leukemia... 298
   4.3 Prostate cancer 299
   4.4 Ovarian Cancer 300
   4.5 Pancreatic Cancer 301
   4.6 Brain Cancer... 301
   4.7 Liver Cancer.. 302
   4.8 Lung Cancer 302
   4.9 Gastric Cancer 303
   4.10 Colorectal Cancer.. 303
   4.11 Melanoma 304
   4.12 Renal Cell Carcinoma.. 304
   4.13 Osteosarcoma 304
   4.14 Neuroblastoma 305
5. Conclusion 305
References 306

Diagnostic and Therapeutic Applications of Exosomes in Cancer. https://doi.org/10.1016/B978-0-12-812774-2.00016-X

1. Introduction

Cancer is the second worldwide cause of death, exceeded only by cardiovascular diseases (Perez-Herrero & Fernandez-Medarde, 2015). Only 70 years have passed since Goodman and his colleagues introduced chemotherapy as a form of cancer treatment (Goodman et al., 1946). Alongside surgical resection and radiotherapy, chemotherapy has been utilized as an antitumor treatment for decades. Unfortunately, drug resistance to chemotherapy is a stubborn impediment for the effective treatment of malignant diseases. In over 90% of patients with metastatic cancer, drug resistance is the cause of treatment failures (Longley & Johnston, 2005). Recently, new therapeutic modalities have been developed, such as targeted cancer therapy agents. In general, targeted therapy increases efficacy and decreases toxicity in comparison with traditional chemotherapy (Hojjat-Farsangi, 2016). However, the results of targeted therapy include drug resistance (Kaiser, 2011; Perez-Herrero & Fernandez-Medarde, 2015). Today, drug resistance in tumors is a serious and hitherto unresolved problem, and the overcoming of this hurdle is essential for clinical efficacy.

The mechanisms of drug resistance underlying conventional chemotherapy and targeted therapy in human cancers share many characteristics, such as an inadequate accumulation of agents intracellularly, enhanced drug metabolism, detoxification, and activated antiapoptotic pathways. Instead of ascribing a single factor—such as a specific gene or protein—to the responsibility of causing the variety of drug-resistant mechanisms in a tumor, drug resistance to cancer therapies most likely involves multiple mechanisms that act in parallel or unison. Emerging evidence suggests that drug-resistant tumors rely on robust biological interaction networks, such as intracellular and intercellular interactions that can even incorporate networks at distant sites or in the surrounding environment. These complex biological interactions are maintained by the ceaseless bidirectional exchange of molecular information across the nuclear and plasma membranes of cells. In addition to well-known biological transport systems, the involvement and role of other novel mechanisms are becoming apparent, such as those facilitated by extracellular vesicles (Andaloussi et al., 2013; Miyawaki, 2011; Sugano et al., 2010). Among the extracellular vesicles, exosomes are of peak interest due to their central role in cellular communication (Andaloussi et al., 2013; Milane et al., 2015). The primary role of exosomes in tumor biology is to transport bioactive molecules between cells to subsequently induce malignant functions such as angiogenesis, immunosuppression, proliferation, and the transfer of genetic material that aids in resistance (Andaloussi et al., 2013). Indeed, exosomes are known to transfer microRNAs to target locations (Zhang et al., 2016). Recent studies report that exosomes are released in greater concentrations by drug-resistant tumor cells than by drug-sensitive cells; a finding which indicates that exosomes may be key in the transfer of drug resistance in cancer (Andaloussi et al., 2013; Min et al., 2016).

Insight into how exosomes confer drug resistance in tumors has been revealed by recent investigations. Moreover, preclinical and clinical trials for cancer treatments are currently evaluating the benefits of manipulating exosomal contents. Exosome-based therapeutics hold the potential to overcome the obstacle of drug resistance in cancer treatment.

2. Mechanisms of Resistance to Current Drug Therapies

Based on the genetic or epigenetic alterations, resistance to drug therapies can be divided into two main categories, namely: intrinsic drug resistance and acquired drug resistance (also known as multidrug resistance [MDR]) (Holohan et al., 2013). Intrinsic drug resistance exists naturally and prior to any drug or treatment exposure. Factors associated with intrinsic drug resistance are as follows: an increased efflux of drugs from the tumor cell membrane, heterogeneity of tumor cells, and extracellular vesicles, including exosomes (Table 1) (Bagrodia, Smeal, & Abraham, 2012). Genetic variation is the basis of the theory of evolution by natural selection, and cancer treatments can be viewed as applying a selective pressure that only the most resistant cells within a heterogeneous tumor cell population can survive. Acquired drug resistance or MDR is a gradual process through which cancer cells exposed to drug therapies undergo genetic or epigenetic changes to express drug-resistant phenotypes (Table 1) (Bagrodia et al., 2012). Most cancer patients develop drug resistance during their course of treatment. In most cases, a long exposure to a single chemotherapeutic drug can extend resistance to multiple structurally and functionally unrelated agents, leading to MDR. The development of MDR in vitro can be attributed to an overexpression of the ATP-binding cassette (ABC) gene, the multidrug resistance gene 1 (ABC subfamily B member 1, ABCB1/MDR1), and the coded drug efflux pump P-glycoprotein (P-gp) (Gottesman, Fojo, & Bates, 2002, 2016). However, the contribution of MDR1/P-gp to clinical drug resistance remains controversial. MDR may

Table 1   Mechanisms of Chemoresistance in Cancer

Types	Mechanisms of Chemotherapy	References
Intrinsic resistance	• Heterogeneity of tumor cell population before treatment • Cancer stem cells • Increased efflux of drugs from the tumor cell membrane and subsequent drug inactivation • Expression of drug-resistant proteins • Extracellular vesicles • Low/overexpression of the target protein • Inefficient or dysregulation of apoptotic machinery • DNA damage repair • Bone marrow microenvironment • Steric hindrance	Bagrodia et al. (2012)
Acquired resistance	• Heterogeneity of tumor cell population after treatment • Mutation and methylation of drug targets • Mutations in the target gene after treatment • Hyperactivation of prosurvival signaling pathways • Up/downregulation of miRNAs • Cross-talk between intracellular signaling pathways • Chronopharmacology	Bagrodia et al. (2012)

miRNAs, microRNAs.

also develop due to mutations and the methylation of genetic drug targets, as well as the cross-talk between intracellular signaling pathways (Ansari, Shackelford, & El-Osta, 2016; Turtoi, Blomme, & Castronovo, 2015). MDR has also been linked with tumor cell heterogeneity. Recently, emerging evidence indicates that other mechanisms—such as the deregulation of exosomes—may also contribute significantly to chemotherapeutic resistance (Tables 2–4).

Table 2 Exosome-Mediated Direct Drug Efflux in Different Cancers

Tumor	Exosomal Drugs	Sample Source	Influencing Factors	References
Glioblastoma	Temozolomide	Medium of cells	–	Munoz et al. (2013)
Melanoma	Cisplatin	Medium of cells	pH, proton pump inhibitors	Federici et al. (2014)

pH, potential of hydrogen.

Table 3 The Roles of Exosomal Influx in Drug Resistance for Different Cancers

Tumor	Exosomal Contents	Sample Source	Drugs	The Roles in Drug Resistance	References
Breast cancer	HER2 (as an antigen)	Medium of cells, serum	Trastuzumab, lapatinib	Upregulated in trastuzumab No function in lapatinib	Ciravolo et al. (2012)
Breast cancer	miR-221/222	Medium of cells	Tamoxifen	Upregulated	Wei et al. (2014)
Breast cancer	miR-23a, miR-1246, miR-29a, miR-222, miR-452, and so on	Medium of cells	Docetaxel, Adriamycin	Upregulated	Chen et al. (2014a,b), Mao et al. (2016), Yu et al. (2016)
Breast cancer	P-gp	Medium of cells	Docetaxel	Upregulated	Lv et al. (2014)
Breast cancer	BRCP	Medium of cells	Doxorubicin	Upregulated	Kong et al. (2015)
Breast cancer	miR-134	Medium of cells	Anti-Hsp90 drugs	Downregulated	O'Brien et al. (2015)
Breast cancer	miR-34a, miR-452, β-elemene	Medium of cells	Adriacin, docetaxel	Upregulated (miR-34a, miR-452) Downregulated (β-elemene)	Zhang et al. (2015)
Breast cancer	miR-4443, miR-574, miR-7847 and so on	Medium of cells	Docetaxel, epirubicin, vinorelbine	Upregulated	Zhong et al. (2016)
Multiple myeloma	miR-16-5p, miR-15a-5p, miR-20a-5p, miR-17-5p	Serum	Bortezomib	Upregulated	Zhang et al. (2016)
Lymphoma	CD20	Medium of cells	Rituximab	Downregulated	Aung et al. (2011)
Leukemia	–	Medium of cells	Imatinib, dasatinib	Downregulated	Liu et al. (2016)

Table 3   The Roles of Exosomal Influx in Drug Resistance for Different Cancers—cont'd

Tumor	Exosomal Contents	Sample Source	Drugs	The Roles in Drug Resistance	References
Prostate cancer	MDR-1/P-gp	Medium of cells, serum	Docetaxel	Upregulated	Corcoran et al. (2012)
Prostate cancer	miR-485-3p	Medium of cells	Fludarabine	Upregulated	Lucotti et al. (2013)
Prostate cancer	miR-34a	Medium of cells	Docetaxel	Downregulated	Corcoran, Rani, and O'Driscoll (2014)
Prostate cancer	MDR-1, MDR-3, endophilin-A2, and PABP4	Medium of cells, serum	Docetaxel	Upregulated	Kharaziha et al. (2015)
Prostate cancer	P-gp	Medium of cells, serum	Docetaxel	Upregulated	Kato et al. (2015)
Prostate cancer	miR3176 and other 12 miRNAs	Medium of cells	Paclitaxel	4 miRNAs downregulated 9 miRNAs upregulated	Li et al. (2016)
Ovarian cancer	Annexin A3	Medium of cells	Platinum	Upregulated	Yin et al. (2012)
Ovarian cancer	miR-433	Medium of cells	Paclitaxel	Upregulated	Weiner-Gorzel et al. (2015)
Pancreatic cancer	Survivin T34A	Medium of cells	Gemcitabine	Downregulated	Aspe et al. (2014)
Glioma	miR-221	Medium of cells	TMZ	Upregulated	Yang et al. (2016)
Glioblastoma	miR-9	Medium of cells	TMZ	Upregulated	Munoz et al. (2013, 2015)
Glioblastoma	MGMT and APNG related mRNAs	Medium of cells, blood	TMZ	Downregulated	Shao et al. (2015)
Liver cancer	miR122	Medium of cells	Sorafenib	Downregulated	Lou et al. (2015)
Liver cancer	–	Medium of cells	Sorafenib	Upregulated	Qu et al. (2016)
Lung cancer	miR-21, miR-98, miR-133b, miR-138, miR-181a, miR-200c; mRNAs: ERCC1, BRCA1, and RRM1	Medium of cells	Cisplatin	Upregulated (miR-21, ERCC1, BRCA1, and RRM1) Downregulated (miR-98, miR-133b, miR-138, miR-181a, and miR-200c)	Xiao et al. (2014)
Lung cancer	–	Medium of cells	Cisplatin	Upregulated	Li et al. (2016)
Renal cell carcinoma	lncARSR	Medium of cells	Sunitinib	Upregulated	Qu et al. (2016), Stone (2016)
Osteosarcoma	MDR-1 mRNA and its product P-gp	Medium of cells	Doxorubicin	Upregulated	Torreggiani et al. (2016)

APNG, alkylpurine-DNA-*N*-glycosylase; *BCRP*, breast cancer resistance protein; *CDDP*, cisplatin; *HER2*, human epidermal growth factor receptor 2; *Hsp*, heat shock protein; *lncARSR*, lncRNA activated in RCC with sunitinib resistance; *MDR-1*, multidrug resistance protein-1; *MGMT*, O6-methylguanine DNA methyltransferase; *miR*, microRNA; *P-gp*, P-glycoprotein; *TMZ*, temozolomide.

Table 4 Cross-Talk Between Tumor Cells and Stroma Through Exosomes in Different Cancers

Tumor	Exosomal Contents	Microenvironment	Drugs	Mechanism of Exosomes	References
Breast cancer	5′-triphosphate viral RNA	Fibroblasts	–	Increase drug resistance through chemoresistant gene-dependent antiviral signaling	Boelens et al. (2014)
Breast cancer	miR-222/223	MSCs	Carboplatin	Increase drug resistance due to the contribution of miRNAs in exosome to the cell dormancy	Bliss et al. (2016)
Multiple myeloma	–	BMSCs	Bortezomib	Induce chemoresistance through blocking reduction of Bcl-2, cleaved caspase-9, caspase-3, and PARP expression	Wang et al. (2014)
Lymphoma	miRNAs	Fibroblasts	Rapamycin, dexamethasone, ruxolitinib	Target the mismatch repair pathway and BRCA1 pathway	Habiel et al. (2016)
Leukemia	Galectin-3	Fibroblasts	Vincristine, nilotinib, trametinib	Protected pre-B ALL cells through auto-induction of Galectin-3 mRNA and tonic NF-κB pathway activation	Fei et al. (2015)
Leukemia	miR-155, miR-375, TGFB1	BMSCs	–	Downregulate the promoters of apoptosis or cell differentiation	Viola et al. (2016)
Ovarian cancer	miR21	Adipocytes, fibroblasts	Paclitaxel	miR21 binding to APAF1	Au Yeung et al. (2016)
Pancreatic cancer	–	Fibroblasts	Gemcitabine	Increase chemoresistance-inducing factor, snail	Richards et al. (2016)
Gastric cancer	–	MSCs	5-fluorouracil	Induce chemoresistance partly through CaM-Ks/Raf/MEK/ERK pathway	Ji et al. (2015)
Colorectal cancer	–	Fibroblast	5-fluorouracil, oxaliplatin	Decrease chemoresistance through priming cancer stem cells via Wnt signaling pathway before chemotherapy	Hu et al. (2015)
Neuroblastoma	miR-21, miR-155	Human monocytes	CDDP	Increase drug resistance through exosomic miR-21/TLR8-NF-κB/exosomic miR-155/TERF1 signaling pathway	Challagundla et al. (2015)

ALL, acute lymphoblastic leukemia; BMSCs, bone marrow stroma cells; CDDP, cisplatin; EGF, epidermal growth factor; EMT, epithelial-to-mesenchymal transition; ERK, extracellular signal–regulated kinase; HER2, human epidermal growth factor receptor 2; HRG, heregulin; MEK, mitogen-activated protein kinase/ERK kinase; miR, microRNA; MSCs, mesenchymal stem cells; NF-κB, nuclear factor kappa-light-chain-enhancer of activated B cells; TERF1, telomeric repeat-binding factor 1; TLR8, toll-like receptor 8.

3. Mechanisms of Exosome in Drug Resistance

Exosomes are the most important members of extracellular vesicles. Exosomes are 40–100 nm-wide extracellular vesicles that consist of a lipid bilayer membrane surrounding a small cytosol and are devoid of cellular organelles. Exosomes contain various molecular constituents, including proteins, DNAs (single- and double-stranded), and RNAs (message RNA [mRNA], microRNA [miR], and long noncoding RNA [lncRNA]) (Gusachenko, Zenkova, & Vlassov, 2013; Kowal, Tkach, & Thery, 2014; Kruger et al., 2014; Takahashi et al., 2014; Thakur et al., 2014). The contents of exosomes are dependent on several conditions and the donor cell type. Owing to the selective sorting of exosomes within the donor cells, exosomes can also vary in their gene expression profiles. Exosomes are known to confer drug resistance to nonresistant cancer cells through a variety of mechanisms. Substantial data have demonstrated that exosomes play a key role in the intrinsic and acquired drug resistance of different cancers.

3.1 Exosome-Mediated Direct Drug Efflux

Exosomes were initially discovered to be responsible for the release of cellular waste and toxins, which is a mechanism utilized by tumor-derived exosomes (Johnstone et al., 1987). Analysis of exosome shedding in response to drug treatment in various cancer cell lines has revealed the consistent correlation between exosome shedding and drug sensitivity (Kreger et al., 2016; Li et al., 2016; Qu et al., 2016). To negate the cytotoxic effects of drugs, tumor cells can simply sequestrate the drugs and their metabolites and encapsulate them for export (Shedden et al., 2003) (Fig. 1).

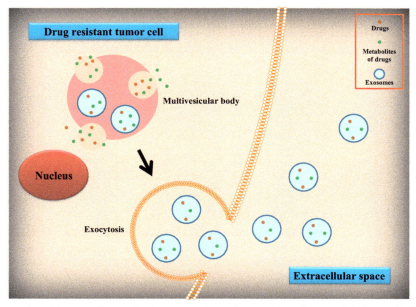

FIGURE 1 Drug-resistant tumor cells can simply sequestrate and encapsulate drugs and their metabolites into tumor-derived exosomes and then directly export these exosomes out into their environment.

3.2 Transportations of Drug Resistance via Exosomal Contents

Drug-resistant tumor cells can transmit resistance to sensitive cells via exosomes, creating new pools of resistant tumor cells. Exosomes are an efficient delivery system for the bioactive molecules that they transport between the donor and recipient cells, and increasing evidence suggests that this leads to the exchange of genetic information and reprogramming of the recipient cells (Kalluri, 2016a, 2016b; Min et al., 2016). Tumor-derived exosomes contain the biomolecular information of RNA in the form of mRNAs, mitochondrial RNAs (mtRNAs), miRs, lncRNAs, and some other noncoding RNAs (van Balkom et al., 2015; Chen et al., 2014; Falcone, Felsani, & D'Agnano, 2015; Milane et al., 2015; Shao et al., 2015; Takahashi et al., 2014). Additionally, exosomes have been confirmed to carry antigens, which counteract antibody-dependent cell cytotoxicity and therefore contribute to treatment failure by neutralizing tumor-reactive antibodies (Aung et al., 2011; Battke et al., 2011) (Fig. 2).

Moreover, tumor-derived exosomes can indirectly confer drug resistance by transporting efflux pump proteins associated with chemoresistance. Indirect drug efflux depends on the tumor cell membrane and the presence of specific membrane-bound proteins. Among the several membrane-bound drug efflux pump families, the most important transmembrane proteins are of the ABC, solute carrier (SLC), and xenobiotic metabolizing enzyme

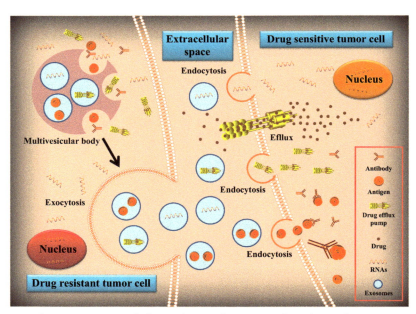

FIGURE 2 There are three ways to transmit drug resistance via exosomes from drug-resistant tumor cells to drug-sensitive tumor cells. First, the biomolecular information contained in exosomes can be in the form of RNA, including messenger RNAs (mRNAs), mitochondrial RNAs (mtRNAs), microRNAs (miRNAs), long noncoding RNAs (lncRNA), and some other noncoding RNAs. Second, exosomes carrying antigens can counteract antibody-dependent cell cytotoxicity. Third, exosomes can also transport chemoresistant-related drug efflux pumps for indirect drug efflux.

(XME) families (Holohan et al., 2013). The ABC transporter family is composed of more than 49 genes that are categorized into 7 subfamilies (A to G) (Khamisipour et al., 2016). Of these, the most involved proteins in drug resistance are ABCB1 (also known as P-gp and MDR1), ABC subfamily G member 2 (ABCG2, also known as breast cancer resistance protein (BCRP)/mitoxantrone resistance protein/placenta-specific ABC protein), and ABC subfamily C member 1 (ABCC1, also known as multidrug resistance–associated protein 1) (Cascorbi & Haenisch, 2010). Among the 52 families of SLC transporter system, the major contributors for cancer drug efflux include SLC21 (SLCO), SLC22, and SLC15 (He, Vasiliou, & Nebert, 2009; Khamisipour et al., 2016). Active XME enzymes facilitate the biotransformation of anticancer drugs to subsequently induce drug resistance (Khamisipour et al., 2016). Cytochrome P-450 isoform families, uridine diphosphate glucuronosyltransferase families, and glutathione S-transferases are among the most prominent families (Khamisipour et al., 2016). Moreover, dysfunctional transporters can increase toxicity by impeding the export of drugs from the cells (Beck, 1991).

3.3 Cross-Talk Between Tumor Cells and the Stroma via Exosomes

Tumor tissues consist of cancer cells and various nontumor cell types including stromal cells, which are sustained by the extracellular matrix along with a vascular network (Joyce & Pollard, 2009). Stromal cells include connective tissue cells, such as fibroblasts and pericytes. The main differences between tumor-associated stromal cells and normal tissues include an increased concentration of fibroblasts, an altered extracellular matrix, and the most common tumor-associated macrophages that are involved in tumorigenesis (Grivennikov, Greten, & Karin, 2010). Fibroblasts are characterized by their high resilience, as they are the only wild-type cell that can be live-cultured from decaying tissue. Cancer-associated fibroblasts are responsible for producing a variety of tumor components (Kalluri, 2016a, 2016b). The tumor stroma is a critical determinant of drug resistance due to the following mechanisms: the reduction of drug distribution throughout the tumor and the restriction of drug delivery due to vascular disorganization (Andre Mdo, Pedro, & Lyden, 2016). Recently, exosomes have been discovered to play a key role in the interactions between tumor and stromal cells. Exosomes derived from tumor cells can provide stromal cells with tumor-promoting activities (Andre Mdo et al., 2016; Battke et al., 2011) (Fig. 3).

4. Exosome-Associated Drug Resistance in Cancers
4.1 Breast Cancer

Breast cancer is one of the three most common cancers worldwide, alongside lung and colon cancer (Harbeck & Gnant, 2016). Breast cancer is the most prevalent cancer that afflicts women, and the World Health Organization estimates that breast cancer represents 16% of all female cancers, with a higher occurrence in developing countries (Malvezzi et al., 2014). Although prognosis has been highly improved by endocrine therapy, antihuman

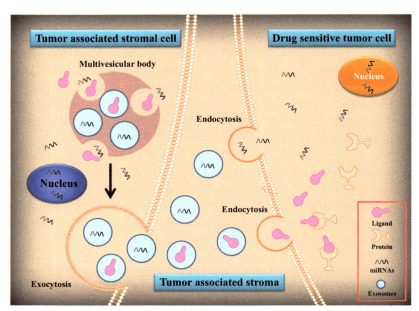

FIGURE 3 The tumor stroma can confer drug resistance to drug-sensitive tumor cells via the exosomes of stromal cells, such as fibroblasts. Exosomal miRNAs and ligands—of cancer-associated fibroblasts, for example—can endow drug resistance.

epidermal growth factor receptor 2 (HER2) targeting, and chemotherapy, some forms of the disease continue to be resistant to current treatments (Cardoso et al., 2016; Harbeck & Gnant, 2016). Recently, exosomes filled with proteins and miRNAs have been observed to influence drug resistance in breast cancer.

Approximately 25% of invasive primary breast cancer tumors exhibit HER2 gene amplification, which enhances the following cellular characteristics: proliferation, migration, and metastasis (Ciravolo et al., 2012; Tagliabue et al., 2010). High levels of activated HER2 were found in HER2-positive exosomes derived from either HER2-positive breast cancer cell lines or from the serum of breast cancer patients with HER2-overexpressed (Ciravolo et al., 2012). The HER2-positive exosomes were found to bind efficiently to trastuzumab (monoclonal antibody targeting HER2) but not lapatinib (a dual tyrosine kinase inhibitor), which decreased the sensitivity of trastuzumab—especially in patients with advanced disease (Ciravolo et al., 2012). Hence, patients with advanced disease and high levels of circulating HER2-positive exosomes could be treated with oral lapatinib (Ciravolo et al., 2012).

Current studies have alluded to the potential responsibility that specific miRNAs may have in influencing drug resistance in breast cancer. For instance, there were significant differences in the concentration and size distribution of exosomes between tamoxifen-sensitive and tamoxifen-resistant breast cancer cells. Exosomes from tamoxifen-resistant breast cancer cells entered tamoxifen-sensitive cells and then released miR-221/222 (Wei et al., 2014). This exosomal miR-221/222 transferred tamoxifen resistance to the tamoxifen-sensitive cells (Wei et al., 2014). Furthermore, enriched miRs have been found in the exosomes of adriamycin- and docetaxel-resistant breast cancer cells (Chen et al., 2014). The exosomes from docetaxel-resistant breast cancer cells were proven to contain different miR profiles when compared with their parent cell (Chen et al., 2014). The 30 most abundant miRs found within the exosomes of adriamycin-resistant breast cancer cells were confirmed to be involved in crucial biological processes (Mao et al., 2016). These processes include the production and release of exosomes; participation in activities of the Wnt signaling pathway, which generally promotes drug resistance when upregulated; and the alteration of gene expression in recipient cells after exosome delivery (Mao et al., 2016). Additionally, selective miRs in adriamycin- and docetaxel-resistant breast cancer cells were found to be transferable to drug-sensitive breast cancer cells via exosomes, which effectively increases chemoresistance (Chen et al., 2014; Mao et al., 2016). After extracting exosomes from adriamycin-resistant breast cancer cells and applying them to adriamycin-sensitive breast cancer cells, the concentration of miR-222 increased in recipient cells (Yu et al., 2016). Meanwhile, miR-222 mimics were able to induce adriamycin resistance, whereas miR-222 inhibitors were able to reverse drug resistance (Yu et al., 2016). Exosome-associated miR-134 was found to be the most downregulated miR in aggressive breast cancer cells (O'Brien et al., 2015). Moreover, to enhance the sensitivity of breast cancer cells to antiheat shock protein 90 (Hsp90) drugs, miR-134 was capable to be delivered through exosomes into the breast cancer cells (O'Brien et al., 2015). Exosomes extracted from the adriamycin- and docetaxel-resistant breast cancer cells raised chemoresistance in the recipient breast cancer cells (Zhang et al., 2015). After a miR profile analysis of the exosomes from adriamycin- and docetaxel-resistant breast cancer cells, 31 miRs correlated with the constant changes associated with MDR (Zhang et al., 2015). Among these miRs, miR-34a and miR-452 were able to be mediated by an MDR-reversing agent in the cells (Zhang et al., 2015). In each breast cancer cell line exhibiting docetaxel-, epirubicin- or vinorelbine resistance, most exosomal miRs had a lower expression level than that of their parental cells, yet a subset of miRs were concentrated in exosomes (Zhong et al., 2016). Among the dysregulated miRs, 22 miRs were upregulated in exosomes and their parental cells (Zhong et al., 2016). Upon further investigation of the miR profiles in preneoadjuvant chemotherapy biopsies and their paired postoperative specimens for 23 breast cancer patients, 12 of the 22 miRs were shown to be significantly upregulated after preneoadjuvant chemotherapy (Zhong et al., 2016).

Tumor-derived exosomes can also transport drug efflux pump proteins that support chemoresistance in breast cancer cells. P-gp, a drug efflux pump protein, can expel drug substrates out of a cell to maintain a nontoxic intracellular environment. Docetaxel resistance has also been shown to be transmittable from docetaxel-resistant to docetaxel-sensitive breast cancer cells via exosomes (Lv et al., 2014). P-gp was detected in exosomes from docetaxel-resistant breast cancer cells, indicating that exosomal P-gp may be a mechanism for inducing drug resistance (Lv et al., 2014). After docetaxel-sensitive breast cancer cells were incubated with exosomes from docetaxel resistance cells, the expression of P-gp increased in the docetaxel-sensitive cells (Lv et al., 2014). Breast cancer cells can also achieve multidrug resistance through other ABC transporters, such as BCRP. BCRP is an ABC family member that exists in the plasma membrane and is known to induce drug resistance in tumor cells (Krishnamurthy & Schuetz, 2005). High concentrations of BCRP were found in the exosomes of breast cancer cells (Kong et al., 2015). After treatment with compounds—bexarotene and guggulsterone—that induce the exosomal secretion of BCRP, chemosensitivity toward doxorubicin increased in breast cancer cells that were originally resistant to doxorubicin (Kong et al., 2015).

Stromal cell communication with breast cancer cells is also a determinant of treatment response. The interaction between stromal and breast cancer cells was able to result in stroma-mediated resistance via paracrine signaling events (Boelens et al., 2014). Through the paracrine signaling pathway, exosomal 5′-triphosphate viral RNA was capable to be transferred from fibroblasts to breast cancer cells, which subsequently stimulated the development of chemoresistant genes associated with antiviral signaling (Boelens et al., 2014). Prolonged dormancy of breast cancer cells in the bone marrow has been proven a cause for recurrence, and dormancy was directly caused by interactions between breast cancer cells and their stromal environment (Bliss et al., 2016; Patel et al., 2014). Naïve mesenchymal stem cells (MSCs) and MSCs exposed to breast cancer cells (primed MSCs) had exosomes containing distinct miR profiles—including miR222/223—which suggests that the cancer cells may influence the MSCs in favor of their survival (Bliss et al., 2016). Furthermore, the exosomes derived from primed MSCs initiated more mitotic quiescence than the exosomes derived from naïve MSCs, which influenced breast cancer cell dormancy (Bliss et al., 2016). To prevent dormancy and chemosensitize the breast cancer cells, antagomirs derived from MSC exosomes were able to target key miRs such as miR-222/223 (Bliss et al., 2016). Antagomirs were known to increase the sensitivity of carboplatin in the mice model of breast cancer (Bliss et al., 2016).

4.2 Hematologic Malignancies

4.2.1 Multiple Myeloma

The second most common hematological malignancy is multiple myeloma, which is characterized by the following: uncontrolled proliferation and accumulation of monoclonal plasma cells in the bone marrow—the monoclonal immunoglobulin fraction exists in the serum or urine; renal failure; and osteolytic bone lesions (Borrello, 2012; Di Marzo et al., 2016; Xu et al., 2012). Currently, the median survival of multiple myeloma has been improved by 5–6 years due to chemotherapy paired with the treatment of proteasome inhibitors, immunomodulatory drugs, corticosteroids, and/or alkylating agents (Kumar et al., 2014). However, drug resistance is still a major challenge to ensure the effectiveness of treatment in multiple myeloma. Emerging studies are suggesting the role of exosomes in inducing drug resistance in multiple myeloma.

Recent studies have uncovered a biomarker for drug resistance in multiple myeloma, which exists in the form of circulating exosome-associated miRs. In both bortezomib-resistant and bortezomib-sensitive patients, the concentrations of exosome-derived RNA were higher within exosomes than in circulation (Zhang et al., 2016). However, the concentration of exosomal RNA was higher in bortezomib-resistant patients than in bortezomib-sensitive patients (Zhang et al., 2016). There was also a significant difference in the exosomal RNA profiles exhibited in bortezomib-resistant and bortezomib-sensitive patients (Zhang et al., 2016). Upon microarray analysis and polymerase chain reaction confirmation, bortezomib resistance was able to be enhanced by the synergistic downregulation of exosomal miR-16-5p, miR-15a-5p, miR-20a-5p, and miR-17-5p (Zhang et al., 2016).

The communication between stromal and multiple myeloma cells can also influence treatment response. Within the bone marrow microenvironment, an exchange of specific cytokines through exosomes was observed between stromal cells of the bone marrow and multiple myeloma cells (Wang et al., 2014). Exosomes from bone marrow stromal cells were able to induce drug resistance to bortezomib by blocking mechanisms that reduce the expression of Bcl-2, cleaved caspase-9, caspase-3, and poly ADP-ribose polymerase (PARP) in multiple myeloma cells (Wang et al., 2014). Moreover, multiple myeloma cell survival and proliferation were inducible by the growth factors and antiapoptotic signaling pathways activated by bone marrow stromal cells (Wang et al., 2014). Exosomes from both human bone marrow stromal cells of multiple myeloma patients and normal donors were found to promote cell survival, replication, and viability in multiple myeloma cells, which increased their bortezomib resistance (Wang et al., 2014).

4.2.2 Lymphoma

Lymphoma causes blood cell tumors developed from lymphocytes. The two main categories of lymphomas are Hodgkin's and non-Hodgkin (Mugnaini & Ghosh, 2016). Drug resistance is a serious and currently unresolved obstacle in lymphoma treatment. Accumulative research on exosomes and their role in lymphomas can provide enough insight to present mechanisms to overcome the hurdle of drug resistance.

Targeted immunotherapy has become a cornerstone of lymphoma therapy. In fact, for most lymphoma patients, anti-CD20 chimeric antibody—one of the first antibodies for malignant B-cell lymphoma—can provide a cure or prolong survival (Aung et al., 2011). However, the prognosis for patients with primary resistance or relapsed aggressive lymphoma remains unfavorable (Aung et al., 2011). The mechanism granting immunotherapy resistance is associated with the release of exosomes from lymphoma cells that contain CD20, which binds to the anti-CD20 antibody and subsequently protects the target cells from an antibody attack (Aung et al., 2011). Additionally, an ABC transporter known to contribute to chemotherapy resistance, namely A3 (ABCA3), was verified to be essential in the enhancement of exosome release from B-cell lymphoma cells (Aung et al., 2011).

T-cell acute lymphoblastic leukemia/lymphoma (T-ALL/LBL) is a specific type of leukemia that overlaps in its characteristics with some types of lymphoma. A significant enrichment of profibrotic transcripts was found in T-LBL biopsies relative to T-ALL biopsies, which suggests that T-ALL and T-LBL may be altered by the stromal cells within the microenvironment (Habiel et al., 2016). Furthermore, T-ALL and T-LBL cells demonstrated an affinity for metastasis to the lung, which suggests a potential interaction between these cancer cells and lung fibroblasts of the stromal microenvironment that can facilitate pulmonary remodeling (Habiel et al., 2016). After coculturing T-ALL cells with fibroblasts, T-ALL cells were found to develop drug resistance to rapamycin, dexamethasone, and ruxolitinib through exosomes (Habiel et al., 2016). Unlike proliferating fibroblasts, senescent fibroblasts were confirmed to release exosomal miRs that target the mismatch repair pathway and BRCA1 pathways (Habiel et al., 2016).

4.2.3 Leukemia

Leukemia is a group of cancers that usually arises in the bone marrow and results in a high number of abnormal white blood cells. Standard treatment includes the combination of chemotherapy, radiation therapy, targeted therapy, a bone marrow transplant, and both supportive and palliative care. Owing to the intrinsic and acquired drug resistance exhibited by these hematologic malignancies, the average 5-year survival rate is below 60% (Bakker et al., 2016). Exosomes released from stromal cells have been reported to reprogram the bone marrow microenvironment to create a niche for leukemia cells, which promotes drug resistance.

The stromal cells in the bone marrow of B-cell precursor acute lymphoblastic leukemia (pre-B ALL) cells were found to contain high levels of Galectin-3, a multifunctional galactose-binding lectin related to drug resistance (Fei et al., 2015). When pre-B ALL cells were cocultured with stromal cells, the stromal cells—especially fibroblasts—were able to transfer Galectin-3 to the pre-B ALL cells via exosomes (Fei et al., 2015). During the development of drug resistance in pre-B ALL cells, the involved stromal cells increased the exosomal secretion of the soluble form of Galectin-3 (Fei et al., 2015). These soluble and exosomal Galectin-3 were internalized by pre-B ALL cells and granted resistance via the autoinduction of Galectin-3 mRNA and the activation of the tonic nuclear factor kappa-light-chain-enhancer of activated B cells (NF-κB) pathway (Fei et al., 2015). In acute myeloid leukemia, the surrounding bone marrow stromal cells also released exosomes that carried miRs and cytokines, which granted chemoresistance by downregulating promoters of apoptosis or cell differentiation (Viola et al., 2016).

Dasatinib—the second generation of SRC family protein tyrosine kinase inhibitors—has superior efficacy in imatinib-resistant chronic myeloid leukemia (CML) patients when compared with its predecessor, but drug resistance is still prevalent (Liu et al., 2016). Imatinib resistance in CML cells was able to be correlated with the following: an increase in released exosomes; activation of the phosphoinositide 3-kinase (PI3K)/protein kinase B (Akt)/mammalian target of rapamycin (mTOR) signaling pathway; and autophagic activity (Liu et al., 2016). In addition, the exosomal release was proven to be augmented by the mTOR-independent beclin-1/Vps34 signaling pathway. Interestingly, this signaling pathway was linked to the observed reduction of phosphatase and tensin homolog due to Notch1 activation (Liu et al., 2016). However, dasatinib was able to enhance apoptosis by targeting the Akt pathway, which prevents mTOR activation. Dasatinib was capable to also decrease exosome release by inhibiting beclin-1/Vps34-dependent autophagy (Liu et al., 2016).

4.3 Prostate cancer

Prostate cancer is the most commonly diagnosed noncutaneous malignancy in men and the second most common cause of death by cancer worldwide (Shukla et al., 2015; Siegel et al., 2014). Although the early detection and treatment of localized prostate cancer has been improved, many patients still die from metastatic disease. The hormone-refractory, castration-resistant, and drug-resistant prostate cancer is especially malevolent. Increasing evidence indicates that exosomes play important roles in inducing castration and drug resistance in prostate cancer.

In hormone-refractory prostate cancer cells, a set of prostate cancer secretory miRs were spontaneously released via exosomes (Lucotti et al., 2013). Notably, in fludarabine-treated prostate cancer cells, there was a reduction in the release of exosomes and exosomal miR-485-3p, which suggests that their uptake by the surviving cell population is beneficial for fludarabine resistance (Lucotti et al., 2013). Fludarabine-resistant prostate cancer cells are dependent upon high levels of intracellular miR-485-3p because it modulates the transcriptional repressor nuclear factor-Y, which triggers the transcription of the following genes: topoisomerase IIα, multidrug resistance gene 1 and cyclin B2 prosurvival (Lucotti et al., 2013). Four miRs were selected after the global miRNA profiling analysis of the docetaxel-resistant prostate cancer cells and their exosomes (Corcoran et al., 2014). Among these miRs, miR-34a was found to reduce the malignancy of prostate cancer by regulating BCL-1 (Corcoran et al., 2014). If miR-34a were introduced into docetaxel-resistant prostate cancer cells, docetaxel sensitivity could be increased (Corcoran et al., 2014). After comparing the exosomal miRNA profiles between the naïve prostate cancer cells and chemoresistant hormone-refractory prostate cancer cells, 29 significantly deregulated miRNAs—including 19 upregulated and 10 downregulated—were identified to exist within the chemoresistant hormone-refractory prostate cancer cells (Li et al., 2016). Further studies have implied that exosomal miRs are capable of affecting genes related to chemoresistance in prostate cancer (Li et al., 2016).

In support of indirect drug efflux, tumor-derived exosomes can also transport drug efflux pump proteins related to chemoresistance in prostate cancer. Exosomes derived from docetaxel-resistant prostate cancer cells were able to confer resistance to docetaxel-sensitive cells partly through the transfer of exosomal MDR-1/P-gp (Corcoran et al., 2012). Furthermore, after longer treatment of docetaxel, the exosomes from patient sera showed greater capacity to induce docetaxel resistance in docetaxel-sensitive prostate cancer cells (Corcoran et al., 2012). Comparative proteomic analysis revealed that different concentrations and molecular compositions existing within the exosomes of docetaxel-sensitive and docetaxel-resistant prostate cancer cells (Kharaziha et al., 2015). Among these proteins, MDR-1 was only enriched in the exosomes of prostate cancer cells and patients sera that exhibited docetaxel resistance (Kharaziha et al., 2015). The exosomes of docetaxel-resistant prostate cancer cells were capable to disseminate docetaxel resistance to the docetaxel-sensitive cells (Kharaziha et al., 2015). P-gp is encoded by the MDR1 gene. The P-gp concentration was higher in exosomes and cell lysates from docetaxel-resistant prostate cancer cells compared with naïve prostate cancer cells (Kato et al., 2015). Similarly, in the docetaxel-resistant patients, P-gp level was also higher in the circulating exosomes than in those analyzed from docetaxel response patients (Kato et al., 2015). Furthermore, MDR1 knockdown improved the sensitivity of docetaxel-resistant prostate cancer cells to docetaxel (Kato et al., 2015).

4.4 Ovarian Cancer

Ovarian cancer is the second most commonly diagnosed gynecological malignancy. In fact, on a global scale, ovarian cancer annually causes 230,000 new cases and 150,000 deaths (Siegel, Miller, & Jemal, 2015). Moreover, the 5-year survival rate for ovarian cancer patients is approximately 45% (Siegel et al., 2015). Drug resistance presents in ovarian cancer and remains a major challenge. Exosomes that are released by drug-resistant ovarian cancer or stromal cells can be recognized and taken up by other cancer cells; in this way, exosomes facilitate the intercellular communication supportive of drug resistance.

Upregulated expression of annexin A3—a Ca^{2+} and phospholipid-binding protein— was a mechanism for platinum resistance in ovarian cancer cells (Yin et al., 2012). In platinum-resistant ovarian cancer patients, the amount of annexin A3 in sera was significantly higher than the concentration present in platinum-sensitive patients, which suggests that the secretion of annexin A3 is positively correlated with platinum resistance (Yin et al., 2012). Meanwhile, the annexin A3 was proven to be released from ovarian cancer cells through exosomes (Yin et al., 2012).

High-grade serous ovarian cancer is the most common and aggressive ovarian cancer subtype that exhibits chemoresistance (Weiner-Gorzel et al., 2015). In high-grade serous ovarian cancer, a high expression of miR-433 was detected (Weiner-Gorzel et al., 2015). Overexpression of miR-433 in ovarian cancer cells could increase paclitaxel resistance via the inhibition of apoptosis and induction of cellular senescence (Furlong et al., 2012; Weiner-Gorzel et al., 2015). miR-433 induced cellular senescence by downregulating cyclin-dependent kinases 6 (Weiner-Gorzel et al., 2015). The ovarian cancer cells with high miR-433 expression successfully transferred paclitaxel resistance to neighboring cancer cells by the release of miR-433 packaged into exosomes (Weiner-Gorzel et al., 2015).

Aggressive ovarian cancer usually spreads to the visceral adipose tissue of the omentum (Au Yeung et al., 2016). Exosome and tissue lysates were isolated from ovarian cancer cells and cancer-associated adipocytes and fibroblasts. These lysates were analyzed and compared with next-generation sequencing technology, which revealed a significantly higher expression of miR21 in the exosomes and tissue of cancer-associated adipocytes and fibroblasts (Au Yeung et al., 2016). miR21 enables paclitaxel resistance as miR21 binds to paclitaxel's direct target, APAF1 (Weiner-Gorzel et al., 2015). Moreover, exosomes transferred miR21 from cancer-associated adipocytes and fibroblasts to ovarian cancer cells (Au Yeung et al., 2016).

4.5 Pancreatic Cancer

Pancreatic cancer, with a 5-year survival rate of less than 7%, is the fourth leading cause of cancer-related deaths (Rahib et al., 2014; Stathis & Moore, 2010). The poor prognosis of this disease is partially attributed to patient response to available therapies. Increasing data suggest that exosomes are also a culprit for poor prognosis.

The protein survivin inhibits apoptosis and is thus a key contributing factor to cellular resistance of apoptosis. The dominant-negative mutant of survivin, survivin T34A, has been shown to block survivin to subsequently induce caspase activation and apoptosis (Aspe et al., 2014). Scientists repackaged exosomes isolated from a melanoma cell line to carry survivin T34A (Aspe et al., 2014). Pancreatic cancer cells containing exosomes carrying survivin T34A have been found to significantly augment sensitivity to gemcitabine (Aspe et al., 2014).

Pancreatic ductal adenocarcinomas are mostly comprised of cancer-associated fibroblasts, which maintain an intrinsic resistance to gemcitabine (Richards et al., 2016). Under the treatment of gemcitabine, cancer-associated fibroblasts increase their release of exosomes, which protects pancreatic cancer cells by promoting drug resistance (Richards et al., 2016). Further research indicated that the exosomes from cancer-associated fibroblasts targeted the chemoresistance-inducing factor, Snail (Richards et al., 2016).

4.6 Brain Cancer

Gliomas are the most common primary tumors of the central nervous system, and they have a high incidence of metastasis and drug resistance, which endows them with a reputation for being one of the most aggressive and deadliest of cancers (Louis, 2006; Omuro & DeAngelis, 2013). Among all gliomas, glioblastoma constitutes 55% all primary gliomas, and patients afflicted with glioblastoma survive less than 1 year, on average (Yang et al., 2016). Drug resistance in gliomas arises due to genetic cellular heterogeneity in combination with an academically underdeveloped understanding of the precise mechanisms endowing the glioma's degree of aggression. However, recent studies in brain cancer have linked heightened drug resistance with increased exosomal secretion.

Elevated miR-221 levels in tumor tissue and exosomes were proven to be positively correlated with temozolomide resistance in glioma (Yang et al., 2016). Further analyses revealed that exosomal miR-221 can be mediated by RELA, a well-documented regulator of a variety of oncogenic pathways. Exosomal miR-221 directly targeted and inhibited the expression of the DNM3 gene, a tumor suppressive gene (Yang et al., 2016; Zhang et al., 2016). The RELA/miR-221 axis is a therapeutic target for overcoming drug resistance (Yang et al., 2016). In temozolomide-resistant glioblastoma cells, a high expression of exosomal miR-9 was observed, which indicates that chemoresistance in glioblastoma may be attributable to the upregulation of exosomal miR-9 (Munoz et al., 2013). In addition, an increased level of temozolomide was detected in the exosomes derived from the temozolomide-dresistant glioblastoma cells, confirming the mechanism of direct drug efflux via exosomes (Munoz et al., 2013). After the transmission of exosomal anti-miR-9 from MSCs to temozolomide-resistant glioblastoma cells, temozolomide resistance was significantly downregulated. This event suggests that miR-9 can induce temozolomide resistance via the activation of the sonic hedgehog pathway (Munoz et al., 2013, 2015). O6-methylguanine DNA methyltransferase and alkylpurine-DNA-N-glycosylase are key enzymes that repair temozolomide-induced DNA damages, and their expression levels are inversely related to treatment efficacy in glioblastoma (Shao et al., 2015). The levels of exosomal mRNAs related to these two enzymes were associated with those of the original glioblastoma cells (Shao et al., 2015). Compared with temozolomide-sensitive glioblastoma cells, higher levels of exosomal mRNAs related to these two enzymes were detected in the temozolomide-resistant glioblastoma cells (Shao et al., 2015). The mRNAs associated with these two enzymes were reliably detected in the exosomes isolated from the blood of glioblastoma patients, and its expression levels were positively correlated with temozolomide resistance (Shao et al., 2015).

4.7 Liver Cancer

The fifth most common cancer and second leading cause of cancer-related deaths at 700,000 deaths per year on a global scale is hepatocellular carcinoma (Torre et al., 2015). Currently, surgery gives hope to early-stage patients; however, late diagnosis and chemoresistance causes the 5-year survival rate to be less than 20% (El-Serag, 2011). Recently, exosomes have attracted much interest for their potential to combat drug resistance to chemotherapies.

miR-122 has been reported to promote chemosensitivity of hepatocellular carcinoma cells (Xu et al., 2011). The MSCs derived from adipose tissue were used to produce large amounts of exosomes packaged with miR-122, via transfection with a miR-122 expression plasmid (Lou et al., 2015). Through in vitro and in vivo studies, the exosomes containing miR-122 were verified to be transferred from adipose tissue–derived MSCs to hepatocellular carcinoma cells, which significantly increased sensitivity to sorafenib in hepatocellular carcinoma cells (Lou et al., 2015).

Further investigation revealed that exosomes derived from hepatocellular carcinoma cells induced sorafenib resistance in vitro by activating the hepatocyte growth factor/c-Met/Akt signaling pathway and inhibiting sorafenib-induced apoptosis (Qu et al., 2016). Meanwhile, the exosomes derived from hepatocellular carcinoma cells induced sorafenib resistance in vivo by the inhibition of sorafenib-induced apoptosis (Qu et al., 2016). Moreover, the degree of invasiveness observed in hepatocellular carcinoma cells can cause the spread of sorafenib resistance observed in tumor-derived exosomes (Qu et al., 2016).

4.8 Lung Cancer

Lung cancer is the leading cause of cancer-related mortality in the world. Non–small-cell lung cancer is the most common type of lung cancer, and it has a poor 5-year survival rate of less than 20%, which is partially due to drug resistance (Zhang, 2016; Quintanal-Villalonga et al., 2016). Exosomes have been recently considered for their potential to enable a novel way of overcoming drug resistance in lung cancer.

In lung cancer cells, cisplatin is verified to strengthen exosomal secretion (Xiao et al., 2014). Moreover, exosomes derived from cisplatin-treated lung cancer cells can increase chemoresistance of naïve lung cancer cells (Xiao et al., 2014). When naïve lung cancer cells were treated with exosomes derived from cisplatin-treated lung cancer cells, expression levels of miR-21 specifically along with other mRNAs increased. Also, the expression levels of some miRs decreased in these exosomally treated naïve lung cancer cells, which indicates that exosomal contents can change the miRs and mRNAs expressed in the recipient cancer cells (Xiao et al., 2014).

Gefitinib is a widely used treatment for non–small-cell lung cancer patients who harbor the epidermal growth factor receptor mutation (Li et al., 2016). Exosomes derived from gefitinib-treated lung cancer cells were capable of decreasing the anti-tumor effects of cisplatin by inhibiting apoptosis while inducing autophagy (Li et al., 2016). However, exosomes derived from cisplatin-treated lung cancer cells did not significantly change the antitumor effects of gefitinib (Li et al., 2016). In addition, inhibition of exosome secretion led to a modest synergistic effect when cisplatin and gefitinib were administered in combination (Li et al., 2016). These results suggest that in lung cancer cells, exosomes are critical for cisplatin resistance after the pretreatment of gefitinib (Li et al., 2016).

4.9 Gastric Cancer

Gastric cancer is the fourth most common malignant tumor worldwide and the second most frequent cause of cancer death after lung cancer (Orditura et al., 2014; Sun et al., 2013). Chemotherapy with 5-fluorouracil and cisplatin has improved survival rates, yet the development of resistance is one of the most significant obstacles to effective gastric cancer therapy (Orditura et al., 2014; Sun et al., 2013). Recent studies have shown that the exosomes derived from stromal cells can communicate with gastric cancer cells to influence the treatment response. MSCs, a distinct group of cells with self-renewal and multilineage differentiation abilities, participate in the formation of the tumor microenvironment (Ji et al., 2015). Exosomes derived from MSCs were proven to induce chemoresistance of 5-fluorouracil in gastric cancer cells in vivo and ex vivo (Ji et al., 2015). Rather than the direct transfer of drug resistance via proteins or miRNAs, the underlying mechanism of drug resistance was at least partially due to the activation of a pathway in gastric cancer cells called CaM-Ks/Raf/mitogen-activated protein kinase/ERK kinase/extracellular-signal-regulated kinase (Ji et al., 2015).

4.10 Colorectal Cancer

Recent estimations indicate that colorectal cancer is the third most common cancer in western countries (Siegel et al., 2014). Although the survival rate has been improved, colorectal cancer still accounts for approximately 9% of cancer-related deaths due to drug resistance (Siegel et al., 2014). The role of exosomes in drug resistance via communication between stromal cells and colorectal cancer cells has been investigated. Colorectal cancer stem cells isolated from either primary colorectal cancer tissues or colorectal cancer xenografts were found to have an inherent resistance to 5-fluorouracil or oxaliplatin (Hu et al., 2015). Exosomes derived from carcinoma-associated fibroblasts were demonstrated to transfer and enhance drug resistance to colorectal cancer stem cells by the Wnt signaling pathway (Hu et al., 2015).

4.11 Melanoma

Melanoma, as one of the leading causes of death from skin cancer, is developed through the malignant transformation of melanocytes (Wong et al., 2013). Despite recent advances in chemotherapy and immunotherapy, therapeutic options for chemoresistant melanoma cells remain limited. Nowadays, exosomes have been recognized as an important factor in the regulation of chemoresistance in melanoma. In cisplatin-resistant melanoma cells, exosomes were proven to contribute to the elimination of drugs, which protected melanoma cells (Federici et al., 2014). Furthermore, the pretreatment of proton pump inhibitors not only decreased the concentration of cisplatin in the exosomes of melanoma cells but also significantly inhibited the release of exosomal cisplatin from melanoma cells, which suggests that proton pump inhibitors may increase chemosensitivity in melanoma (Federici et al., 2014).

4.12 Renal Cell Carcinoma

Renal cell carcinoma accounts for more than 90% of all kidney cancer cases, and is the 10th most common malignancy (Garcia, Cowey, & Godley, 2009). Kidney tumors maintain a high potential for exhibiting drug resistance leading to a poor prognosis. For example, sunitinib resistance is one of the major challenges in the treatment of advanced renal cell carcinoma (Stone, 2016; Qu et al., 2016). Increasing evidence indicates that exosomes maintain a key role in the development of sunitinib resistance in renal cell carcinoma. In sunitinib-resistant renal carcinoma cells, lncARSR (lncRNA activated in renal cell carcinoma (RCC) with sunitinib resistance) was found to correlate with clinically poor sunitinib response by the competitive binding of miR-34/miR-449, which facilitates AXL and c-MET expression (Stone, 2016; Qu et al., 2016). Importantly, the lncARSR in sunitinib-resistant renal carcinoma cells was packaged into exosomes and transmitted to drug-sensitive cells, which conferred sunitinib resistance (Stone, 2016; Qu et al., 2016).

4.13 Osteosarcoma

Osteosarcoma is a high-grade primary skeletal malignancy of mesenchymal origin (Durfee, Mohammed, & Luu, 2016). It accounts for less than 1% of all newly diagnosed cancers in adults and 3%–5% of those in children, but it is the most common primary malignancy in adolescents outside of leukemia and lymphoma (Durfee et al., 2016). In the past 30 years, neoadjuvant chemotherapy has highly improved the prognosis of osteosarcoma patients; however, chemoresistance has slowed the progress in recent years. Exosomes, as a potential means to overcome chemoresistance, have been given high attention. In doxorubicin-resistant osteosarcoma cells, the exosomes from these cells were capable of detection (Torreggiani et al., 2016). Incubation of doxorubicin-sensitive osteosarcoma cells with exosomes extracted from doxorubicin-resistant osteosarcoma cells increased the chemoresistance of the originally sensitive cells (Torreggiani et al., 2016). Mechanically, the exosome-mediated transfer of drug resistance was proven to be caused by MDR-1 mRNA and its product P-gp from the exosomes of doxorubicin-resistant osteosarcoma cells (Torreggiani et al., 2016).

4.14 Neuroblastoma

Neuroblastoma is the most common solid malignancy in children outside of the skull (Cohn et al., 2009). The prognosis of patients with high-risk neuroblastoma is still poor due to chemoresistance (Pearson et al., 2008). Exosomes may provide a potential mechanism for reversing chemoresistance in neuroblastoma.

miR-21 and miR-155 are both regarded as oncogenic miRNAs. The neuroblastoma cells are able to transfer exosomal miR-21 to neighboring human monocytes and affect monocytes by binding to miR-21 with the toll-like receptor 8 (TLR8), which subsequently activates the pathway called NF-κB (Challagundla et al., 2015). Afterward, miR-155 was upregulated due to the miR-21/TLR8-NF-κB/miR-155 pathway in the affected monocytes and transferred back to neuroblastoma cells via exosomes (Torreggiani et al., 2016). This finding suggests that a high level of miR-155 in neuroblastoma cells is ectogenic, as it is produced in a cell other than that expected (Challagundla et al., 2015). Then, in neuroblastoma cells, miR-155 directly targeted telomeric repeat-binding factor 1 and subsequently increased drug resistance (Challagundla et al., 2015; Deville et al., 2011; Diotti & Loayza, 2011; Yamada et al., 2011). Interfering with exosomal secretion in neuroblastoma cells may be a potential method for overcoming drug resistance.

5. Conclusion

The development of intrinsic and acquired drug resistance in cancer is a major obstacle for effective therapeutic treatment. There are multiple mechanisms causing drug resistance in cancer cells among and within cancer types. Specifically, exosomes have emerged as previously unrecognized vehicles that have key roles in the progression of drug resistance. A large amount of evidence supports that drug-resistant cancer-derived exosomes confer drug resistance to drug-sensitive cells. For instance, exosomes have been shown to contribute to drug resistance of tumors from various cancer types. All in all, further basic and clinical studies on drug-resistant tumor-derived exosomes hold great potential to prolong the efficacy of chemotherapy as well as provide mechanisms for additional targeted therapies, which can improve the clinical outcome of cancer patients.

References

Andaloussi, S. E. L., et al. (2013). Extracellular vesicles: Biology and emerging therapeutic opportunities. *Nature Reviews Drug Discovery, 12*(5), 347–357.

Andre Mdo, R., Pedro, A., & Lyden, D. (2016). Cancer exosomes as mediators of drug resistance. *Methods in Molecular Biology, 1395,* 229–239.

Ansari, J., Shackelford, R. E., & El-Osta, H. (2016). Epigenetics in non-small cell lung cancer: From basics to therapeutics. *Translational Lung Cancer Research, 5*(2), 155–171.

Aspe, J. R., et al. (2014). Enhancement of Gemcitabine sensitivity in pancreatic adenocarcinoma by novel exosome-mediated delivery of the Survivin-T34A mutant. *Journal of Extracellular Vesicles, 3.*

Au Yeung, C. L., et al. (2016). Exosomal transfer of stroma-derived miR21 confers paclitaxel resistance in ovarian cancer cells through targeting APAF1. *Nature Communications, 7,* 11150.

Aung, T., et al. (2011). Exosomal evasion of humoral immunotherapy in aggressive B-cell lymphoma modulated by ATP-binding cassette transporter A3. *Proceedings of the National Academy of Sciences of the United States of America, 108*(37), 15336–15341.

Bagrodia, S., Smeal, T., & Abraham, R. T. (2012). Mechanisms of intrinsic and acquired resistance to kinase-targeted therapies. *Pigment Cell & Melanoma Research, 25*(6), 819–831.

Bakker, E., et al. (2016). The role of microenvironment and immunity in drug response in leukemia. *Biochimica et Biophysica Acta, 1863*(3), 414–426.

Battke, C., et al. (2011). Tumour exosomes inhibit binding of tumour-reactive antibodies to tumour cells and reduce ADCC. *Cancer Immunology, Immunotherapy, 60*(5), 639–648.

Beck, W. T. (1991). Modulators of P-glycoprotein-associated multidrug resistance. *Cancer Treatment and Research, 57,* 151–170.

Bliss, S. A., et al. (2016). Mesenchymal stem cell-derived exosomes stimulate cycling quiescence and early breast cancer dormancy in bone marrow. *Cancer Research, 76*(19), 5832–5844.

Boelens, M. C., et al. (2014). Exosome transfer from stromal to breast cancer cells regulates therapy resistance pathways. *Cell, 159*(3), 499–513.

Borrello, I. (2012). Can we change the disease biology of multiple myeloma? *Leukemia Research, 36*(Suppl. 1), S3–S12.

Cardoso, F., et al. (2016). Research needs in breast cancer. *Annals of Oncology, 28*(2).

Cascorbi, I., & Haenisch, S. (2010). Pharmacogenetics of ATP-binding cassette transporters and clinical implications. *Methods in Molecular Biology, 596,* 95–121.

Challagundla, K. B., et al. (2015). Exosome-mediated transfer of microRNAs within the tumor microenvironment and neuroblastoma resistance to chemotherapy. *Journal of the National Cancer Institute, 107*(7).

Chen, W. X., et al. (2014a). Exosomes from docetaxel-resistant breast cancer cells alter chemosensitivity by delivering microRNAs. *Tumor Biologyogy, 35*(10), 9649–9659.

Chen, W. X., et al. (2014b). Exosomes from drug-resistant breast cancer cells transmit chemoresistance by a horizontal transfer of microRNAs. *PLoS One, 9*(4), e95240.

Ciravolo, V., et al. (2012). Potential role of HER2-overexpressing exosomes in countering trastuzumab-based therapy. *Journal of Cellular Physiology, 227*(2), 658–667.

Cohn, S. L., et al. (2009). The International Neuroblastoma Risk Group (INRG) classification system: An INRG Task Force report. *Journal of Clinical Oncology, 27*(2), 289–297.

Corcoran, C., et al. (2012). Docetaxel-resistance in prostate cancer: Evaluating associated phenotypic changes and potential for resistance transfer via exosomes. *PLoS One, 7*(12), e50999.

Corcoran, C., Rani, S., & O'Driscoll, L. (2014). miR-34a is an intracellular and exosomal predictive biomarker for response to docetaxel with clinical relevance to prostate cancer progression. *The Prostate*, *74*(13), 1320–1334.

Deville, L., et al. (2011). hTERT promotes imatinib resistance in chronic myeloid leukemia cells: therapeutic implications. *Molecular Cancer Therapeutics*, *10*(5), 711–719.

Di Marzo, L., et al. (2016). Microenvironment drug resistance in multiple myeloma: Emerging new players. *Oncotarget*, *7*(37).

Diotti, R., & Loayza, D. (2011). Shelterin complex and associated factors at human telomeres. *Nucleus*, *2*(2), 119–135.

Durfee, R. A., Mohammed, M., & Luu, H. H. (2016). Review of osteosarcoma and current management. *Rheumatology and Therapy*, *3*(2), 221–243.

El-Serag, H. B. (2011). Hepatocellular carcinoma. *The New England Journal of Medicine*, *365*(12), 1118–1127.

Falcone, G., Felsani, A., & D'Agnano, I. (2015). Signaling by exosomal microRNAs in cancer. *Journal of Experimental & Clinical Cancer Research*, *34*, 32.

Federici, C., et al. (2014). Exosome release and low pH belong to a framework of resistance of human melanoma cells to cisplatin. *PLoS One*, *9*(2), e88193.

Fei, F., et al. (2015). B-cell precursor acute lymphoblastic leukemia and stromal cells communicate through Galectin-3. *Oncotarget*, *6*(13), 11378–11394.

Furlong, F., et al. (2012). Low MAD2 expression levels associate with reduced progression-free survival in patients with high-grade serous epithelial ovarian cancer. *The Journal of Pathology*, *226*(5), 746–755.

Garcia, J. A., Cowey, C. L., & Godley, P. A. (2009). Renal cell carcinoma. *Current Opinion in Oncology*, *21*(3), 266–271.

Goodman, L. S., Wintrobe, M. M., et al. (1946). Nitrogen mustard therapy; use of methyl-bis (beta-chloroethyl) amine hydrochloride and tris (beta-chloroethyl) amine hydrochloride for Hodgkin's disease, lymphosarcoma, leukemia and certain allied and miscellaneous disorders. *The Journal of the American Medical Association*, *132*, 126–132.

Gottesman, M. M., Fojo, T., & Bates, S. E. (2002). Multidrug resistance in cancer: Role of ATP-dependent transporters. *Nature Reviews Cancer*, *2*(1), 48–58.

Gottesman, M. M., et al. (2016). Toward a better understanding of the complexity of cancer drug resistance. *Annual Review of Pharmacology and Toxicology*, *56*, 85–102.

Grivennikov, S. I., Greten, F. R., & Karin, M. (2010). Immunity, inflammation, and cancer. *Cell*, *140*(6), 883–899.

Gusachenko, O. N., Zenkova, M. A., & Vlassov, V. V. (2013). Nucleic acids in exosomes: Disease markers and intercellular communication molecules. *Biochemistry*, *78*(1), 1–7.

Habiel, D. M., et al. (2016). Senescent stromal cell-induced divergence and therapeutic resistance in T cell acute lymphoblastic leukemia/lymphoma. *Oncotarget*, *7*(50).

Harbeck, N., & Gnant, M. (2017). Breast cancer. *Lancet*, *389*(10074), 1134–1150.

He, L., Vasiliou, K., & Nebert, D. W. (2009). Analysis and update of the human solute carrier (SLC) gene superfamily. *Human Genomics*, *3*(2), 195–206.

Hojjat-Farsangi, M. (2016). Targeting non-receptor tyrosine kinases using small molecule inhibitors: An overview of recent advances. *Journal of Drug Targeting*, *24*(3), 192–211.

Holohan, C., et al. (2013). Cancer drug resistance: An evolving paradigm. *Nature Reviews Cancer*, *13*(10), 714–726.

Hu, Y., et al. (2015). Fibroblast-derived exosomes contribute to chemoresistance through priming cancer stem cells in colorectal cancer. *PLoS One*, *10*(5), e0125625.

Ji, R., et al. (2015). Exosomes derived from human mesenchymal stem cells confer drug resistance in gastric cancer. *Cell Cycle, 14*(15), 2473–2483.

Johnstone, R. M., et al. (1987). Vesicle formation during reticulocyte maturation. Association of plasma membrane activities with released vesicles (exosomes). *Journal of Biological Chemistry, 262*(19), 9412–9420.

Joyce, J. A., & Pollard, J. W. (2009). Microenvironmental regulation of metastasis. *Nature Reviews Cancer, 9*(4), 239–252.

Kaiser, J. (2011). Combining targeted drugs to stop resistant tumors. *Science, 331*(6024), 1542–1545.

Kalluri, R. (2016a). The biology and function of exosomes in cancer. *Journal of Clinical Investigation, 126*(4), 1208–1215.

Kalluri, R. (2016b). The biology and function of fibroblasts in cancer. *Nature Reviews Cancer, 16*(9), 582–598.

Kato, T., et al. (2015). Serum exosomal P-glycoprotein is a potential marker to diagnose docetaxel resistance and select a taxoid for patients with prostate cancer. *Urologic Oncology, 33*(9), 385 e15-20.

Khamisipour, G., et al. (2016). Mechanisms of tumor cell resistance to the current targeted-therapy agents. *Tumor Biology, 37*(8), 10021–10039.

Kharaziha, P., et al. (2015). Molecular profiling of prostate cancer derived exosomes may reveal a predictive signature for response to docetaxel. *Oncotarget, 6*(25), 21740–21754.

Kong, J. N., et al. (2015). Guggulsterone and bexarotene induce secretion of exosome-associated breast cancer resistance protein and reduce doxorubicin resistance in MDA-MB-231 cells. *International Journal of Cancer, 137*(7), 1610–1620.

Kowal, J., Tkach, M., & Thery, C. (2014). Biogenesis and secretion of exosomes. *Current Opinion in Cell Biology, 29*, 116–125.

Kreger, B. T., et al. (2016). The enrichment of Survivin in exosomes from breast cancer cells treated with paclitaxel promotes cell survival and chemoresistance. *Cancers, 8*(12).

Krishnamurthy, P., & Schuetz, J. D. (2005). The ABC transporter ABCG2/BCRP: Role in hypoxia mediated survival. *Biometals, 18*(4), 349–358.

Kruger, S., et al. (2014). Molecular characterization of exosome-like vesicles from breast cancer cells. *BMC Cancer, 14*, 44.

Kumar, S. K., et al. (2014). Continued improvement in survival in multiple myeloma: Changes in early mortality and outcomes in older patients. *Leukemia, 28*(5), 1122–1128.

Li, J., et al. (2016). Exosome-derived microRNAs contribute to prostate cancer chemoresistance. *International Journal of Oncology, 49*(2), 838–846.

Li, X. Q., et al. (2016). Exosomes derived from gefitinib-treated EGFR-mutant lung cancer cells alter cisplatin sensitivity via up-regulating autophagy. *Oncotarget, 7*(17), 24585–24595.

Liu, J., et al. (2016). Distinct dasatinib-induced mechanisms of apoptotic response and exosome release in imatinib-resistant human chronic myeloid leukemia cells. *International Journal of Molecular Sciences, 17*(4), 531.

Longley, D. B., & Johnston, P. G. (2005). Molecular mechanisms of drug resistance. *The Journal of Pathology, 205*(2), 275–292.

Lou, G., et al. (2015). Exosomes derived from miR-122-modified adipose tissue-derived MSCs increase chemosensitivity of hepatocellular carcinoma. *Journal of Hematology & Oncology, 8*, 122.

Louis, D. N. (2006). Molecular pathology of malignant gliomas. *Annual Review of Pathology, 1*, 97–117.

Lucotti, S., et al. (2013). Fludarabine treatment favors the retention of miR-485-3p by prostate cancer cells: Implications for survival. *Molecular Cancer, 12*(1), 52.

Lv, M. M., et al. (2014). Exosomes mediate drug resistance transfer in MCF-7 breast cancer cells and a probable mechanism is delivery of P-glycoprotein. *Tumor Biologyogy, 35*(11), 10773–10779.

Malvezzi, M., et al. (2014). European cancer mortality predictions for the year 2014. *Annals of Oncology*, *25*(8), 1650–1656.

Mao, L., et al. (2016). Exosomes decrease sensitivity of breast cancer cells to adriamycin by delivering microRNAs. *Tumor Biologyogy*, *37*(4), 5247–5256.

Milane, L., et al. (2015). Exosome mediated communication within the tumor microenvironment. *Journal of Controlled Release*, *219*, 278–294.

Min, L., et al. (2016). The roles and implications of exosomes in sarcoma. *Cancer and Metastasis Reviews*, *35*(3), 377–390.

Miyawaki, A. (2011). Proteins on the move: Insights gained from fluorescent protein technologies. *Nature Reviews Molecular Cell Biology*, *12*(10), 656–668.

Mugnaini, E. N., & Ghosh, N. (2016). Lymphoma. *Primary Care*, *43*(4), 661–675.

Munoz, J. L., et al. (2013). Delivery of functional anti-miR-9 by mesenchymal stem cell-derived exosomes to glioblastoma multiforme cells conferred chemosensitivity. *Molecular Therapy – Nucleic Acids*, *2*, e126.

Munoz, J. L., et al. (2015). Temozolomide resistance and tumor recurrence: Halting the Hedgehog. *Cancer Cell Microenvironment*, *2*(2).

O'Brien, K., et al. (2015). miR-134 in extracellular vesicles reduces triple-negative breast cancer aggression and increases drug sensitivity. *Oncotarget*, *6*(32), 32774–32789.

Omuro, A., & DeAngelis, L. M. (2013). Glioblastoma and other malignant gliomas: A clinical review. *JAMA*, *310*(17), 1842–1850.

Orditura, M., et al. (2014). Treatment of gastric cancer. *World Journal of Gastroenterology*, *20*(7), 1635–1649.

Patel, S. A., et al. (2014). Treg/Th17 polarization by distinct subsets of breast cancer cells is dictated by the interaction with mesenchymal stem cells. *The Journal of Cancer Stem Cell Research*, *2014*(2).

Pearson, A. D., et al. (2008). High-dose rapid and standard induction chemotherapy for patients aged over 1 year with stage 4 neuroblastoma: A randomised trial. *The Lancet Oncology*, *9*(3), 247–256.

Perez-Herrero, E., & Fernandez-Medarde, A. (2015). Advanced targeted therapies in cancer: Drug nano-carriers, the future of chemotherapy. *European Journal of Pharmaceutics and Biopharmaceutics*, *93*, 52–79.

Qu, L., et al. (2016). Exosome-transmitted lncARSR promotes sunitinib resistance in renal cancer by acting as a competing endogenous RNA. *Cancer Cell*, *29*(5), 653–668.

Qu, Z., et al. (2016). Exosomes derived from HCC cells induce sorafenib resistance in hepatocellular carcinoma both in vivo and in vitro. *Journal of Experimental & Clinical Cancer Research*, *35*(1), 159.

Quintanal-Villalonga, A., et al. (2016). Tyrosine kinase receptor landscape in lung Cancer: Therapeutical implications. *Disease Markers*, *2016*, 9214056.

Rahib, L., et al. (2014). Projecting cancer incidence and deaths to 2030: The unexpected burden of thyroid, liver, and pancreas cancers in the United States. *Cancer Research*, *74*(11), 2913–2921.

Richards, K. E., et al. (2016). Cancer-associated fibroblast exosomes regulate survival and proliferation of pancreatic cancer cells. *Oncogene*, *36*(13).

Shao, H., et al. (2015). Chip-based analysis of exosomal mRNA mediating drug resistance in glioblastoma. *Nature Communications*, *6*, 6999.

Shedden, K., et al. (2003). Expulsion of small molecules in vesicles shed by cancer cells: Association with gene expression and chemosensitivity profiles. *Cancer Research*, *63*(15), 4331–4337.

Shukla, M. E., et al. (2015). Evaluation of the current prostate cancer staging system based on cancer-specific mortality in the surveillance, epidemiology, and end results database. *Clinical Genitourinary Cancer*, *13*(1), 17–21.

Siegel, R., et al. (2014). Cancer statistics, 2014. *CA: A Cancer Journal for Clinicians*, *64*(1), 9–29.

Siegel, R. L., Miller, K. D., & Jemal, A. (2015). Cancer statistics, 2015. *CA: A Cancer Journal for Clinicians*, *65*(1), 5–29.

Stathis, A., & Moore, M. J. (2010). Advanced pancreatic carcinoma: Current treatment and future challenges. *Nature Reviews Clinical Oncology*, *7*(3), 163–172.

Stone, L. (2016). Kidney cancer: Exosome transmission of sunitinib resistance. *Nature Reviews Urology*, *13*(6), 297.

Sugano, K., et al. (2010). Coexistence of passive and carrier-mediated processes in drug transport. *Nature Reviews Drug Discovery*, *9*(8), 597–614.

Sun, J., et al. (2013). Clinical significance of palliative gastrectomy on the survival of patients with incurable advanced gastric cancer: A systematic review and meta-analysis. *BMC Cancer*, *13*, 577.

Tagliabue, E., et al. (2010). HER2 as a target for breast cancer therapy. *Expert Opinion on Biological Therapy*, *10*(5), 711–724.

Takahashi, K., et al. (2014). Extracellular vesicle-mediated transfer of long non-coding RNA ROR modulates chemosensitivity in human hepatocellular cancer. *FEBS Open Bio*, *4*, 458–467.

Thakur, B. K., et al. (2014). Double-stranded DNA in exosomes: A novel biomarker in cancer detection. *Cell Research*, *24*(6), 766–769.

Torre, L. A., et al. (2015). Global cancer statistics, 2012. *CA: A Cancer Journal for Clinicians*, *65*(2), 87–108.

Torreggiani, E., et al. (2016). Multimodal transfer of MDR by exosomes in human osteosarcoma. *International Journal of Oncology*, *49*(1), 189–196.

Turtoi, A., Blomme, A., & Castronovo, V. (2015). Intratumoral heterogeneity and consequences for targeted therapies. *Bulletin Du Cancer*, *102*(1), 17–23.

van Balkom, B. W., et al. (2015). Quantitative and qualitative analysis of small RNAs in human endothelial cells and exosomes provides insights into localized RNA processing, degradation and sorting. *Journal of Extracellular Vesicles*, *4*, 26760.

Viola, S., et al. (2016). Alterations in acute myeloid leukaemia bone marrow stromal cell exosome content coincide with gains in tyrosine kinase inhibitor resistance. *British Journal of Haematology*, *172*(6), 983–986.

Wang, J., et al. (2014). Bone marrow stromal cell-derived exosomes as communicators in drug resistance in multiple myeloma cells. *Blood*, *124*(4), 555–566.

Wei, Y., et al. (2014). Exosomal miR-221/222 enhances tamoxifen resistance in recipient ER-positive breast cancer cells. *Breast Cancer Research and Treatment*, *147*(2), 423–431.

Weiner-Gorzel, K., et al. (2015). Overexpression of the microRNA miR-433 promotes resistance to paclitaxel through the induction of cellular senescence in ovarian cancer cells. *Cancer Medicine*, *4*(5), 745–758.

Wong, J. R., et al. (2013). Incidence of childhood and adolescent melanoma in the United States: 1973-2009. *Pediatrics*, *131*(5), 846–854.

Xiao, X., et al. (2014). Exosomes: Decreased sensitivity of lung cancer A549 cells to cisplatin. *PLoS One*, *9*(2), e89534.

Xu, Y., et al. (2011). MicroRNA-122 sensitizes HCC cancer cells to adriamycin and vincristine through modulating expression of MDR and inducing cell cycle arrest. *Cancer Letters*, *310*(2), 160–169.

Xu, S., et al. (2012). Bone marrow-derived mesenchymal stromal cells are attracted by multiple myeloma cell-produced chemokine CCL25 and favor myeloma cell growth in vitro and in vivo. *Stem Cells*, *30*(2), 266–279.

Yamada, O., et al. (2011). Activation of STAT5 confers imatinib resistance on leukemic cells through the transcription of TERT and MDR1. *Cellular Signalling*, *23*(7), 1119–1127.

Yang, J. K., et al. (2016). Exosomal miR-221 targets DNM3 to induce tumor progression and temozolomide resistance in glioma. *Journal of Neuro-oncology, 131*(2).

Yin, J., et al. (2012). Secretion of annexin A3 from ovarian cancer cells and its association with platinum resistance in ovarian cancer patients. *Journal of Cellular and Molecular Medicine, 16*(2), 337–348.

Yu, D. D., et al. (2016). Exosomes from adriamycin-resistant breast cancer cells transmit drug resistance partly by delivering miR-222. *Tumor Biologyogy, 37*(3), 3227–3235.

Zhang, H. (2016). Osimertinib making a breakthrough in lung cancer targeted therapy. *OncoTargets and Therapy, 9*, 5489–5493.

Zhang, J., et al. (2015). Beta-elemene reverses chemoresistance of breast cancer cells by reducing resistance transmission via exosomes. *Cellular Physiology and Biochemistry, 36*(6), 2274–2286.

Zhang, L., et al. (2016). Potential role of exosome-associated microRNA panels and in vivo environment to predict drug resistance for patients with multiple myeloma. *Oncotarget, 7*(21), 30876–30891.

Zhang, Z., et al. (2016). DNM3 attenuates hepatocellular carcinoma growth by activating P53. *Medical Science Monitor, 22*, 197–205.

Zhong, S., et al. (2016). MicroRNA expression profiles of drug-resistance breast cancer cells and their exosomes. *Oncotarget, 7*(15), 19601–19609.

17

Exosomes in Cancer Immunotherapy

Yuki Takahashi, Yoshinobu Takakura

KYOTO UNIVERSITY, KYOTO, JAPAN

CHAPTER OUTLINE

1. Introduction 313

2. Effect of Tumor Exosomes on T Cells 314

3. Effect of Tumor Exosomes on Natural Killer Cells 316

4. Effect of Tumor Exosomes on Myeloid Lineage Cells 317

5. Other Effects of Exosomes in Cancer Immunotherapies 318

6. Cancer Immunotherapies Using Exosomes 319

7. Conclusion 321

References 321

1. Introduction

Mammalian immune systems, comprising innate and acquired immunity, play important roles in protecting a body (Alberts et al., 2002; Kawai & Akira, 2006). Leukocytes such as macrophages and natural killer (NK) cells are major types of cells responsible for innate immunity. In acquired immunity, T and B cells are primarily responsible for cellular and humoral immunity, respectively, which are subclasses of acquired immunity (Janeway, 2005). Dendritic cells (DCs) are important immune cells that link innate and acquired immunity by presenting antigens to T cells (Ito, Connett, Kunkel, & Matsukawa, 2013; Schreibelt et al., 2010).

The immune system plays important roles in the development and progression of cancer. To avoid immune surveillance, tumors have developed a variety of immune evasion methods to actively protect themselves. Because the mechanisms that cancers use for immune evasion differ among tumor types, the degree of immune dysfunction caused by cancers varies widely, although the more advanced the tumor stage, the more dysfunction tends to increase (Pitt et al., 2016). Therefore, the degree of immune dysfunction correlates with the outcome of cancer patients (Whiteside, 2013).

In conventional cancer immunotherapies, manipulation of both innate and acquired immunity has been used to suppress tumor growth. In the case of acquired immunity, the therapeutic effect is specific to tumors, whereas with innate immunity, tumor destruction is nonspecific. For innate immune cancer immunotherapies, ligands for pattern recognition receptors such as toll-like receptors (TLRs) and nod-like receptors have frequently been used (Maisonneuve, Bertholet, Philpott, & De Gregorio, 2014). In the case of cancer immunotherapies based on acquired immunity, tumor-specific immune responses are mediated by memory and effector T and B cells. Cancer vaccine is the most commonly investigated cancer immunotherapy because cancer vaccines can induce a tumor-specific immune response through the induction of effector T cells specific to tumor antigens. With cancer vaccines, immune adjuvants such as TLR ligands have frequently been coadministered with tumor antigens to elicit a strong tumor-specific immune response in conjunction with innate immunity. However, because of cancer immune evasion described above, cancer immunotherapies have had very limited success.

In the last decade, the concept of cancer therapy has largely changed from the standard treatments of surgery, radiation, and chemotherapy to one that includes immunotherapy. As symbolized by a breakthrough of the year 2013 in *Science*, cancer immunotherapy has become a standard strategy to fight cancer (Couzin-Franke, 2013). In the first breakthrough, monoclonal antibodies targeting T cell checkpoint molecules such as CTLA-4 and PD-1 showed great therapeutic effects in patients with advanced cancers, including melanoma, non–small cell lung cancer, and renal cancer (Pardoll, 2012; Sharma & Allison, 2015). In most patients who responded to the treatment, durable remission was observed. In a second breakthrough in cancer immunotherapy, adoptive transfer of T cells engineered ex vivo to express chimeric antigen receptors eliminated tumor cells in cancer patients (Grupp et al., 2013). Of these novel cancer immunotherapies, inhibition of immune checkpoint molecules, especially PD-1/PD-L1, is the most promising approach.

As discussed in other chapters of this book, exosomes play various roles in delivering cargoes to recipient cells for intercellular communication. There are important roles for exosomes in the immune system as well. Exosomes secreted from immune cells have been shown to have immunoregulatory effects (Robbins, Dorronsoro, & Booker, 2016). In addition, exosomes secreted from tumor cells facilitate intercellular communication between tumor cells and normal cells in various capacities, including immune evasion. In this chapter, the use of exosomes, especially exosomes from tumor cells, in cancer immunotherapies, mostly in immune evasion, is discussed. Then, the possibility of developing exosome-based cancer immunotherapies other than exosome-based vaccines is discussed based on the literature.

2. Effect of Tumor Exosomes on T Cells (Fig. 1)

Effective CD4+ and CD8+ T cells are very important for antitumor immunity; therefore, it is not surprising that exosomes derived from tumor cells directly and indirectly suppress these T cells. Reports so far indicate that T cells are more susceptible to the suppressive

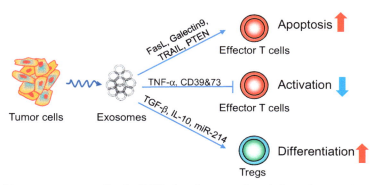

FIGURE 1 Effect of tumor exosomes on T cells. *PTEN*, phosphatase and tensin homologue; *TGF-β*, transforming growth factor β; *TRAIL*, tumor necrosis factor–related apoptosis-inducing ligand.

effects of tumor-derived exosomes than other types of immune cells. Molecules on the surfaces of exosomes, such as ligands and cytokines, rather than molecules in the exosomes, such as miRNAs, frequently interact with T cells because T cells are less able to internalize exosomes than other types of immune cells such as macrophages. The effects of tumor exosomes on T cells include induction of apoptosis, inhibition of proliferation and differentiation, and inhibition of function. In addition, tumor exosomes have been shown to suppress antigen-specific immunity using the model antigen ovalbumin (OVA). In one study, an OVA-specific immune response was suppressed by exosomes secreted from OVA-expressing melanoma (Yang, Kim, Bianco, & Robbins, 2011).

In vitro experiments have shown that tumor exosomes induce T cell apoptosis through molecules such as Fas ligand (FasL), tumor necrosis factor (TNF)-related apoptosis-inducing ligand, and galectin-9 on exosomal membranes (Andreola et al., 2002; Frängsmyr et al., 2005; Huber et al., 2005; Keryer-Bibens et al., 2006; Klibi et al., 2009; Taylor, Gercel-Taylor, Lyons, Stanson, & Whiteside, 2003). In addition to the induction of apoptosis by surface ligands, tumor exosomes containing phosphatase and tensin homologue (PTEN) were found to regulate the PI3K/AKT pathway in activated CD8+ T cells, resulting in AKT dephosphorylation and an increase in proapoptotic BAX and decrease in antiapoptotic Bcl-2, Bcl-xL, and MCL-1 in these cells (Czystowska et al., 2009; Putz et al., 2012).

Tumor exosomes have also been shown inhibit T cell reactivity to interleukin (IL)-2, inhibiting T cell proliferation (Clayton, Mitchell, Court, Mason, & Tabi, 2007). Moreover, administration of exosomes collected from a GL26 glioblastoma cell line to mice reduced the number of CD8+ T cells and decreased the expression of interferon (IFN)-γ and granzyme in these cells (Liu, Wang, & Yuan, 2013).

Activation of CD4+ and CD8+ T cells was found to be suppressed by tumor exosomes in the presence of TNF-α by affecting the CD3-T cell receptor complex through redox signaling (Soderberg, Barral, Soderstrom, Sander, & Rosen, 2007). The effects of adenosine produced by the exosome-associated ectonucleotidases CD39 and CD73 on effector T cells were demonstrated by Clayton, Al-Taei, Webber, Mason, and Tabi (2011). Immunosuppression by adenosine is caused by ligation of adenosine to its receptor on effector T cells, which increases cAMP levels in these cells to inhibit their function (Muller-Haegele, Muller, & Whiteside, 2014).

Because regulatory T cells (Tregs) are immune cells that negatively regulate the immune system, Tregs help cancer cells escape immunity. Therefore, the relationship between tumor exosomes and Tregs has been investigated. Transforming growth factor (TGF)-β1 and IL-10 associated with tumor exosomes have been shown to promote the differentiation of CD4$^+$ CD25$^-$ T cells into Tregs and promote the proliferation of Tregs by increasing the amount of phosphorylated SMAD2/3 and STAT3 in Tregs (Szajnik, Czystowska, Szczepanski, Mandapathil, & Whiteside, 2010). Moreover, the addition of tumor exosomes increased the immunosuppressive function of Tregs by upregulating the expression levels of FasL, TGF-β, IL-10, CTLA-4, granzyme B, and perforin (Mrizak et al., 2015). In addition, tumor exosomes containing miR-214 converted peripheral CD4$^+$ T cells into Tregs by reducing PTEN levels in the T cells (Yin et al., 2014). In the case of nasopharyngeal carcinoma (NPC), miRNAs contained in exosomes secreted from NPC cells were found to inhibit T cell proliferation and differentiation into Th1 and Th17 cells and promote Treg differentiation. In this study, hsa-miR-106a-5p, hsa-miR-891a, hsa-miR-20a-5p, hsa-miR-24-3p, and hsa-miR-1908 in the exosomes suppressed MAPK1 signaling to decrease phosphorylation of ERK, STAT1, and STAT3 and increase phosphorylation of STAT5 in T cells (Ye et al., 2014).

3. Effect of Tumor Exosomes on Natural Killer Cells (Fig. 2)

In addition to T cells, NK cells are important immune cells in cancer immunotherapies, as they enact various strategies to kill tumor cells (Vivier, Ugolini, Blaise, Chabannon, & Brossay, 2012). Although anticancer drug–treated carcinoma cells have been reported to release exosomes rich in heat shock proteins (HSPs) that promote NK cell activity, tumor-derived exosomes are more likely to suppress the function of NK cells (Lv et al., 2012), as the number of reports on the suppressive effects of tumor exosomes on NK cells is greater than the number of reports of the opposite effect. Clayton et al. (2008) showed that tumor exosomes downregulated NK cell surface expression of NKG2D, an NK cell–activating receptor that is important for the cytotoxic activity of NK cells. Another study indicated a similar effect of tumor exosomes on NKG2D (Szczepanski, Szajnik, Welsh, Whiteside, & Boyiadzis, 2011). TGF-β associated with tumor exosomes might be the main reason for

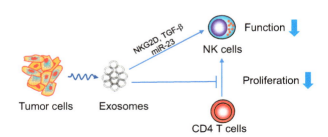

FIGURE 2 Effect of tumor exosomes on NK cells. *NK*, natural killer; *TGF-β*, transforming growth factor β.

NKG2D downregulation because TGF-β neutralizing antibody was found to inhibit the NKG2D-reducing effect of tumor exosomes (Szczepanski et al., 2011). In addition to TGF-β, NKG2D ligands expressed on tumor exosomes might also be involved in tumor exosome suppression of NK cells because the ligands can downregulate NKG2D expression and might serve as decoys for the receptor (Hedlund, Nagaeva, Kargl, Baranov, & Mincheva-Nilsson, 2011). Moreover, a recent study showed that miR-23, in addition to TGF-β, contained in exosomes from lung carcinoma cells could suppress NK cell function (Berchem et al., 2015). In addition to suppressing NK cell activity, tumor exosomes suppressed proliferation of NK cells stimulated by CD4+ T cells (Clayton et al., 2007).

4. Effect of Tumor Exosomes on Myeloid Lineage Cells (Fig. 3)

DCs are cells that link innate and acquired immunity through their antigen presentation function. In the bone marrow, differentiation of DCs from myeloid precursors is reported to be hindered by tumor exosomes. In an in vitro experiment, tumor exosomes were shown to inhibit the differentiation of myeloid precursor cells into DCs by the induction of IL-6 (Liu et al., 2010). In addition, maturation of DCs was suppressed by the uptake of tumor exosomes by immature DCs. Specifically, Yang et al. demonstrated inhibition of DC maturation by TGF-β associated with tumor exosomes (2011). Moreover, an antitumor response mediated by DCs can be inhibited by tumor exosome regulation of TLR4 expression on DCs (Zhou et al., 2014). Finally, miRNA-203 in pancreatic cancer cell–derived exosomes downregulated the TLR4 expression level in DCs, resulting in decreased secretion of cytokines such as IL-12 and TNF-α from DCs and suppression of the immune response mediated by DCs.

Tumor exosomes not only suppress the differentiation of myeloid precursors into DCs but also promote the differentiation of precursors into myeloid-derived suppressor cells (MDSCs), which show immunosuppressive effects (Filipazzi, Burdek, Villa, Rivoltini, & Huber, 2012). Xiang et al. (2009) demonstrated the promotion of MDSC differentiation through TGF-β and prostaglandin E2 in tumor exosomes. Moreover, pretreatment of mice with tumor exosomes was found to increase the number of MDSCs in the spleen, lung, and

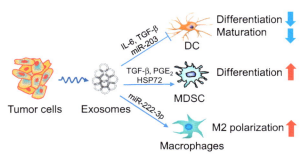

FIGURE 3 Effect of tumor exosomes on myeloid lineage cells. *DC*, dendritic cell; *HSP*, heat shock protein; *IL*, interleukin; *MDSC*, myeloid-derived suppressor cells; *PGE2*, prostaglandin E2; *TGF-β*, transforming growth factor β.

peripheral blood. In addition, HSP72 on the surfaces of tumor exosomes stimulated TLR2/MyD88 and activated STAT3 to produce IL-6, which promoted the immunosuppression by MDSCs (Chalmin et al., 2010).

Macrophages are important cells in an innate immune response. Because macrophages actively take up exosomes (Charoenviriyakul et al., 2017; Imai et al., 2015; Takahashi et al., 2013), macrophages are largely affected by tumor exosomes. Tumor exosomes have been shown to affect macrophages, inducing M2-type macrophages, which have immunosuppressive functions (Marton et al., 2012; de Vrij et al., 2015). Recently, Ying et al. (2016) found that uptake of ovarian cancer–derived exosomes by macrophages induced the production of macrophages with a tumor-associated macrophage-like phenotype. They also found that miR-222-3p contained in the tumor exosomes induced M2 polarization of macrophages by reducing the expression of suppressor of cytokine signaling 3 in macrophages.

5. Other Effects of Exosomes in Cancer Immunotherapies (Fig. 4)

Because tumor exosomes contain tumor antigens derived from exosome-producing cells, i.e., tumor cells, tumor exosomes suppress antitumor immunity mediated by tumor antigen–specific antibodies by acting as a decoy for the antibodies. Ciravolo et al. (2012) showed that HER2[+] exosomes collected from the plasma of patients with breast cancer cells bound to a HER2-specific antibody, trastuzumab, and that the HER2[+] exosomes suppressed the inhibitory effect of trastuzumab on the proliferation of HER2-overexpressing tumor cells. In addition, antibody binding by tumor exosomes was found to suppress antibody-dependent cell-mediated cytotoxicity, which is a major antitumor mechanism of tumor antigen–specific antibodies (Battke et al., 2011). In addition, exosomes from B cell lymphoma displaying CD20 bound to rituximab, a CD20-specific antibody, protecting tumor cells from complement-dependent cytotoxicity by binding and consuming the complement (Aung et al., 2011).

In addition to tumor exosomes, exosomes from other cell types have been applied in cancer immunotherapies. Exosomes secreted from Tregs suppressed effector T cells by delivering miRNAs associated with proapoptotic and antiproliferative functions (Okoye et al., 2014). Moreover, Xie, Zhang, Zhao, Li, and Xiang (2013) reported that natural Treg-secreted exosomes suppressed cytotoxic T lymphocyte activity in a B16 murine melanoma model. In addition to those from Tregs, exosomes secreted from CD11b[+] cells suppressed tumor antigen–specific

FIGURE 4 Effect of tumor exosomes on antibody-dependent cancer immunotherapy. *ADCC*, antibody-dependent cell-mediated cytotoxicity.

immune responses by a major histocompatibility complex (MHC) class II–dependent mechanism (Yang, Ruffner, Kim, & Robbins, 2012). The existence of PD-L1 on tumor exosomes has been reported, although there has been no study showing the direct effect of exosome-associated PD-L1 on the immune system (Hong, Funk, Muller, Boyiadzis, & Whiteside, 2016). Considering that placenta-derived exosomes associated with PD-L1 showed an immunomodulatory effect (Sabapatha, Gercel-Taylor, & Taylor, 2006), further study is needed to elucidate the role of tumor exosome–associated PD-L1 in cancer immunotherapies.

6. Cancer Immunotherapies Using Exosomes (Fig. 5)

Because of their nature as endogenous intercellular vehicles of various molecules, including nucleic acids and proteins, in addition to their biological activities discussed above, various applications of exosomes in cancer immunotherapies have been reported.

The most successful cancer immunotherapy using exosomes is exosome-based vaccination. Because tumor exosomes contain endogenous tumor antigens, tumor exosomes have occasionally been used as cancer vaccines (Hao, Moyana, & Xiang, 2007). Moreover, exosomes collected from DCs pulsed with tumor antigens have frequently been used in cancer vaccines because the exosomes contain tumor antigens, MHC Class I and II molecules, and costimulatory molecules (Pitt et al., 2014). In tumor vaccines, including exosome-based tumor vaccines, ligands for TLRs, such as CpG DNA (TLR9), poly I:C (TLR3), and lipopolysaccharides (LPS; TLR4) have been frequently used as adjuvants (Goutagny, Estornes, Hasan, Lebecque, & Caux, 2012). In the application of TLR ligands as adjuvants, delivery of the ligands to their target cells, i.e., immune cells, is important for the induction of an effective immune response. Because exosomes are endogenous delivery vehicles, exosomes can be used to deliver TLR ligands to immune cells. We have used exosomes

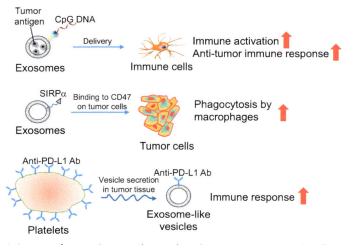

FIGURE 5 Schematic images of cancer immunotherapy by using exosomes. *SIRPα*, signal regulatory protein α.

secreted from B16BL6 murine melanoma cells to deliver CpG DNA, as well as tumor antigens contained in exosomes, to DCs (Morishita, Takahashi, Matsumoto, Nishikawa, & Takakura, 2016). In this study, plasmid DNA encoding a fusion protein composed of streptavidin (SAV; a protein that binds to biotin with high affinity) and lactadherin (LA; an exosome-tropic protein) was transfected into B16BL6 cells to produce genetically engineered SAV-LA-expressing exosomes (SAV-exo). Then, SAV-exo was incubated with biotinylated CpG DNA to prepare CpG DNA-conjugated exosomes. The uptake efficiency of CpG DNA by a murine DC line (DC2.4 cells) was significantly increased by conjugation to exosomes, which resulted in the increased secretion of immunostimulatory cytokines from DC2.4 cells and enhanced presentation of tumor antigens by these cells. Moreover, intratumoral injection of CpG DNA conjugated to exosomes was shown to suppress tumor growth in tumor-bearing mice much more effectively than simple coadministration of exosomes and CpG DNA. Our result clearly indicated that exosomes can be used as a codelivery system for TLR ligands and tumor antigens. Detail on exosomal cancer vaccines are presented in other chapters of this book.

In addition, there have been a limited number of reports on exosome-based cancer immunotherapies other than vaccines. Exosomes derived from DCs (Dex) were shown to increase the immunostimulatory activity of LPS. Sobo-Vujanovic, Munich, and Vujanovic (2014) reported that Dex bound LPS and that the immunostimulatory activity of LPS bound to Dex was approximately 40- to 100-fold greater than that of LPS alone. As TLR4, a receptor for LPS, is present on the outside of the cellular membrane, increased uptake is not likely to be the reason for this increased activity. The authors hypothesized that cytokines and costimulatory molecules associated with Dex are the reason for the enhanced activity of LPS bound to Dex compared with that of LPS alone. In addition to TLR ligands, cytokines can be delivered by exosomes in cancer immunotherapies. Cossetti et al. (2014) showed that IFN-γ bound to its receptor on exosomes stimulated target cells more efficiently than IFN-γ alone, which suggests the possibility of cancer immunotherapies based on cytokines delivered by exosomes.

Koh et al. (2017) developed a cancer immunotherapy that targets CD47, a "don't eat me" signal that is overexpressed on the surfaces of most tumors, using exosomes that harbor signal regulatory protein α (SIRPα), a protein that interacts with CD47 and is expressed on phagocytes. Because the interaction of CD47 with SIRPα was shown to suppress the phagocytic activity of macrophages, antagonism of CD47 by exosomes harboring SIRPα resulted in an increase in phagocytosis of tumor cells, followed by the induction of an antitumor T cell response. Therefore, increasing tumor phagocytosis by antagonizing the CD47 signal with exosomes may be an effective cancer immunotherapy approach.

Recently, inhibition of the immune checkpoint molecule PD-L1 using anti-PD-L1 monoclonal antibodies conjugated to the surfaces of platelets was shown to reduce postsurgical tumor recurrence (Chao et al., 2017). In this study, anti-PD-L1 antibody was effectively released from platelet-derived microparticles on platelet activation, which occurs frequently in tumor microenvironments. Administration of platelet-bound anti-PD-L1 antibody greatly increased the number of infiltrating CD8$^+$ and CD4$^+$ T cells, decreased the number of CD4$^+$ Foxp3$^+$ T cells, and promoted the proliferation of CD8$^+$ and CD4$^+$ effector T cells in the tumor. Moreover, the administration of platelet-bound anti-PD-L1 antibody significantly prolonged the survival of mice after primary tumor resection by reducing cancer regrowth and metastatic spread in tumor-bearing mice. These results indicate that microparticles, which are similar to exosomes, secreted from platelets can be served as effective delivery vehicles for the inhibition of immune checkpoint molecules.

7. Conclusion

Cancer immunotherapy has become an effective approach to treat cancer patients. Because exosomes play important roles in the immune system, understanding their roles can aid in the development of more effective cancer immunotherapies.

References

Alberts, B., Johnson, A., Lewis, J., Raff, M., Roberts, K., & Walter, P. (2002). *Molecular biology of the cell* (4th ed.). Nigel Chaffey.

Andreola, G., Rivoltini, L., Castelli, C., Huber, V., Perego, P., Deho, P., ... Fais, S. (2002). Induction of lymphocyte apoptosis by tumor cell secretion of FasL-bearing microvesicles. *Journal of Experimental Medicine, 195,* 1303–1316.

Aung, T., Chapuy, B., Vogel, D., Wenzel, D., Oppermann, M., Lahmann, M., ... Wulf, G. G. (2011). Exosomal evasion of humoral immunotherapy in aggressive B-cell lymphoma modulated by ATP-binding cassette transporter A3. *Proceedings of the National Academy of Sciences, 108*(37), 15336–15341.

Battke, C., Ruiss, R., Welsch, U., Wimberger, P., Lang, S., Jochum, S., & Zeidler, R. (2011). Tumour exosomes inhibit binding of tumour-reactive antibodies to tumour cells and reduce ADCC. *Cancer Immunology, Immunotherapy, 60*(5), 639–648.

Berchem, G., Noman, M. Z., Bosseler, M., Paggetti, J., Baconnais, S., Le Cam, E., ... Chouaib, S. (2015). Hypoxic tumor-derived microvesicles negatively regulate NK cell function by a mechanism involving TGF-beta and miR23a transfer. *OncoImmunology, 5,* e1062968.

Chalmin, F., Ladoire, S., Mignot, G., Vincent, J., Bruchard, M., Remy-Martin, J. P., ... Ghiringhelli, F. (2010). Membrane-associated Hsp72 from tumor-derived exosomes mediates STAT3-dependent immunosuppressive function of mouse and human myeloid-derived suppressor cells. *Journal of Clinical Investigation, 120,* 457–471.

Chao, W., Sun, W., Ye, Y., Hu, Q., Bomba, H. N., & Guo, Z. (2017). In situ activation of platelets with checkpoint inhibitors for post-surgical cancer immunotherapy. *Nature Biomedical Engineering, 1*:11.

Charoenviriyakul, C., Takahashi, Y., Morishita, M., Matsumoto, A., Nishikawa, M., & Takakura, Y. (2017). Cell type-specific and common characteristics of exosomes derived from mouse cell lines: Yield, physicochemical properties, and pharmacokinetics. *European Journal of Pharmaceutical Sciences, 96,* 316–322.

Ciravolo, V., Huber, V., Ghedini, G. C., Venturelli, E., Bianchi, F., Campiglio, M., ... Pupa, S. M. (2012). Potential role of HER2-overexpressing exosomes in countering trastuzumab-based therapy. *Journal of Cellular Physiology, 227*(2), 658–667.

Clayton, A., Al-Taei, S., Webber, J., Mason, M. D., & Tabi, Z. (2011). Cancer exosomes express CD39 and CD73, which suppress t cells through adenosine production. *The Journal of Immunology, 187,* 676–683.

Clayton, A., Mitchell, J. P., Court, J., Linnane, S., Mason, M. D., & Tabi, Z. (2008). Human tumor-derived exosomes down-modulate NKG2D expression. *The Journal of Immunology, 180,* 7249–7258.

Clayton, A., Mitchell, J. P., Court, J., Mason, M. D., & Tabi, Z. (2007). Human tumor-derived exosomes selectively impair lymphocyte responses to interleukin-2. *Cancer Research, 67,* 7458–7466.

Cossetti, C., Iraci, N., Mercer, T. R., Leonardi, T., Alpi, E., Drago, D., … Vega, B. (2014). Extracellular vesicles from neural stem cells transfer IFN-γ via Ifngr1 to activate Stat1 signaling in target cells. *Molecular Cell, 56*(2), 193–204.

Couzin-Franke, J., (2013). Cancer immunotherapy. *Science, 342,* 1432–1433.

Czystowska, M., Han, J., Szczepanski, M. J., Szajnik, M., Quadrini, K., Brandwein, H., … Whiteside, T. L. (2009). IRX-2, a novel immunotherapeutic, protects human T cells from tumor-induced cell death. *Cell Death and Differentiation, 16,* 708–718.

Filipazzi, P., Burdek, M., Villa, A., Rivoltini, L., & Huber, V. (2012). Recent advances on the role of tumor exosomes in immunosuppression and disease progression. *Seminars in Cancer Biology, 22,* 342–349.

Frängsmyr, L., Baranov, V., Nagaeva, O., Stendahl, U., Kjellberg, L., & Mincheva-Nilsson, L. (2005). Cytoplasmic microvesicular form of Fas ligand in human early placenta: Switching the tissue immune privilege hypothesis from cellular to vesicular level. *Molecular Human Reproduction, 11,* 35–41.

Goutagny, N., Estornes, Y., Hasan, U., Lebecque, S., & Caux, C. (2012). Targeting pattern recognition receptors in cancer immunotherapy. *Targeted Oncology, 7*(1), 29–54.

Grupp, S. A., Kalos, M., Barrett, D., Aplenc, R., Porter, D. L., Rheingold, S. R., … June, C. H. (2013). Chimeric antigen receptor-modified T cells for acute lymphoid leukemia. *New England Journal of Medicine, 368,* 1509–1518.

Hao, S., Moyana, T., & Xiang, J. (2007). Review: Cancer immunotherapy by exosome-based vaccines. *Cancer Biotherapy & Radiopharmaceuticals, 22*(5), 692–703.

Hedlund, M., Nagaeva, O., Kargl, D., Baranov, V., & Mincheva-Nilsson, L. (2011). Thermal- and oxidative stress causes enhanced release of NKG2D ligand-bearing immunosuppressive exosomes in leukemia/lymphoma T and B cells. *PLoS One, 6,* e16899.

Hong, C. S., Funk, S., Muller, L., Boyiadzis, M., & Whiteside, T. L. (2016). Isolation of biologically active and morphologically intact exosomes from plasma of patients with cancer. *Journal of Extracellular Vesicles, 5,* 29289.

Huber, V., Fais, S., Iero, M., Lugini, L., Canese, P., Squarcina, P., … Rivoltini, L. (2005). Human colorectal cancer cells induce T-cell death through release of proapoptotic microvesicles: Role in immune escape. *Gastroenterology, 128,* 1796–1804.

Imai, T., Takahashi, Y., Nishikawa, M., Kato, K., Morishita, M., Yamashita, T., … Takakura, Y. (2015). Macrophage-dependent clearance of systemically administered B16BL6-derived exosomes from the blood circulation in mice. *Journal of Extracellular Vesicles, 4,* 26238.

Ito, T., Connett, J. M., Kunkel, S. L., & Matsukawa, A. (2013). The linkage of innate and adaptive immune response during granulomatous development. *Frontiers in Immunology, 4,* 10.

Janeway, C. A. (2005). *Immunobiology.* Garland Science.

Kawai, T., & Akira, S. (2006). Innate immune recognition of viral infection. *Nature Immunology, 7,* 131–137.

Keryer-Bibens, C., Pioche-Durieu, C., Villemant, C., Souquère, S., Nishi, N., Hirashima, M., … Busson, P. (2006). Exosomes released by EBV-infected nasopharyngeal carcinoma cells convey the viral latent membrane protein 1 and the immunomodulatory protein galectin 9. *BMC Cancer, 6,* 283.

Klibi, J., Niki, T., Riedel, A., Pioche-Durieu, C., Souquere, S., Rubinstein, E., … Busson, P. (2009). Blood diffusion and Th1-suppressive effects of galectin-9-containing exosomes released by Epstein-Barr virus-infected nasopharyngeal carcinoma cells. *Blood, 113,* 1957–1966.

Koh, E., Lee, E. J., Nam, G. H., Hong, Y., Cho, E., Yang, Y., & Kim, I. S. (2017). Exosome-SIRPα, a CD47 blockade increases cancer cell phagocytosis. *Biomaterials, 121,* 121–129.

Liu, Z. M., Wang, Y. B., & Yuan, X. H. (2013). Exosomes from murine-derived GL26 cells promote glioblastoma tumor growth by reducing number and function of CD8+ T cells. *Asian Pacific Journal of Cancer Prevention, 14,* 309–314.

Liu, Y., Xiang, X., Zhuang, X., Zhang, S., Liu, C., Cheng, Z., … Zhang, H. G. (2010). Contribution of MyD88 to the tumor exosome-mediated induction of myeloid derived suppressor cells. *American Journal of Pathology, 176*, 2490–2499.

Lv, L. H., Wan, Y. L., Lin, Y., Zhang, W., Yang, M., Li, G. L., … Min, J. (2012). Anticancer drugs cause release of exosomes with heat shock proteins from human hepatocellular carcinoma cells that elicit effective natural killer cell antitumor responses in vitro. *Journal of Biological Chemistry, 287*, 15874–15885.

Maisonneuve, C., Bertholet, S., Philpott, D. J., & De Gregorio, E. (2014). Unleashing the potential of NOD- and Toll-like agonists as vaccine adjuvants. *Proceedings of the National Academy of Sciences, 111*, 12294–12299.

Marton, A., Vizler, C., Kusz, E., Temesfoi, V., Szathmary, Z., Nagy, K., … Tubak, V. (2012). Melanoma cell-derived exosomes alter macrophage and dendritic cell functions in vitro. *Immunology Letters, 148*(1), 34–38.

Morishita, M., Takahashi, Y., Matsumoto, A., Nishikawa, M., & Takakura, Y. (2016). Exosome-based tumor antigens–adjuvant co-delivery utilizing genetically engineered tumor cell-derived exosomes with immunostimulatory CpG DNA. *Biomaterials, 111*, 55–65.

Mrizak, D., Martin, N., Barjon, C., Jimenez-Pailhes, A. S., Mustapha, R., Niki, T., … Delhem, N. (2015). Effect of nasopharyngeal carcinoma-derived exosomes on human regulatory T cells. *Journal of the National Cancer Institute, 107*(1), 363.

Muller-Haegele, S., Muller, L., & Whiteside, T. L. (2014). Immunoregulatory activity of adenosine and its role in human cancer progression. *Expert Review of Clinical Immunology, 10*(7), 897–914.

Okoye, I. S., Coomes, S. M., Pelly, V. S., Czieso, S., Papayannopoulos, V., Tolmachova, T., … Wilson, M. S. (2014). MicroRNA-containing T-regulatory-cell-derived exosomes suppress pathogenic T helper 1 cells. *Immunity, 41*(1), 89–103.

Pardoll, D. M. (2012). The blockade of immune checkpoints in cancer immunotherapy. *Nature Reviews Cancer, 12*, 252–264.

Pitt, J. M., Charrier, M., Viaud, S., André, F., Besse, B., Chaput, N., & Zitvogel, L. (2014). Dendritic cell–derived exosomes as immunotherapies in the fight against cancer. *The Journal of Immunology, 193*(3), 1006–1011.

Pitt, J. M., Marabelle, A., Eggermont, A., Soria, J. C., Kroemer, G., & Zitvogel, L. (2016). Targeting the tumor microenvironment: Removing obstruction to anticancer immune responses and immunotherapy. *Annals of Oncology, 27*, 1482–1492.

Putz, U., Howitt, J., Doan, A., Goh, C. P., Low, L. H., Silke, J., & Tan, S. S. (2012). The tumor suppressor PTEN is exported in exosomes and has phosphatase activity in recipient cells. *Science Signalling, 5*(243), ra70.

Robbins, P. D., Dorronsoro, A., & Booker, C. N. (2016). Regulation of chronic inflammatory and immune processes by extracellular vesicles. *Journal of Clinical Investigation, 126*, 1173–1180.

Sabapatha, A., Gercel-Taylor, C., & Taylor, D. D. (2006). Specific isolation of placenta-derived exosomes from the circulation of pregnant women and their immunoregulatory consequences. *American Journal of Reproductive Immunology, 56*(5–6), 345–355.

Schreibelt, G., Tel, J., Sliepen, K. H., Benitez-Ribas, D., Figdor, C. G., Adema, G. J., & de Vries, I. J. (2010). Toll-like receptor expression and function in human dendritic cell subsets: Implications for dendritic cell-based anti-cancer immunotherapy. *Cancer Immunology, Immunotherapy, 59*, 1573–1582.

Sharma, P., & Allison, J. P. (2015). The future of immune checkpoint therapy. *Science, 348*, 56–61.

Sobo-Vujanovic, A., Munich, S., & Vujanovic, N. L. (2014). Dendritic-cell exosomes cross-present Toll-like receptor-ligands and activate bystander dendritic cells. *Cellular Immunology, 289*(1), 119–127.

Soderberg, A., Barral, A. M., Soderstrom, M., Sander, B., & Rosen, A. (2007). Redox-signaling transmitted in trans to neighboring cells by melanoma-derived TNF-containing exosomes. *Free Radical Biology and Medicine, 43*, 90–99.

Szajnik, M., Czystowska, M., Szczepanski, M. J., Mandapathil, M., & Whiteside, T. L. (2010). Tumor-derived microvesicles induce, expand and up-regulate biological activities of human regulatory T cells (Treg). *PLoS One, 5*(7), e11469.

Szczepanski, M. J., Szajnik, M., Welsh, A., Whiteside, T. L., & Boyiadzis, M. (2011). Blast-derived microves-icles in sera from patients with acute myeloid leukemia suppress natural killer cell function via mem-brane-associated transforming growth factor-beta1. *Haematologica, 96,* 1302–1309.

Takahashi, Y., Nishikawa, M., Shinotsuka, H., Matsui, Y., Ohara, S., Imai, T., & Takakura, Y. (2013). Visualization and in vivo tracking of the exosomes of murine melanoma B16-BL6 cells in mice after intravenous injection. *Journal of Biotechnology, 165*(2), 77–84.

Taylor, D. D., Gercel-Taylor, C., Lyons, K. S., Stanson, J., & Whiteside, T. L. (2003). T-cell apoptosis and sup-pression of T-cell receptor/CD3-zeta by Fas ligand-containing membrane vesicles shed from ovarian tumors. *Clinical Cancer Research, 9,* 5113–5119.

Vivier, E., Ugolini, S., Blaise, D., Chabannon, C., & Brossay, L. (2012). Targeting natural killer cells and natu-ral killer T cells in cancer. *Nature Reviews Immunology, 12,* 239–252.

de Vrij, J., Maas, S. L., Kwappenberg, K. M., Schnoor, R., Kleijn, A., Dekker, L., … van Strien, M. E. (2015). Glioblastoma-derived extracellular vesicles modify the phenotype of monocytic cells. *International Journal of Cancer, 137*(7), 1630–1642.

Whiteside, T. L. (2013). Immune responses to cancer: Are they potential biomarkers of prognosis? *Frontiers in Oncology, 3,* 107.

Xiang, X., Poliakov, A., Liu, C., Liu, Y., Deng, Z. B., Wang, J., … Zhang, H. G. (2009). Induction of myeloid-derived suppressor cells by tumor exosomes. *International Journal of Cancer, 124,* 2621–2633.

Xie, Y., Zhang, X., Zhao, T., Li, W., & Xiang, J. (2013). Natural CD8+ 25+ regulatory T cell-secreted exo-somes capable of suppressing cytotoxic T lymphocyte-mediated immunity against B16 melanoma. *Biochemical and Biophysical Research Communications, 438*(1), 152–155.

Yang, C., Kim, S. H., Bianco, N. R., & Robbins, P. D. (2011). Tumor-derived exosomes confer antigen-specific immunosuppression in a murine delayed-type hypersensitivity model. *PLoS One, 6,* e22517.

Yang, C., Ruffner, M. A., Kim, S. H., & Robbins, P. D. (2012). Plasma-derived MHC class II+ exosomes from tumor-bearing mice suppress tumor antigen-specific immune responses. *European Journal of Immunology, 42*(7), 1778–1784.

Ye, S. B., Li, Z. L., Luo, D. H., Huang, B. J., Chen, Y. S., Zhang, X. S., … Li, J. (2014). Tumor-derived exosomes promote tumor progression and T-cell dysfunction through the regulation of enriched exosomal microRNAs in human nasopharyngeal carcinoma. *Oncotarget, 5,* 5439–5452.

Yin, Y., Cai, X., Chen, X., Liang, H., Zhang, Y., Li, J., … Zhang, C. Y. (2014). Tumor-secreted miR-214 induces regulatory T cells: A major link between immune evasion and tumor growth. *Cell Research, 24,* 1164–1180.

Ying, X., Wu, Q., Wu, X., Zhu, Q., Wang, X., Jiang, L., … Wang, X. (2016). Epithelial ovarian cancer-secreted exosomal miR-222-3p induces polarization of tumor-associated macrophages. *Oncotarget, 7*(28), 43076–43087.

Zhou, M., Chen, J., Zhou, L., Chen, W., Ding, G., & Cao, L. (2014). Pancreatic cancer derived exosomes regu-late the expression of TLR4 in dendritic cells via miR-203. *Cellular Immunology, 292,* 65–69.

18

Extracellular Vesicles as Vehicles of B Cell Antigen Presentation: Implications for Cancer Vaccine Therapies[1]

Michael W. Graner

UNIVERSITY OF COLORADO DENVER, AURORA, CO, UNITED STATES

CHAPTER OUTLINE

1. Introduction 325

2. Extracellular Vesicles: Classifications and Nomenclature 326

3. B Cells, Extracellular Vesicles, and Associated Impacts 329
   3.1 Antigen-Presenting Cells in Immune Responses 329
   3.2 B Cells and Antigen Presentation 331
   3.3 B Cell Extracellular Vesicles and Follicular Dendritic Cells............................ 335
   3.4 B Cell Extracellular Vesicles and the Promotion of Dendritic Cells–Mediated T Cell
       Immune Responses 335
   3.5 B Cell Extracellular Vesicles as Cancer Vaccines/Adjuvants.......................... 336

4. Conclusions.. 336

References 338

1. Introduction

Cells communicate with each other proximally and distally in various modalities such as direct cell-surface molecular contact, gap junctions, tunneling nanotubes (Gerdes, Rustom, & Wang, 2013), and autocrine/paracrine and endocrine secretions of soluble and less-than-soluble moieties (Kim & Moustaid-Moussa, 2000; Zhang & Yang, 2017). Extracellular vesicles (EVs) are of the latter class. Here we discuss the roles of B cell–derived

---

[1] This work was supported by the University of Colorado Neurosurgery Department and by National Institutes of Health 4R01EB016378-04.

Diagnostic and Therapeutic Applications of Exosomes in Cancer. https://doi.org/10.1016/B978-0-12-812774-2.00018-3

325

EVs in forms of antigen presentation and the provocation of immune responses, along with an outlook for potential clinical utility.

2. Extracellular Vesicles: Classifications and Nomenclature

Exosomes and microvesicles are categories of EVs, membrane bilayer–delineated nanoparticles released from essentially all cell types. Apoptotic bodies are the third form of EV (Gyorgy et al., 2011; Yanez-Mo et al., 2015). These EVs differ in mode of production and, to some extent, by content. Exosomes are generated in the endosomal system as late endosomes undergo "reverse budding," with invaginations at their limiting membrane surfaces. This results in the formation of multivesicular bodies (MVBs), and the smaller interior vesicles are called intraluminal vesicles (ILVs). MVBs canonically proceed to lysosomal degradation (Luzio, Pryor, & Bright, 2007); however, under certain circumstances (Mittelbrunn, Vicente Manzanares, & Sanchez-Madrid, 2015), MVBs may dock with the plasma membrane, which fuses the MVB-limiting membrane with the plasma membrane, releasing the ILVs outside the cell as exosomes. Microvesicles (also called "ectosomes," "shed membrane vesicles," or "membrane blebs") are released directly from the cell surface via similar molecular mechanisms as used for the formation of ILVs, but with disruption of the actin cytoskeleton and membrane reorganization. Apoptotic bodies form during cellular dismantling amid programmed cell death, involving protease (largely caspase) activation, DNA fragmentation, and externalization of phosphatidylserine. Illustrations of the three types of EV formation are shown in Fig. 1.

---

shown). Endosomal trafficking results in changes in internal lumenal pH (usually decreasing) and alterations in phospholipids and RAB GTPases, which define the character of the endosomes to bestow "identities." As the endosomes mature, they may interact with vesicles of the trans-Golgi network (not shown) and/or progress toward late endosomes/multivesicular bodies (MVBs). Here, changes in pH, membrane lipids, and functions of the ESCRT (endosomal *sorting complexes required for transport*, not shown) may result in invaginations or reverse budding of the endosome-limiting membrane. Cytosolic components (proteins, nucleic acids, metabolites) may be sorted into these invaginations. As these bud off into the lumenal interior, the small vesicles are referred to as "intraluminal vesicles" (ILVs, or sometimes, intravesicular bodies). Note that these vesicles, to some extent, replicate the topography of the cell itself, with lipids and membrane proteins facing "outward," and with cytosolic components contained within. Canonically, late endosomes/MVBs are destined to the fuse with lysosomes for content degradation and recycling (not shown); however, an alternate route is for the MVBs to fuse with the plasma membrane to release the ILVs outside the cell. *These extracellular vesicles are now known as "exosomes."* Another form of EV arises from "membrane blebbing" or outward protrusion of the cellular membrane (utilizing some of the molecular machinery to form ILVs in an MVB). Often this is facilitated by a gathering of membrane domains ("lipid rafts") or characterized by the "flipping" of phosphatidylserine to the external surface of the plasma membrane. As with the generation of MVBs, cytosolic components may enter into the membrane extensions. These blebs pinch off, releasing membrane vesicles extracellularly. *Such EVs have been called microvesicles, ectosomes, or shed vesicles,* among other terms. Again, there is a recapitulation of the cellular topography, as with exosomes, although a variety of sorting mechanisms for both proteins and lipids do not allow for an exact replication. As cells undergo apoptosis (Bottom panel), signaling events drive protease (often caspase) activity, degrading cellular proteins, while characteristic chromosomal condensation occurs in the nucleus. Organelles degenerate, and a variety of cytoplasmic blebs may arise, altering the shape of the cell, and leading to the release of *a third type of EV, apoptotic bodies.* These are often amorphously shaped and sometimes contain significant amounts of DNA. For relative sizes of the various EVs, see Fig. 2. *Note that the sizes of cells, organelles, and vesicles are not drawn to scale at all.*

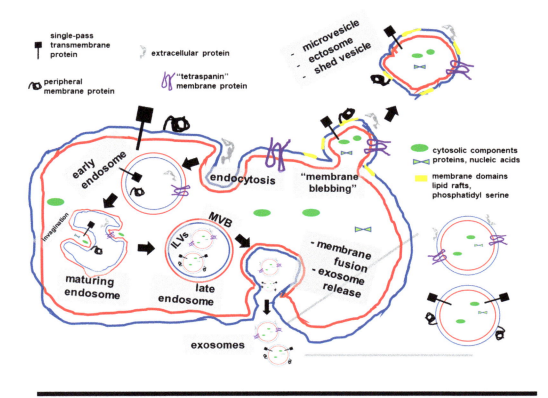

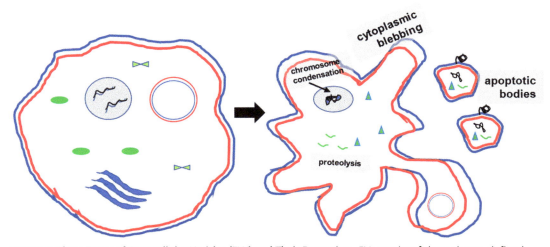

FIGURE 1 Three Types of Extracellular Vesicles (EVs) and Their Formation. EVs consist of three classes, defined by their biogenesis but with frequent overlap of molecular content. The top panel depicts a generic cell with various types of membrane or extracellular proteins (the extracellular protein—which could be a serum protein—may be bound electrostatically or by specific receptors). Other cytosolic proteins or biomolecules (nucleic acids, metabolites) are also represented. Endocytic events (typically phagocytic, fluid-phase pinocytic, receptor-mediated, or caveolar) pull membrane and extracellular proteins into the endocytic pathway, such as into early endosomes. These may consist of recycling endosomes, which may return components to the cell surface (not

There are several excellent reviews on the topic of EV types and subtypes, their formation, and their biologic functions, and the reader is referred to those for better explanations (Gyorgy et al., 2011; Kalra, Drummen, & Mathivanan, 2016; Yanez-Mo et al., 2015).

EV subclasses have been defined by their diameters (30–130 nm for exosomes, 100–1000 nm for microvesicles, and 50–5000 nm for apoptotic bodies) (Gyorgy et al., 2011; Yanez-Mo et al., 2015). A portrayal of the relative sizes of such vesicles (and comparisons to other biologic entities) is shown in Fig. 2. Obviously, there is substantial overlap in these size ranges, as well as marked molecular heterogeneity, making clear distinctions essentially impossible. Thus the term "extracellular vesicles" has become the default term to use (Gould & Raposo, 2013), along with guides for EV isolation and characterization (Consortium et al., 2017; Lotvall et al., 2014). Most of the publications we will cite herein frequently use the term

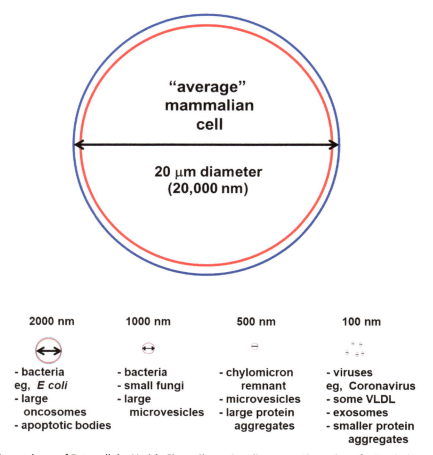

FIGURE 2 Comparisons of Extracellular Vesicle Sizes. Shown is a diagrammatic version of a "typical" mammalian cell, represented in two dimensions as a circle; it has a diameter of ~20 μm (2×10^{-5} m; 20,000 nm). Shown below the cell are circles with diameters representing 2000 nm, 1000 nm, 500 nm, and 100 nm drawn proportionately, along with short lists of other biologic entities of similar diameters.

"exosomes" in describing the EVs used in those studies. We recognize that other EV subtypes may exist in these populations and will continue to use the term EVs for the rest of the chapter. Nonetheless, the reader should be aware that the materials referenced would largely fall in the "exosome" size range and possess several such molecular makers.

3. B Cells, Extracellular Vesicles, and Associated Impacts

3.1 Antigen-Presenting Cells in Immune Responses

B cells are both antigen-presenting cells (APCs) and effector lymphocytes in the cellular immune system, being the cellular basis for humoral (antibody-based) immunity (Cooper, 2015). While originally there was a disconnect between the existence of serum antibodies (gamma globulins) and B cells, plasma cells—and eventually their B cell origins—were recognized as the source of immunoglobulins (LeBien & Tedder, 2008). The roles of B cells as central players in overall immune functions have been further recognized (Youinou, 2007), along with their roles in various disease states ranging from autoimmunity to oncology (Suryani & Tangye, 2010).

A key feature of the mammalian immune response is the generation of adaptive immunity, the ability to drive cellular (T cell) and humoral (antibody/B cell) immune responses against specific antigens derived from invading pathogens, or even tumors. Those outputs require the acquisition, processing, and presentation of potentially antigenic materials from pathogens/tumors to those effector cells (T cells and B cells). "Professional" APCs, including dendritic cells (DCs) (Guermonprez, Valladeau, Zitvogel, Thery, & Amigorena, 2002), macrophage (Unanue, 1984), and, as mentioned, B cells (Rodriguez-Pinto, 2005), are the primary promoters and stimulators that drive the initial phases of the adaptive immune response.

Technically, essentially all mammalian cells are APCs, in that cells display endogenously derived materials (usually small peptides of 8–10 amino acids in length) on major histocompatibility complex class I (MHC I) molecules. This occurs as a mechanism for surveillance by T cells, which may be searching for pathogenically affected cells (tumorous, viral or bacterial infections, etc.). The professional APCs are also capable of expressing MHC class II molecules (presenting peptides of 15–25 amino acids in length, which are derived from exogenous sources) (Blum, Wearsch, & Cresswell, 2013) and can enhance activation of T cells by expressing costimulatory molecules such as CD80 (B7-1) and CD86 (B7-2) (Chen & Flies, 2013). Thus such APCs are capable of "educating" and stimulating both CD8$^+$ and CD4$^+$ T cells (via presentation of peptides on MHC I and MHC II, respectively). This is explained in more detail in Fig. 3.

Most antigen presentation phenomenon by professional APCs occur in secondary lymphoid organs, such as lymph nodes and spleen (Millington, Zinselmeyer, Brewer, Garside, & Rush, 2007). Here, APCs such as macrophage, DCs, and follicular dendritic cells (FDCs) contact effector lymphocytes such as T cells and B cells in a unique architectural environment (Junt, Scandella, & Ludewig, 2008). The "intrinsic" presentation pathway (Fig. 3A) utilizes obsolete or unfolded/misfolded proteins found within the cell itself. These

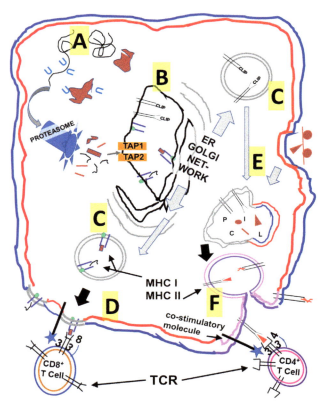

FIGURE 3 Generic Professional Antigen-Presenting Cell (APC) in an Overview of Antigen Presentation. A highly diagrammatic APC is depicted. The "intrinsic" antigen presentation pathway starts at (A), where senescent or improperly folded intracellular proteins undergo poly-ubiquitination (blue U's), marking the proteins for degradation via the proteasome. In this scenario, the red protein is of pathogenic origin, such as viral, bacterial, or even a mutated tumor protein. Peptides are generated by the proteasome, which leads to the passage of peptides into the endoplasmic reticulum (ER) lumen via the TAP (transporters associated with antigen processing) transporters. Inside the ER, major histocompatibility complex (MHC) class I and II molecules are assembled (B). MHC I molecules are loaded with peptides that bind with sufficient affinity into their peptide-binding clefts, while MHC II molecules have class II invariant chain peptide, CLIP, associated with the MHC II peptide-binding cleft (see Fig. 4). The MHC molecules are packed into transport vesicles exiting the ER/Golgi network (C). These intracellular vesicles fuse with the cellular plasma membrane to display the MHC I molecules with peptides bound for presentation to CD8+ T cells (D). Such T cells may have a T cell receptor (TCR) that recognizes the peptide "in the context" of the MHC I molecule. Costimulatory molecules (rods with stars), such as CD80/CD86, are necessary for full activation of the naïve T cells. Moving to the "extrinsic" antigen presentation pathway, extracellular proteins (red shapes, right side of figure) are internalized and enter the endosomal system (E). Antigen-containing endosomes fuse with the MHC II–containing secretory vesicles, and proteolysis removes CLIP from the MHC II peptide–binding pocket, allowing the binding of the internalized antigen species (F). The vesicles fuse with the plasma membrane, displaying the MHC II–peptide complex for presentation to an antigen-specific CD4+ T cell, with accompanying costimulation. As mentioned for Fig. 1, none of the entities depicted are drawn to scale.

are often multiply ubiquitinated, targeting the dysfunctional proteins for proteasomal degradation (Kraft, Peter, & Hofmann, 2010). The peptides generated by the proteasome enter the endoplasmic reticulum (ER) via the transporters associated with antigen processing (TAP 1 and 2), where the peptides are loaded into the binding grooves of appropriate MHC I molecules (i.e., different MHC I molecules have different binding affinities for various peptides, e.g., note the red peptide in Fig. 3B). The stabilized MHC I structure now passes into the trans-Golgi network for vesicular transport to the cell surface (Fig. 3C). Also assembling in the ER are MHC II molecules, which have their peptide-binding clefts blocked by a portion of the invariant chain (Ii) called class II invariant chain peptide (CLIP). This prevents self-peptide association with MHC II and promotes passage through the ER/Golgi network and packaging into transport vesicles (Fig. 3C). The MHC I–containing vesicles may fuse with the cell's plasma membrane, resulting in display of the MHC I-peptide complex, that can be recognized by CD8+ T cells with a T cell receptor (TCR) that fits the MHC I–presented peptide (Fig. 3D). Along with costimulation provided by the APC, the T cell will now be activated to proliferate and move into the peripheral environment on a "search-and-destroy" mission to eliminate cells that display the same CD8+ TCR-reactive pathogenic peptide on their MHC I molecules. The APC itself does not have to endogenously express the "offending" protein/peptide itself; there are numerous ways for professional APCs to acquire, process, and present exogenous materials into the endogenous/intrinsic pathways, mechanisms grouped under "cross-presentation" (Brode & Macary, 2004; Joffre, Segura, Savina, & Amigorena, 2012). Thus pathogen-derived antigenic materials arriving in a secondary lymphoid structure such as a draining lymph node may enter into the APC intrinsic presentation pathways and be presented on MHC I molecules, leading to activation of cytotoxic T lymphocytes (CD8+ T cells).

APCs that are resident in secondary lymphoid organs, or those "scavenging" in potential areas of infection or damage, may also acquire pathogenic antigens via specific or nonspecific cell surface interactions (Mak & Jett, 2014). Such proteins are internalized into endosomal bodies that result in degradation of the proteins. Fusion with the MHC II–containing vesicles results in the cleavage of CLIP and replacement in the MHC II–binding clefts with potentially antigenic peptides (Fig. 3E) (see also Fig. 4). These MHC II–peptide complexes are displayed for presentation to CD4+ T cells (Fig. 3F); costimulation promulgates the proliferation of the "helper" T cells, which secrete cytokines and growth factors that aid in CD8+ "killer" T cell responses, as well as activation of B cells (Zhu & Paul, 2008).

3.2 B Cells and Antigen Presentation

The role of B cells as APCs has been known for some time (Pierce et al., 1988), and they are included in the repertoire of APCs with efficient MHC class II presentation capabilities (Roche & Furuta, 2015). B cell APC activity is frequently overlooked in background of their antibody-producing roles but also in that the outcomes of B cell antigen presentation are not entirely straightforward (Chen & Jensen, 2008). The rest of this chapter proceeds from the hypothesis that B cell exosomes are important contributors to B cell antigen presentation capacities.

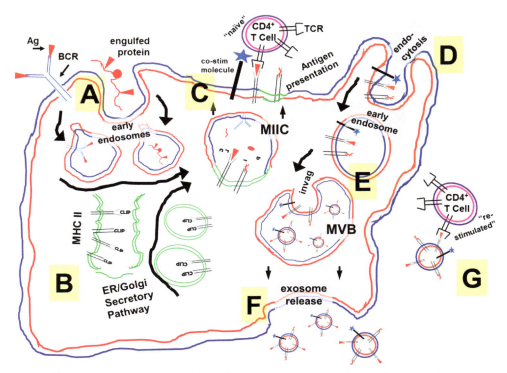

FIGURE 4 B Cell Antigen Presentation to T Cells and Activation Maintenance by B Cell Extracellular Vesicles (EVs). Depicted is a highly diagrammatic representation of a B cell in antigen-presentation mode. (A) Shows the B cell receptor (BCR) binding an antigen (Ag) specific for the paratope of the antibody, formed by the N-termini of the variable regions of the heavy and light chains of the immunoglobulin. The BCR on a B cell is a transmembrane receptor that can internalize by endocytic mechanisms, much like those that also take in (engulf) proteins in the extracellular milieu. These engulfed proteins and antigen-bound BCRs end up in early endosomes. In (B), different major histocompatibility class II (MHC II) molecules consisting of alpha and beta chains are assembled within the endoplasmic reticulum (ER), with a portion of the invariant chain (class II invariant chain peptide, CLIP) in the MHC peptide-binding cleft. This prevents binding of self-peptides (or other peptides that may be destined to bind MHC I molecules) and promotes passage of the MHC II molecules through the Golgi apparatus for packaging into vesicles of the secretory pathway (in C). As these vesicles fuse with incoming endosomes (with material scavenged from the cell membrane), the invariant chain is proteolyzed and CLIP is removed from the MHC peptide-binding groove, allowing for the binding of potentially antigenic peptides arriving in the endosomal vesicle. These are often called MHC class II–enriched compartments (MIICs). As the membranes from these vesicles fuse with the cellular plasma membrane, the membrane-bound MHC II molecules can now display the peptides acquired in the MIIC, often accompanied by costimulatory molecules (co-stim) such as CD80 and/or CD86. CD4+ T cells with the appropriate T cell receptor (TCR) recognize their cognate peptide antigens presented in the context of the B cell's MHC II molecules (antigen presentation). The formerly naïve T cells are now activated and will proliferate and/or secrete cytokines, influencing the overall immune response, including stimulation of B cells. The ligation of MHC II by the T cell TCR promotes exosome/EV release that begins with the endocytosis of B cell membrane components (D). The trajectory of such endosomes is to progress from early to late endosomes (E). Late endosomes develop inward budding (invaginations) of the endosomal, limiting membrane that eventually results in the formation of vesicles within the lumen of the endosome ("intraluminal vesicles," or "intravesicular bodies"). The endosome is now called a multivesicular body (MVB), and the vesicles within it have, to some extent, recapitulated the topography of the cell itself. The canonical fate of MVBs is often lysosomal degradation. However, intra- and extracellular factors can promote fusion of the MVB membranes with the plasma membrane, now releasing the interior vesicles outside the cell, where they are now called "exosomes" (F). Outside the cell, the peptide antigen–MHC II complex in the exosome membrane can "restimulate" or maintain the activated status of already-primed CD4+ T cells (G). As mentioned for Fig. 1, none of the entities depicted are drawn to scale.

Some of the earlier works on exosomes/EVs were from B cells, describing the EVs as antigen-presenting vesicles (Raposo et al., 1996). The EVs/"exosomes" believed to be derived from MHC class II-enriched compartments (MIICs) had peptide-bound MHC II in their membranes (whether such MIICs were "conventional" or "unconventional" endocytic compartments was, and remains, of some debate (Kleijmeer, Morkowski, Griffith, Rudensky, & Geuze, 1997; Pierre et al., 1996; Rocha & Neefjes, 2008)). When derived from B cell lines and pulsed with known antigens, the EVs themselves were able to promote proliferation and IL2 release from CD4$^+$ T cell clones that specifically recognized those antigens. Additionally, T cells specific for a given antigen may activate B cells (via ligation of the specific peptide-MHC II complex with the cognate TCR), leading to release of B cell EVs. Those EVs could further stimulate antigen-primed T cells, suggesting that the EVs are necessary to maintain the activated status of the T cells (Muntasell, Berger, & Roche, 2007). In situations where expansion or maintenance of clonal populations of T cells are driven by alloantigens, these studies may provide mechanisms for the reduced response thresholds of memory T cells versus their naïve counterparts (London, Lodge, & Abbas, 2000). This rather complicated scenario is depicted in Fig. 4.

One important point to note about Fig. 4A and C is the role of the B cell receptor (BCR). The BCR of a given B cell is essentially the antibody that the cell would secrete (i.e., with the same antigen-binding sites), but with a membrane-spanning domain. The BCR links to numerous signaling molecules, starting with the CD79A/B structure (Seda & Mraz, 2015). Stimulation of the BCR by antigen promotes cell proliferation, metabolic changes, and cytoskeletal rearrangements. Endocytosis of the BCR/antigen complex may lead to a more protected degradation of the antigen, in that epitopes may be preserved when the antigen is in embraced in the binding pocket formed by the antibody's heavy and light chain variable regions. Combined with the high efficiency (affinity) of antibody/antigen binding, and the selectivity of the altered endosomal degradation, the BCR can concentrate small amounts of particular antigenic epitopes to be parlayed into antigen presentation pathways to CD4$^+$ T cells (Rodriguez-Pinto, 2005). As shown, T cells responsive to those epitopes then promote the release of B cell EVs with MHC II–peptide complexes capable of restimulating the T cells in a feedback loop.

Another mode of antigen presentation that likely involves B cell EVs is "cross-dressing," which is the transfer of preformed and presumably functional MHC–peptide complexes from one cell to another by some form of intercellular transfer (Campana, De Pasquale, Carrega, Ferlazzo, & Bonaccorsi, 2015). If such passaged MHC molecules end up on a professional APC, that APC may gain an operational capacity outside of its own presentation machinery (Bonaccorsi, Pezzino, Morandi, & Ferlazzo, 2014). Three mechanisms have been described that may accomplish this transfer of MHC molecules. These include trogocytosis, the exchange of membrane patches between cells that had prior physical contact (Ahmed, Munegowda, Xie, & Xiang, 2008; Joly & Hudrisier, 2003). This is a recognized means of immune plasticity (Caumartin, Lemaoult, & Carosella, 2006) that could potentially create APCs out of effector T cells. Another mechanism of intercellular communication involves tunneling nanotubes, intercellular bridges up to 200 nm in diameter, and possibly >300 μm in length (Gerdes et al., 2013). These allow the transfer of both membrane and cytosolic materials, including metabolites, proteins, and RNAs (Gallagher & Benfey, 2005) via tubules that can contact numerous cells across distances exceeding many cell diameters. A third mechanism of cross-dressing is the passage of EVs to APCs, with EVs essentially perching intact on the APC membrane surface. The external face of the EV appears almost as a protrusion of the APC, but with potentially different transmembrane molecules on the external face (Campana et al., 2015). These three forms of MHC passage are shown in Fig. 5.

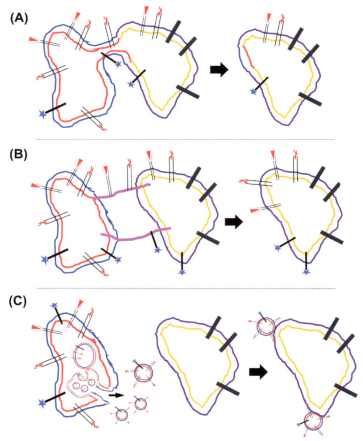

FIGURE 5 Trogocytosis, Tunneling Nanotubes, and Extracellular Vesicles (EVs) in Passive Antigen Presentation. Depicted are schematics of the transfer of major histocompatibility class II (MHC II) molecules (and/or costimulatory molecules) from one cell to another. (A) Shows a form of trogocytosis with the cell on the left (blue and red plasma membrane) contacting and briefly fusing membranes with the cell on the right (purple and orange membrane). Passage of the MHC II molecules (as redrawn from Figs. 3 and 4) and membrane domains are shown. Tunneling nanotubes are shown in (B) as lavender extensions from the left cell to the right cell, leading to transfer of MHC II and costimulatory molecules to the surface of the right cell. (C) shows a continuum of endosome maturation into a multivesicular body (see Figs. 1, 3, and 4) and extracellular release of exosomes/EVs from the cell on the left with eventual docking of the EVs on the cell to the right. The EVs are capable of displaying MHC and costimulatory molecules, acting as a functional unit on the recipient cell. As mentioned for Fig. 1, none of the entities depicted are drawn to scale, particularly the surface proteins in relation to the nanotubes.

3.3 B Cell Extracellular Vesicles and Follicular Dendritic Cells

FDCs do not appear to be bone marrow–derived but are instead stromal-like, probably from mesenchymal stem cell origin (van Nierop & de Groot, 2002). FDCs form critical networks in primary and secondary lymphoid organs as components of germinal centers, leading to clonal expansion of antigen-activated B cells (Liu, Grouard, de Bouteiller, & Banchereau, 1996). FDCs themselves are proficient at acquiring native antigen forms from antigen–antibody complexes that filter through the germinal center. These antigen–antibody complexes are bound at the FDC surface and stimulate the BCR of B cells that have sufficiently high affinity for the antigens (van Nierop & de Groot, 2002). FDCs use a variety of surface receptors to collect circulating antigen to present to B cells, while also passing "iccosomes" (immune complex coated bodies), which could be considered a very distinct form of EV. These "beads-on-a-string" structures are FDC membrane–derived vesicles of 0.3–0.7 µm that contain surface antigen complexes that transfer to B cells (Szakal, Kosco, & Tew, 1988). Those antigens can then be processed (i.e., Fig. 4) for presentation to CD4[+] T cells, which provide "help" in the form of cytokines and cell surface as receptor/ligand interactions to further activate the B cells (Tew, Wu, Fakher, Szakal, & Qin, 2001). Curiously, FDCs themselves do not synthesize MHC II molecules and yet are capable of stimulating T cells (Gray, Kosco, & Stockinger, 1991). One explanation for this comes from the discovery of presumed B cell EVs with MHC II on their surfaces, docking on FDCs as a unique form of antigen presentation (Denzer et al., 2000). This scenario is essentially that of Fig. 5C, where EVs from one cell may stably adhere to another cell, extending the repertoire of surface molecules and membrane lipids on the recipient cell. In the case of B cell EVs and FDCs in the germinal center, this binding appears rather specific for the FDCs (Denzer et al., 2000).

3.4 B Cell Extracellular Vesicles and the Promotion of Dendritic Cells– Mediated T Cell Immune Responses

As already noted, it appears that B cell EVs are capable of presenting antigen to CD4[+] T cells, leading to activation of the T cells, which then may reciprocate and further stimulate B cells. Other recent work has shown that B cells may also be conduits for antigen supply to DCs to promote antigen presentation to T cells (Henningsson et al., 2011). In that scenario, IgE–antigen complexes are captured by the low-affinity IgE receptor, CD23, on B cells, with endocytosis of the complex and presumed antigen presentation on B cell MHC II. However, this process requires DCs and does not seem to occur as direct presentation from B cells to T cells (Henningsson et al., 2011). Later studies (Martin, Brooks, Henningsson, Heyman, & Conrad, 2014) implicated B cell EVs or exosomes (there called "bexosomes") from CD23[+] B cells (and IgE–antigen complexes) in antigen transfer to DCs in the periphery. Those DCs were believed to proceed to secondary lymphoid organs where the DCs could not only stimulate antigen-specific T cells but also drive antibody production against the antigen. To further complicate the matter, DC exosomes/EVs, presumably with the same antigens that may have been acquired from B cell EVs, also contribute to B cell antibody responses to those antigens and, indeed, require B cells for Th1-skewed/proinflammatory CD4[+] T cell responses (Qazi, Gehrmann, Domange Jordo, Karlsson, & Gabrielsson, 2009).

3.5 B Cell Extracellular Vesicles as Cancer Vaccines/Adjuvants

Tumor EVs have been utilized as anticancer vaccines to promote T cell and B cell responses in preclinical settings (Graner et al., 2009; Kunigelis & Graner, 2015), and EVs from antigen-pulsed DCs have progressed into cancer clinical trials (Besse et al., 2016; Pitt et al., 2014). One could imagine settings where B cells could be harvested from a patient and grown in short-term culture for autologous EV preparations (there is a significant portion, ~60+%, of the naïve B cell compartment that expresses CD23 (Perez-Andres et al., 2010) which could be selected from the total population). Alternatively, there are numerous CD23[+] B cell lines that may be potentially useful (Okumura, Kishi, Tagoh, Minowada, & Muraguchi, 1995). One envisions taking tumor antigens (reference lists here for T cell–defined antigens/epitopes (Vigneron, 2015; Vigneron, Stroobant, Van den Eynde, & van der Bruggen, 2013) and an active database here http://cvc.dfci.harvard.edu/tadb/) and chemically cross-linking them to IgE molecules. These antibody–antigen complexes (single or multiple complexes) could then be incubated with the B cell cultures; EVs could be harvested by clinically relevant protocols (Lener et al., 2015). The treated B cell–derived EVs could then be pulsed on patient autologous DCs (there are several preclinical models that demonstrate effectiveness of exogenous EVs as promoters of DC-driven immunotherapy (Dai et al., 2006; Napoletano et al., 2009; Rao et al., 2016; Yao et al., 2014)). Alternatively, such EVs may be employed as antigen-specific treatment vehicles themselves, with injections into recipient "patients" (animal models or humans) as vaccines. This is outlined in Fig. 6. The goal would be to target the dermal DCs (generally, Langerhans cells) to promote an immune response (Kunigelis & Graner, 2015). Measurements of efficacy in preclinical (e.g., murine in vitro or in vivo, or human in vitro) models would be antigen-specific T cell responses (measured by proliferation, cytokine secretion, MHC-antigen-based tetramer responses, cytotoxicity against antigen-expressing cells), or in murine models with antitumor activities in tumor-bearing mice. These surrogates are recognized as incomplete in terms of monitoring immune responses for cancer immunotherapy paradigms, but they are what we have available (Macchia, Urbani, & Proietti, 2013; Whiteside, 2013). Success in the preclinical settings would bode favorably for continuation into human cancer patient clinical trials.

4. Conclusions

B cell EVs play important roles in immune responses to cancer antigens. Those roles amount to various forms of T cell stimulation via MHC II expression on the B cells/B cell EVs, and presentation of T cell–specific peptides, or stimulation of T cells in the context of docking on FDCs. Additionally, B cell EVs may deliver antigen to other DCs in the organismic periphery, or in the primary/secondary lymphoid organs. The BCR aids in collection and concentration of antigen, enabling potent stimulation of T cells with a potentially high-affinity antigen, but perhaps one of low prevalence. Passage of antigenic materials to MHC II molecules that may then exit the cell in the form of EVs provides for distance

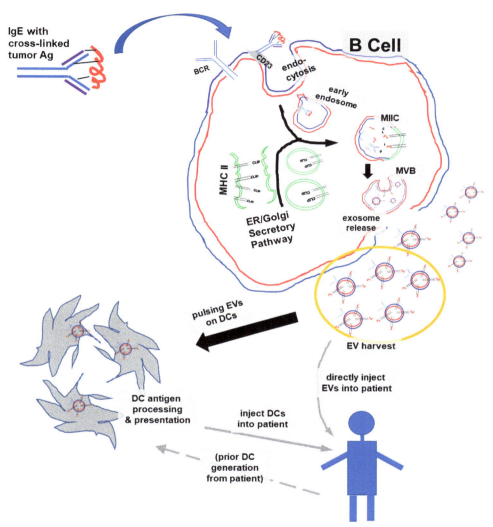

FIGURE 6 IgE-Linked Antigen as an Antigen Source for B Cell Extracellular Vesicles (EVs), and Their Use as DC Antigen Sources. IgE can be purified in large-scale batches and assessed for quality control. Known tumor antigens may be chemically cross-linked to IgE by many standard compatible chemistries. These antigen/antibody complexes could then be incubated with CD23+ B cells, which would undergo antigen processing as described in Figs. 3 and 4, with preferential processing of the IgE-linked antigen. Eventually the endosomal system would lead to exosome/EV release. Those EVs could be harvested by regulatory agency-approved means and then used as antigen sources when pulsed onto patient-derived DCs. The DCs would process and present antigen (Fig. 1) and would be infused/injected to the patient as autologous cellular vaccines. Alternatively, the EVs could be directly injected into patients as vaccines, where they would presumably stimulate local/dermal DCs and deliver antigen to those DCs, who would then traverse to draining lymph nodes for antigen presentation and activation of CD4+ (but also CD8+) T cells. As mentioned for Fig. 1, none of the entities depicted are drawn to scale.

signaling mechanisms and continued stimulation necessary to keep T cells in an activation status. The possibility of using B cell EVs as sources of antigen for stimulation of DCs, or of T cells directly, may show promise by virtue of the preferential processing of IgE-linked antigens within CD23+ B cells. This mechanism leads to efficacious MHC II presentation of IgE-bound antigens, and thus CD4+ T cell stimulation. B cell EVs could thus be used clinically as effective antigen sources in a cancer vaccine setting, either by direct use of the B cell EVs, or by employing the B cell EVs as antigen sources and stimulatory agents in DC-based cellular vaccinations.

References

Ahmed, K. A., Munegowda, M. A., Xie, Y., & Xiang, J. (2008). Intercellular trogocytosis plays an important role in modulation of immune responses. *Cellular & Molecular Immunology*, 5(4), 261–269. https://doi.org/10.1038/cmi.2008.32.

Besse, B., Charrier, M., Lapierre, V., Dansin, E., Lantz, O., Planchard, D., ... Chaput, N. (2016). Dendritic cell-derived exosomes as maintenance immunotherapy after first line chemotherapy in NSCLC. *Oncoimmunology*, 5(4), e1071008. https://doi.org/10.1080/2162402X.2015.1071008.

Blum, J. S., Wearsch, P. A., & Cresswell, P. (2013). Pathways of antigen processing. *Annual Review of Immunology*, 31, 443–473. https://doi.org/10.1146/annurev-immunol-032712-095910.

Bonaccorsi, I., Pezzino, G., Morandi, B., & Ferlazzo, G. (2014). Drag cells in immunity: Plasmacytoid DCs dress up as cancer cells. *OncoImmunology*, 3, e28184. https://doi.org/10.4161/onci.28184.

Brode, S., & Macary, P. A. (2004). Cross-presentation: Dendritic cells and macrophages bite off more than they can chew!. *Nature Reviews Immunology*, 112(3), 345–351. https://doi.org/10.1111/j.1365-2567.2004.01920.x.

Campana, S., De Pasquale, C., Carrega, P., Ferlazzo, G., & Bonaccorsi, I. (2015). Cross-dressing: An alternative mechanism for antigen presentation. *Immunology Letters*, 168(2), 349–354. https://doi.org/10.1016/j.imlet.2015.11.002.

Caumartin, J., Lemaoult, J., & Carosella, E. D. (2006). Intercellular exchanges of membrane patches (trogocytosis) highlight the next level of immune plasticity. *Transplant Immunology*, 17(1), 20–22. https://doi.org/10.1016/j.trim.2006.09.032.

Chen, L., & Flies, D. B. (2013). Molecular mechanisms of T cell co-stimulation and co-inhibition. *Nature Reviews Immunology*, 13(4), 227–242. https://doi.org/10.1038/nri3405.

Chen, X., & Jensen, P. E. (2008). The role of B lymphocytes as antigen-presenting cells. *Archivum immunologiae et therapiae experimentalis*, 56(2), 77–83. https://doi.org/10.1007/s00005-008-0014-5.

Consortium, E.-T., Van Deun, J., Mestdagh, P., Agostinis, P., Akay, O., Anand, S., ... Hendrix, A. (2017). EV-TRACK: Transparent reporting and centralizing knowledge in extracellular vesicle research. *Nature Methods*, 14(3), 228–232. https://doi.org/10.1038/nmeth.4185.

Cooper, M. D. (2015). The early history of B cells. *Nature Reviews Immunology*, 15(3), 191–197. https://doi.org/10.1038/nri3801.

Dai, S., Zhou, X., Wang, B., Wang, Q., Fu, Y., Chen, T., ... Cao, X. (2006). Enhanced induction of dendritic cell maturation and HLA-A*0201-restricted CEA-specific CD8(+) CTL response by exosomes derived from IL-18 gene-modified CEA-positive tumor cells. *Journal of Molecular Medicine (Berlin)*, 84(12), 1067–1076. https://doi.org/10.1007/s00109-006-0102-0.

Denzer, K., van Eijk, M., Kleijmeer, M. J., Jakobson, E., de Groot, C., & Geuze, H. J. (2000). Follicular dendritic cells carry MHC class II-expressing microvesicles at their surface. *The Journal of Immunology*, 165(3), 1259–1265.

Gallagher, K. L., & Benfey, P. N. (2005). Not just another hole in the wall: Understanding intercellular protein trafficking. *Genes & Development, 19*(2), 189–195. https://doi.org/10.1101/gad.1271005.

Gerdes, H. H., Rustom, A., & Wang, X. (2013). Tunneling nanotubes, an emerging intercellular communication route in development. *Mechanisms of Development, 130*(6–8), 381–387. https://doi.org/10.1016/j.mod.2012.11.006.

Gould, S. J., & Raposo, G. (2013). As we wait: Coping with an imperfect nomenclature for extracellular vesicles. *Journal of Extracellular Vesicles, 2*. https://doi.org/10.3402/jev.v2i0.20389.

Graner, M. W., Alzate, O., Dechkovskaia, A. M., Keene, J. D., Sampson, J. H., Mitchell, D. A., & Bigner, D. D. (2009). Proteomic and immunologic analyses of brain tumor exosomes. *The FASEB Journal, 23*(5), 1541–1557. https://doi.org/10.1096/fj.08-122184.

Gray, D., Kosco, M., & Stockinger, B. (1991). Novel pathways of antigen presentation for the maintenance of memory. *International Immunology, 3*(2), 141–148.

Guermonprez, P., Valladeau, J., Zitvogel, L., Thery, C., & Amigorena, S. (2002). Antigen presentation and T cell stimulation by dendritic cells. *Annual Review of Immunology, 20*, 621–667. https://doi.org/10.1146/annurev.immunol.20.100301.064828.

Gyorgy, B., Szabo, T. G., Pasztoi, M., Pal, Z., Misjak, P., Aradi, B., … Buzas, E. I. (2011). Membrane vesicles, current state-of-the-art: Emerging role of extracellular vesicles. *Cellular and Molecular Life Sciences, 68*(16), 2667–2688. https://doi.org/10.1007/s00018-011-0689-3.

Henningsson, F., Ding, Z., Dahlin, J. S., Linkevicius, M., Carlsson, F., Gronvik, K. O., … Heyman, B. (2011). IgE-mediated enhancement of CD4+ T cell responses in mice requires antigen presentation by CD11c+ cells and not by B cells. *PLoS One, 6*(7), e21760. https://doi.org/10.1371/journal.pone.0021760.

Joffre, O. P., Segura, E., Savina, A., & Amigorena, S. (2012). Cross-presentation by dendritic cells. *Nature Reviews Immunology, 12*(8), 557–569. https://doi.org/10.1038/nri3254.

Joly, E., & Hudrisier, D. (2003). What is trogocytosis and what is its purpose? *Nature Immunology, 4*(9), 815. https://doi.org/10.1038/ni0903-815.

Junt, T., Scandella, E., & Ludewig, B. (2008). Form follows function: Lymphoid tissue microarchitecture in antimicrobial immune defence. *Nature Reviews Immunology, 8*(10), 764–775. https://doi.org/10.1038/nri2414.

Kalra, H., Drummen, G. P., & Mathivanan, S. (2016). Focus on extracellular vesicles: Introducing the next small big thing. *International Journal of Molecular Sciences, 17*(2), 170. https://doi.org/10.3390/ijms17020170.

Kim, S., & Moustaid-Moussa, N. (2000). Secretory, endocrine and autocrine/paracrine function of the adipocyte. *The Journal of Nutrition, 130*(12), 3110S–3115S.

Kleijmeer, M. J., Morkowski, S., Griffith, J. M., Rudensky, A. Y., & Geuze, H. J. (1997). Major histocompatibility complex class II compartments in human and mouse B lymphoblasts represent conventional endocytic compartments. *Journal of Cell Biology, 139*(3), 639–649.

Kraft, C., Peter, M., & Hofmann, K. (2010). Selective autophagy: Ubiquitin-mediated recognition and beyond. *Nature Cell Biology, 12*(9), 836–841. https://doi.org/10.1038/ncb0910-836.

Kunigelis, K. E., & Graner, M. W. (2015). The dichotomy of tumor exosomes (TEX) in cancer immunity: Is it all in the ConTEXt? *Vaccines (Basel), 3*(4), 1019–1051. https://doi.org/10.3390/vaccines3041019.

LeBien, T. W., & Tedder, T. F. (2008). B lymphocytes: How they develop and function. *Blood, 112*(5), 1570–1580. https://doi.org/10.1182/blood-2008-02-078071.

Lener, T., Gimona, M., Aigner, L., Borger, V., Buzas, E., Camussi, G., … Giebel, B. (2015). Applying extracellular vesicles based therapeutics in clinical trials – an ISEV position paper. *Journal of Extracellular Vesicles, 4*, 30087. https://doi.org/10.3402/jev.v4.30087.

Liu, Y. J., Grouard, G., de Bouteiller, O., & Banchereau, J. (1996). Follicular dendritic cells and germinal centers. *International Review of Cytology, 166*, 139–179.

London, C. A., Lodge, M. P., & Abbas, A. K. (2000). Functional responses and costimulator dependence of memory CD4+ T cells. *The Journal of Immunology, 164*(1), 265–272.

Lotvall, J., Hill, A. F., Hochberg, F., Buzas, E. I., Di Vizio, D., Gardiner, C., … Thery, C. (2014). Minimal experimental requirements for definition of extracellular vesicles and their functions: A position statement from the International Society for Extracellular Vesicles. *Journal of Extracellular Vesicles, 3*, 26913. https://doi.org/10.3402/jev.v3.26913.

Luzio, J. P., Pryor, P. R., & Bright, N. A. (2007). Lysosomes: Fusion and function. *Nature Reviews Molecular Cell Biology, 8*(8), 622–632. https://doi.org/10.1038/nrm2217.

Macchia, I., Urbani, F., & Proietti, E. (2013). Immune monitoring in cancer vaccine clinical trials: Critical issues of functional flow cytometry-based assays. *BioMed Research International, 2013*, 726239. https://doi.org/10.1155/2013/726239.

Mak, T. W., & Jett, B. D. (2014). Antigen processing and presentation. In T. W. Mak, M. E. Saunders, & B. D. Jett (Eds.), *Primer to the immune response* (2nd ed.) (pp. 161–179). Amsterdam: Elsevier.

Martin, R. K., Brooks, K. B., Henningsson, F., Heyman, B., & Conrad, D. H. (2014). Antigen transfer from exosomes to dendritic cells as an explanation for the immune enhancement seen by IgE immune complexes. *PLoS One, 9*(10), e110609. https://doi.org/10.1371/journal.pone.0110609.

Millington, O. R., Zinselmeyer, B. H., Brewer, J. M., Garside, P., & Rush, C. M. (2007). Lymphocyte tracking and interactions in secondary lymphoid organs. *Inflammation Research, 56*(10), 391–401. https://doi.org/10.1007/s00011-007-7017-2.

Mittelbrunn, M., Vicente Manzanares, M., & Sanchez-Madrid, F. (2015). Organizing polarized delivery of exosomes at synapses. *Traffic, 16*(4), 327–337. https://doi.org/10.1111/tra.12258.

Muntasell, A., Berger, A. C., & Roche, P. A. (2007). T cell-induced secretion of MHC class II-peptide complexes on B cell exosomes. *The EMBO Journal, 26*(19), 4263–4272. https://doi.org/10.1038/sj.emboj.7601842.

Napoletano, C., Rughetti, A., Landi, R., Pinto, D., Bellati, F., Rahimi, H., … Nuti, M. (2009). Immunogenicity of allo-vesicle carrying ERBB2 tumor antigen for dendritic cell-based anti-tumor immunotherapy. *International Journal of Immunopathology & Pharmacology, 22*(3), 647–658. https://doi.org/10.1177/039463200902200310.

Okumura, A., Kishi, H., Tagoh, H., Minowada, J., & Muraguchi, A. (1995). Expression of 18.6/CD23 antigen on human lymphoid progenitor cell lines and phorbol 12-myristate 13-acetate (PMA)-induced microglia-shaped cells. *Microbiology and Immunology, 39*(11), 879–890.

Perez-Andres, M., Paiva, B., Nieto, W. G., Caraux, A., Schmitz, A., Almeida, J., … Primary Health Care Group of Salamanca for the Study of MBL (2010). Human peripheral blood B-cell compartments: A crossroad in B-cell traffic. *Cytometry B Clinical Cytometry, 78*(Suppl. 1), S47–S60. https://doi.org/10.1002/cyto.b.20547.

Pierce, S. K., Morris, J. F., Grusby, M. J., Kaumaya, P., van Buskirk, A., Srinivasan, M., … Smolenski, L. A. (1988). Antigen-presenting function of B lymphocytes. *Immunological Reviews, 106*, 149–180.

Pierre, P., Denzin, L. K., Hammond, C., Drake, J. R., Amigorena, S., Cresswell, P., & Mellman, I. (1996). HLA-DM is localized to conventional and unconventional MHC class II-containing endocytic compartments. *Immunity, 4*(3), 229–239.

Pitt, J. M., Charrier, M., Viaud, S., Andre, F., Besse, B., Chaput, N., & Zitvogel, L. (2014). Dendritic cell-derived exosomes as immunotherapies in the fight against cancer. *The Journal of Immunology, 193*(3), 1006–1011. https://doi.org/10.4049/jimmunol.1400703.

Qazi, K. R., Gehrmann, U., Domange Jordo, E., Karlsson, M. C., & Gabrielsson, S. (2009). Antigen-loaded exosomes alone induce Th1-type memory through a B-cell-dependent mechanism. *Blood, 113*(12), 2673–2683. https://doi.org/10.1182/blood-2008-04-153536.

Rao, Q., Zuo, B., Lu, Z., Gao, X., You, A., Wu, C., … Yin, H. (2016). Tumor-derived exosomes elicit tumor suppression in murine hepatocellular carcinoma models and humans in vitro. *Hepatology, 64*(2), 456–472. https://doi.org/10.1002/hep.28549.

Raposo, G., Nijman, H. W., Stoorvogel, W., Liejendekker, R., Harding, C. V., Melief, C. J., & Zitvogel, L. (1996). B lymphocytes secrete antigen-presenting vesicles. *The Journal of Experimental Medicine, 183*(3), 1161–1172.

Rocha, N., & Neefjes, J. (2008). MHC class II molecules on the move for successful antigen presentation. *The EMBO Journal, 27*(1), 1–5. https://doi.org/10.1038/sj.emboj.7601945.

Roche, P. A., & Furuta, K. (2015). The ins and outs of MHC class II-mediated antigen processing and presentation. *Nature Reviews Immunology, 15*(4), 203–216. https://doi.org/10.1038/nri3818.

Rodriguez-Pinto, D. (2005). B cells as antigen presenting cells. *Cellular Immunology, 238*(2), 67–75. https://doi.org/10.1016/j.cellimm.2006.02.005.

Seda, V., & Mraz, M. (2015). B-cell receptor signalling and its crosstalk with other pathways in normal and malignant cells. *European Journal of Haematology, 94*(3), 193–205. https://doi.org/10.1111/ejh.12427.

Suryani, S., & Tangye, S. G. (2010). Therapeutic implications of advances in our understanding of transitional B-cell development in humans. *Expert Review of Clinical Immunology, 6*(5), 765–775. https://doi.org/10.1586/eci.10.55.

Szakal, A. K., Kosco, M. H., & Tew, J. G. (1988). A novel in vivo follicular dendritic cell-dependent iccosome-mediated mechanism for delivery of antigen to antigen-processing cells. *The Journal of Immunology, 140*(2), 341–353.

Tew, J. G., Wu, J., Fakher, M., Szakal, A. K., & Qin, D. (2001). Follicular dendritic cells: Beyond the necessity of T-cell help. *Trends in Immunology, 22*(7), 361–367.

Unanue, E. R. (1984). Antigen-presenting function of the macrophage. *Annual Review of Immunology, 2*, 395–428. https://doi.org/10.1146/annurev.iy.02.040184.002143.

van Nierop, K., & de Groot, C. (2002). Human follicular dendritic cells: Function, origin and development. *Seminars in Immunology, 14*(4), 251–257.

Vigneron, N. (2015). Human tumor antigens and cancer immunotherapy. *BioMed Research International, 2015*, 948501. https://doi.org/10.1155/2015/948501.

Vigneron, N., Stroobant, V., Van den Eynde, B. J., & van der Bruggen, P. (2013). Database of T cell-defined human tumor antigens: The 2013 update. *Cancer Immunity, 13*, 15.

Whiteside, T. L. (2013). Immune responses to cancer: Are they potential biomarkers of prognosis? *Frontiers in Oncology, 3*, 107. https://doi.org/10.3389/fonc.2013.00107.

Yanez-Mo, M., Siljander, P. R., Andreu, Z., Zavec, A. B., Borras, F. E., Buzas, E. I., … De Wever, O. (2015). Biological properties of extracellular vesicles and their physiological functions. *Journal of Extracellular Vesicles, 4*, 27066. https://doi.org/10.3402/jev.v4.27066.

Yao, Y., Wang, C., Wei, W., Shen, C., Deng, X., Chen, L., … Hao, S. (2014). Dendritic cells pulsed with leukemia cell-derived exosomes more efficiently induce antileukemic immunities. *PLoS One, 9*(3), e91463. https://doi.org/10.1371/journal.pone.0091463.

Youinou, P. (2007). B cell conducts the lymphocyte orchestra. *Journal of Autoimmunity, 28*(2–3), 143–151. https://doi.org/10.1016/j.jaut.2007.02.011.

Zhang, G., & Yang, P. (2017). A novel cell-cell communication mechanism in the nervous system: Exosomes. *Journal of Neuroscience Research.* https://doi.org/10.1002/jnr.24113.

Zhu, J., & Paul, W. E. (2008). CD4 T cells: Fates, functions, and faults. *Blood, 112*(5), 1557–1569. https://doi.org/10.1182/blood-2008-05-078154.

19

Translational Potential of Tumor Exosomes in Diagnosis and Therapy

Naureen Javeed[1], Debabrata Mukhopadhyay[2]

[1]MAYO CLINIC, ROCHESTER, MN, UNITED STATES; [2]MAYO CLINIC, JACKSONVILLE, FL, UNITED STATES

CHAPTER OUTLINE

1. Introduction.. 343

2. Exosomes and Cancer 344

3. Exosomes as Cancer Biomarkers 345

4. Therapeutic Potential of Exosomes 346

5. Conclusions.. 348

References 349

1. Introduction

As the number of cancer incidences arises, what is urgently needed are effective novel methodologies for improving cancer detection, monitoring, and treatment. Despite decades of intensive research in the field, cancer therapies have failed to show promising efficacy and complete reversal from the tumorigenic state. The heterogeneity and complexity of tumors alone are just a few of the several confounding variables that contribute to the elusiveness of the disease. What is needed now more than ever are reliable biomarkers for early diagnosis and prognosis, and more effective treatment regimens for cancer.

The understanding of the complexity and heterogeneity of malignant tumors is a topic of much discussion. Tumors consist of cancer cells, fibroblasts, immune cells, endothelial cells, and the surrounding stroma (Hanahan & Weinberg, 2011). Collectively, these cells are known as the tumor microenvironment, which is critical for the initiation, progression, and dissemination of the tumor as well as its response to therapy (Hanahan & Weinberg, 2011). Molecular/cellular events and factors secreted by tumor cells have been shown to impact the tumor microenvironment to favor neoplastic transformation (Hanahan & Weinberg, 2011; Swartz et al., 2012). Understanding the interplay between tumor cells and their microenvironment will be critical for the development of novel biomarkers and therapies to treat the disease.

For over a decade, significant research efforts have been made in the field of cancer biomarker development and novel therapeutic treatments. One potential avenue that has garnered much attention is the use of exosomes for cancer detection and treatment. Exosomes are 30–150 nm endocytic-derived nanovesicles, which are secreted by almost all cell types in the body (Javeed & Mukhopadhyay, 2017; Mathivanan, Ji, & Simpson, 2010). Originally viewed as a mechanism to rid the cell of waste and toxins, significant strides have been made to show the relevance of these nanoparticles as important mediators of intercellular communication through the exchange of bioactive cargo. Interestingly, although both normal and tumor-derived exosomes share importance in the tumorigenesis process, the concentration of exosomes shed from tumors has been shown to be substantially higher (Rabinowits, Gercel-Taylor, Day, Taylor, & Kloecker, 2009; Taylor & Gercel-Taylor, 2008). Several reports have shown the contribution of exosomes to tumorigenesis through aiding in the formation of the premetastatic niche, promotion of angiogenesis, and modulation of tumor immune responses (Katoh, 2013; Psaila & Lyden, 2009; Sceneay, Smyth, & Moller, 2013; Zhang & Grizzle, 2011). The cell specificity, stability, and bioavailability of exosomes make them attractive biomarkers and therapeutic agents for both the treatment and prevention of cancer.

2. Exosomes and Cancer

The formation and progression of cancer is a complex process, which consists of several "hallmark" characteristics leading to the development of the tumor (Hanahan & Weinberg, 2011). This multistep process not only involves the evolution of normal cells to those of a neoplastic state but also incorporates surrounding tumor-associated stroma (including immune cells, endothelial cells, and fibroblasts) and blood vessels thus comprising the "tumor microenvironment" (Hanahan & Weinberg, 2011). Communication between cells is an important factor in the transmission of biological signals. Cells that comprise the tumor microenvironment exchange information via secretion of soluble factors or by direct cell-to-cell interactions. Intercellular communication in this environment can therefore promote tumor initiation, dissemination, and even the response to therapies (Wang, Chen, Liu, & Tian, 2016). For example, soluble factors secreted by the tumor can aid in resisting apoptosis, inducing cell proliferation, and promoting angiogenesis (Hanahan & Weinberg, 2011). Understanding the interplay between the cells that make up the tumor microenvironment and the factors they secrete has great implications for cancer prevention, biomarker development, and novel therapies to treat the disease.

Intercellular communication through the secretion and uptake of extracellular vesicles (EVs) has been extensively studied in the last three decades. Much progress has been made in understanding the fundamentals of EV biogenesis, release, uptake, content, and in the pathophysiology of disease states (Javeed & Mukhopadhyay, 2017; Mathivanan, et al., 2010). In recent years, exosomes, a subtype of EVs, have garnered much attention because of their capacity to house important biologically active cargoes, transfer their contents to target cells, and thereby aid in promotion of disease states. Exosomes have been shown to be mediators of tumorigenesis, neurodegenerative diseases, such as Parkinson's and Alzheimer's disease (Baixauli, Lopez-Otin, & Mittelbrunn, 2014; Bolmont et al., 2007; Danzer et al., 2012), cardiovascular diseases (Amabile, Rautou, Tedgui, & Boulanger, 2010; Arslan et al., 2013; Bang et al., 2014), and infectious diseases, such as HIV (Narayanan et al., 2013). Perhaps, the most well-understood role of exosomes in disease pathology is their role in the progression of tumorigenesis. Exosomes have been shown to induce cell proliferation (Al-Nedawi et al., 2008), while inhibiting apoptosis (Khan et al., 2009), promote angiogenesis (Kucharzewska et al., 2013; Umezu et al., 2014; Zhou et al., 2014), and aid in establishing the premetastatic niche (Peinado et al., 2012; Zhou, et al., 2014).

3. Exosomes as Cancer Biomarkers

The identification of novel cancer biomarkers and therapeutic regimes has proven to be a challenge in cancer research. Owing to the molecular and pathophysiological complexity of the disease, there is a dire need for more definitive biomarkers, better avenues for monitoring disease progression, as well as more effective treatment regimes. The identification of novel biomarkers is of great importance to improve patient diagnosis and prognosis. Emerging evidence has shown the utility of tumor-derived exosomes as promising biomarkers for the disease. The heterogeneity of the content in exosomes is often reflective of the particular disease state. Several reports have now shown the utility of tumor-derived exosomes as potential biomarkers through their characterization, and function both in vivo and in vitro.

Proteomic profiling of two human colorectal cancer cell lines revealed a selective enrichment of key metastatic factors, signal transduction molecules, and various lipid raft components in these exosomes (Ji et al., 2013). In pancreatic cancer, a cell surface proteoglycan, glypican-1 (GPC1) was shown to be enriched in circulating pancreatic cancer–derived exosomes along with mutant Kras mRNA (Melo et al., 2015). These GPC1+ circulating exosomes were found to be detected in the serum of pancreatic cancer patients and shown to be an overall more reliable serum biomarker than the current standard, CA19-9 (Melo et al., 2015).

The discovery of nucleic acids as potential biomarkers in exosomes has been suggested for the identification of various cancers. For example, variants of mRNA and miRNAs shown in gliomas were detected in exosomes isolated from the serum of glioblastoma patients (Skog et al., 2008). miRNAs found in exosomes have been shown to be prognostic biomarkers for various cancers. Studies have shown that miRNAs in exosomes are protected from RNase degradation and can be detected in the serum of patients (Rabinowits, et al., 2009; Takeshita et al., 2013). In one study, the expression of exosomal miR-19α in the serum of colorectal cancer patients was shown to be significantly elevated compared with healthy donors. Additionally, elevated exosomal miR-19α correlated to a poorer prognosis for the patient (Matsumura et al., 2015). In ovarian cancer, eight diagnostic miRNAs were consistently found in the sera of patients (Taylor & Gercel-Taylor, 2008). Not only exosomal mRNAs and miRNAs have been identified as potential biomarkers for cancer, but it has been shown that double-stranded DNA has also been identified particularly in pancreatic cancer exosomes (Kahlert et al., 2014). Interestingly, genomic DNA from both cell lines and serum of pancreatic cancer patients showed the existence of mutant KRAS and p53 (Kahlert et al., 2014). These findings suggest that the exosomal DNA from pancreatic cancer represents the entirety of the genome and therefore sheds light on the mutational status of parental tumor cells (Kahlert et al., 2014).

In addition to nucleic acids, exosomes have been shown to house small peptides. A 52-amino-acid polypeptide adrenomedullin was found in pancreatic cancer tumor-derived exosomes (PC-Exo) isolated from both conditioned media of pancreatic cancer cell lines and patient plasma (Javeed et al., 2015). These PC-Exo not only contained both adrenomedullin and CA19-9 (a clinical marker for the disease) but were also shown to be functionally active by inducing β-cell dysfunction in vitro (Javeed, et al., 2015). Collectively, this work demonstrated the functional role of tumor-derived exosomes in mediating new-onset diabetes in pancreatic cancer patients. As new-onset of diabetes was shown to precede cancer diagnosis in a subset of pancreatic cancer patients, PC-Exo may serve as an early prognostic marker for the disease (Javeed, et al., 2015).

4. Therapeutic Potential of Exosomes

The potential use of exosomes as vehicles for targeted anticancer drug delivery has garnered significant attention in recent years. The role that exosomes take on as cellular carriers/deliverers of important biologically active cargoes to recipient cells makes these vesicles optimal candidates for drug delivery. Not only are exosomes abundantly secreted by the majority of cell types, but they are well tolerated, stable both in vivo and in vitro, membrane permeable, and can be targeted to various tissue types (Javeed & Mukhopadhyay, 2017). Collectively, these qualities make exosomes prime candidates for antitumor therapies as they have the potential to serve as vehicles for targeted delivery of anticancer compounds/molecules, proteins, nucleic acids, and miRNA/siRNAs. To exploit the therapeutic potential of exosomes, several strategies can be employed such as: (1) drug loading and delivery of antitumor compounds/molecules; (2) immune modulation; (3) modulation of exosomal content; and (4) complete removal of tumor-derived exosomes from the body.

One potential avenue for the use of exosomes in cancer therapeutics involves the loading and delivery of antitumor agents, nucleic acids, miRNAs, and siRNAs. The utility of exosomes as a means for drug delivery holds several advantages. As exosomes contain several transmembrane/membrane-anchored proteins, this could facilitate their ability to fuse with plasma membranes with ease, thus allowing for efficient cellular internalization of exosomal cargo. In addition, exosomes may be better tolerated than artificial delivery vehicles because of their endogenous production in the body and their ability to resist phagocytosis impart by their expression of CD47 (Kaur et al., 2014; Kibria et al., 2016).

The potential use of exosomes as deliverers of tissue-targeting siRNAs to the mouse brain has been explored. In one study, dendritic cell–derived exosomes were isolated and loaded with siRNA via electroporation (Alvarez-Erviti et al., 2011). The loaded exosomes were able to traverse the blood brain barrier and target the mouse neurons in the brain resulting in an efficient knockdown (~60%) of their gene of interest (Alvarez-Erviti et al., 2011). In a recent report, exosomes derived from normal mesenchymal cells (termed iExosomes) were engineered to contain either siRNA or shRNA to oncogenic Kras^{G12D}, a common mutation found in pancreatic cancer (Kamerkar et al., 2017). iExosomes were shown to effectively suppress pancreatic cancer in vivo and increase overall survival potential (Kamerkar, et al., 2017).

The loading of tumor-derived antigens onto exosomes has been explored as a means to modulate immune responses. Over two decades ago, it was shown that exosomes secreted from B lymphocytes carried major histocompatibility complex (MHC) class II molecules and that these exosomes could elicit a T cell response (Raposo et al., 1996). Soon after, a report demonstrated that exosomes released from mouse dendritic cells contained both class I and II MHC molecules and could elicit a potent immune response in vivo as shown by decreased tumor growth or complete eradication of the tumor (Zitvogel et al., 1998). Work from these two seminal reports has provided a foundation for researchers to further investigate the role of exosomes in immune modulation. Three Phase I clinical trials went underway using both dendritic cell (DC)- and ascite-derived exosomes; however, only modest improvements were shown in the patients (Dai et al., 2008; Escudier et al., 2005; Morse et al., 2005). Furthermore, in 2009, a Phase II clinical trial was conducted using peptide-loaded DC-derived exosomes (termed dexosomes) (Viaud et al., 2009).

As exosomes have been highly implicated in tumorigenesis, inhibition of these vesicles has been viewed as a potential therapeutic intervention for the disease. It is known that the level of circulating exosomes increases during cancer progression (Logozzi et al., 2009). Thereby, therapies to reduce the overall production, uptake, and/or secretion of exosomes will be important to exploit. Specific inhibition strategies to prevent exosome formation are underway but have mostly not been tested in disease models in vivo. The exact mechanism(s) involved in exosome internalization is currently up for debate. There are several potential means of exosome internalization, which include clathrin-mediated endocytosis, caveolin-mediated endocytosis, macropinocytosis, phagocytosis, and plasma membrane fusion (Mulcahy, Pink, & Carter, 2014). In one study, tumor-derived exosome production was reduced using the blood pressure–reducing drug, dimethyl amiloride, which was correlated to enhance in vivo antitumor efficacy of a chemotherapeutic drug (Chalmin et al., 2010). Another study showed that a potential methodology to target tetraspanins, which are important for the formation of exosomes, has been implicated in the induction of angiogenesis in tumors (Nazarenko et al., 2010).

Targeting the secretion of exosomes is also a plausible therapeutic target; however, the exact mechanism of exosome release remains unknown. It has been suggested in tumor cells that the small GTPase Rab27a is required for exosome release and that its knockdown (via RNAi) reduced the rate of tumor growth and metastasis (Bobrie et al., 2012; Ostrowski et al., 2010). Other Rab GTPases have also been implicated in the release of exosomes, specifically Rab11 and Rab35. However, inhibition of these GTPases has been shown to impair the docking/fusion of multivesicular bodies to the plasma membrane, thereby inhibiting exosomal release (Hsu et al., 2010; Savina, Fader, Damiani, & Colombo, 2005).

Although inhibition of exosome release, internalization, and formation show much promise for the treatment of cancers, it should be noted as a cautionary approach that disrupting any of these pathways could lead to unfavorable off-target effects. Many of the targetable molecules/pathways suggested are also critical for maintaining normal cellular processes. As novel approaches continue to be developed, it will be important to assess methodologies that can specifically target the exosomal machinery and pathways.

5. Conclusions

Interest in the diagnostic and therapeutic potential of exosomes has been exponentially growing as researchers begin to unlock the complexities that surround the field of EV biology. Although compelling data in the field has demonstrated the importance of exosomes in tumor modulation, there remains several unopened questions about their role in tumorigenesis and how exosomes can be effectively used as biomarkers and therapeutic agents to treat the disease. This is in part due to the many technical challenges that face this type of research such as efficient exosome isolation procedures, which will yield high concentrations of exosomes, and the identification of exosomal content that impacts the tumorigenesis process. Additionally, a more clear understanding of the differences between exosomal signaling in both normal and the tumorigenic state will be imperative in designing novel therapies to treat the disease. Despite these setbacks, researchers are making significant strides toward understanding the biological relevance and utility of exosomes. Because of their bioavailability, ease of uptake, and ability to target recipient cells, exosomes have the potential to revolutionize patient care through the development of noninvasive therapeutic approaches to treat cancers. This is exemplified by several ongoing clinical trials in which exosomes are used as both biomarkers and therapeutic agents to treat various cancers (Table 1).

Table 1   Ongoing Clinical Trials Involving Exosomes and Cancer

Disease Condition	Study Title	NCT ID
Pancreatic cancer	Interrogation of Exosome-mediated Intercellular Signaling in Patients With Pancreatic Cancer	NCT02393703
	Diagnostic Accuracy of Circulating Tumor Cells (CTCs) and Onco-exosome Quantification in the Diagnosis of Pancreatic Cancer - PANC-CTC	NCT03032913
	A Study of Blood Based Biomarkers for Pancreas Adenocarcinoma	NCT03334708

Table 1   Ongoing Clinical Trials Involving Exosomes and Cancer—cont'd

Disease Condition	Study Title	NCT ID
Breast cancer	Effects of MK-3475 (Pembrolizumab) on the Breast Tumor Microenvironment in Triple Negative Breast Cancer	NCT02977468
Prostate cancer	Detection of ARv7 in the Plasma of Men With Advanced Metastatic Castrate Resistant Prostate Cancer (MCRP)	NCT03236688
	Clinical Evaluation of the 'ExoDx Prostate IntelliScore' (EPI)	NCT03031418
	Decision Impact Trial of the ExoDx Prostate (IntelliScore)	NCT03235687
	Localising Occult Prostate Cancer Metastases With Advanced Imaging TEchniques	NCT02935816
	Molecular Studies and Clinical Correlations in Human Prostatic Disease	NCT00578240
Colon cancer	Study Investigating the Ability of Plant Exosomes to Deliver Curcumin to Normal and Colon Cancer Tissue	NCT01294072
Lung cancer	Circulating Exosome RNA in Lung Metastases of Primary High-Grade Osteosarcoma	NCT03108677
Thyroid cancer	Metformin Hydrochloride in Mitigating Side Effects of Radioactive Iodine Treatment in Patients With Differentiated Thyroid Cancer	NCT03109847
	Anaplastic Thyroid Cancer and Follicular Thyroid Cancer-derived Exosomal Analysis Via Treatment of Lovastatin and Vildagliptin and Pilot Prognostic Study Via Urine Exosomal Biological Markers in Thyroid Cancer Patients	NCT02862470

ClinicalTrials.gov.

References

Al-Nedawi, K., Meehan, B., Micallef, J., Lhotak, V., May, L., Guha, A., & Rak, J. (2008). Intercellular transfer of the oncogenic receptor EGFRvIII by microvesicles derived from tumour cells. [Research Support, Non-U.S. Gov't] *Nature Cell Biology, 10*(5), 619–624. https://doi.org/10.1038/ncb1725.

Alvarez-Erviti, L., Seow, Y., Yin, H., Betts, C., Lakhal, S., & Wood, M. J. (2011). Delivery of siRNA to the mouse brain by systemic injection of targeted exosomes. [Research Support, Non-U.S. Gov't] *Nature Biotechnology, 29*(4), 341–345. https://doi.org/10.1038/nbt.1807.

Amabile, N., Rautou, P. E., Tedgui, A., & Boulanger, C. M. (2010). Microparticles: Key protagonists in cardio-vascular disorders. [Review] *Seminars in Thrombosis and Hemostasis, 36*(8), 907–916. https://doi.org/10.1055/s-0030-1267044.

Arslan, F., Lai, R. C., Smeets, M. B., Akeroyd, L., Choo, A., Aguor, E. N., … de Kleijn, D. P. (2013). Mesenchymal stem cell-derived exosomes increase ATP levels, decrease oxidative stress and activate PI3K/Akt pathway to enhance myocardial viability and prevent adverse remodeling after myocardial ischemia/reperfusion injury. [Research Support, Non-U.S. Gov't] *Stem Cell Research, 10*(3), 301–312. https://doi.org/10.1016/j.scr.2013.01.002.

Baixauli, F., Lopez-Otin, C., & Mittelbrunn, M. (2014). Exosomes and autophagy: Coordinated mechanisms for the maintenance of cellular fitness. [Review] *Frontiers in Immunology, 5*, 403. https://doi.org/10.3389/fimmu.2014.00403.

Bang, C., Batkai, S., Dangwal, S., Gupta, S. K., Foinquinos, A., Holzmann, A., … Thum, T. (2014). Cardiac fibroblast-derived microRNA passenger strand-enriched exosomes mediate cardiomyocyte hypertrophy. [Research Support, Non-U.S. Gov't] *The Journal of Clinical Investigation, 124*(5), 2136–2146. https://doi.org/10.1172/JCI70577.

Bobrie, A., Krumeich, S., Reyal, F., Recchi, C., Moita, L. F., Seabra, M. C., … Thery, C. (2012). Rab27a supports exosome-dependent and -independent mechanisms that modify the tumor microenvironment and can promote tumor progression. [Research Support, Non-U.S. Gov't] *Cancer Research, 72*(19), 4920–4930. https://doi.org/10.1158/0008-5472.CAN-12-0925.

Bolmont, T., Clavaguera, F., Meyer-Luehmann, M., Herzig, M. C., Radde, R., Staufenbiel, M., … Jucker, M. (2007). Induction of tau pathology by intracerebral infusion of amyloid-beta-containing brain extract and by amyloid-beta deposition in APP x Tau transgenic mice. [Research Support, N.I.H., Extramural Research Support, Non-U.S. Gov't] *The American Journal of Pathology, 171*(6), 2012–2020. https://doi.org/10.2353/ajpath.2007.070403.

Chalmin, F., Ladoire, S., Mignot, G., Vincent, J., Bruchard, M., Remy-Martin, J. P., … Ghiringhelli, F. (2010). Membrane-associated Hsp72 from tumor-derived exosomes mediates STAT3-dependent immunosuppressive function of mouse and human myeloid-derived suppressor cells. [Research Support, Non-U.S. Gov't] *The Journal of Clinical Investigation, 120*(2), 457–471. https://doi.org/10.1172/JCI40483.

Dai, S., Wei, D., Wu, Z., Zhou, X., Wei, X., Huang, H., & Li, G. (2008). Phase I clinical trial of autologous ascites-derived exosomes combined with GM-CSF for colorectal cancer. [Clinical Trial, Phase I Randomized Controlled Trial Research Support, Non-U.S. Gov't] *Molecular Therapy, 16*(4), 782–790. https://doi.org/10.1038/mt.2008.1.

Danzer, K. M., Kranich, L. R., Ruf, W. P., Cagsal-Getkin, O., Winslow, A. R., Zhu, L., … McLean, P. J. (2012). Exosomal cell-to-cell transmission of alpha synuclein oligomers. [Research Support, N.I.H., Extramural] *Molecular Neurodegeneration, 7*, 42. https://doi.org/10.1186/1750-1326-7-42.

Escudier, B., Dorval, T., Chaput, N., Andre, F., Caby, M. P., Novault, S., … Zitvogel, L. (2005). Vaccination of metastatic melanoma patients with autologous dendritic cell (DC) derived-exosomes: Results of the first phase I clinical trial. *Journal of Translational Medicine, 3*(1), 10. https://doi.org/10.1186/1479-5876-3-10.

Hanahan, D., & Weinberg, R. A. (2011). Hallmarks of cancer: The next generation. [Research Support, N.I.H., Extramural Review] *Cell, 144*(5), 646–674. https://doi.org/10.1016/j.cell.2011.02.013.

Hsu, C., Morohashi, Y., Yoshimura, S., Manrique-Hoyos, N., Jung, S., Lauterbach, M. A., … Simons, M. (2010). Regulation of exosome secretion by Rab35 and its GTPase-activating proteins TBC1D10A-C. [Research Support, Non-U.S. Gov't] *The Journal of Cell Biology, 189*(2), 223–232. https://doi.org/10.1083/jcb.200911018.

Javeed, N., & Mukhopadhyay, D. (2017). Exosomes and their role in the micro-/macro-environment: A comprehensive review. *Journal of Biomedical Research, 31*(5), 386–394. https://doi.org/10.7555/JBR.30.20150162.

Javeed, N., Sagar, G., Dutta, S. K., Smyrk, T. C., Lau, J. S., Bhattacharya, S., … Mukhopadhyay, D. (2015). Pancreatic cancer-derived exosomes cause paraneoplastic beta-cell dysfunction. [Research Support, N.I.H., Extramural] *Clinical Cancer Research, 21*(7), 1722–1733. https://doi.org/10.1158/1078-0432.CCR-14-2022.

Ji, H., Greening, D. W., Barnes, T. W., Lim, J. W., Tauro, B. J., Rai, A., … Simpson, R. J. (2013). Proteome profiling of exosomes derived from human primary and metastatic colorectal cancer cells reveal differential expression of key metastatic factors and signal transduction components. [Research Support, Non-U.S. Gov't] *Proteomics, 13*(10–11), 1672–1686. https://doi.org/10.1002/pmic.201200562.

Kahlert, C., Melo, S. A., Protopopov, A., Tang, J., Seth, S., Koch, M., … Kalluri, R. (2014). Identification of double-stranded genomic DNA spanning all chromosomes with mutated KRAS and p53 DNA in the serum exosomes of patients with pancreatic cancer. [Clinical Trial Research Support, N.I.H., Extramural Research Support, Non-U.S. Gov't] *The Journal of biological chemistry, 289*(7), 3869–3875. https://doi.org/10.1074/jbc.C113.53226.

Kamerkar, S., LeBleu, V. S., Sugimoto, H., Yang, S., Ruivo, C. F., Melo, S. A., … Kalluri, R. (2017). Exosomes facilitate therapeutic targeting of oncogenic KRAS in pancreatic cancer. [Research Support, N.I.H., Extramural Research Support, Non-U.S. Gov't] *Nature, 546*(7659), 498–503. https://doi.org/10.1038/nature22341.

Katoh, M. (2013). Therapeutics targeting angiogenesis: Genetics and epigenetics, extracellular miRNAs and signaling networks (review). [Research Support, Non-U.S. Gov't Review] *International Journal of Molecular Medicine*, 32(4), 763–767. https://doi.org/10.3892/ijmm.2013.1444.

Kaur, S., Singh, S. P., Elkahloun, A. G., Wu, W., Abu-Asab, M. S., & Roberts, D. D. (2014). CD47-dependent immunomodulatory and angiogenic activities of extracellular vesicles produced by T cells. [Research Support, N.I.H., Extramural Research Support, N.I.H., Intramural] *Matrix Biology*, 37, 49–59. https://doi.org/10.1016/j.matbio.2014.05.007.

Khan, S., Aspe, J. R., Asumen, M. G., Almaguel, F., Odumosu, O., Acevedo-Martinez, S., … Wall, N. R. (2009). Extracellular, cell-permeable survivin inhibits apoptosis while promoting proliferative and metastatic potential. [Research Support, N.I.H., Extramural Research Support, Non-U.S. Gov't] *British Journal of Cancer*, 100(7), 1073–1086. https://doi.org/10.1038/sj.bjc.6604978.

Kibria, G., Ramos, E. K., Lee, K. E., Bedoyan, S., Huang, S., Samaeekia, R., … Liu, H. (2016). A rapid, automated surface protein profiling of single circulating exosomes in human blood. *Scientific Reports*, 6, 36502. https://doi.org/10.1038/srep36502.

Kucharzewska, P., Christianson, H. C., Welch, J. E., Svensson, K. J., Fredlund, E., Ringner, M., … Belting, M. (2013). Exosomes reflect the hypoxic status of glioma cells and mediate hypoxia-dependent activation of vascular cells during tumor development. [Research Support, Non-U.S. Gov't] *Proceedings of the National Academy of Sciences of the United States of America*, 110(18), 7312–7317. https://doi.org/10.1073/pnas.1220998110.

Logozzi, M., De Milito, A., Lugini, L., Borghi, M., Calabro, L., Spada, M., … Fais, S. (2009). High levels of exosomes expressing CD63 and caveolin-1 in plasma of melanoma patients. [Research Support, Non-U.S. Gov't] *PLoS One*, 4(4), e5219. https://doi.org/10.1371/journal.pone.0005219.

Mathivanan, S., Ji, H., & Simpson, R. J. (2010). Exosomes: Extracellular organelles important in intercellular communication. [Research Support, Non-U.S. Gov't Review] *Journal of Proteomics*, 73(10), 1907–1920. https://doi.org/10.1016/j.jprot.2010.06.006.

Matsumura, T., Sugimachi, K., Iinuma, H., Takahashi, Y., Kurashige, J., Sawada, G., … Mimori, K. (2015). Exosomal microRNA in serum is a novel biomarker of recurrence in human colorectal cancer. [Research Support, Non-U.S. Gov't] *British Journal of Cancer*, 113(2), 275–281. https://doi.org/10.1038/bjc.2015.201.

Melo, S. A., Luecke, L. B., Kahlert, C., Fernandez, A. F., Gammon, S. T., Kaye, J., … Kalluri, R. (2015). Glypican-1 identifies cancer exosomes and detects early pancreatic cancer. [Research Support, N.I.H., Extramural Research Support, Non-U.S. Gov't] *Nature*, 523(7559), 177–182. https://doi.org/10.1038/nature14581.

Morse, M. A., Garst, J., Osada, T., Khan, S., Hobeika, A., Clay, T. M., … Lyerly, H. K. (2005). A phase I study of dexosome immunotherapy in patients with advanced non-small cell lung cancer. *Journal of Translational Medicine*, 3(1), 9. https://doi.org/10.1186/1479-5876-3-9.

Mulcahy, L. A., Pink, R. C., & Carter, D. R. (2014). Routes and mechanisms of extracellular vesicle uptake. [Review] *Journal of Extracellular Vesicles*, 3. https://doi.org/10.3402/jev.v3.24641.

Narayanan, A., Iordanskiy, S., Das, R., Van Duyne, R., Santos, S., Jaworski, E., … Kashanchi, F. (2013). Exosomes derived from HIV-1-infected cells contain trans-activation response element RNA. [Research Support, N.I.H., Extramural Research Support, U.S. Gov't, Non-P.H.S.] *The Journal of Biological Chemistry*, 288(27), 20014–20033. https://doi.org/10.1074/jbc.M112.438895.

Nazarenko, I., Rana, S., Baumann, A., McAlear, J., Hellwig, A., Trendelenburg, M., … Zoller, M. (2010). Cell surface tetraspanin Tspan8 contributes to molecular pathways of exosome-induced endothelial cell activation. [Research Support, Non-U.S. Gov't] *Cancer Research*, 70(4), 1668–1678. https://doi.org/10.1158/0008-5472.CAN-09-2470.

Ostrowski, M., Carmo, N. B., Krumeich, S., Fanget, I., Raposo, G., Savina, A., … Thery, C. (2010). Rab27a and Rab27b control different steps of the exosome secretion pathway. [Research Support, Non-U.S. Gov't] *Nature Cell Biology*, 12(1), 19–30. https://doi.org/10.1038/ncb2000. sup pp. 11–13.

Peinado, H., Aleckovic, M., Lavotshkin, S., Matei, I., Costa-Silva, B., Moreno-Bueno, G., … Lyden, D. (2012). Melanoma exosomes educate bone marrow progenitor cells toward a pro-metastatic phenotype through MET. [Research Support, N.I.H., Extramural Research Support, Non-U.S. Gov't Research Support, U.S. Gov't, Non-P.H.S.] *Nature Medicine, 18*(6), 883–891. https://doi.org/10.1038/nm.2753.

Psaila, B., & Lyden, D. (2009). The metastatic niche: Adapting the foreign soil. [Research Support, N.I.H., Extramural Research Support, Non-U.S. Gov't Review] *Nature Reviews Cancer, 9*(4), 285–293. https://doi.org/10.1038/nrc2621.

Rabinowits, G., Gercel-Taylor, C., Day, J. M., Taylor, D. D., & Kloecker, G. H. (2009). Exosomal microRNA: A diagnostic marker for lung cancer. *Clinical Lung Cancer, 10*(1), 42–46. https://doi.org/10.3816/CLC.2009.n.006.

Raposo, G., Nijman, H. W., Stoorvogel, W., Liejendekker, R., Harding, C. V., Melief, C. J., & Geuze, H. J. (1996). B lymphocytes secrete antigen-presenting vesicles. [Research Support, Non-U.S. Gov't Research Support, U.S. Gov't, P.H.S.] *The Journal of Experimental Medicine, 183*(3), 1161–1172.

Savina, A., Fader, C. M., Damiani, M. T., & Colombo, M. I. (2005). Rab11 promotes docking and fusion of multivesicular bodies in a calcium-dependent manner. [Research Support, Non-U.S. Gov't] *Traffic, 6*(2), 131–143. https://doi.org/10.1111/j.1600-0854.2004.00257.x.

Sceneay, J., Smyth, M. J., & Moller, A. (2013). The pre-metastatic niche: Finding common ground. [Research Support, Non-U.S. Gov't Review] *Cancer Metastasis Reviews, 32*(3–4), 449–464. https://doi.org/10.1007/s10555-013-9420-1.

Skog, J., Wurdinger, T., van Rijn, S., Meijer, D. H., Gainche, L., Sena-Esteves, M., … Breakefield, X. O. (2008). Glioblastoma microvesicles transport RNA and proteins that promote tumour growth and provide diagnostic biomarkers. [Research Support, N.I.H., Extramural Research Support, Non-U.S. Gov't] *Nature Cell Biology, 10*(12), 1470–1476. https://doi.org/10.1038/ncb1800.

Swartz, M. A., Iida, N., Roberts, E. W., Sangaletti, S., Wong, M. H., Yull, F. E., … DeClerck, Y. A. (2012). Tumor microenvironment complexity: Emerging roles in cancer therapy. [Congresses Research Support, N.I.H., Extramural Research Support, Non-U.S. Gov't] *Cancer Research, 72*(10), 2473–2480. https://doi.org/10.1158/0008-5472.CAN-12-0122.

Takeshita, N., Hoshino, I., Mori, M., Akutsu, Y., Hanari, N., Yoneyama, Y., … Matsubara, H. (2013). Serum microRNA expression profile: miR-1246 as a novel diagnostic and prognostic biomarker for oesophageal squamous cell carcinoma. [Comparative Study] *British Journal of Cancer, 108*(3), 644–652. https://doi.org/10.1038/bjc.2013.8.

Taylor, D. D., & Gercel-Taylor, C. (2008). MicroRNA signatures of tumor-derived exosomes as diagnostic biomarkers of ovarian cancer. [Research Support, N.I.H., Extramural Research Support, Non-U.S. Gov't] *Gynecologic Oncology, 110*(1), 13–21. https://doi.org/10.1016/j.ygyno.2008.04.033.

Umezu, T., Tadokoro, H., Azuma, K., Yoshizawa, S., Ohyashiki, K., & Ohyashiki, J. H. (2014). Exosomal miR-135b shed from hypoxic multiple myeloma cells enhances angiogenesis by targeting factor-inhibiting HIF-1. [Research Support, Non-U.S. Gov't] *Blood, 124*(25), 3748–3757. https://doi.org/10.1182/blood-2014-05-576116.

Viaud, S., Terme, M., Flament, C., Taieb, J., Andre, F., Novault, S., … Chaput, N. (2009). Dendritic cell-derived exosomes promote natural killer cell activation and proliferation: A role for NKG2D ligands and IL-15Ralpha. [Research Support, Non-U.S. Gov't] *PLoS One, 4*(3), e4942. https://doi.org/10.1371/journal.pone.0004942.

Wang, Z., Chen, J. Q., Liu, J. L., & Tian, L. (2016). Exosomes in tumor microenvironment: Novel transporters and biomarkers. [Review] *Journal of Translational Medicine, 14*(1), 297. https://doi.org/10.1186/s12967-016-1056-9.

Zhang, H. G., & Grizzle, W. E. (2011). Exosomes and cancer: A newly described pathway of immune suppression. [Research Support, N.I.H., Extramural Research Support, Non-U.S. Gov't] *Clinical Cancer Research, 17*(5), 959–964. https://doi.org/10.1158/1078-0432.CCR-10-1489.

Zhou, W., Fong, M. Y., Min, Y., Somlo, G., Liu, L., Palomares, M. R., … Wang, S. E. (2014). Cancer-secreted miR-105 destroys vascular endothelial barriers to promote metastasis. [Research Support, N.I.H., Extramural Research Support, Non-U.S. Gov't] *Cancer Cell, 25*(4), 501–515. https://doi.org/10.1016/j.ccr.2014.03.007.

Zitvogel, L., Regnault, A., Lozier, A., Wolfers, J., Flament, C., Tenza, D., … Amigorena, S. (1998). Eradication of established murine tumors using a novel cell-free vaccine: Dendritic cell-derived exosomes. [Research Support, Non-U.S. Gov't] *Nature Medicine, 4*(5), 594–600.

Index

Note: Page numbers followed by "f" indicate figures, "t" indicate tables.

A

ABC gene. *See* ATP-binding cassette gene (ABC gene)
ABC subfamily G member 2 (ABCG2), 292–293
ABCB1, 275–276, 287–288, 292–293
ABCG2. *See* ABC subfamily G member 2 (ABCG2)
Acidic culture microenvironment, 225
Acquired drug resistance, 287–288, 298
Acquired immunity, 313–314
Activated antiapoptotic pathways, 286
Activation protein kinase B (Akt), 263–265, 273, 298
"Active-sorting" machinery, 41–42
Adaptive dialysis-like affinity platform technology (ADAPT), 275
Adaptor protein 2 (AP2), 196–197
ADCC. *See* Antibody-dependent cell-mediated cytotoxicity (ADCC)
Adenosine, 315
Adhesion molecules, 37
Adipose stem cells (ASCs), 42
Adrenomedullin, 346
Aerobic glycolysis–based energy generation, 270–271
Aethlon Hemopurifier, 275
Affinity-based lectin and antibody, 275
AFM. *See* Atomic force microscopy (AFM)
Aggressive breast cancer cell lines, 267
Aggressive ovarian cancer, 300
Agilent Bioanalyzer pico chip, 102
Ago. *See* Argonaute (Ago)
Akt. *See* Activation protein kinase B (Akt)
AKT. *See* Protein kinase B (AKT)
ALG-2-interacting protein X (ALIX), 30
ALIX. *See* ALG-2-interacting protein X (ALIX)

Alkylpurine-DNA-N-glycosylase (APNG), 124–125
Alzheimer's disease, 5–6, 72, 344–345
52-Amino-acid polypeptide adrenomedullin, 346
Amphiregulin (AREG), 267
Angiogenesis, 6–7, 12, 220, 286
    in cancer, 235–236
    exosome role in, 265–266
Angiogenic factors, 265–266
Angiogenic inducers, 236
Annexin II, 265–266
Anti-PD-L1 antibody, 320–321
Anti-PD-L1 monoclonal antibodies, 320–321
Antibody-dependent cell-mediated cytotoxicity (ADCC), 318, 318f
Antibody/antibodies, 274
    antibody-based capture method, 98–99
    binding, 318
Anticancer drug–treated carcinoma cells, 316–317
Antigen editing, 65
Antigen presentation, B cells and, 331–334, 332f
    trogocytosis, tunneling nanotubes, and EVs in passive antigen, 334f
Antigen processing, 329–331
Antigen-presenting cells (APCs), 329
    in immune responses, 329–331
    sgeneric professional APCs, 330f
Antigen-specific T cell responses, 336
Antihuman epidermal growth factor receptor 2 (HER2), 293–294
Antitumor effect of neoadjuvant chemotherapy, 239
Antitumorigenic effects, 6

AP2. *See* Adaptor protein 2 (AP2)

APCs. *See* Antigen-presenting cells (APCs)

APNG. *See* Alkylpurine-DNA-N-glycosylase (APNG)

Apoptotic bodies, 219–220

Arachidonic acid–derived eicosanoids, 46–47

AREG. *See* Amphiregulin (AREG)

Arf6 in exosome production, 15–16

Argonaute (Ago), 95–96, 132–133

Ascite-derived exosomes, 347

ASCs. *See* Adipose stem cells (ASCs)

Astrocyte-derived exosomes, 268, 272

Atomic force microscopy (AFM), 47

ATP-binding cassette gene (ABC gene), 287–288

Autophagosomes, 17

Autophagy pathways, 17

B

B cell receptor (BCR), 333

B cells, 313
   B cell–derived EVs, 325–326
   EVs, and associated impacts, 329–336

B-cell precursor acute lymphoblastic leukemia (Pre-B ALL), 298

B16 murine melanoma model, 318–319

B16BL6 cells, 319–320

B16BL6–derived exosomes, 274

Bafilomycin A, 196

BAL. *See* Bronchoalveolar lavage (BAL)

Base-catalyzed hydrolysis, 100–101

Basic fibroblast growth factors (bFGFs), 236

BCR. *See* B cell receptor (BCR)

"Beads-on-a-string" structures, 335

Bexosomes, 335

bFGFs. *See* Basic fibroblast growth factors (bFGFs)

Bioanalyzer 2100 chip, 102

Bioanalyzer Pico chip, 102

Bioanalyzer software, 100–101

Bioassay, 119

Biochemical products, 60

Biodistribution of cancer-derived exosomes
   cancer-derived exosome functions, 176–177
   diverse factors on evaluation of exosome biodistribution in vivo, 179–182

exosome, 175–176
   mechanisms involving in tumor-derived exosomes homing to specific organs, 177–179

Biofunctional proteins, 1–2

Biogenesis, 250–251
   EV biogenesis pathways, 13f
   exosome, 14–17
   MV, 13–14

Biology
   biological fluids, 97, 110
   biological processes, 176
   of exosomes, 236–237

Bioluminescent marker, 181

Biomarkers, 6, 119, 130, 134
   development, 130
   discovery, 7, 69–70
   exosome RNAs as, 130
   exosomes as cancer biomarkers, 345–346
   nucleic acids as potential biomarkers, 345–346
   structural biomarkers of TNT formation in cancer, 227–229

Biosensing applications, 120

Biotinylated antibodies, 123

Bladder cancer, 80
   cell–derived exosomes, 272

Blood
   blood-based biomarker tests, 161–162
   blood-based diagnosis, 161–162
   blood/plasma secretome, 72
   brain barrier, 346
   capillaries, 235–236
   coagulation, 5–6
   pressure–reducing drug, 347
   vessels, 344

BM. *See* Brain metastatic (BM)

BMDCs. *See* Bone marrow–derived cells (BMDCs)

Body fluids, 98

Bone marrow, 317
   progenitor cells, 248–250

Bone marrow–derived cells (BMDCs), 247–248

Brain
   cancer, 301–302
   function, 5–6
   tumor, 81

Brain metastatic (BM), 72
BRCA2 reversion mutation, 64–65
Breast cancer, 74–78, 293–296
   cell–derived extracellular vesicles, 270–271
   exosomal annexin II, 269–270
Breast cancer resistance protein (BCRP). *See* ABC subfamily G member 2 (ABCG2)
Breast milk, exosomes in, 73
Bronchoalveolar lavage (BAL), 135

C
CA 19–9. *See* Carbohydrate antigen 19–9 (CA 19–9)
CA-125 antigen, 165–166
Cadherin, 39
CAFs. *See* Cancer-associated fibroblasts (CAFs)
Cancer, 6–7, 12, 133–134, 161–162, 191, 286, 344–345
   angiogenesis in, 235–236
   biological function, 5–7
   biology, 220
   brain, 301–302
   cancer-derived neoantigens, 65
   cells, 31, 34, 266
      cancer cell–derived exosomes role in initiation, 248–250
      lines, 45–46
      physiology, 142
      types, 36–37
   colorectal, 303–304
   cross-talk between tumor cells and stroma through exosomes, 290t
   diagnosis, 119
      tumor exosomes, 164–170
   EV biogenesis
      CD63-dependent vesicle production, 19–20
      EV biogenesis pathways, 13f
      exosome biogenesis, 14–17
      increasing importance of examining diverse vesicles, 17–19
      MV biogenesis, 13–14
      vesicle secretion to intercellular signaling, 20
   exosomal influx roles in drug resistance, 288t–289t

exosome-associated drug resistance, 293–305
exosome-mediated direct drug efflux, 288t
exosomes, 6, 27–28
   as cancer biomarkers, 345–346
      particle size and zeta potential of exosomes, 49t–50t
      proteins in, 36–41
      size and charge, 47–51
   gastric, 303
   lung, 302–303
   mechanisms of chemoresistance in, 287t
   metastasis, 269
      TME cell–derived exosomes in, 268
   molecular and structural biomarkers of TNT formation, 227–229
   ovarian, 300
   pancreatic, 301
   prostate, 299–300
   proteomic content of exosomes, 74–82, 75t–77t
      bladder cancer, 80
      brain tumor, 81
      breast cancer, 74–78
      colorectal cancer, 82
      head and neck cancer, 79
      leukemia, 79
      liver cancer, 80–81
      lung cancer, 82
      ovarian cancer, 78–79
      pancreatic cancer, 81
      prostate cancer, 78
   proteomics, 72
   therapeutics, 346
   therapy, 12, 314
   treatments and diagnostic tools, 71
   vesicle contents, 17
Cancer immunotherapy, exosomes in, 314, 318–319
   using exosomes, 318–319, 319f
   myeloid lineage cells, tumor exosomes effect on, 317–318
   natural killer cells, tumor exosomes effect on, 316–317
   T cells, tumor exosomes effect on, 314–316

Cancer vaccine, 314, 319–320
  B cell EVs as cancer vaccines/adjuvants, 336
    IgE-linked antigen, 337f
  implications for cancer vaccine therapies
    B cells, EVs, and associated impacts,
      329–336
    comparisons of EV sizes, 328f
    types of EVs and formation, 327f
Cancer-associated fibroblasts (CAFs), 251–252,
    263–265
  CAF-derived exosomes, 270–271
  exosomes in transformation of normal
    fibroblast cells into, 267–268
Cancer-derived exosomes, 71, 176
  functions, 176–177
  in tumor angiogenesis, 237–239
Cancer-specific exosomes (CSEs), 64–65
Canonical endocytic pathway, 29
Carbohydrate antigen 19–9 (CA 19–9), 61–62
  biomarker, 161–162, 345–346
Carcinogenesis, 62, 204–205
Cardio protection, 5–6
Cardiolipin, 45
Cardiovascular diseases, 286, 344–345
Cargo sorting, 12
Casein kinase II (CK2), 39–40, 263–265
Catalog of Somatic Mutations in Cancer
    (COSMIC), 64
Caveolin, 29
Caveolin-1, 197–198
CD151 protein, 37, 64–65, 82, 265
CD171 protein, 82
CD23 cell line, 336
CD3-T cell receptor, 315
CD4+ T cells, 314–315
CD44 receptor, 227–228, 265, 267
CD47 receptor, 320
CD63 protein, 4, 60–61, 265
  CD63-dependent LMP1 secretion, 20
  CD63-dependent mechanism, 16–17
  CD63-dependent vesicle production, 19–20
  in vesicle cargo sorting, 19–20
CD66 b (neutrophil marker), 43–44
CD8+ "killer" T cell responses, 331
CD8+ T cells, 314–315, 329–331

CD81 proteins, 4, 31–33, 265
CD9 protein, 4, 265
CD90+ liver cancer cells, 266
Cell-to-cell communication, 69–70
Cellular/cell(s), 344
  adhesion, 227–228
  communication, 130, 221–223, 325–326
  culture additives, 97–98
  culture supernatants, 97–98
  endocytic pathway–derived extracellular
    nanovesicles, 162
  homeostasis, 131
  internalization mechanism, 191
  processes, 95
  senescence, 34–36
  source of exosomes, 180
  waste, 291
Central nervous system diseases, 5–6
Centrifugation parameters, 97–98
Ceramide, 33
Cerebrospinal fluid (CSF), 71, 224
  exosomes in, 72
cfDNA. See Circulating cell free DNA (cfDNA)
CFPAC-1, 275–276
cfRNA, 65
Charged multivesicular body proteins
    (CHMPs), 14–15
Chemokine CCL2, 250–251
Chemokine receptors, 37
Chemotherapeutics, 142
Chemotherapy, 286, 303. See also Cancer
    immunotherapy, exosomes in
Chimeric antigen receptors, 314
Chloroquine, 196
Chlorpromazine (CPZ), 196–197
CHMPs. See Charged multivesicular body
    proteins (CHMPs)
Cholesterol, 16, 45–46
Chronic inflammation, 252–253
Chronic lymphocytic leukemia (CLL), 251–252
  cells, 267–268
Chronic myeloid leukemia (CML), 223,
    265–266, 298
  CML–derived exosomes, 79
Circulating cancer-derived exosomes, 179

Circulating cell free DNA (cfDNA), 60–61
  ddPCR analysis, 62
Circulating exosomes, double-stranded
      genomic DNA in, 60–61
Circulating tumor cells (CTCs), 61–62, 162
Circulating tumor DNA (ctDNA), 60
Cisplatin, 303
CK2. *See* Casein kinase II (CK2)
Class II invariant chain peptide (CLIP).
      See Invariant chain (Ii)
Clathrin, 29
  clathrin-dependent endocytosis, 191
  clathrin-independent endocytosis, 191
  clathrin-mediated endocytosis, 197–198
Claudin-7, 40
Claudin-containing exosomes, 78–79
CLDN4 protein, 64–65
CLL. *See* Chronic lymphocytic leukemia (CLL)
CML. *See* Chronic myeloid leukemia (CML)
Colorectal cancer (CRC), 82, 136, 248–250, 266,
      303–304
  cells, 266
Constitutive exocytosis, 187–188
Conventional analytical methods, 120
Conventional cancer immunotherapies, 314
Conventional imaging techniques, 119
Conventional semiconductor fabrication
      processes, 121–122
COSMIC. *See* Catalog of Somatic Mutations in
      Cancer (COSMIC)
Costimulatory molecules, 329
CpG DNA (TLR9), 319–320
CPZ. *See* Chlorpromazine (CPZ)
CRC. *See* Colorectal cancer (CRC)
CRISPR/Cas9 system, 19–20
"Cross-dressing", 333–334
Cross-talk between tumor cells and stroma via
      exosomes, 293
CSEs. *See* Cancer-specific exosomes (CSEs)
CSF. *See* Cerebrospinal fluid (CSF)
CTCs. *See* Circulating tumor cells (CTCs)
ctDNA. *See* Circulating tumor DNA (ctDNA)
CTLA-4, 314, 316
Culture systems, 97–98
Curcumin

curcumin-induced exosomal secretion, 273
  treatment, 273
Cytokines, 265–266, 320
Cytonemes, 224–225
Cytoplasmic protein, 5
Cytoskeletal
  actin–myosin machinery, 14
  components, 3
  degradation, 265
Cytosolic RNAs, 96

D
Dasatinib, 273, 298
Database for annotation, visualization, and
      integrated discovery (DAVID), 151
DCs. *See* Dendritic cells (DCs)
ddPCR. *See* Droplet digital polymerase chain
      reaction (ddPCR)
Deep sequencing, 107
  platforms, 103–104
Degree of immune dysfunction, 313–314
DEL-1. *See* Developmental endothelial locus-1
      (DEL-1)
Delta-like ligand 4 (Dll4), 265–266
Dendritic cells (DCs), 16–17, 194–195, 313,
      317, 329–331, 347
  B cell EVs and promotion of DCs–mediated
      T cell immune responses, 335
  DC-derived exosomes, 274, 346–347
Density gradient
  centrifugation, 4
  media, 97–98
Detoxification, 286
Developmental endothelial locus-1 (DEL-1),
      71
Dexosomes. *See* Peptide-loaded DC-derived
      exosomes
DHA. *See* Docosahexaenoic acid (DHA)
Diagnostic potential of tumor exosomes, 170
  circulating tumor antigens, exosomes, and
      CTCs, 162f
  methods and tools for exosome isolation,
      163–164
  tumor exosomes for cancer diagnosis,
      164–170

Differential ultracentrifugation process, 98–99
Digital PCR, 107
Dimethyl amiloride, 347
Direct fusion, 191–198
    macropinocytosis, 195–196
    phagocytosis, 194–195
    receptor-and raft-mediated endocytosis, 196–198
    soluble signaling and Juxtacrine signaling, 198
Diverse factors on evaluation of exosome biodistribution in vivo, 179–182
    influence of cellular source of exosomes, 180
    kinetics of exosome distribution, 180
    labeling of exosomes for tracking studies, 181–182
    route of injection, 179–180
Diverse vesicles, 17–19
Dll4. *See* Delta-like ligand 4 (Dll4)
DLS. *See* Dynamic light scattering (DLS)
DNA, 262
    isolation techniques for RNA/DNA from exosomes, 97–100
    presence of, 110
Docetaxel resistance, 295–296
Docosahexaenoic acid (DHA), 34
"Double-edged" miRNA, 206–207
Double-stranded DNA, 345–346
    genomic DNA in circulating exosomes, 60–61
Doxorubicin (Dox), 142
    Dox-treated PC-3 cells, 34
Droplet digital polymerase chain reaction (ddPCR), 62
Drosha-independent miRNA processing pathways, 95–96
Drug resistance, 286, 288t–289t
    exosome-associated drug resistance in cancers, 293–305
    mechanisms of exosome in, 291–293
    mechanisms of resistance to drug therapies, 287–288
Drug therapies, mechanisms of resistance to, 287–288
Drug-resistant tumor cells, 291f

Dynamic light scattering (DLS), 47, 97–98
Dynamic processing of RNA, 44–45
Dynamin 2, 196–197

E
E2F1 gene, 227–228
EBV. *See* Epstein–Barr virus (EBV)
ECM. *See* Extracellular matrix (ECM)
"Ectosomes". *See* Microvesicles
Effector cells, 329
Effector lymphocytes, 329–331
EGFR. *See* Epidermal growth factor receptor (EGFR)
Eicosanoids, 46–47
Electrochemical sensing, 122–124
Electron microscopy (EM), 152, 220–221
ELISA method, 80–82, 120
EM. *See* Electron microscopy (EM)
EMMPRIN. *See* Extracellular matrix metallo-proteinase inducer (EMMPRIN)
EMT. *See* Epithelial-to-mesenchymal transition (EMT)
Endocrine, 130
Endoplasmic reticulum (ER), 188, 329–331
Endosomal sorting complex required for transport (ESCRT), 3, 13–14, 29, 188
    ESCRT-0 complexes, 14–15
    ESCRT-associated proteins, 30
    ESCRT-dependent mechanism, 15–16
    ESCRT-dependent pathway, 3
    ESCRT-I complexes, 14–15
    ESCRT-II complexes, 14–15
    ESCRT-III complexes, 14–15
    ESCRT-independent pathways, 33
    machinery, 30
Endosomal system, 17, 188
Endosomal vesicle pathways, 96–97
Endothelial cells, 236–239, 343–344
Engineered T-cell receptors, 65
Enhanced drug metabolism, 286
EOC. *See* Epithelial ovarian cancer (EOC)
EPCAM protein, 64–65, 165–166
Epidermal growth factor receptor (EGFR), 63, 74–78, 199–201, 263–265, 267, 269–270
    EGFRvIII, 263–265
    EGFRvIII-regulated genes, 41

Epidermal growth factor receptor pathway substrate clone 15 (EPS15), 196–197
Epithelial ovarian cancer (EOC), 136
Epithelial-to-mesenchymal transition (EMT), 6–7, 27–28, 39, 253–255, 254f, 263–265
    EMT-inducing proteins, 263–265
    EMT-related proteins, 39–40
    EMT-transducing signaling molecules, 253–255
    exosomes in, 263–265
    hallmark of premetastatic niche formation, 253–255
EPS15. *See* Epidermal growth factor receptor pathway substrate clone 15 (EPS15)
Epstein–Barr virus (EBV), 253–255
    LMP1, 16–17
ER. *See* Endoplasmic reticulum (ER)
ERK. *See* Extracellular signal–regulated protein kinase (ERK)
Erythroleukemia, 194–195
ESCART protien, 273
ESCRT. *See* Endosomal sorting complex required for transport (ESCRT)
esRNA. *See* Exosomal shuttle RNA (esRNA)
EVpedia, 109
evRNA. *See* RNA component of exosomes (evRNA)
Exo. *See* Exosome (Exo)
"EXO-motif" sequences, 41–42
ExoCarta database, 41, 74, 109
Exocytosis, 187–188
exoDNA. *See* Exosome-derived DNA (exoDNA)
exo*KRAS*. *See* Exosomal *KRAS* (exo*KRAS*)
EXOmotif. *See* Exosome-sorting RNA motifs (EXOmotif)
ExoQuick kit, 47–48, 74, 163
exoRNA. *See* RNA within exosomes (exoRNA)
ExoSearch chip, 163–164, 165f, 167f, 168–170
Exosomal
    biomarkers, 164–165
    cargo packaging inhibition, 273
    content
        modulate tumor angiogenesis, 201–202
        regulate tumor immunity, 202–203
    exosomal miR-221/222 transferred tamoxifen resistance, 294–295

    exosomal noncoding RNAs role, 250–251
    genomic molecular profiling of exosomal cargo, 63–65
    microRNA content modulation affecting cellular reprogramming, 207–210
        effect of miR-155/miR-125b-2 pDNA transfection, 208f–209f
    miR-19α, 345–346
    miR-21, 273
    miRNA, 138, 250–251, 345–346
        content affects cellular reprogramming, 207–210
        microarray and sequencing for exosomal content profiling, 145–150
        in silico analysis, 150–151
        technology for study, 147f, 148t
    proteins, 28
        role in priming metastasis, 251–252
    proteomics, 80–81
    RNA in cancer, 41–45
        small and total RNA analyses, 43f
Exosomal *KRAS* (exo*KRAS*), 62
Exosomal shuttle RNA (esRNA), 131
Exosome (Exo), 1–2, 2f, 5, 13–15, 27–28, 47, 59–60, 69–70, 94–95, 142, 175–176, 198–203, 219–220, 247–248, 262, 326–328, 344–345
    activity, 225–226
    additional diagnostic potential, 170
    as agents for early detection, diagnosis, and stratification, 61–62
    from B cell lymphoma, 318
    biodistribution in vivo evaluation, 179–182
    biogenesis, 3, 14–19, 29–33
        biogenesis pathways of cancer exosomes, 33f
        inhibition of, 273
        proteins in exosome secretion, 32t
        in tumor cells and uptake by recipient, 187–191
    biology, 236–237
    from CAF cell, 268
    as cancer biomarkers, 345–346
    in cancer immunotherapy, 314, 318–319
        cancer immunotherapies using, 318–319, 319f

Exosome (Exo) *(Continued)*
  tumor exosomes effect on myeloid
    lineage cells, 317–318
  tumor exosomes effect on natural killer
    cells, 316–317
  tumor exosomes effect on T cells, 314–316
  composition, 36–47, 95
    exosomal RNA in cancer, 41–45
    lipids in exosomes, 45–47
    proteins in cancer exosomes, 36–41
  content, 5
  exosome-associated GPC1, 28
  exosome-based delivery of therapeutic
    miRNAs, 140–142
  exosome-enriched miRNAs, 250–251
  exosome-intrinsic organotropic
    mechanism, 178
  exosome-mediated events at metastatic
    sites, 269–272
  exosomes mediate cancer–stroma
    interactions, 242f
  exosomes/EVs, 333, 335
  genesis, 3
  for imaging and chemotherapeutics, 142
  internalization, 347
    inhibition of, 274
  interplay between tunneling nanotubes
    and, 220–223
  isolation
    and characterization, 4–5
    and RNA/DNA, 98–99
  labeling for tracking studies, 181–182
  lipids in, 45–47
  lncRNA in, 142–144
  markers, 4–5
  mechanisms in drug resistance, 291–293
    cross-talk between tumor cells and
      stroma, 293
    exosome-mediated direct drug efflux, 291
    transportations of drug resistance,
      292–293
  in metastasis progression, 262–268
  at metastatic sites, 270f
  methods and tools for exosome isolation,
    163–164
    ExoQuick, 163

    filtration, 163
    microfluidic, 164
    ultracentrifugation, 163
  from normal mesenchymal cells, 347
  nucleic acid loading into, 95–97
  perspective, 276
  at primary tumor sites, 264f
  production, 15, 255
  proteins, 80–81
  proteomes, 71
  RNA history in, 130–131
  secreted from cancer cells
    exosome secretion from healthy and
      cancer cells, 35t
    factors affecting release of, 34–36
    size and charge of cancer exosomes,
      47–51
  secretion, 17
    immune cell, 314
    tumor cells, 314
  sources, 71–73, 144–145
    in breast milk, 73
    in CSF, 72
    in plasma, 72
    in saliva, 73
    in urine, 72–73
  strategies against exosome-mediated
    metastasis, 272–276
  therapeutic potential, 346–348
  transport oncoproteins, 199–201
  in tumor angiogenesis, 237–241
  tunneling nanotube–like nanostructures
    and interaction, 223–225
  uptake mechanism, 191
Exosome RNAs as biomarkers and targets for
    cancer therapy
  challenges and perspective, 151–152
  history of RNA in exosomes, 130–131
  RNA species present in exosomes, 132–144
  sources of exosomes, 144–145
  technology for exosomal miRNA study,
    145–151
Exosome-associated drug resistance in
    cancers
  brain cancer, 301–302
  colorectal cancer, 303–304

gastric cancer, 303
hematologic malignancies, 296–298
liver cancer, 302
lung cancer, 302–303
melanoma, 304
neuroblastoma, 305
osteosarcoma, 304–305
ovarian cancer, 300
pancreatic cancer, 301
prostate cancer, 299–300
RCC, 304
Exosome-derived DNA (exoDNA), 60–62
Exosome-mediated communication
direct fusion, 191–198
exosome biogenesis in tumor cells and
uptake, 187–191
exosome uptake and cellular internalization
mechanisms, 191
exosome-mediated cellular reprogramming,
203–210
microRNA-mediated reprogramming of
tumor cells, 204–205
microRNAs role on reprogramming
tumor-associated macrophages,
205–210
exosomes and tumor microenvironment,
198–203
future prospects, 210
Exosome-mediated metastasis
pharmacologic and genetic strategies
targeting, 255
strategies against, 272–276
exosome as therapeutic vehicle, 275–276
inhibition of exosome biogenesis, 273
inhibition of exosome internalization, 274
inhibition of packaging exosomal cargo, 273
removal of tumor exosomes, 275
TLRs role in exosome-mediated metastasis,
252–253
Exosome-sorting RNA motifs (EXOmotif), 96
Exportin-5 (Exp5), 132–133
Extracellular DNA, 60–61, 110
Extracellular matrix (ECM), 6, 27–28, 262, 267
deposition, 177
Extracellular matrix metalloproteinase
inducer (EMMPRIN), 267

Extracellular microvesicles, 94
Extracellular signal–regulated protein kinase
(ERK), 272
Extracellular vesicles (EVs), 11–12, 29, 59–60,
72, 175, 198, 199f, 219–220, 325–329,
344–345
analysis, 122
B cells, EVs, and associated impacts
antigen-presenting cells in immune
responses, 329–331
B cell EVs and follicular dendritic cells,
335
B cell EVs and promotion of dendritic
cells–mediated T cell immune
responses, 335
B cell EVs as cancer vaccines/adjuvants,
336
B cells and antigen presentation, 331–334
biogenesis in cancer
CD63-dependent vesicle production, 19–20
EV biogenesis pathways, 13f
exosome biogenesis, 14–17
increasing importance of examining
diverse vesicles, 17–19
MV biogenesis, 13–14
vesicle secretion to intercellular signaling,
20
comparisons of EV sizes, 328f
production, 19–20
proteins, 124–125
proteomic analyses, 120
research, 1
secretion, 20
source of, 97–98
body fluids, 98
cell culture supernatants, 97–98
types and formation, 327f

F
Fas ligand (FasL), 198, 315
fas ligand–positive exosomes, 271
FDCs. *See* Follicular dendritic cells (FDCs)
"Fexosomes", 142
Fibroblasts, 6–7, 240, 293
exosomes in transformation of normal
fibroblast cells into CAF, 267–268

Fibronectin deposits, 269–270
Fibrosarcoma cells, 40–41
FIGO. *See* International Federation of Gynecology and Obstetrics (FIGO)
Filipin, 197–198
Filtration, 162–163
First generation nPLEX sensor, 121
ΔFL. *See* Increased fluorescence signals (ΔFL)
Flotillins, 16, 197–198
Flow cytometry, 97–98
Fludarabine-resistant prostate cancer cells, 299
Fluorescent markers, 181
Fluorescent RNA-tracking dyes, 108
Follicular dendritic cells (FDCs), 329–331, 335
    B cell EVs and, 335
Foscan, 34
Fusion machinery, 3

G
Galectin-3, 298
Galectin-9, 315
Gamma delta T cells (γδ T cells), 194–195
Gastric cancer (GC), 137, 248–250, 303
    gastric cancer cell–derived exosomes, 269–270
Gastrointestinal stromal cells secrete exosomes, 268
GBM. *See* Glioblastoma multiforme (GBM)
GBM cell–derived exosomes, 274
GC. *See* Gastric cancer (GC)
Gefitinib, 303
Gemcitabine (GEM), 140
GEMs. *See* Glycolipid-enriched microdomains (GEMs)
Gene ontology (GO), 37
Gene transcript expression, 17–18
Genetic material, 119
Genetic variation, 287–288
Genomic DNA-enriched exosomes, mutational signatures detection in, 61
Genomic molecular profiling of exosomal cargo, 63–65
    clinical progression of metastatic pancreatic cancer patient, 63f

Glioblastoma (SF-539), 266
Glioblastoma multiforme (GBM), 124, 266
Gliomas, 199–201, 301
Glycolipid-enriched microdomains (GEMs), 40
Glycoproteins, 37
Glypican1 (GPC1), 28, 122, 345
    in exosomes, 6
GNPs. *See* Gold nanoparticles (GNPs)
GO. *See* Gene ontology (GO)
Gold nanoparticles (GNPs), 142
GPC1. *See* Glypican1 (GPC1)
Growth factors, 130, 235–236
Guanosine triphosphatases (GTPases), 189–190, 273
    GTPase Rab27a, 347–348
GW4869 inhibitor, 273

H
H1299. *See* NSCLC (H1299)
Hairpin-shaped precursor miRNAs, 132–133
HCC. *See* Hepatocellular carcinoma cells (HCC)
HE4. *See* Human epididymis protein 4 (HE4)
Head and neck cancer, 79
Heat shock proteins (HSPs), 4–5, 316–317
    HSP72, 317–318
    HSP70, 4–5, 17, 30, 80–81
    HSP90, 4–5, 48–51, 294–295
Hedgehog (Hh), 224–225
HEK293 cells, 60, 180
Hematologic malignancies
    leukemia, 298
    lymphoma, 297–298
    multiple myeloma, 296–297
Heparin sulfate proteoglycan, 274
Hepatic stellate cells, 251–252, 269–270
Hepatocellular carcinoma cells (HCC), 144, 206–207, 302
Hepatocyte growth factor (HGF), 269–270
    receptor, 48–51
Hepatocyte growth factor receptor substrate (Hrs), 188–189
HER2. *See* Antihuman epidermal growth factor receptor 2 (HER2)

HER2⁺ exosomes, 318
Heterogeneity of tumor exosomes
    double-stranded genomic DNA in
        circulating exosomes, 60–61
    exosomes as agents for early detection,
        diagnosis, and stratification, 61–62
    genomic molecular profiling of exosomal
        cargo, 63–65
    mutational signatures detection in genomic
        DNA-enriched exosomes, 61
    transcriptomic characterization of tumors, 65
Heterogeneous nuclear ribonucleoproteins
        A2/B1 (hnRNPA2B1), 96
HGF. See Hepatocyte growth factor (HGF)
Hh. See Hedgehog (Hh)
HIF-1. See Hypoxia-induced factor 1 (HIF-1)
High-grade serous ovarian cancer, 300
High-resolution flow cytometry analysis, 4–5,
        108
High-throughput chip fabrication, 121–122
HIST2H2BE protein, 64–65
Histone deacetylase inhibitor (MS-275), 273
HIV, 13–14, 344–345
    HIV-1, 5–6
hnRNPA2B1. See Heterogeneous nuclear
        ribonucleoproteins A2/B1
        (hnRNPA2B1)
Hormone
    hormone-refractory prostate cancer cells, 299
    signaling molecules, 130
    treatment, 190
Horseradish peroxidase (HRP), 123
Host components, 71
Housekeeping genes, 107
HRP. See Horseradish peroxidase (HRP)
Hrs. See Hepatocyte growth factor receptor
        substrate (Hrs)
HSP90. See Heat shock protein 90 (HSP90)
HSPs. See Heat shock proteins (HSPs)
Human bladder carcinoma cell lines, 39–40
Human epididymis protein 4 (HE4), 167–168
Human lung adenocarcinoma cells (A549), 36
Human salivary proteome, 79
Human umbilical vein endothelial cells
        (HUVECs), 223, 273

Hypoxia, 34–36
    hypoxia-induced exosomes, 266
    hypoxia/HIF-1-signaling axis, 241
Hypoxia-induced factor 1 (HIF-1), 241
    HIF1-α, 267, 270–271
Hypoxic Glioma secretome, 81
Hypoxic myeloma cell–derived exosomes, 266

I
IAPs. See Inhibitors of apoptosis proteins
        (IAPs)
ICAMs. See Intercellular adhesion molecules
        (ICAMs)
"Iccosomes", 335
IExosomes, 347
IFN. See Interferon (IFN)
IL. See Interleukins (IL)
IL8-mediated VCAM-1 activation, 265–266
ILK-1. See Integrin-linked kinase 1 (ILK-1)
Illumina HiSeq system, 103–104
ILVs. See Intraluminal vesicles (ILVs)
Imatinib, 273
iMER. See Immunomagnetic exosome RNA
        (iMER)
iMEX. See Integrated magnetic-electro-
        chemical exosome (iMEX)
Immune
    adjuvants, 314
    antigen-presenting cells in immune
        responses, 329–331
    cells, 240, 319–320
    checkpoint molecule PD-L1, 320–321
    function and surveillance, 5–6
    system, 71, 313–314
Immunomagnetic exosome RNA (iMER), 124,
        125f
    analysis, 124–125
Immunomagnetic tumor EV enrichment, 124
Immunosuppression, 271–272, 286, 315
Immunotherapy, 65, 314
In silico analysis, 150–151
    contents of EVs, 151t
    DAVID, 151
    IPA, 151
In vitro cell-based models, 177

Increased fluorescence signals (ΔFL), 166–167
Induce neo-angiogenesis, 265–266
Infectious
    cargo, 70
    diseases, 344–345
Inflammatory cytokines, 252–253
Ingenuity pathway analysis (IPA), 151
Inhibitors of apoptosis proteins (IAPs), 41
Injection route, 179–180
Innate immune cancer immunotherapies, 314
Insulin-like growth factor–binding protein 3, 81
Insulin-like growth factor–binding protein 6, 81
Integral membrane proteins, 37
Integrated magnetic-electrochemical exosome
    (iMEX), 122, 123f
Integrin alpha 6 (ITGα6), 39
Integrin beta 5 (ITGβ5), 39
Integrin-linked kinase 1 (ILK-1), 253–255
Integrins, 37
    integrin beta-1, 18–19
    protein family members, 229–230
    repertoire of tumor-derived exosomes, 178
Intercellular adhesion molecules (ICAMs), 37
    ICAM-1, 39
Intercellular communication, 219–220,
    333–334, 344–345
Intercellular highways, TNT, 225–226
Intercellular milieu, 272
Interferon (IFN), 206
    IFN-γ, 315, 320
Interferon regulatory factor 4 (IRF-4), 206–207
Interleukins (IL), 265–266
    IL-2, 315
    IL-6, 253–255, 317–318
    IL-10, 316
International Federation of Gynecology and
    Obstetrics (FIGO), 136
International Society for Extracellular Vesicles
    (ISEV), 152, 176
Intracellular lipid raft, 17–18
Intracellular signaling, 20
Intraluminal vesicles (ILVs), 3, 14–15, 29,
    188–189, 326–328
Intratumoral injection of CpG DNA
    conjugated to exosomes, 319–320
Intrinsic drug resistance, 287–288

Intrinsic presentation pathway, 329–331
Invariant chain (Ii), 329–331
Invasion, exosomes in, 267
IPA. See Ingenuity pathway analysis (IPA)
IRF-4. See Interferon regulatory factor 4
    (IRF-4)
ISEV. See International Society for Extracellular
    Vesicles (ISEV)
Isolation, 47–48
    and characterization, 4–5
    methods and tools for exosome, 163–164
    techniques for RNA/DNA from exosomes
        exosomes isolation and RNA/DNA,
            98–99
        isolation of microRNAs/noncoding RNAs
            specifically, 99–100
        source of extracellular vesicles, 97–98
ITGα6. See Integrin alpha 6 (ITGα6)
ITGβ5. See Integrin beta 5 (ITGβ5)
iTRAQ
    iTRAQ-basedquantitative proteomics, 79
    quantitative iTRAQ proteomic analysis,
        39–40
    quantitative proteomic analyses, 81

J
JAK-STAT pathway, 266
Juxtacrine, 130
    signaling, 198

K
Kaposi sarcoma–associated herpesvirus
    (KSHV), 253–255
Kinetics of exosome distribution, 180
KRAS mutant, 263–265
KRAS mutations, 61–63
KRAS^G12D mutated, 270–271
KSHV. See Kaposi sarcoma–associated
    herpesvirus (KSHV)
Kupffer cells, 269–270

L
Lactadherin (LA), 319–320
LAMP-1. See Lysosomal-associated membrane
    protein 1 (LAMP-1)
"Large-scale omics" approaches, 19–20

Latent membrane protein 1 (LMP1), 16–17, 19–20

LC-MS. *See* Liquid chromatography–mass spectrometry (LC-MS)

Leucine-rich α-2-glycoprotein (LRG1), 82

Leukemia, 79, 298

Leukocytes, 313

Lewis lung carcinoma (LLC), 252–253

LGALS3BP protein, 64–65

Library preparations from evRNA samples, 104–106

   small-RNA-sequencing library preparation, 104t

   sources of bias in RNA isolation and sequencing methods, 105t

Ligands, 96

lincRNA. *See* Long intergenic noncoding RNA (lincRNA)

Lipid(s), 1–2, 119

   domains, 33

   in exosomes, 45–47, 46f

   lipid-dependent ceramide/nSMase pathway, 33

   lipid-signaling molecules, 46–47

   rafts, 16, 59–60

      lipid rafts–exosomes–TNTs, 221–223

Lipopolysaccharides (LPS), 206, 319–320

Lipoproteins, 109–110

Liposomes, 255

Liquid biopsies

   assays, 63

   transcriptomic characterization of tumors, 65

Liquid chromatography–mass spectrometry (LC-MS), 131

Liver cancer, 80–81, 302

LLC. *See* Lewis lung carcinoma (LLC)

LMP1. *See* Latent membrane protein 1 (LMP1)

lncRNA. *See* Long noncoding RNA (lncRNA)

Long intergenic noncoding RNA (lincRNA), 131

Long noncoding RNA (lncRNA), 95, 107, 131, 132f, 251, 266, 291

   lncRNA ZFAS1, 267

LPS. *See* Lipopolysaccharides (LPS)

LRG1. *See* Leucine-rich α-2-glycoprotein (LRG1)

Lung cancer, 82, 134–135, 302–303

lung cancer cell (A-549)–derived exosomes, 271

lung cancer–derived exosomes secrete miRs, 202–203

Lymphoma, 297–298

Lysobisphosphatidic acid, 3

Lysosomal-associated membrane protein 1 (LAMP-1), 195

Lysosome, 3, 188

M

M2-type macrophages, 318

Macrophage(s), 205, 313, 318, 329–331

   macrophage-derived exosomes, 268

   migration inhibitory factor, 81

   repolarization, 202–203

Macropinocytosis, 195–196

MAF. *See* Mutant allele frequency (MAF)

Magnetic nanoparticles (MNPs), 142, 143f

Magnetic resonance imaging (MRI), 142, 143f, 161–162, 182

Major histocompatibility complex (MHC), 347

   class I molecules, 329

   class II molecules, 329

   class II–dependent mechanism, 318–319

   MHC-II-bearing dendritic cell exosomes, 30

MALAT1. *See* Metastasis-associated lung adenocarcinoma transcript 1 (MALAT1)

Malignant melanoma cells (B16–F10), 48–51

Mammalian cells, 95

Mammalian immune

   response, 329

   systems, 313

Mammalian target of rapamycin (mTOR), 298

MAPK, 263–265

   pathways, 252–253

   and PI3K signaling proteins, 17

MARCKS. *See* Myristoylated alanine-rich C-kinase substrate (MARCKS)

Mass spectrometry–based approaches, 73

Mass spectrometry–based proteomics, 109

Mass spectroscopic analysis of bladder cancer exosomes, 270–271

Matrix metalloproteinases (MMPs), 40, 236, 253–255, 263–265

   MMP-9, 40

Matrix metalloproteinases (MMPs) *(Continued)*
 MMP13, 253–255
  MMP13-containing exosomes, 79
 MMP-14, 40
Matrix remodeling, 40–41
Mature endosome, 3
MbCD. *See* Methyl-b-cyclodextrin (MbCD)
MDA-MB-231
 cells, 274
 MDA-MB-231-derived exosomes, 40–41
MDR1. *See* ABCB1
MDSCs. *See* Myeloid-derived suppressor cells
  (MDSCs)
Melanoma, 266, 304, 314
 cells, 33
 exosomes, 269–270
 melanoma-bearing SCID mice, 34
 melanoma-derived exosomes, 177, 265–266
Membrane
 endosome, 252–253
 membrane-derived subcellular structures,
  162
 proteins, 37
"Membrane blebs". *See* Microvesicles
Mesenchymal stem cells (MSCs), 45, 267–268,
  296
 MSC-derived exosomes, 268
Mesenchymal-to-epithelial transition (MET),
  272
 survival, and maintenance of metastasized
  cells, 272
Mesothelin, 227–228
Mesothelioma cell–derived exosomes, 267–268
Messenger RNA (mRNA), 42, 65, 202–203, 272,
  291
MET. *See* Mesenchymal-to-epithelial
  transition (MET)
Metabolic conditions, 225–226
Metabolic shift, exosomes in, 270–271
Metastasis, 177, 201–202, 223–224, 236,
  247–248, 262, 347–348
 cancer cell–derived exosomes role in
  initiation and progression, 248–250
 exosomal noncoding RNAs role in initiation
  and priming, 250–251
 exosomal proteins role, 251–252

 pathways, 81
 TLRs role in exosome-mediated metastasis,
  252–253
Metastasis progression, 248–250
 exosomes in, 262–268
  angiogenesis, 265–266
  EMT, 263–265
  invasion and migration, 267
  microRNA, lncRNA, and DNA transferred
   via exosomes, 263t
  proteins transferred via exosomes and
   involving, 263t
  TME cell–derived exosomes in cancer
   metastasis, 268
  transformation of normal fibroblast cells
   into CAF, 267–268
Metastasis-associated lung adenocarcinoma
  transcript 1 (MALAT1), 251
Metastasized cells, 272
Metastatic
 exosome-mediated events at metastatic
  sites, 269–272
  exosomes in metabolic shift, 270–271
  exosomes in metastatic niche formation,
   269–270
  immunosuppression, 271–272
  MET, survival, and maintenance of
   metastasized cells, 272
 gastric cancer cells, 138
 process, 176–177
Methyl-b-cyclodextrin (MbCD), 197–198
MGMT. *See* O^6-methylguanine DNA methyltransferase (MGMT)
MHC. *See* Major histocompatibility complex
  (MHC)
MHC class II-enriched compartments
  (MIICs), 333
Microarray(s)
 analysis of RNA from exosome, 107
 method, 101
 microarray-based transcriptomics, 109
 and sequencing for exosomal content
  profiling, 145–150
  bioanalyzer assay, 150f
  miRNAs analyzed using SB biosciences
   PCR array system, 148t–149t

Microautophagy, 17
Microenvironment-derived exosomes in
        tumor angiogenesis, 239–240
Microfluidic(s), 108–109, 162
    approach, 164
    microfluidic-based systems, 4–5
    platform, 124
    technology, 164
Microparticles, 219–220, 320–321
MicroRNAs (miRNAs), 34, 41–42, 44, 95,
        187–188, 225–226, 250–253, 267, 291,
        314–315
    biogenesis, 132–133
    isolation, 99–100
    loading into exosomes, 95–96
    microRNA-mediated reprogramming of
        tumor cells, 204–205
    miR-9, 266
    miR-21, 255
    miR-21–5p, 267
    miR-100–5p, 267
    miR-105, 255
    miR-122, 270–271
    miR-139–5p, 267
    miR-155, 206
    miR-210-enriched exosomes, 238–239
    miR-222–3p, 318
    miR-1246, 266
    miRNA-203, 317
    role on reprogramming tumor-associated
        macrophages, 205–210
        exosome-mediated cellular communi-
            cation, 207f
        modulation of exosomal microRNA,
            207–210
    therapeutic application, 138–140
Microscopy-based methods, 47
"Microtentacles", 223–224
Microtubules (MTs), 226
Microvesicles (MVs), 13–14, 18–19, 59–60, 94,
        187–188, 219–220, 326–328
    biogenesis, 13–14
MIF. *See* Migration inhibitory factor (MIF)
Migration, exosomes in, 267
Migration inhibitory factor (MIF), 251–252,
        269–270

Migratory phenotypes, 12
MIICs. *See* MHC class II-enriched
        compartments (MIICs)
Miniaturized electrochemical sensors, 122
miR. *See* MicroRNAs (miRNAs)
miR-125b. *See* MiRNA-125b (miR-125b)
miRBase, 107
miRCURY kit, 99–100
miRNA recognition elements. *See* Three prime
        untranslated region (3′ UTR)
MiRNA-125b (miR-125b), 206–207
miRNAs. *See* MicroRNAs (miRNAs)
Mitochondrial RNAs (mtRNAs), 292
Mitoxantrone resistance protein. *See* ABC
        subfamily G member 2 (ABCG2)
MMP13. *See* Matrix-metalloproteinase 13
        (MMP13)
MMPs. *See* Matrix metalloproteinases
        (MMPs)
MNPs. *See* Magnetic nanoparticles (MNPs)
Molecular motors, 3
Molecular switches, 3
Monensin, 196
Morphogens, 235–236
Motility, 227–228
Mouse
    brain, 347
    model, 271
    neurons, 347
MRI. *See* Magnetic resonance imaging (MRI)
mRNA. *See* Messenger RNA (mRNA)
MS-275. *See* Histone deacetylase inhibitor
        (MS-275)
MSCs. *See* Mesenchymal stem cells (MSCs)
MSTO-211H malignant mesothelioma cells,
        227–228
mTOR. *See* Mammalian target of rapamycin
        (mTOR)
mtRNAs. *See* Mitochondrial RNAs (mtRNAs)
MTs. *See* Microtubules (MTs)
MUC1 proteins, 270–271
Multicellular system, 130
Multidrug resistance (MDR). *See* Acquired
        drug resistance
Multidrug resistance–associated protein 1,
        292–293

Multifunctional messengers with mixed
    intentions
  angiogenesis in cancer, 235–236
  exosomes
    biology, 236–237
    exosomes in tumor angiogenesis, 237–241
  future perspectives, 241–243
Multiple myeloma, 248–250, 296–297
Multiple sclerosis, 5–6
Multiplexed biomarker detection, 167–168
Multivesicular body (MVB), 3, 14–15,
    29, 69–70, 187–188, 192f, 236–237,
    326–328
  formation, 96
  proteins, 4–5
Murine DC line (DC2. 4 cells), 319–320
Mutant allele frequency (MAF), 62
Mutational signatures, 64
  detection in genomic DNA-enriched
    exosomes, 61
MVB. See Multivesicular body (MVB)
MVs. See Microvesicles (MVs)
Myeloid cell lineage, 205
Myeloid lineage cells, tumor exosomes effect
    on, 317–318, 317f
Myeloid-derived suppressor cells (MDSCs),
    203, 317–318
Myofibroblasts, 201
Myristoylated alanine-rich C-kinase substrate
    (MARCKS), 240
  gene, 250–251

N
N-SMase. See Neutral sphingomyelinase
    (nSMase)
Naïve MSCs, 296
Named nano-plasmonic exosome system
    (nPLEX system), 120–121
Nanochip, 102
NanoDrop spectrophotometer, 101–102
Nanoparticle tracking analysis (NTA), 4–5, 47,
    97–98, 165–166
Nanoplasmonic sensing, 120–122
Nanoplasmonic sensor (NPS), 121–122, 121f
NanoString nCounter platform, 108

Nanotechnology platforms for cancer
    exosome analyses
  iMER analysis, 124–125
  sensing
    electrochemical, 122–124
    nanoplasmonic, 120–122
Nasopharyngeal carcinoma (NPC), 316
  metastasis, 253–255
  progressions, 79
National Cancer Institute-60 panel (NCI-60
    panel), 17
National Institute of Allergy and Infectious
    Diseases (NIAID), 151
Natural antisense RNAs, 95
Natural killer cells (NK cells), 195, 271, 313
  NK cell–activating receptor, 316–317
  tumor exosomes effect on, 316–317, 316f
NCI-60 panel. See National Cancer Institute-60
    panel (NCI-60 panel)
nCounter NanoString technology, 108
ncRNAs. See Noncoding RNAs (ncRNAs)
Near-infrared (NIR), 181
Neural plasticity, 5–6
Neuroblastoma, 305
  cells, 36–37
Neurodegenerative diseases, 344–345
Neurological disorders, 72
Neutral sphingomyelinase (nSMase), 33,
    188–189
  nSMase2, 16
Neutrophil mobilization, 202–203
Next generation sequencing (NGS), 60–61, 95
Next-generation massive sequencing
    platforms, 147–150
NF-κB. See Nuclear factor kappa-light-chain-
    enhancer of activated B cells (NF-κB)
NGS. See Next generation sequencing (NGS)
NHDFs. See Normal human dermal fibroblasts
    (NHDFs)
NIAID. See National Institute of Allergy and
    Infectious Diseases (NIAID)
NIR. See Near-infrared (NIR)
NK cells. See Natural killer cells (NK cells)
NKG2D receptor, 316–317
Nod-like receptors, 314

Non-BM. *See* Nonbrain metastatic (Non-BM)
Non-EV-associated RNA and other lab-derived contaminations, 109–110
Non-small cell lung cancer (NSCLC), 82, 266, 314
  patients, 251
Nonbrain metastatic (Non-BM), 72
Noncoding RNAs (ncRNAs), 36, 238–239
  isolation, 99–100
Noninvasive liquid biopsy strategies, 61–62
Nontumor cells, 191–193, 293
Normal human dermal fibroblasts (NHDFs), 36
Notch signaling, 238, 265–266
NPC. *See* Nasopharyngeal carcinoma (NPC)
nPLEX system. *See* Named nano-plasmonic exosome system (nPLEX system)
NPS. *See* Nanoplasmonic sensor (NPS)
NSCLC. *See* Non-small cell lung cancer (NSCLC)
nSMase. *See* Neutral sphingomyelinase (nSMase)
NTA. *See* Nanoparticle tracking analysis (NTA)
Nuclear factor kappa-light-chain-enhancer of activated B cells (NF-κB), 265–266
Nucleases, 98
Nucleic acid, 130–131, 151, 345–346
Nucleic acid profiling in tumor exosomes.
    See also Proteomic profiling of tumor exosomes
  exosomes, 94–95
    composition of exosomes, 95
  experimental artifacts and contaminants affecting evRNA analysis, 109–110
  isolation techniques for RNA/DNA from exosomes, 97–100
  nucleic acid loading into exosomes, 95–97
    endosomal vesicle pathways, 96–97
    miRNA loading into exosomes, 95–96
  nucleic acid profiling—RNA, 100–108
  single EV analysis, 108–109
  statistical and bioinformatic software for evRNA characterization, 109
Nutilin-3-induced radiosensitization, 36
Nutrients, 96

O
O^6-methylguanine DNA methyltransferase (MGMT), 124–125
Oligodendrocytes, 31, 33
Oligomerization of caveolins, 197–198
Omeprazole. *See* Proton pump inhibitor
Oncosomes, 13–14
Organotropism, 176–177
Osteopontin, 227–228
Osteosarcoma, 304–305
Ovalbumin (OVA), 314–315
Ovarian cancer, 78–79, 300, 345–346
    ovarian cancer–derived exosomes, 78–79
    patients, 167–168
Ovarian tumor–associated markers, 165–166
Ovarian tumor–derived exosomes, 167–168
"Overdiagnosis" phenomenon, 62

P
P-glycoprotein (P-gp). *See* ABCB1
P-selectin glycoprotein ligand-1 (PSGL-1), 191–193
p38MAPK, 265–266, 269–270
P66Shc regulates mast cell activation, 224
PA. *See* Phosphatidic acid (PA)
Pancreatic cancer (Panc-1), 47, 266
Pancreatic cancer, 81, 255, 270–271, 301, 345
    exosomes, 269–270
Pancreatic cancer tumor-derived exosomes (PC-Exo), 346
Pancreatic ductal adenocarcinoma (PDAC), 63, 140, 301
Pancreatic stellate cell–derived exosomes, 268
Paracrine, 130
Parasites, 5–6
Parkinson's diseases, 5–6, 344–345
Pattern recognition receptors (PRRs), 252–253
PC-3. *See* Prostate cancer cells (PC-3)
PC-Exo. *See* Pancreatic cancer tumor-derived exosomes (PC-Exo)
PCR. *See* Polymerase chain reaction (PCR)
PD-1, 314
PD-L1, 318–319
PDAC. *See* Pancreatic ductal adenocarcinoma (PDAC)

PDGFs. *See* Platelet-derived growth factors (PDGFs)

PDZ binding motifs, 31–33

Peptide-loaded DC-derived exosomes, 347

Peptides, 130

Perforin, 316

Periostin, 80

Phagocytosis, 194–195

Pharmacologic suppression of TNTs, 226–227

Phosphatase and tensin homologue (PTEN), 268, 272, 315

Phosphatases, 105

Phosphatidic acid (PA), 16

Phosphatidylinositol, 45

Phosphatidylinositol 3-kinase (PI3K), 195–196, 274, 298

Phosphatidylserine (PS), 14, 190–191

Phosphoinositide 3-kinase. *See* Phosphatidylinositol 3-kinase (PI3K)

Phospholipase D2 (PLD2), 15–16

Phospholipids, 46–47

Phosphorylation, 266

Physicochemical properties of exosomes, 28

Physiologic factors, 225–226

PI3K. *See* Phosphatidylinositol 3-kinase (PI3K)

Piwi-interacting RNA (piRNA), 131

Placenta-derived exosomes, 318–319

Placenta-specific ABC protein. *See* ABC subfamily G member 2 (ABCG2)

Plasma, 4
  exosomes, 72

Plasma membrane (PM), 13–14, 29, 224
  components, 96
  protrusions, 196

Plasmid DNA, 319–320

Platelet dust, 1

Platelet-bound anti-PD-L1 antibody, 320–321

Platelet-derived growth factors (PDGFs), 253–255

PLD2. *See* Phospholipase D2 (PLD2)

PM. *See* Plasma membrane (PM)

Poly (ADP-ribose) polymerase inhibitor therapy, 64–65

Poly I:C (TLR3), 319–320

Polymer-based precipitation reagents, 98–99

Polymerase chain reaction (PCR), 105, 120

PPIs. *See* Proton pump inhibitors (PPIs)

Pre-B ALL. *See* B-cell precursor acute lymphoblastic leukemia (Pre-B ALL)

Precancerous adenoma, 137

Precipitation reagent ExoQuick, 47–48

Precision medicine, exoDNA role in, 60
  double-stranded genomic DNA in circulating exosomes, 60–61
  exosomes as agents for early detection, diagnosis, and stratification, 61–62
  genomic molecular profiling of exosomal cargo, 63–65
  mutational signatures detection in genomic DNA-enriched exosomes, 61
  transcriptomic characterization of tumors, 65

Premetastatic niche(s), 176–177, 247–248, 252–253
  exosomes in development, 249f
    cancer cell–derived exosomes role in initiation and progression, 248–250
    EMT, hallmark of premetastatic niche formation, 253–255
    exosomal noncoding RNAs role in initiation and priming, 250–251
    exosomal proteins role in priming metastasis, 251–252
    pharmacologic and genetic strategies, 255
    TLRs role in exosome-mediated metastasis, 252–253
  formation, 247–248

pri-miRNAs. *See* Primary miRNAs (pri-miRNAs)

Primary miRNAs (pri-miRNAs), 95–96, 132–133

Prions, 70

"Professional" APCs, 329

Proliferation, 286

Prostaglandin E2, 317–318

Prostate cancer, 78, 145, 299–300
  prostate cancer–derived exosomes, 267

Prostate cancer cells (PC-3), 34, 36, 45–46, 48, 263–265, 267

Prostate-specific antigen (PSA), 145

Protein(s), 119
  in cancer exosomes, 36–41
    Venn diagram of proteins, 38f
  cargoes, 40–41
  survivin inhibits apoptosis, 301
Proteolipid protein, 188–189
Proteome, 70
Proteomic profiling of tumor exosomes, 70.
    See also Nucleic acid profiling in tumor
    exosomes
  exosomes
    proteomic content in cancers, 74–82
    sources, 71–73
  representation of exosomal content and
    secretion, 70f
Proteomic(s)
  approach, 17
  data corroborate, 39–40
  profiling, 345
Proto-oncogenes, 272
Proton pump inhibitors (PPIs), 190, 274
Prototypical ESCRT biogenesis mechanism, 31
PRRs. *See* Pattern recognition receptors (PRRs)
PS. *See* Phosphatidylserine (PS)
PSA. *See* Prostate-specific antigen (PSA)
PSGL-1. *See* P-selectin glycoprotein ligand-1
    (PSGL-1)
PTEN. *See* Phosphatase and tensin homologue
    (PTEN)

Q
qPCR. *See* Quantitative real-time PCR (qPCR)
qRT-PCR. *See* Quantitative reverse
    transcription-polymerase chain
    reaction (qRT-PCR)
qTRS. *See* Tunable resistive pulse sensing (qTRS)
Quant-iT RiboGreen RNA Assay Kit, 102
Quantitative iTRAQ proteomic analysis, 39–40
Quantitative mass spectroscopic-and Western
    blot analysis, 269–270
Quantitative real-time PCR (qPCR), 106–107
  method, 147
  validation by, 106–107
Quantitative reverse transcription-polymerase
    chain reaction (qRT-PCR), 95, 101
  profiling of miRNA, 74

Quantum nanocrystal-tagged oligonucle-
    otides, 108–109
Qubit RNA HS assay, 101–102

R
Rab GTPase, 34, 347–348
Rab27, 267
Rab27a, 34, 273
Rac GTPase protein, 31–33
Rac1, GTPase of Rho family, 196
Radiation-induced 217 exosomal protein
    expression, 79
Radiotherapy, 286
Raft-mediated endocytosis, 196–198
Ras GTPase family members, 18–19
Rat adenocarcinoma cell line, 40
RBPs. *See* RNA-binding proteins (RBPs)
RCC. *See* Renal cell carcinoma (RCC)
Recalcitrant cancers, 61–62
Receiver operating characteristic (ROC),
    168–170
Receptor-mediated endocytosis, 196–198
Receptor–ligand interaction, 94, 130–131
Recipient cells, 138–139, 251
  exosome biogenesis in tumor cells and
    uptake by, 187–191
  exosomes communicate with, 262
  internalization of exosomes into, 274
Regulation of exosomes during tumor
    angiogenesis, 240–241
Regulatory small RNAs, 140–142
Regulatory T cells (Tregs), 203, 316
RELA/miR-221 axis, 301–302
Renal cancer, 314
Renal cell carcinoma (RCC), 304
Resistance mechanisms to drug therapies,
    287–288
Retroviruses, 70
Reverse budding, 326–328
Reverse intracellular budding process, 188
Ribonucleases (RNases), 100–101
Ribonucleoprotein complexes (RNPs),
    109–110
RISC. *See* RNA-induced silencing complex
    (RISC)

RNA
    analysis, 120
    dyes, 108–109
    exosomes isolation and RNA/DNA, 98–99
    history in exosomes, 130–131
    isolation techniques for RNA/DNA from
        exosomes, 97–100
    marker detection, 124
    nucleic acid profiling
        analysis of quality, quantity, and diversity
            of evRNA, 100–102
        characterizing evRNA, 103–108
    packaging, 41
    purification, 124
    RNA-binding fluorescent protein probes,
        108–109
    RNA-sequencing experiments, 105
    species, 1–2
    species present in exosomes
        lncRNA in exosomes, 142–144
        small noncoding RNAs, 132–142
RNA component of exosomes (evRNA), 95
    analysis of quality, quantity, and diversity,
        100–102
        assessing evRNA quality, 100–101
        evaluating quantity, 101–102
        methods for determining evRNA purity
            and integrity, 101t
        RNA detection methods for evRNA
            quantification, 103t
    characterization, 103–108
        deep sequencing platforms, 103–104
        library preparations from evRNA samples,
            104–106
        microarray analysis of RNA from
            exosome, 107
        NanoString, 107–108
        statistical and bioinformatic software,
            109
        validation by qRT-PCR, 106–107
        validation of deep sequencing data, 106
    experimental artifacts and contaminants
        affecting analysis
        non-EV-associated RNA and other
            lab-derived contaminations, 109–110
        presence of DNA, 110

RNA interference (RNAi), 224–225
RNA within exosomes (exoRNA), 65
RNA-binding proteins (RBPs), 41–42, 95–96,
    263–265
RNA-induced silencing complex (RISC),
    132–133
RNAi. See RNA interference (RNAi)
RNase, 98
    III enzymes, 95–96
    RNaseA, 98
RNases. See Ribonucleases (RNases)
RNPs. See Ribonucleoprotein complexes
    (RNPs)
ROC. See Receiver operating characteristic
    (ROC)
Roche 454 pyrosequencing, 103–104
rRNAs, 95

S
S100-A9 protein, 251–252
Saliva, exosomes in, 73
Sample Preparation Kit, 108
SAV. See Streptavidin (SAV)
SAV-exo. See SAV-LA-expressing exosomes
    (SAV-exo)
SAV-LA-expressing exosomes (SAV-exo),
    319–320
"Search-and-destroy" mission, 329–331
Secretion by exocytosis, 190
Secretory autophagy, 17
"Seed and soil hypothesis", 247–248
Sensitive techniques, 102
Sequencing, 101
    sequencing-based transcriptomics,
        109
Serum antibodies, 329
SF-539. See Glioblastoma (SF-539)
"Shed membrane vesicles". See Microvesicles
Short interfering RNA (siRNA), 12
Signal regulatory protein α (SIRPα), 320
Signaling molecules, 130
SILAC-based approach, 74–78
Single EV analysis, 108–109
siRNA. See Short interfering RNA (siRNA)
SIRPα. See Signal regulatory protein α (SIRPα)
Size-exclusion chromatography, 4

SLC. *See* Solute carrier (SLC)
Small noncoding RNAs, 95, 132–142
    exosomes for imaging and chemotherapeutics, 142
    miRNA
        exosome-based delivery of therapeutic, 140–142
        therapeutic application, 138–140
Small nuclear RNAs (snRNAs), 41–44, 131, 251
Small nucleolar RNA (snoRNA), 131
Small-RNA
    chip, 102
    small-RNA-sequencing library preparation, 104t
Smaller interior vesicles, 326–328
SNAREs, 3
snoRNA. *See* Small nucleolar RNA (snoRNA)
snRNAs. *See* Small nuclear RNAs (snRNAs)
Soil, 247–248
SOJ6 cell line, 45–46
SOLiD system, 103–104
Solid tumors, 201, 266
Soluble signaling, 198
Solute carrier (SLC), 292–293
Solutes, 96
SP1. *See* Sphingosine 1-phosphate (SP1)
Specific molecular and structural biomarkers of TNT formation in cancer, 227–229
Sphingomyelin, 46–47
Sphingosine 1-phosphate (SP1), 16
Spin column–based method, 98–99
SPION. *See* Superparamagnetic iron oxide nanoparticle (SPION)
SPR sensors. *See* Surface plasmon resonance sensors (SPR sensors)
SRC-family kinases, 263–265
Stanniocalcin 1 (STC1), 81
Stanniocalcin 2, 81
STAT3, 269–270
Statistical and bioinformatic software for evRNA characterization, 109
STC1. *See* Stanniocalcin 1 (STC1)
Stem cell maintenance and function, 5–6
Streamlined exosome molecular analysis, 164
Streptavidin (SAV), 319–320

Stroma, cross-talk with tumor cells via exosomes and, 293
Stromal cell(s), 6–7
    communication, 296
    phenotype manipulation, 12
Structural biomarkers of TNT formation in cancer, 227–229
Sugar chains, 48
Superparamagnetic iron oxide, 142
Superparamagnetic iron oxide nanoparticle (SPION), 182
Surface charge of Exosomes, 28, 48
Surface plasmon resonance sensors (SPR sensors), 120
Surface protein, 5
"Surfaceome", 64–65
Survivin, 74–78
Sustained angiogenesis, 236
Synaptic pathway, 130
Syndecan, 30
Syntenin, 30
    functions, 15–16
    overexpression, 30
    syntenin-1, 183

T
T cell receptor (TCR), 329–331
T cell(s), 313, 329
    B cell EVs and promotion of dendritic cells–mediated T cell immune responses, 335
    cytotoxicity, 12
    leukemia cells, 194–195
    tumor exosomes effect on, 314–316, 315f
T-cell acute lymphoblastic leukemia/lymphoma (T-ALL/LBL), 297–298
T-helper-1 (Th1), 205
TACSTD2. *See* Tumor-associated calcium-signal transducer 2 (TACSTD2)
TAMs. *See* Tumor-associated macrophages (TAMs)
Target cell, 5, 191, 198
Targeted cancer therapy, 286
TBC1 domain family member 10A-C (TBC1D10A-C), 189–190
TCGA. *See* The cancer genome atlas (TCGA)
TCR. *See* T cell receptor (TCR)

TEM. *See* Transmission electron microscopy
(TEM)
Temoprofin, 34
TeMs. *See* Tetraspanin-enriched
microdomains (TEMs)
TEMs. *See* Tetraspanin-enriched
microdomains (TEMs)
Tetraspanin 6 (Tspan6), 16–17
Tetraspanin 8 (Tspan8), 82
knockout animals, 37
Tetraspanin-enriched microdomains (TEMs),
16–17, 31
Tetraspanins, 4–5, 37, 193–194, 265, 347
protein, 16–17
TfR. *See* Transferrin receptor (TfR)
TGF. *See* Transforming growth factor (TGF)
TGM2 protein, 80–81
Th1. *See* T-helper-1 (Th1)
The cancer genome atlas (TCGA), 134
Therapeutic application of miRNA, 138–140,
139t, 141t
Therapeutic potential of exosomes, 346–348
Therapeutic vehicle, exosome as, 275–276
Three prime untranslated region (3′ UTR),
132–133
Thrombospondins, 37
Tissue
biopsy, 119
origins, 17–18
repair, 5–6
tissue-specific immune-suppressive effects,
178–179
TLRs. *See* Toll-like receptors (TLRs)
TME. *See* Tumor microenvironment (TME)
TMs. *See* Tumor microtubes (TMs)
TNF. *See* Tumor necrosis factor (TNF)
TNF-related apoptosis-inducing ligand
(TRAIL), 198
TNTs. *See* Tunneling nanotubes (TNTs)
Toll-like receptors (TLRs), 194–195, 252–253,
271, 314
ligands, 314
role in exosome-mediated metastasis,
252–253
TLR3, 138–139, 252–253

TLR4, 320
TLR7, 252–253
TLR8, 252–253, 305
Toxins, 291
*TP*53 mutations, 61
TRAIL. *See* TNF-related apoptosis-inducing
ligand (TRAIL)
Transcriptomic characterization of tumors, 65
Transfer RNAs (tRNAs), 41–42, 95–97
Transferrin receptor (TfR), 30
Transforming growth factor (TGF), 316
TGFβ, 236, 251–255, 269–270
neutralizing antibody, 316–317
TGF-β-mediated SMAD-dependent
signaling process, 201
TGF-β1, 316
Translational research perspective, 120
Transmembrane protein CD63, 80–81
Transmission electron microscopy (TEM), 4–5,
47, 145, 146f, 165–166, 166f, 236–237
Transport epidermal growth factor receptor,
238
Transportations of drug resistance via
exosomal contents, 292–293, 292f
Trastuzumab, 318
Tregs. *See* Regulatory T cells (Tregs)
Triple negative breast cancer cell–derived
exosomes, 267–268
tRNAs. *See* Transfer RNAs (tRNAs)
TRP2. *See* Tyrosinase-related protein 2 (TRP2)
TSAP6. *See* Tumor suppressor–activated
pathway 6 (TSAP6)
TSG101 protein, 30, 255
Tspan6. *See* Tetraspanin 6 (Tspan6)
TTF. *See* Tumor-treating fields (TTF)
Tumor exosomes, 269–270
for cancer diagnosis, 164–170
CCD images of multiplexed three-color
fluorescence detection, 169f
diagnostic accuracy analysis, 169t
effect
on myeloid lineage cells, 317–318, 317f
on natural killer cells, 316–317, 316f
on T cells, 314–316, 315f
removal, 275

translational potential in diagnosis and
    therapy
    exosomes and cancer, 344–345
    exosomes as cancer biomarkers, 345–346
    therapeutic potential of exosomes,
        346–348
Tumor immunity, exosomal content
    regulating, 202–203
Tumor microenvironment (TME), 71, 198–203,
        200t, 241, 248–250, 255, 262, 343–344
    cell–derived exosomes in cancer metastasis,
        268
    exosomal content
        modulating tumor angiogenesis and
            metastasis, 201–202
        regulating tumor immunity, 202–203
    exosomes transport oncoproteins, 199–201
    formation, 12
Tumor microtubes (TMs), 223–224
Tumor necrosis factor (TNF), 198, 315
    TNFaip2, 227–228
    TNFα, 236, 253–255, 315
Tumor suppressor–activated pathway 6
        (TSAP6), 189–190
Tumor-associated calcium-signal transducer 2
        (TACSTD2), 80
Tumor-associated macrophages (TAMs), 205
Tumor-associated macrophages repro-
        gramming, MicroRNAs role on, 205–210
Tumor-treating fields (TTF), 227
Tumor(s), 286, 313–314, 343
    antigens, 318
        presentation, 12
    biopsy, 161–162
    cells, 253–255, 292, 318
        cross-talk between stroma via exosomes
            and, 293
        exosome biogenesis in, 187–191
        invasion, 267
        microRNA-mediated reprogramming of,
            204–205
        tumor cell–derived exosomal, 250–251, 272
        tumor cell–derived exosomes, 164,
            267–268
    EVs, 336

biogenesis, 12
exosomes in tumor angiogenesis
    cancer-derived exosomes in, 237–239
    microenvironment-derived exosomes in,
        239–240
    regulation of exosomes during tumor
        angiogenesis, 240–241
phagocytosis, 320
stroma, 293, 294f
suppressors, 204–205
transcriptomic characterization, 65
tumor-associated stroma, 344
tumor-derived exosomes, 6–7, 41, 176,
        248–250, 271, 292–293, 295–296
    homing to specific organs, 177–179
    production, 347
tumor-derived vesicle concentrations, 119
tumor-specific immune responses, 314
vaccines, 319–320
Tumorigenesis process, 19–20, 252–253, 344
Tunable resistive pulse sensing (qTRS), 97–98
Tunneling nanotubes (TNTs), 219–220,
        221f–222f
    formation, 225–226
    interplay between exosomes and, 220–223
    pharmacologic suppression, 226–227
    specific molecular and structural biomarkers
        of formation in cancer, 227–229
    tunneling nanotube–like nanostructures
        and interaction with exosomes, 223–225
Tyrosinase-related protein 2 (TRP2), 48–51

U
U6 snRNA detection, 99–100
U87 cell line, 45–46
Ultracentrifugation, 162–163
Ultrafiltration techniques, 4
Ultrasensitive mutation detection
        methodology, 62
Urinary Exosome Protein Database, 80
Urine, 4, 145
    exosomes, 72–73
Urothelial cells, 263–265
3′ UTR. *See* Three prime untranslated region
        (3′ UTR)

V

V-Raf murine sarcoma viral oncogene
    homolog B, 63
Vaccines, 65
Validation
    of deep sequencing data, 106
    by qRT-PCR, 106–107
VAMP3. *See* Vesicle-associated membrane
    protein 3 (VAMP3)
Vascular endothelial growth factor (VEGF), 81,
    236, 255, 265–266, 273
Vascular endothelial growth factor receptor 1
    (VEGFR1), 247–248
Vascular proliferation, 177
Vav and paxillin, 224
VEGF. *See* Vascular endothelial growth factor
    (VEGF)
VEGFR1. *See* Vascular endothelial growth
    factor receptor 1 (VEGFR1)
Vesicle
    production, 18
    secretion to intercellular signaling, 20
    vesicle–cell fusion process, 191–193

Vesicle-associated membrane protein 3
    (VAMP3), 18–19
Vesicular lipid raft contents, 17–18
Vesicular transport regulation GTPases, 3
Vimentin, 39–40
Viral infection, 17
Visceral cancers, 63
VPS33 B protein, 79

W

Wafer-scale batch process, 121–122
Western blotting method, 120, 224
Wnt signaling pathway, 294–295
World Health Organization, 293–294

X

Xenobiotic metabolizing enzyme (XME),
    292–293

Z

Zeta potential of exosomes, 48
ZFAS1, 142–144, 267
Zonula occludens 1 (ZO-1), 250–251

CPI Antony Rowe
Chippenham, UK
2018-12-05 02:04